Crustacean and Mollusk Aquaculture in the United States

CRUSTACEAN AND MOLLUSK AQUACULTURE IN THE UNITED STATES

Edited by

Jay V. Huner

Center for Small Farm Research
Southern University
Baton Rouge, Louisiana

E. Evan Brown

Department of Agricultural Economics
The University of Georgia
Athens, Georgia

AVI PUBLISHING COMPANY, INC.
Westport, Connecticut

Cover illustration courtesy of J. Haamer.
Reprinted from "Mussel Culture and Harvest"
edited by R. A. Lutz, Elsevier Scientific
Publishing Company, Amsterdam, 1980, with
permission.

250 Post Road East
P.O. Box 831
Westport, Connecticut 06881

Library of Congress Cataloging in Publication Data

Main entry under title:
Crustacean and mollusk aquaculture in the United
States.

Bibliography: p.
Includes index.
1. Shellfish culture—United States. I. Huner, Jay V.
II. Brown, E. Evan.
SH365.A3C78 1985 639'.5 84-24302
ISBN 0-87055-468-9

Printed in the United States of America
A B C D 4321098765

To Arlene H. Brown (1927–1983)

and to John and Edith Huner, Kade and Virginia McInnis, Ossi, Leena, and Lasse Lindqvist and my wife Judy

Contents

Contributors xi
Preface xiii

1 Crawfish Culture in the United States 1
James W. Avault, Jr. and Jay V. Huner
Introduction 2
Species of Importance 3
Basic Biology 7
Diseases and Parasites 15
Pond Culture 18
Soft-Shelled Crawfish 38
Intensive Crawfish Culture 40
Status of Crawfish Culture 44
Crawfish Processing 47
Economics and Markets 49
Summary 52
Literature Cited 54

2 Freshwater Prawns 63
Paul A. Sandifer and Theodore I. J. Smith
Introduction 63
Basic Biology 65
Culture Techniques 89
Diseases, Parasites, and Predators 104
Constraints on Development of Prawn Farming 107
Status of Culture in the United States 109
Economic Overview and Outlook 111
Literature Cited 118

3 Penaeid Shrimp Culture 127
A. L. Lawrence, J. P. McVey, and J. V. Huner
Introduction 127
Species of Importance 130
Basic Biology and Culture Requirements 135
Culture Techniques 137
Diseases and Parasites 145
Status of Shrimp Farming in the United States 147

Economic Overview and Outlook 150
Literature Cited 152

4 Lobster Aquaculture 159
Louis R. D'Abramo and Douglas E. Conklin
Introduction 159
General Biology—Selected Aspects 162
Culture 167
Diets for Larger Juveniles 181
Diseases 189
Site Selection and Design of Facilities 191
Marketing and Associated Economics 193
Future Prospects 194
Literature Cited 196

5 Other Crustacean Species 203
Michael J. Oesterling and Anthony J. Provenzano
Introduction 203
Blue Crab 204
Cancer Crabs 215
Spot Prawn 219
Spiny Lobsters 225
The Smaller Shrimps 228
Literature Cited 230

6 Oyster Culture 235
Victor B. Burrell, Jr.
Introduction 236
Natural History 236
The American Oyster *(Crassostrea virginica)* 239
The Pacific Oyster *(Crassostrea gigas)* 247
The European Oyster *(Ostrea edulis)* 252
The Olympia Oyster *(Ostrea lurida)* 256
Oyster Hatcheries 259
Problems Facing the Oyster Culture Industry 266
Summary 268
Literature Cited 269

7 Clam Aquaculture 275
John J. Manzi
Introduction 275
Clam Fisheries and Aquaculture Production 276
Basic Biology 280
Culture Techniques 282
Parasites and Diseases 295
Constraints 297
Status and Economic Overview 300
Summary 303
Literature Cited 304

8 Mussel Aquaculture in the United States **311**
Richard A. Lutz
Introduction 311
European Mussel Culture Technology 314
Feasibility of European Culture Techniques in the United States 319
Current Mussel Aquaculture Production 320
Recent Research Efforts 327
Life Cycle of the Blue Mussel 328
Experimental Culture—East Coast 331
Experimental Culture—West Coast 337
Presence of Pearls 338
Predation, Parasitism, and Accumulation of Algal Biotoxins 342
Mussel Culture in Heated Effluents 344
Aquaculture Carrying-Capacity Model 345
Economics 350
Summary 354
Literature Cited 356

9 Abalone: The Emerging Development of Commercial Cultivation in the United States **365**
Neal Hooker and Daniel E. Morse
Introduction 366
Principal U.S. Species 371
Biology 376
Conventional Cultivation Technology 385
Barriers to Efficient Production Using Conventional Technology 390
Ocean Ranching 404
Summary and Prospects for Future Industrial Development 407
Literature Cited 409

10 Water Quality **415**
Robert P. Romaire
Introduction 415
Physical Variables 416
Chemical Variables 422
Biological Variables 444
Pesticides 448
Water Analysis 449
Literature Cited 451

Appendix: The Brine Shrimp, Genus *Artemia* 457
Index **463**

List of Contributors

James W. Avault, Jr., School of Forestry and Wildlife Management, Louisiana State University Agricultural Center, Baton Rouge, Louisiana 70803

Victor G. Burrell, Jr., South Carolina Marine Resources Research Institute, Charleston, SC 29412

Douglas E. Conklin, Bodega Marine Laboratory, University of California, Bodega, California 94923

*Louis R. D'Abramo,** Bodega Marine Laboratory, University of California, Bodega, California 94923

Neal Hooker, Department of Biological Sciences and The Marine Science Institute, University of California, Santa Barbara, California 93106

Jay V. Huner, Center for Small Farm Research, College of Agriculture, Southern University, Baton Rouge, Louisiana 70813

A. L. Lawrence, Texas Agricultural Experiment Station and Department of Animal Science and Wildlife and Fisheries Sciences, Port Aransas, Texas 78373

Richard A. Lutz, Department of Oyster Culture, New Jersey Agricultural Experiment Station, and Center for Coastal and Environmental Studies, Rutgers University, New Brunswick, New Jersey 08903

J. P. McVey, American Embassy, Jakarta, Indonesia

John J. Manzi, Marine Resources Research Institute, Charleston, South Carolina 29412

Daniel E. Morse, Department of Biological Sciences and The Marine Science Institute, University of California, Santa Barbara, California 93106

Michael J. Oesterling, Virginia Institute of Marine Science and School of Marine Science, College of William and Mary, Gloucester Point, Virginia 23062

Anthony J. Provenzano, Department of Oceanography, Old Dominion University, Norfolk, Virginia 23508

Robert P. Romaire, School of Forestry and Wildlife Management, Louisiana Agricultural Experiment Station, Louisiana State University Agricultural Center, Baton Rouge, Louisiana 70803

Paul A. Sandifer, Marine Resources Research Institute, Charleston, South Carolina 29412

Theodore I. J. Smith, Marine Resources Research Institute, Charleston, South Carolina 29412

*Present Address: Department of Wildlife and Fisheries, P.O. Drawer LW, Mississippi State University, Mississippi State, Mississippi 39762

Preface

Crustaceans and mollusks are the glamor species of the seafood industry. Interest in aquaculture[1] of these invertebrates, frequently referred to as shellfishes, has resulted from increased exploitation of wild stocks. This has driven prices up to levels very attractive to investors. Much attention has been focused on crustaceans such as freshwater crawfishes, freshwater prawns, penaeid shrimps, and homarid lobsters, and mollusks such as oysters, clams, mussels, scallops, and abalone. Success has been greatest with low trophic level species such as freshwater crawfishes and bivalve mollusks where nature subsidizes the aquaculturist in many ways, providing feed at low or no cost, seed for culture systems, waste removal, and so forth. Species such as homarid lobsters, penaeid shrimps, and abalone may have complicated life cycles, relatively slow growth rates, and other problems that have, so far, limited development of aquaculture of these high value, high visibility species.

All taxa listed initially can be cultured from egg to egg in captivity, but many factors influence the commercial profitability of raising them in the United States. Investment in aquaculture of crustaceans and mollusks in the United States or by U. S. companies abroad has been extensive. Yet, only freshwater crawfishes and oysters have been cultured on a truly large-scale, profitable basis to date in the United States proper. Other taxa such as penaeid shrimps and abalones are being cultured in the Orient where demand justifies expensive, labor-intensive culture systems. Penaeid shrimps are also being cultured quite successfully in Latin America where costs of land and labor are low and year-round growing seasons occur. In Europe, mussels have been cultured for centuries using very simple techniques. These can be and are now being applied in the United States, but demand for the product does not yet begin to meet that in Europe.

[1] Aquaculture is a general term that applies to the culture of any aquatic organism. Mariculture is a much more specific term refering to culture of marine (saltwater) organisms.

In this text, we have drawn together specialists to address the status of the culture of freshwater crawfishes, freshwater prawns, penaeid shrimps, homarid lobsters, oysters, marine clams, marine mussels, and abalones in the United States. Additional contributions deal with culture of miscellaneous crustaceans like soft-shelled blue crabs and spiny lobsters and the monitoring and controlling of water quality.

The primary reason for culturing of all taxa is to produce food for human consumption. Because all species are natural fish foods, a secondary reason for culturing some of them is for fish bait, and this is noted where appropriate. A few species may find their way, because of some peculiarity, into the aquarium pet market.

There are many general phases of aquaculture that could be singled out for emphasis in this text after discussing the culture of specific taxa. We have chosen, however, to emphasize only one, water quality. Water is the medium in which aquatic species live and are cultured. Most commercial failures are directly or indirectly related to failure to meet the water quality requirements of the target species. Thus, we end this text with a chapter on water quality.

Authors have utilized the same basic outline in preparing their chapters, but variations in the number of species within a taxon, degree of commercialization to date, and cultural techniques—not to mention individual style—have resulted in a potpourri of presentations. We feel that this is a positive aspect of this text. Where one crustacean chapter may treat a subject like molting lightly, another will fill in the gap, as molting is a common process with all crustaceans. In addition, a cultural technique or product form commonly associated with one taxon may have application with another. For example, soft-shelled crabs and crawfishes are especially high value products at sizes of 25 g (crawfishes) to 100 g (crabs). Growing time of lobsters, a major economic impediment to their successfull culture, would be significantly reduced if this product form could be successfully pursued. Thus, this text should serve not only as a status report on development of commercially oriented crustacean and mollusk aquaculture in the United States, but also to promote the exchange of technology and product form ideas.

We trust that you will find this text useful and informative.

Jay V. Huner
E. Evan Brown

1

Crawfish Culture in the United States

James W. Avault, Jr.
Jay V. Huner

Introduction
Species of Importance
Basic Biology
  Life Cycles
  Reproduction
  Growth and Molting
  Food Habits
  Genetics
  Environmental Requirements
  Relationship between *P. acutus* and *P. clarkii*
Diseases and Parasites
Pond Culture
  Ponds and Pond Construction
  Water and Its Management
  Stocking Crawfish Ponds
  Forages
  Harvesting
  Population Monitoring
  Predator Control
  Rice/Crawfish Double Cropping
  Polyculture of Crawfish with Other Species
Soft-Shelled Crawfish
  Production
  Molt Induction
Intensive Crawfish Culture
Status of Crawfish Culture
  Climatic Restrictions on Culture of Crawfish
  Current Acreage under Culture
  Expansion Possibilities

Crawfish Processing
Economics and Markets
Summary
Literature Cited

INTRODUCTION

Freshwater crawfish* culture represents the only large-scale, profitable crustacean culture in the continental United States. In 1983, about 45,700 ha (112,879 Ac) were devoted to crawfish production, with about 90% of the total acreage centered in south Louisiana where ponds yielded at least 807 kg/ha/year (700 lb/Ac/year) (L. de la Bretonne, Louisiana Cooperative Extension Service, Baton Rouge, LA, personal communication, 1983). Other production areas, in order of decreasing acreage, are southeast Texas, northwestern Mississippi, South Carolina, and Arkansas. (See Status of Crawfish Culture section for details.) In the Mississippi River valley and California most established finfish farms, especially those specializing in fry, fingerling, minnow, and goldfish production, have feral crawfish crops which may or may not be managed depending on the profit margin involved (Huner 1976, 1978a).

Crawfish cultured in Louisiana, Texas, Mississippi, and South Carolina are generally sold for food. In other areas, they normally find their way into the fish bait trade where very small crawfish, 2.5–3.5 cm (1–1.5 in.), are excellent bait for yellow perch *(Perca flavescens)* and bluegill *(Lepomis macrochirus)* and larger, 4.0–7.0 cm (1.5–2.5 in.), crawfish species are used to catch largemouth bass *(Micropterus salmoides)* and various catfishes (*Ictalurus* spp.).

Crawfish are much easier to culture than most commercially cultured aquatic species for several reasons. Once stocked, they establish sustaining populations, and restocking is usually not necessary. Hatchery production of young crawfish and stocking of known numbers is not practiced or required on a commercial scale. Water depths need not exceed 0.5 m (19 in.); thus, less water and less expensive levees are required than in fin fish culture. Little effort is required to convert rice acreage to crawfish production. In the late 1970s crawfish generated more revenues than those received for other crops, including rice and soybeans, with which they may be rotated. The feeding of crawfish, although effective, is normally not practiced. Rather, vol-

*The terms crawfish and crayfish are interchangeable but crawfish is preferred in the industry.

unteer or intentionally cultivated cover crops—grasses (including rice and/or millet), sedges, and semiaquatic plants—are grown during the dry summer period while crawfish remain quiescent and/or spawn in burrows. This vegetation provides the basis for a detritus food chain that can routinely generate crawfish crops of 500–1500 kg/ha (445–1335 lb/Ac) if proper management is practiced. Harvesting is generally labor intensive, requiring the use of wire mesh traps, but improved harvesting devices have been developed and the use of family labor often reduces labor expenses.

There are several general books and bulletins available on crawfish culture. Most deal with production of *Procambarus* spp. in Louisiana (Hill and Cancienne 1966; Visoca 1966; Gary 1974; Gooch and Huner 1980; LaCaze 1981; Huner and Avault 1981; Huner and Barr 1981). Alon and Dean (1980) discuss culture of *Procambarus clarkii* in South Carolina. Other notable publications include Forney's (1968) bulletin on the culture of *Orconectes immunis* in New York state, Biggs' (1980) general review on the suitability of various North American species for culture, and Arrignon's (1981) book on the biology and culture of the North American species *Pacifastacus leniusculus* and European crawfishes in Europe.

SPECIES OF IMPORTANCE

The species of crawfish most suited for culture are residents of temporary lentic habitats, primarily swamps and marshes. They dig burrows during dry periods that are rarely more than 1 m below the surface. They are classified as tertiary burrowers (Hobbs 1981) and are not especially well adapted for burrowing. Other species are classified as secondary and primary burrowers; these are morphologically and behaviorally more oriented than tertiary burrowers for a life below the surface of the earth in intricate burrows. No secondary or primary burrower shows any potential as an aquaculture candidate.

The following commentary deals almost exclusively with one family, Cambaridae, and one species, the red swamp crawfish, *Procambarus clarkii* (Fig. 1.1). This species alone accounts for at least 85% by volume of all cultured crawfish in the United States (Huner and Barr 1981) even though there are two crawfish families (Astacidae being the other) and well over 350 species of crawfishes in the continental United States (Hobbs 1972, 1974b). *P. clarkii* is native to the Mississippi River Valley from Texas into southern Illinois and has been successfully introduced into the Southwest, the Far West (California and Oregon), the Pacific (Hawaii), the Midwest (Ohio), and the

FIG. 1.1 Four important U.S. crawfish species, all large males (color is faded because animals were preserved). Left to right: *Orconectes rusticus, Pacifastacus leniusculus, Procambarus clarkii,* and *Procambarus acutus acutus.*
Courtesy of J. E. Barr.

Atlantic Coast states of Maryland, South Carolina, and North Carolina (Huner and Barr 1981).

Other species that have been cultured profitably include the white river crawfish, *Procambarus acutus acutus* (Fig. 1.1), and the paper shell crawfish, *Orconectes immunis* (Huner 1976). *Procambarus acutus* shares much of the same natural range as *P. clarkii* but is native to the Atlantic Coast and occurs naturally in Wisconsin. *Orconectes immunis* is restricted to the Midwest and Northeast. Interestingly, one of the first bulletins on crawfish culture available to the public dealt with the culture of *O. immunis* in New York state (Forney 1968). Both *P. clarkii* and *P. acutus* can be present in the same culture pond. In many cases, *P. acutus* is present in greater numbers than *P. clarkii* in new ponds, but its relative percentage declines in successive years. The relationship between the two species is not well understood and will be discussed later (Fig. 1.2). Two other *Orconectes, O. rusticus* (Fig. 1.1) and *O. nais,* are found in midwestern finfish culture ponds (Huner 1976). They are harvested when economics are favorable. Langlois (1935) and Rickett (1974) reported harvest of *O. rusticus* and *O. nais,* respectively, from largemouth bass fingerling ponds. Both *P.*

FIG. 1.2 Male *Procambarus clarkii* (top) and *Procambarus acutus acutus* (bottom).
Courtesy of J. E. Barr.

clarkii and *P. acutus* grow to more than 10 cm (4 in.) in total length in 3 to 6 months. *O. immunis* rarely exceeds 6.5 cm (2.6 in.) in a comparable growing period. Crayfish of this size are marginal as food, but this size and smaller ones are ideal for fish bait.

Orconectes virilis is among the larger members of its genus and is harvested for bait and food in Wisconsin (Threinen 1958; Huner 1978a). It is discontinuously distributed across North America from Maine to California, and northward from the Mississippi Valley into the Canadian provinces of Ontario, Manitoba, Saskatchewan, and east-central Alberta (Aiken 1968). It is a lentic species but does not do well in the temporary habitats frequented by species that are now cultured (Bovbjerg 1970). This undoubtedly accounts for the absence of reports about its presence in aquaculture operations that invariably involve annual dewatering. This species can be cultured readily in closed systems (Leonard 1981) and may someday be cultured commercially.

One other species warrants comment. This is the signal crawfish, *Pacifastacus leniusculus,* native to the Far West and a member of the family Astacidae (Fig. 1.1). Wild stocks are commercially exploited in

Washington, Oregon, and California, with most of the 200,000–300,000 kg (440,000–666,000 lb) of annual production being exported to the Scandinavian countries (Huner 1978a). Size limits are about 9.3 cm (3.7 in.) total length, but this species requires stable, lentic culture systems, cool water, and a growing season of about 18 months or longer for successful culture (Cabantous 1975). There is limited interest in culturing this species on the West Coast; however, *P. leniusculus* shows its greatest culture potential in Europe where hatcheries produce young animals for stocking lakes that have lost their native stocks of *Astacus astacus* to the fungal crawfish "plague" *Aphanomyces astaci. Pacifastacus leniusculus* is not affected by the pathogen (Abrahamsson 1973; Cabantous 1975; Arrignon 1981).

There are other large cambarid crawfishes that may be suitable for culture. These include *Procambarus hayi, Procambarus troglodytes,* and *Procambarus fallax. P. hayi* and *P. fallax* are large members of the subgenus *Ortmannicus,* the same subgenus to which *P. acutus* belongs (Hobbs 1974a). *P. hayi* is native to Mississippi and Tennessee; *P. fallax* to Florida and Georgia. *P. troglodytes* is a large member of the subgenus *Scapulicambarus,* the same subgenus to which *P. clarkii* has been assigned (Hobbs 1974a). It is native to the southeastern United States. There is some controversy as to whether or not it is a "good" species and distinct from *P. clarkii,* but Hobbs (1981) still contends that the two species are distinct. Because there are so many species of crawfish in the United States, there may be others with culture potential. No species that is found consistently in fish culture systems should be ruled out.

It is appropriate at this time to mention the so-called "Australian" freshwater crawfishes. These are members of the family Parastacidae (Hobbs 1974a). Because their weight can reach 4 kg (8.8 lb), much attention has been directed to them. Two smaller species, *Cherax tenuimanus* (Marron) and *Cherax destructor* (Yabbie), are cultured in Australia (Frost 1975; Crook 1981). *C. tenuimanus,* which reaches 1 kg (2.2 lb) is twofold larger than *C. destructor* but requires a permanent lentic habitat for successful culture, whereas *C. destructor* is a hardier burrowing species. Although introductions of species to new regions are tempting, they are fraught with potential ecological danger. One major objection to the introduction of Australian crawfishes into North America is that they show no resistance to the crawfish plague, *A. astaci.* North American crawfishes are apparently vectors for this pathogen (Unestam 1969). Thus, a substantial investment could be lost in a very short period of time (Huner and Barr 1981) if introduced Australian crawfish were infected by the crawfish plague.

BASIC BIOLOGY

Life Cycles

Cultured cambarid crawfishes are characteristically found in temporary lentic habitats, such as swamps and marshes. These tertiary burrowers are abundant in open water during wet periods and retreat to simple burrows during dry seasons. This annual watering/dewatering cycle reduces the impact of finfish predation on crawfish since few fishes survive with virtually no water. In the deep South, *P. clarkii* and *P. acutus* are active during the late fall to late spring wet season. As one moves northward, these species, as well as other cambarids, become dormant during the winter, and the growing season extends well into the summer. Eventually a point is reached in the Midwest where the various cambarids are active in the middle to late spring through early to middle fall. Thus, the kind of management practiced in Louisiana with its fall–winter–spring wet period may not be appropriate in regions such as Ohio; however, it may be necessary to hold water in ponds to prevent freezing in shallow burrows.

For all practical purposes, *P. leniusculus,* the important astacid species, is not cultured in the United States. It is typically active during the spring–summer–fall period and inactive during the winter. The duration of the inactive period is governed by altitude and latitude.

Selected studies of the life cycles of commercially important species include the following: *Orconectes immunis* in Illinois (Tack 1941); *Orconectes rusticus* in Kentucky (Prins 1968); *Orconectes virilis* in Michigan (Momot and Gowing 1977); *Procambarus clarkii* in Louisiana (Penn 1943; Konikoff 1977) and Illinois (Brown 1955); *Procambarus acutus* in Louisiana (Penn 1956) and Illinois (Brown 1955); *Procambarus hayi* in Mississippi (Payne 1972); *Procambarus troglodytes* in South Carolina (Wishart and Loyacano 1974); *Pacifastacus leniusculus* in California (Abrahamsson and Goldman 1970) and Oregon (Mason 1975).

Reproduction

The sexes are distinct in crawfishes (Fig. 1.3). In males, the first two paris of abdominal appendages, called *swimmerets,* are modified to transfer sperm to a seminal receptacle located between the female's walking legs. (Female astacid crawfish have no seminal receptacles. Spermatophores are attached where the seminal receptacle is found

FIG. 1.3 *Procambarus clarkii,* form I male (left), immature male (center), and mature female (right). Note differences in claw sizes and development of the first two pairs of swimmerets of males for sperm transfer.
Courtesy of J. Huner.

in cambarids.) Sperm travels down the gonopodia after being extruded from openings at the inner bases of the fifth pair of walking legs. Fertilization is not immediate, as sperm is stored for up to 6 months in the female's seminal receptacle. Gonopodia and seminal receptacles have distinct morphology. Male gonopodia serve as the principal taxonomic characters for distinguishing cambarid species. No identification keys have yet been developed for female crawfishes (Hobbs 1972).

Sexually mature cambarid males have inflated claws, hooks at the bases of one or two pairs of walking legs, and cornified gonopodia. Such males are said to be in the form I state. All secondary sexual characters serve to insure successful sperm transfer. Sexually mature females, too, have a form I condition, but their secondary sexual characters are not as well defined as those of the males.

Females generally lay eggs while in burrows, but burrows are not necessary for successful egg laying and incubation. Eggs are extruded from paired oviducts that open at the inner bases of the third pair of walking legs. A sticky substance, called *glair,* is produced by glands on the ventral surfaces of various abdominal segments. Eggs, it is assumed, are fertilized after extrusion. The female assumes a position on her back with her abdomen curled to form a pouch prior to laying eggs. This position facilitates fertilization and attachment of eggs to the paired abdominal appendages (swimmerets).

Once the eggs are firmly attached to the swimmerets, the female rights herself and incubation begins. Oxygen levels in burrow waters are usually too low to support developing embryos (Jaspers and Avault 1969). Generally, this water serves only to maintain a 100% humidity, which assures that oxygen can be absorbed through the gills of the female and the surfaces of the egg cases. The female constantly fans the eggs to aerate them and removes dead eggs. At temperatures of 20°–25°C (68°–77°F), hatching usually occurs in 2–3 weeks. At lower temperatures, development becomes progressively slower and stops altogether below 10°C (50°F).

Fecundity depends on the female's size and species. A 7.5- to 8.5-cm (3- to 3.5-in.) female may lay 100 or more eggs. A 12.5-cm (5-in.) female may well produce 600–700 eggs. The average is about 300 eggs per female. Once eggs hatch, the larvae resemble grotesque adults with exaggerated, yolk-swollen cephalothoraxes. They undergo two molts in the next 2 weeks after which time they are ready to leave the female and fend for themselves. However, after the second molt (3rd instar), they continually return to the mother, if possible, attracted by maternal brooding pheromones. Normally, the young are dispersed when the mother moves about in open water and the detached young are unable to return to her. Astacid crawfishes can leave the mother after one molt, but usually remain through two molts.

P. clarkii is especially successful in culture systems in the South because it may reproduce year-round (Huner 1978b). Other species typically reproduce only once per year. As one progresses northward, two reproductive strategies are apparent: either eggs laid in the fall do not develop until the following spring, or eggs are not laid until the warmer spring period and development proceeds normally. *P. clarkii* uses both strategies at northern latitudes (Suko 1956).

Growth and Molting

It takes a minimum of 11 molts for a young cambarid crawfish to mature. This takes 3 to 9 months depending on environmental conditions; water availability and temperature are the most critical factors. The change from a rapidly growing juvenile form to a slow-growing adult form involves a fairly dramatic metamorphosis, especially in males. Once males and females mature they become form I and usually mate. After young are produced, adults molt back to a quasi-juvenile form called the form II stage. One or two more molts may follow before they return to the form I stage and are capable of reproducing again.

The maximum life span for *Procambarus* spp. is 3–4 years, whereas that for *Orconectes* spp. and *Pacifastacus* spp. is 5–6 years. In pond culture or nature, few individuals of any species live longer than half their maximum age.

As with other arthropods, crawfishes must molt in order to increase in size. Molting is controlled by endocrine secretions mediated directly by the nervous system. Neurosecretions are inhibitory or stimulatory, depending on their source. Generally, removal of both eyestalks eliminates molt-inhibiting hormones produced by these organs. The principal molt-stimulating hormone, beta-ecdysone, is secreted as inactive alpha-ecdysone by the Y-organs located in the mandibullary segment and, perhaps, elsewhere (Huner and Barr 1981; Aiken 1980). Alpha-ecdysone is converted to beta-ecdysone in the hepatopancreas and testes.

Physiologists divide the molt cycle into five distinct stages: A, soft; B, postmolt; C, intermolt; D, premolt; and E, the molt itself. Young crawfish can complete the cycle in 6–10 days but the cycle's duration increases gradually with age. For actively growing crawfish about 60% of the cycle is spent in the premolt stage. Stevenson (1972) and Huner and Avault (1976, 1977) discuss molting in crawfishes at some length.

During the premolt stage, calcium carbonate is gradually removed from the old shell or exoskeleton. Much is lost to surrounding waters but enough is retained in the blood, hepatopancreas, and gastroliths to allow recalcification to about 50% of the intermolt level. The gastroliths ("stomach stones" or "eyes") are paired calcium carbonate concretions located on either side of the stomach slightly below and behind the eye proper. They, along with the uncalcified lining of the stomach, are shed into the lumen of the stomach at the time of molt. When adequate environmental calcium is present, mineralization is completed in 1–3 days depending on species, size, and temperature (Huner *et al.* 1978; Travis 1965).

Food Habits

Crawfish have been described as opportunistic omnivores. Two excellent papers (Momot *et al.* 1978; Lorman and Magnuson 1978) summarize existing literature on crawfish food habits. The bulk of the diet is microbially enriched detritus. Animal matter—including worms, insect larvae, mollusks, and zooplankton—is an especially important source of nutrients for immature crawfish. Vascular plants and epiphytic growths including plants and animals are readily consumed. Living plant material is a valuable source of carotenoids. In the absence of dietary carotenoids, crawfish take on an abnormal pale coloration (Huner and Meyers 1979, 1981).

Genetics

Very little is known about the genetics of crawfishes. Black and Huner (1980) review the status of genetic studies with *P. clarkii*. Virtually no work has been conducted with the exception of studies on the inheritance of simple recessive characters such as "silver or yellow" eye and light blue body color and sex-limited (female only-male lethal) dark blue body color. The great expense associated with genetic studies will, no doubt, limit work in this area. Selection for various traits, including rapid growth, could clearly expand species options. However, it is doubtful that select crawfish could be managed in state-of-the-art pond systems.

An especially nagging concern of southern crawfish culturists is the possibility that current harvesting methods are selecting for smaller crawfishes. That is, once crawfish reach minimum sizes of 6.5–7.5 cm (2.6–3.0 in.) they are susceptible to retention in traps, depending on mesh size. However, since crawfish do mature at smaller sizes, the removal of larger animals may leave smaller ones to produce the next year's crop. Studies are needed to determine the degree of heritability of size, but these will be complicated, no doubt, by the remarkable variation in minimum size at maturity in all species (Fig. 1.4).

Environmental Requirements

Although the various cultured cambarid crawfishes can survive a wide range of environmental extremes, most, regardless of geographic location, do best under similar conditions. These can be summarized as follows: water temperature, 20°–25°C (68°–77°F); dissolved oxygen, greater than 3 ppm (parts per million); total hardness, approximately 100 ppm; salinity, fresh or less than 5 ppt (parts per thousand); water pH, 6.5–8.5; free carbon dioxide, less than 5 ppm; and total ammonia, less than 1 ppm. The cambarids respond to unfavorable conditions by either slowing their molting rate or stopping it altogether. *P. clarkii* and *P. acutus* are both considered to be warm-water species because of their southern distributions, but the most rapid growth for these two species occurs in the late fall and early spring at temperatures of 20°–25°C (68°–77°F) range (Huner and Barr 1981). The shift from a fall–winter–spring growth cycle in the South to a spring–summer–

FIG. 1.4 Variations in size of mature, nongrowing Form I male *Procambarus clarkii*. Stunting is a serious problem, especially late in the season.
Courtesy of J. E. Barr.

fall growth cycle in the North is apparently associated with latitudinal/seasonal shifts in these optimal temperatures. It is not surprising that cultured cambarids can tolerate low dissolved oxygen and high ammonia levels since they are all residents of temporary lentic environments where such extremes occur each year (Fig. 1.5).

It is commonly accepted, based on studies by de la Bretonne *et al.* (1969), that total water hardness in crawfish culture systems should exceed 50 ppm as calcium carbonate for optimal crawfish production. Yet Huner (1978b) noted excellent production in two ponds with total hardness levels below 5 ppm. Total alkalinity did exceed 125 ppm in these ponds so that ionic stress was apparently not a problem. It is clear that the crawfish were able to obtain adequate calcium from food and substrate in the absence of dissolved calcium in the water.

Crawfish can be acclimated to high temperatures, but short-term exposure (12–24 hours) to temperatures greater than 33°C (91.4°F), especially for cool-water species, is not wise as mortality can be great. Fortunately, before temperatures reach these extremes, crawfish will typically retreat to cooler burrows. Most remain active and will enter traps at temperatures around 10°C (50°F), but sustained periods of such low temperatures dramatically decrease catches. Biggs (1980) provides a good review of thermal requirements for crawfishes.

FIG. 1.5 A typical ditch habitat occupied by *Procambarus clarkii* and *Procambarus acutus acutus* in Louisiana.
Courtesy of J. Huner.

When oxygen levels fall below 2 ppm, most crawfish actively move to the surface either by climbing available substrate or moving to the shore. They can utilize atmospheric oxygen as long as the gills remain moist; however, they are stressed and are exposed to all manner of predators under such conditions. While crawfish do seek atmospheric oxygen at lower oxygen levels, most can survive at levels around 1 ppm for up to 24 hours if they are denied access to the surface (Park 1945; Melancon and Avault 1977).

Most commercially important crawfish species tolerate salinities of 15–20 ppt for 4–8 weeks, but molting stops at these high levels. In *P. clarkii,* molting will continue indefinitely at salinities below 10 ppt (Sharfstein and Chafin 1979; Loyacano 1967). Late premolt *P. clarkii* will molt in deionized or distilled water and can survive for 3–6 weeks without food, achieving a hardness of about 50% of the intermolt level (Huner *et al.* 1978; Smith 1940). Biggs (1980) provides a useful review with tabular summaries of the environmental requirements for larger cambarid and astacid crawfishes.

All crawfishes are more susceptible to environmental extremes when they are in the late premolt state, during the molt itself, and immediately after the molt. Stress (transportation, low oxygen, temperatures, etc.) should be minimized in rapidly growing populations.

All crawfishes are sensitive to pesticides, especially insecticides. Huner and Barr (1981) provide a summary of pesticide toxicities for *Procambarus clarkii.*

Relationship Between *P. acutus* and *P. clarkii*

P. acutus and *P. clarkii* share common ranges, are very similar in appearance, and may be found in the same crawfish pond. Though percentages may vary greatly, *P. clarkii* usually accounts for 85 to 95% of the population.

Interest in the association between the species is twofold. First, there seems to be a consumer preference for *P. clarkii* in Louisiana. Second, *P. acutus* has smaller claws than *P. clarkii,* so that the percentage of meat recovered in a peeling operation is somewhat greater because only the abdominal ("tail") meat is retained.

Clearly, there must be some differences in niches for the two species if they are to coexist in the same area, but the differences are not very clear. Bean and Huner (1978) found that young 1-gram specimens of the two species had similar growth rates and survival when stocked separately or together at ratios ranging from 25:75 to 50:50. However, Lutz (1983) found that *P. acutus* grew twice as fast as *P. clarkii* in cool months in the same ponds but the sizes of both species at maturity were comparable.

P. acutus occurs naturally as far north as Wisconsin, whereas *P. clarkii* extends northward only into southern Illinois. In Louisiana, *P. acutus* releases young later in the fall than *P. clarkii* (Huner 1975). This implies differences in temperature tolerance; however, there is no real difference in acute high-temperature tolerance for the two species in Louisiana populations. This does not rule out subtle differences in tolerances to extreme temperatures but does discount major differences.

Differences in tolerance to low oxygen levels has been suggested by Huner and Barr (1981) to explain the relationship between the two species. *P. acutus* apparently prefers a lotic environment and has a reduced branchial area (less gill surface); however, in short-term studies Huner (1982) found no differences in acute oxygen tolerance between the two species. Long-term chronic studies might prove useful in differentiating the preferred habitats of the two species. For example, we have never observed *P. acutus* in marsh- and swamp-type ponds, which have sediments with high organic loads and a biological oxygen demand substantially higher than that of ponds constructed on better drained agricultural lands.

Another identifiable difference between the two species is that *P.*

clarkii spawns spontaneously throughout the year, but *P. acutus* usually spawns in the fall well after the first *P. clarkii* young-of-the-year appear (Huner 1975; Lutz 1983). If *P. clarkii* numbers are low for any reason, *P. acutus* may thrive and make up much of the catch, but it cannot compete if initial *P. clarkii* numbers are high. Then, too, there may be microhabitats in ponds that favor one species over the other regardless of when the young-of-the-year appear. Thus, the proportions of the two species in the catch may depend on the type and relative amounts of microhabitats in any particular pond as long as reproductive success is good.

DISEASES AND PARASITES

No major disease or parasite problems have been associated with North America crawfish species (Amborski 1980). This contrasts sharply with the situation in Europe (Unestam 1973) where *Aphanomyces astaci,* a fungus, has decimated the native astacids. Sporozoans, *Thelonia* spp., periodically reduce astacid populations in Europe, too. Neither malady has proven to be a problem in North America. The North American astacids and cambarids are resistant to *A. astaci* and may have been the source of the infection in Europe (Unestam 1969). These "domesticated" species appear to be remarkably resistant to most of the problems associated with European species and other cultured crustaceans such as prawns and shrimps.

Crawfish generally exhibit considerable resistance to diseases. Unestam (1973) provides a brief summary of the mechanisms involved. The exoskeleton is a formidable barrier to most invading organisms. However, much of the alimentary canal lacks chitin and may be more easily penetrated. Wounds or scratches are some of the most common places for entry of internal parasites. Crawfish blood itself does exhibit resistance to fungi, and bacteria grow very poorly in freshly prepared blood serum. Opsonization of foreign particles followed by phagocytosis occurs. Larger invaders of the blood, like small worms and fungal hyphae, are encapsulated by blood cells. Formation of melanin around the invader in both the blood and cuticle of crawfish may also play a role in their defense.

Two bacterial problems are encountered in *P. clarkii* (Amborski *et al.* 1975a, b), but they do not seem to cause harm in pond-raised crawfish. Cracked and broken areas in the exoskeleton may be attacked by chitinivorous bacteria. Erosion of the cuticle allows secondary invasion by these potentially lethal bacteria present in the surrounding environment. Entire body parts such as the uropods and telson can be lost before death occurs; however, molting eliminates the affected area if erosion has not penetrated the cuticle. Thus, chitinivorous bacteria are not a problem in rapidly growing animals and do not appear to cause problems in wild and domestic populations. Bacteria that can produce pathogenic endotoxins normally occur in crawfish gut flora in low numbers but cause no problems (Amborski *et al.* 1975b); such bacteria can bloom in a nutrient-rich environment such as that found when unpurged crawfish are crowded in shallow, hot water. Endotoxins from *Flavobacterium* sp. can accumulate and kill all crawfish in confined containers in 12–24 hours. Afflicted animals give the appearance of having been exposed to a neurotoxin. Other potentially dangerous bacteria isolated from crawfish guts include *Pseudomonas* sp., *Citrobacter freudii,* and *Aeromonas* sp.

Although sporozoans do not appear to be a problem in the culture of North American crawfish species, protozoan epibionts may cause respiratory problems in nonmolting crawfish. That is, they may, by sheer mass of numbers, interfere with respiration across gill surfaces. This appeared to be a problem in at least one case where adult, nonmolting brood crawfish were taken from Louisiana to Alabama (B. Nelson, Mississippi Cooperative Extension Service, Gulfport, MS, personal communication). Various genera encountered on the gills of *P. clarkii* have included *Zoothamnium, Corthunia, Epistylis,* and *Bodo* (Johnson 1977a, b; Lasher 1975). Modest numbers of these protozoans will always be present, but it is prudent to inspect crawfish before they are stocked.

All crawfishes serve as both intermediate and definitive hosts for parasites, flukes, and acanthocephalans (Table 1.1). It seems reasonable to assume that heavy helminth loads would adversely affect survival and fecundity, but virtually no studies have been conducted to date. *Paragonemus kellcoti* and *P. westermani* are mammalian lung flukes and potentially dangerous to man and his pets. Japanese specialists have suggested that their countrymen not eat *P. clarkii* because it is a vector of *P. westermani* when eaten raw in the traditional Japanese style (Huner and Barr 1981). Raw crawfish or their parts should not be eaten by humans or their pets.

Crawfish exoskeletons serve as substrates for brachiobdellid worms and their eggs, corixid bug eggs, and external epiphytic growths. Heavy infestations affect a crawfish's appearance but probably cause no physical harm. Brachiobdellid worms are tiny segmented annelid worms closely related to leeches and earthworms. They rarely exceed 3 mm in length and can be found attached to crawfish body surfaces, especially those that cannot be reached by grooming appendages. Some specialists feel that they do cause gill damage, but this has not been

TABLE 1.1. HELMINTH PARASITES FOUND IN *PROCAMBARUS CLARKII*

Taxon	Organism/Life-Cycle Stage	Host Loci	Reference
Acanthocephala	*Southwellina dimorpha* (encysted larvae)	Encysted and attached to external lining to gut.	Lantz 1973
Trematoda	*Crepidostomum cornutum*	Encysted in heart, hepatopancreas, pericardial membrane, and muscalature of cephalothorax.	Sogandares-Bernal 1965; Boettcher 1979
	Gorgodera amplicava	Encysted in lower quadrant of stomach wall usually at level of gastric mill.	Sogandares-Bernal 1965
	Macroderoides typicus	Encysted in cephalothoracic and antennal musculature.	Sogandares-Bernal 1965
	Maritrema obstipum	Encysted in central shaft of gills and hepatopancreas.	Sogandares-Bernal 1965
	Microphallus progenetirus	Encysted in cephalothoracic cavity.	Sogandares-Bernal 1965
	Microphallus opacus	Encysted in hepatopancreas.	Sogandares-Bernal 1965; Boettcher 1979
	Ochelosoma sp.	Encysted in abdominal musculature.	Sogandares-Bernal 1965
	Paragonemus kellcoti	Encysted in heart and surrounding membranes.	Sogandares-Bernal 1965
	Paragonemus westermani	Encysted in cephalothoracic region.	Hamajima *et al.* 1976
	Allocorrigia filiformis	Adults in green gland.	Turner and Corkum 1977

well documented. These worms typically lay white eggs in the depression on the dorsal surface of the rostrum. Holt (1975) reviewed brachiobdellid biology.

Water boatmen are small, 1–3 mm in length, hemipteran bugs belonging to the family Corixidae. They lay eggs on any convenient surface including crawfish exoskeletons. These eggs may cover, in extreme cases, most of the cephalothorax and much of the abdomen. Eggs are lost at each molt, but a very minute number of nonmolting adults may be immobilized by the sheer mass of attached corixid eggs. Epiphytes principally algae, will grow rapidly on the shells of crawfish that remain in open water. This, too, is unsightly and is a good indicator of poor growing conditions in a pond, although the growths do not appear to be pathogenic. Crawfish covered by corixid eggs and/or epiphytic algae are usually edible, but consumers may reject them on the basis of appearance. Fortunately, most of the material is physically removed by abrasion when crawfish are packed into sacks or held live in boxes and tanks.

Reproductively spent, senile *P. clarkii* exhibit a syndrome called "wasting disease," described by Lindqvist and Mikkola (1979), who recorded the problem in Kenya. Body tissues literally waste away. This has been known in Louisiana for many years and has been called the "hollow tail" condition by natives (Huner and Barr 1981). The common consensus in Louisiana is that the animals have used up body reserves while in burrows. Lindqvist and Mikkola (1979) report fairly high numbers of emaciated crawfish but a very small percentage is observed in Louisiana.

Crawfish eggs can become infected with fungi such as *Saprolegnia* spp. The fungus normally attacks unfertilized eggs and spreads through the "berry" to infect developing embryos. Fungal infection does not usually become a problem unless water quality is poor because the female crawfish removes dead eggs, thus eliminating the substrate for fungi.

Should concern arise over external parasites and ectocommensals, they may be controlled with theraputic agents used to control the same or similar organisms on finfishes (Williams and Avault 1977). These include sodium chloride, formalin, and potassium permanganate. Insecticides should never be used. Crawfish are treated by dipping them into a specific solution for several minutes. Occasionally increasing the salinity of culture water to 15 ppt for 10 days is said to help control bacterial problems in tank systems (Johnson 1977a, b).

POND CULTURE

Ponds and Pond Construction

Ponds for growing warm-water crawfish may arbitrarily be classified as ricefield, open, wooded, and marsh (Avault *et al.* 1969). In ricefields, crawfish are grown in rotation with rice. Since rice requires only about 10 cm (4 in.) of water, it may be necessary to raise the height of levees to accommodate culture of crawfish, which require 30 to 50 cm of water. Ricefield ponds may be constructed with both a levee plow pulled with a tractor and with a bulldozer. A levee plow throws up interior contour levees, if any are required, and a bulldozer is used to construct those levees requiring traffic.

Open ponds are similar to ricefield ponds, but the only crop is crawfish (Figs. 1.6 and 1.7). Open ponds are typically constructed with a

FIG. 1.6 Open crawfish pond in Texas with rice planted solely for crawfish fodder. Lanes promote water circulation and aid in trapping crawfish.
Courtesy of J. Huner.

FIG. 1.7 Large, open Louisiana crawfish pond with vegetation cut and baled to reduce oxygen problems. This is, in theory, a good practice but is not well worked out.
Courtesy of J. Huner.

bulldozer. The ideal size is 8 ha (20 Ac). Water management is more difficult in ponds larger than this. Open ponds are constructed much like catfish ponds; step by step methodology for building catfish ponds is summarized by Avault (1980). We might note here a few modifications for construction of crawfish ponds. First, levees need be only high enough to allow 30–50 cm (12–20 in.) of water. Deeper ponds, if already available, will do fine for crawfish culture, and some farmers use catfish ponds to grow crawfish. But if one has a choice, lower and thus less expensive levees should be constructed. Drain pipes should have two antiseep collars, to discourage crawfish and rodents from burrowing and causing leaks (Fig. 1.8). The antiseep collars may be

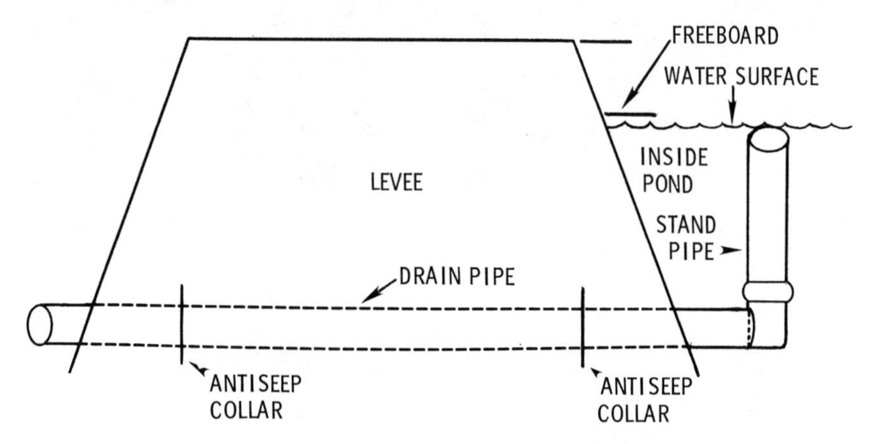

FIG. 1.8 Collars to prevent crawfish from burrowing and causing leaks.

of metal, fiberglass, concrete, or other materials. It is desirable, but not critical, for crawfish ponds to have smooth bottoms. Another suggested modification is to construct baffle levees inside crawfish ponds to aid in proper water circulation (Fig. 1.9). Boat/trapping lanes can also be arranged so as to encourage water flow through the pond (Fig. 1.5). Water pumping and flushing is required, particularly in the fall as vegetation in the pond decomposes. Baffle levees can be constructed with levee plows; in very large ponds of 15–25 ha (37–62 Ac), the levees may be beneficial to crawfish for burrowing requirements.

Wooded ponds contain varied amounts of trees and shrubs (Fig. 1.10). Such ponds consist of a ring levee around a low-lying wooded tract of land. The land is usually too wet and boggy for traditional farm crops and will often not support a bulldozer. A dragline usually is used to construct the levee, and the bottom, except for the borrow ditch, is left undisturbed during construction. Many wooded ponds are never

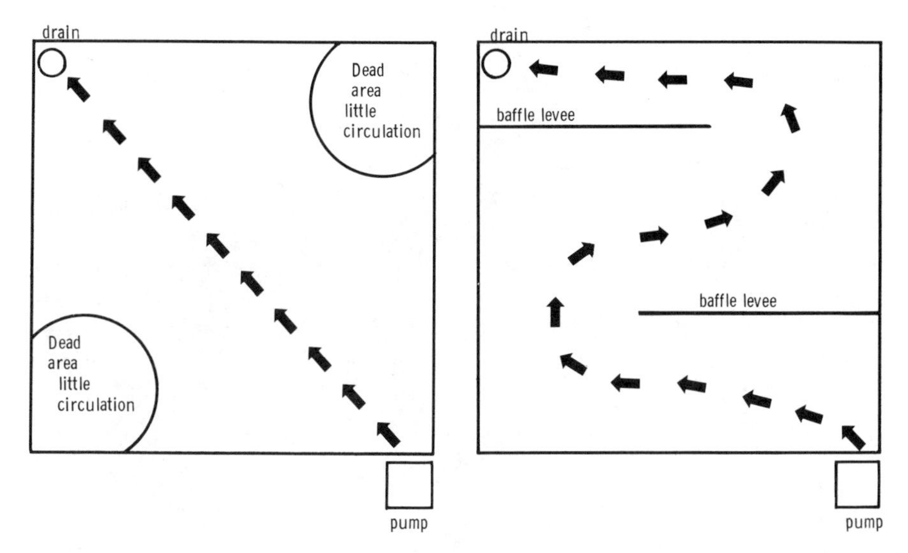

FIG. 1.9 Baffle levees constructed to aid water circulation.

stocked with crawfish; native crawfish move in and subsist on detritus from leaves and other natural vegetation. Poor water circulation and acidic conditions are the rule in such ponds.

Marsh ponds are found along the coastline (Fig. 1.11). High peat soils and brackish water typify these ponds. Levees are constructed

FIG. 1.10 A wooded Louisiana crawfish pond.
Courtesy of J. Huner.

FIG. 1.11 A flooded slough, unsuitable for agriculture but converted to crawfish culture at minimal expense.
Courtesy of J. Huner.

with draglines on mats. To fill and drain ponds, pumps must be used. Techniques for building marsh ponds are covered by Perry (1972).

Water and Its Management

Water Sources. Water for crawfish farming comes from natural streams, water wells, or estuarine bays. Most crawfish farmers in Louisiana, except rice/crawfish farmers, take water from bayous and other streams. In Louisiana, it is legal to pump water from a public stream onto private land (Williams *et al.* 1975), but other states may have regulations prohibiting this. If stream water is used, it should be free from pollutants and toxic substances, particularly insecticides. Wild fish must be screened out as they may become serious predators of crawfish. Saran screen, 21 meshes per cm, is effective but clogs up so much as to be impractical. However, in crawfish farming, it is not critical to screen out fish eggs and fry; only fingerlings and larger fish need be excluded. This is because of the way crawfish are farmed. Ponds are filled with water in the fall and completely drained by June. Stream water is typically low in dissolved oxygen and must be well aerated as it enters ponds. This is best accomplished by allowing water to cascade over two to six mesh plates from 2.5 to 0.6 cm mesh (1 in.–0.25 in.) (Fig. 1.12); in the process larger fish are removed.

Well water is an excellent source of water. It has no wild fish and should be free of toxins and pesticides. It is often chronically low in dissolved oxygen (D.O.) and may be supersaturated with nitrogen and/or high in free carbon dioxide. Aeration will overcome these difficulties. Estaurine waters should be suspected of the same kinds of problems as those associated with stream water. In addition, water that is too saline is not suitable for crawfish. Loyacano (1967) noted that newly hatched young were killed by 15 ppt salinity; sea strength is about 35 ppt. Growth of crawfish was satisfactory up to a salinity of 10 ppt, but data are lacking on the effect of salinity on reproduction. At this point it seems safe to culture crawfish in salinities ranging from 1 to 3 ppt.

Quantity of Water Required. The volume of water needed to fill a pond is usually expressed in acre-feet or cubic meters. A water depth of only 1.5 feet (45 cm) is needed; therefore 1.5 acre-feet (489,000 gallons) of water are required to flood each acre of crawfish. Once craw-

FIG. 1.12 Tiered aerator serves two purposes including aeration of incoming water and removal of larger fish.
Courtesy of J. Huner.

fish ponds are filled, flushing water for approximately one to two water changes is often necessary to offset low dissolved oxygen. Since crawfish are grown primarily during cool months, evaporative loss is minimal. In many instances, winter and spring rains more than compensate for evaporative losses. As a rule, a water well producing 1000 gal/min (3750 liters/min) is considered adequate for a 32-ha (79-Ac) crawfish farm.

Quality of Source Water and Soils. As stated earlier, source water should be free of pollution, toxins, pesticides, and wild fish. Water of the same quality as required for culture of warm-water fish species is generally acceptable for crawfish culture. To review, dissolved oxygen should be 3 ppm or greater (Melancon and Avault 1977); water temperatures, above 20°C (68°F) for optimum growth; water pH, 6.5 to 8.5; alkalinity, 50 to 200 ppm or greater, and optimum bicarbonate alkalinity, above 100 ppm; and total water hardness, 50 to 200 ppm. de la Bretonne and Avault (1971) found that a total hardness of 100 ppm gave best crawfish growth. Above this, crawfish grew no better than they did at 100 ppm. Soils should have enough clay to hold water. Acid soils may adversely affect crawfish production. de la Bretonne and Avault (1971) found, for example, that crawfish production could be more than doubled by liming acid soils. The best soil pH is 6.4. Boyd (1979) reviews liming of fish ponds.

Water Management. A major concern in the water management of crawfish ponds is oxygen depletion (Avault *et al.* 1975). Crawfish ponds contain natural and/or planted vegetation as a food source for crawfish. Filling ponds with water in the fall encourages decomposition of vegetation, especially of terrestrial plants, and oxygen depletion may result. Flushing ponds with fresh water is a common practice to offset low dissolved oxygen (D.O.). Aeration and water circulation also hold promise as techniques to counteract oxygen depletion, but research is lacking. Currently, the best approach to avoid low D.O. includes the following practices: (1) mow circulation access lanes in terrestrial vegetation several weeks before flooding ponds; (2) flood ponds to 15 cm (6 in.) initially, 30 cm (12 in.) 2 to 3 weeks later, and then to 45 cm (18 in.) after another several weeks—this gradual flooding retards decomposition and oxygen depletion; (3) if the weather is very hot, delay flooding of ponds until cooler weather sets in.

Stocking Crawfish Ponds

Brood crawfish are usually stocked into ponds from April through June. At least 2 weeks prior to stocking, ponds should be filled with at least 30 cm (12 in.) of water. Brood crawfish are stocked into ponds at rates varying from 20 to 65 kg/ha (18–59 lb/Ac). If some native crawfish are already present and/or if vegetation is available to provide cover, then the lower rate may be used. If not, then use the higher stocking rate.

The most economical time to stock crawfish in Louisiana is late May or June. Crawfish become tough at this time, and prices are reduced. When stocking late in the season, however, one must be sure that there is about a 50:50 sex ratio. Often females become scarce relative to males. Sexually mature males are identified by hooks on their walking legs. Brood crawfish can be purchased from a processing plant or from a farmer. If purchased from a plant, crawfish should not be held in a cooler beforehand.

When ponds are stocked, brood crawfish are distributed all along the inside of the main levee on all four sides. Crawfish typically burrow underground in the levee rather than in the middle of the field. Once burrowing is complete, water may be removed slowly over a period of about 2 weeks. In newly constructed ponds with no old burrows present, more time may be needed. Many females are bred before stocking, but some mating no doubt occurs in burrows. Bred females store sperm in a seminal receptacle. Though females may lay eggs at any time of the year, they typically do so in the fall. At this time, eggs are fertilized and held on the abdomen with a sticky substance called glair. When water flushes crawfish out of burrows, females release their young, and both young and adults begin foraging.

Forages

Natural volunteer vegetation—including perennials such as alligator weed *(Alternathera philoxeroides),* water primrose (*Ludvigia* spp.), and smartweed (*Polygonum* spp.) and annuals like grasses and sedges—provides good forage for crawfish. Many farmers still rely on natural vegetation. However, research has shown that crawfish production can be increased, over that utilizing natural vegetation, by planting rice. Crawfish feed on green rice plants, but it is the decomposing rice straw, covered with bacteria, fungi, and other microorganisms, which benefits crawfish most.

Seed of a rice variety that produces a lot of foliage, such as Saturn, Mars, or Melrose, should be soaked in water for 12 hours to presprout and then left to drain another 12 hours. After water has been drawn down following crawfish stocking to a depth of about 10 cm (4 in.), rice can be broadcast at an approximate rate of 110 kg/ha (97 lb/Ac). After planting, the remaining water is removed within 24 hours. Although planting rice in June will give low yields of grain, the objective, in this case, is to produce fodder for crawfish. An alternative to wet planting of rice is drill planting on the dry bottom at a seeding rate of about 70 kg/ha (62 lb/Ac). No water is added to the pond except for irrigation during dry periods.

In Louisiana, crawfish ponds are typically reflooded from September 15 to October 15. Although crawfish begin feeding on planktonic and epiphytic organisms and green rice leaves, the detrital food chain, composed of decomposing rice straw or other vegetation, gives highest crawfish production. A number of other materials—millet, sweet potato vines and trimmings, hays, sugar cane wastes, and manures, to name a few—work well as detrital substrate, but the practicality of hauling in these materials must be considered (Goyert *et al.* 1975–76; Goyert and Avault 1977; Clark *et al.* 1975; Rivas *et al.* 1979; Avault *et al.* 1978; Johnson and Avault 1980; Miltner and Avault 1980, 1981). When rice straw first begins to decompose, the carbon:nitrogen ratio can be as high as 60:1. As decomposition progresses, the C:N ratio becomes lower and the detrital substrate becomes more nutritious. Bacteria and other high-protein microorganisms serve as food for crawfish.

Though rice straw is an excellent detrital substrate, it normally is depleted by March in Louisiana. With the food supply dwindling, crawfish may cease to grow and be stunted. To offset this, one may feed a commercial ration, add additional substrate such as hay, plant forages with differing decomposition rates, and harvest intensively. In early studies (Smitherman *et al.* 1967), crawfish grew well on pelleted rations, but at that time the economics of feeding crawfish a commercial ration were questionable. Probably the best option is to start off with rice as a forage then use feed rations as detritus runs out. Cange *et al.* (1982) recorded improved yields when crawfish were fed cattle range pellets, which cost $100 per ton in 1981, after rice forage became depleted.

Adding additional hay to ponds, once the rice forage is nearly depleted, is economically feasible according to farmers and it allows the detrital food chain to continue for a longer period. In Louisiana studies (Day 1983), almost 4000 kg/ha (3520 lb/Ac) of crawfish have been grown in static experimental ponds by addition of rice hay. Once the original rice plants near depletion in late February, bails of rice hay are added in piles within the pond. The quantity of hay to use per hectare should about equal the quantity of hay harvested from a hectare of land, but it should not be added all at one time. To reduce oxygen depletion problems, an initial application and three additional applications 3 to 4 weeks apart are recommended. Wheat straw, alfalfa hay and other available materials also are suitable as detrital substrate.

Though definitive research is lacking, it seems logical that planting two or more forage crops in the same pond with differing C:N ratios (decomposition rates) would stagger the food substrate over the entire growing season. Leafy varieties of sugar cane are being developed in Louisiana to plant in combination with rice. Sugar cane produces high quantities of biomass and has an initial C:N ratio over double that of rice. The fourth option to offset depletion of feed substrate is to harvest intensively. This is probably the most important factor in minimizing stunting. For years Louisiana farmers would run 25 traps/ha (10 traps/Ac) once a day for about 100 trap days. Pfister (1982), however, demonstrated that crawfish catch and production could be greatly increased with improved harvesting techniques.

Harvesting

Crawfish are typically harvested with polyvinyl-coated, 19 mm (0.75 in.) mesh wire traps baited with gizzard shad *(Dorosoma cepedianum)* or other fish; however, smaller mesh (12 and 15 mm) traps are often used early in the season when prices are high and crawfish are scarce. Baited traps are usually run with two men in a boat—one handling the boat, the other running traps—but one-man systems are becoming popular.

Traps and Trapping. Though well over a dozen trap varieties are used, most can be broken down to one of two broad catagories—pillow or stand-up traps (Figs. 1.13 and 1.14). A pillow trap has one or more funnel openings, and the trap lies flat on the bottom submerged. The end opposite the openings is crimped shut. The stand-up trap has two or more funnels at the bottom with the top open and extending above the water surface. A strip of metal, 7.5 cm (3 in.) wide, may be placed around the inside of the trap to prevent crawfish from climbing out. Pfister (1982) set traps at rates varying from 25 to 100/ha (10–40/Ac). Best catch was at a density of 100 traps/ha (40 traps/Ac) but when he considered the economics, 75 traps/ha (30/Ac) was most profitable. He also documented that production of crawfish could be further increased by running traps up to three times a day instead of the customary once a day. Many farmers run traps only 100 days during the harvesting season even though 150 trap days would optimize production. Reasons given by farmers for this low number of harvest days are difficulty in getting bait and its cost, low prices for crawfish, not enough trapping labor, and need to attend to other crops.

Baits. Lack of a good crawfish bait during the crawfish season is a major concern. Heretofore, the preferred bait has been gizzard shad,

FIG. 1.13 Vertical trap with support rod and crawfish retainer ring around top. Two funnel entrances at bottom. *Courtesy of J. Huner.*

Dorosoma cepedianum (Fig. 1.15), but it is not always available, must be kept in a freezer, and requires labor to thaw and chop into proper-sized pieces; in addition, it is often expensive, costing as much as $0.22 per lb (454 g). Other fish, such as menhaden, are used, but they are less effective than shad and have the same drawbacks as shad.

Several artificial baits have been developed (Collazo 1981; Pollock 1982) which show promise. An artificial bait should (1) catch crawfish as well as shad does in both cool water (below 20°C or 68°F) and warm water, (2) require no special freezer or storage space, (3) be portion size for traps, and (4) cost less or no more than shad. Any artificial crawfish bait has three basic components: a carrier, a binder, and an attractant. The carrier forms the bulk of the bait and consists of an animal ingredient such as fish meal and a vegetable ingredient such as soybean meal or cottonseed meal. The binder, such as soya flour,

FIG. 1.14 Cone shaped, self-supporting trap with four funnel entrances and bait well.
Courtesy of J. Huner.

FIG. 1.15 Striped mullet and gizzard shad are used for crawfish bait. Tarp keeps flies away.
Courtesy of J. Huner.

holds all materials together. The attractant might be blood meal, powdered eggs, or fish oil.

Harvesting Techniques. Crawfish are commonly harvested commercially using a crawfish combine, crawfish buggy, or boat (Huner and Barr 1981). The crawfish combine can be operated by one man (Fig. 1.16). A metal wheel mounted on one end of the boat digs into the pond bottom to move the boat along. The motor-powered wheel is guided by foot pedals. The advantage of this device is that only one man is needed to both propel the boat and run traps. The crawfish buggy is is essence a high-boy tractor that stradles traps as it moves through the water. The single operator directs the buggy with foot pedals and also runs and rebaits traps. Boats for harvesting crawfish may be propelled by outboard motors, modified transom-mounted, air-cooled motors (go-devils), or airplane engines (air boats, see Fig. 1.17).

The electro-trawl (Caine and Avault 1983) consists of a surface trawl pushed through the water (Fig. 1.18). An electrical current (pulsed direct current) in front stimulates crawfish to flip off the bottom into the path of the oncoming trawl. This technique requires no traps or bait, and it can harvest soft-shelled crawfish. The electro-trawl is still in the development stage but shows much promise.

FIG. 1.16 Cajun Combine. Crawfish/bait trough is set ahead of operator's seat when in use.
Courtesy of J. Huner.

FIG. 1.17 Airboat used to harvest crawfish. Trough holds crawfish and bait.
Courtesy of J. Huner.

FIG. 1.18 Experimental electro-trawl harvesting system.
Courtesy of J. Avault.

Harvesting in south Louisiana may be started in late November. At this time, usually only a month and a half after flooding, originally stocked brood crawfish are captured (holdover juveniles and some fast-growing young-of-the-year are usually present in older ponds). But a trapper must check two things when he begins harvesting. First, if many captured females contain ovarian eggs or are carrying eggs, a trapper could be removing a potential crop. de la Bretonne and Avault (1977) discuss this and suggest that trapping be delayed until after brood females have spawned. The second thing to look for is emaciated, hollow-tailed crawfish. Brood crawfish, which burrowed during the summer, are often in poor condition in November, and when the tail is removed from the body, the tail meat is stringy. It would be better to wait until crawfish are fattened before harvesting them. Newly molted, shiny crawfish are usually acceptable for market. The best procedure then is to set out only a few test traps to check for eggy females and hollow-tailed crawfish. Assuming neither condition exists, trapping may begin.

Most farmers trap each pond daily, with an average daily harvest per trap of 0.5–1 kg (1.1–2.2 lb). Sometimes in December or January, harvest catch may drop considerably. This can occur if most of the brood crawfish and holdover juveniles have already been removed by trapping and most young-of-the-year are still too small to be held in the 19-mm mesh traps. In addition, the water may have become so cold (10°C or 50°F) that crawfish are inactive. The farmer should then periodically spot check with test traps to determine when it is worth his while to resume normal trapping activities. Once resumed, trapping should continue daily until late May or early June depending on geographic location. In south Louisiana, most farmers stop harvesting in mid-May, but in Mississippi the season runs a month later. Regardless, harvesting should probably cease when crawfish become tough and dark in color, which denotes that growth has stopped, when extensive catch drops drastically, and when burrowing activity is found. When these conditions are noted, water should be slowly removed from the pond to encourage further burrowing by the crawfish. Rice is then replanted as before. The second year, however, brood crawfish need not be stocked. Those crawfish remaining after harvest serve as the next year's brood stock.

Population Monitoring

Population monitoring is an important aspect of crawfish culture because hatching takes place in distinct pulses, usually two to four, in the fall and early winter. At least two age classes, adult brooders

FIG. 1.19 Drop chain-weighted seine can be used to monitor crawfish populations. Note large crab from coastal Louisiana crawfish pond. Large numbers of crabs can decimate crawfish crops.
Courtesy of J. Huner.

and immature juveniles, remain from the preceding year (Huner 1978b). Periodic samples may be taken with small mesh traps and nets to observe growth rates and relative densities. Dip nets are most useful when all pond types are considered (Huner and Barr 1981), but weighted minnow seines work (Fig. 1.19) well in soft-stemmed vegetation and provide better sampling of larger crawfish that would otherwise have to be caught in small mesh traps (Momot and Romaire 1981). Pond monitoring and population dynamics are discussed at some length by Huner (1978b), Huner and Romaire (1979), Huner and Barr (1981), Momot and Romaire (1981), and Lutz (1983).

Predator Control

Fish, aquatic insects, birds, and mammals all prey on crawfish. Blue crabs *(Callinectes sapidus)* can be a problem in coastal areas, too. Fish are best controlled by screening the source water. After crawfish have been harvested and ponds drained, the remaining pools can be treated with 2 to 3 ppm 5% rotenone solution. Antimycin can also be used to control scaled fish but is less effective for control of bullheads and other catfish (Brown and Avault 1975).

Three groups of aquatic, predaceous insects feed upon crawfish young: the hemipterans (true bugs), coleopterans (beetles), and the odonates (dragonflys). The hemipterans and beetles must breathe air, whereas the nymphal stages of dragonflys are truly aquatic and have gills. Hemipterans are very carnivorous and include the water scorpion (*Ranatra* spp.), giant predaceous water bugs (*Belostoma* spp. and *Lethocerus* spp.), and backswimmers (*Notonecta* spp.). Hemipterans have piercing beaks and can inject proteolytic enzymes (digestive juices) that kill and digest their prey. Coleopterans include a wide diversity of species, but the giant predaceous diving battle larva (*Cybister* sp.) is the only major species that threatens young crawfish. Dragonfly nymphs are a threat to crawfish, but adult forms leave water and prey on flying insects.

Ordinarily these various aquatic insects do not appear to seriously damage young crawfish under pond conditions at the current state-of-the-art. If, however, they do cause damage, pond filling can be delayed until after mid-September in Louisiana when some insects such as *Anax junius* and *Belostoma lutarium* are less numerous (Witzig *et al.* 1980). Chemicals used to control insects include 0.25 ppm methyl parathion, 4 ppm Dylox, or 0.25 ppm Baytex®. One must remember, however, that crawfish are easily killed by these insecticides and we do not advocate use of insecticides to control insects. A 75:25 mixture of diesel fuel and motor oil or cottonseed oil can be sprayed on the water surface to form an oil slick. This clogs the breathing tubes of air-breathing insects such as back swimmers. If vegetation is present in ponds, this treatment is not effective because it is difficult to obtain an unbroken oil slick. No intentional efforts are now made to control predaceous insects in Louisiana crawfish ponds (Huner and Barr 1981).

Some aquatic birds, such as herons, egrets, ibises, and ducks, may feed heavily on crawfish (Huner and Abraham 1981). Bird predation can be lessened by flooding ponds rapidly in the fall and by having vegetation as cover in ponds. When the crawfish-harvesting season is complete, ponds should be slowly drained over a week or two to allow crawfish sufficient time to burrow.

Raccoons *(Procyon lotor)* and mink *(Mustela vison)* prey on crawfish and, in some instances, may significantly reduce populations at stocking time when crawfish are most vulnerable. In one instance, a farmer stocked a pond shortly before dark with crawfish that had previously been held in a cooler and were sluggish; raccoons virtually eliminated the crawfish. Daytime stocking with crawfish that have not been held in a cooler overcame this problem.

To sum up, fish, aquatic insects, birds, and mammals all prey on crawfish, but with a few precautions, none should pose a serious problem particularly once crawfish have become established in a pond.

Rice/Crawfish Double Cropping

Techniques for double cropping rice and crawfish are well developed (Chien and Avault 1979, 1980a, b). A general account of double-cropping practices in Louisiana (Avault 1980) is given in this section.

In Louisiana, rice is usually planted in March or April by either wet planting or dry planting. If wet-planted, rice may be presprouted by soaking in water for 12 to 24 hours then placing it under a tap to keep heat in for up to 24 hours. During warmer months there is no need to presprout rice. After several inches of water are added to a previously disked field, rice seed may be broadcast at a rate of 140–151 kg/ha (125–135 lb/Ac). Water is usually drained off the field 1 to 3 days after planting. For dry planting, a well-prepared seed bed is drill-planted at about 112 kg/ha (100 lb/Ac).

A soil test should aid in determining fertilizer requirements. Usually all of the phosphorus and potassium and one-half of the nitrogen are applied at time of planting. The remainder of the nitrogen is top-dressed before the rice is half grown.

Fungicides, herbicides, and insecticides are usually used in growing rice. Fungicides, used as seed protectants, were once thought to be very toxic to crawfish. From Louisiana studies, Cheah *et al.* (1979, 1979–80, 1980) concluded that fungicides as a whole are not very toxic to crawfish except for Arasan 70-S Red (96-hr $LC_{50}=4.3$ mg/liter) and Difolatan (based on field study). At this stage of knowledge, Kocide SD (96-hr $LC_{50}=2918$ mg/liter) appears to be the best choice for rice/crawfish farmers who wet-plant, whereas Captan 80 WP (96-hr $LC_{50}=15{,}631$ mg/liter) appears to be a safe choice for drill planting. Benlate 50 WP (96-hr $LC_{50}=1032$ mg/liter) is another commonly used fungicide that is relatively safe to use.

Propanil, ordram (molinate), and 2,4-D (2,4-dichlorophenoxyacetic acid) are some commonly used herbicides. Propanil (96-hr $LC_{50}=7.9$ mg/liter) is relatively toxic to crawfish. It is usually applied on dry plots at a rate of 2.8–5.6 kg/ha (2.5–5.0 lb/Ac) to control barnyard grass. One way to avoid herbicide toxicity is to wait 2 weeks after treatment before stocking crawfish. If crawfish are already present and a herbicide is normally applied to a wet field, water can be removed until the base of the levee(s) is exposed before treatment and added back several days to a week later. It should be noted, however, that the greatest rice injury, as well as the best weed control, takes place when there is no flood. As the depth of flood increases, rice injury decreases, but so does weed control. An all-around compromise is to apply herbicides at the maximum water depth that does not reach the majority of burrows along levees. This may be an inch or two of water. At this depth, the chance of the chemical being carried into crawfish burrows is reduced. Ordram (96-hr $LC_{50} = 140$ mg/liter) is roughly 17 times less toxic to crawfish as propanil and is usually applied at 2.2–2.8 kg/ha (2.0–2.5 lb/Ac) for control of grasses and certain sedges. Ordram is rather stable under anaerobic conditions and may remain in soils for up to 4 months after application. The herbicide 2,4-D (96-hr $LC_{50} = 1389$ mg/liter), used to control broad-leafed weeds, is relatively nontoxic to crawfish and should pose no problems when used properly.

Insecticides in general are highly toxic to crawfish, with malathion (96-hr $LC_{50} = 50$ mg/liter) an exception. Furadan (96-hr $LC_{50} = 0.5$ mg/liter), commonly applied at a rate of 0.56 kg/ha (0.5 lb/Ac) for control of the rice water weevil, is highly toxic to crawfish. Drawing water off the field before application of furadan lessens the chance of its reaching crawfish burrows. Some farmers opt not to use furadan, figuring that the lower rice production is offset by higher crawfish production. In the Philippines, where fish and rice are grown together simultaneously, furadan is incorporated into soils at planting. Rice plants take up the insecticide through the roots for weevil control; and fish, which are stocked later, are not killed by the insecticide.

Pesticide application must be timed to avoid harm not only to crawfish but also to the rice plant. The seedling stage of rice is perhaps the most delicate stage, and as the plant gets older, it becomes less sensitive through the tillering stage and up to the boot stage. During the boot stage, heading, flowering, and grain development, the sensitivity of rice is again high. Therefore, the best time for treatment is after the tillering stage and before the boot stage, usually 7 to 8 weeks after planting. If insecticides must be applied to control rice stink bugs during grain development, it is probably advisable to drain the pond. However, the rice canopy should minimize the amount of insecticide reaching the soil-water below. Sevin and malathion should be used, if possible, because of their "low" toxicity to crawfish.

Residues of pesticides in crawfish are an important consideration. In preliminary field studies in Louisiana, propanil, ordram, and furadan were not noted in crawfish flesh; however, further tests are required before concrete conclusions can be made.

Most rice varieties take 100 to 120 days to mature, and harvesting usually begins in August. Water that is on the rice field for weed-control purposes is drained about 2 weeks before rice harvest. Between September 15 and October 15 rice stubble is reflooded. Crawfish emerge from burrows, and females release young. Rice stubble should be left standing and not bush-hogged or disked under. Crawfish feed on the ratoon growth (new green plant growth), on plankton and periphyton (organisms attached to rice stubble), and on rice detritus. The rice hay also affords hiding places for molting crawfish. Harvesting of crawfish begins as early as Thanksgiving and continues until the following year at which time the farmer has two options.

The first option is to continue crawfish harvesting until April, at which time water is slowly removed over a week's time to allow crawfish burrowing and then rice is planted. If a large number of crawfish remain, the pond may be overcrowded the next year, resulting in stunted crawfish. This is why it is important to trap heavily when crawfish are going into traps well; it is impossible to trap too intensively. Rice can be wet-planted when about 7.5–10 cm (2–4 in.) of water remain in the dewatering pond. There is no need to drain the pond completely, disk, and then plant rice. The wet-planting method, tested in Louisiana, saves time, labor and money, and weed control is excellent. Moreover, crawfish do a good job of working up the soil. The second year, crawfish will not have to be restocked; those remaining serve as brood stock for the next crop.

The second option is to continue crawfish harvest until late May or early June. This allows a farmer to maximize crawfish production, since peak harvesting occurs in March and April, but it is often too late to plant rice after crawfish harvest. However, in Louisiana farmers can plant soybeans up to June 15 and still realize a good crop. One method for planting soybeans following crawfish, is to pull the water down and then fly (broadcast) soybeans directly on the wet field. For a weed-free stand of soybeans, this technique works well—not to mention the savings in time, labor, and money usually involved in soil preparation.

Polyculture of Crawfish with Other Species

In Louisiana studies, crawfish were cultured in replicated ponds with channel catfish *(Ictalurus punctatus),* bigmouth buffalo *(Ictiobus cyprinellus,* and paddlefish *(Polyodon spathula)* (Tuten and Avault 1981). Buffalo and paddlefish were stocked in the fall after ponds were filled. Catfish were stocked in cages in the spring and later released after the bulk of the crawfish had been harvested. Catfish were fed; other species were not but subsisted on waste feed and on natural pond organisms. Total net production of all species combined averaged 4947 kg/ha (4354 lb/Ac). Crawfish net production averaged 1345 kg/ha (1184 lb/Ac). Production of the fish species in kg/ha were as follows: catfish, 3191; buffalo, 302; and paddlefish 108 (2808, 266, and 96 lb/Ac, respectively). Huner (1976) discusses crawfish–fish polyculture in established finfish systems. Crawfish–fish polyculture is not widely practiced commercially.

SOFT-SHELLED CRAWFISH

Soft-shelled crawfish have long been popular, expensive fish baits in the Midwest. Small, 2- to 8-g crawfish are sold individually for as much as $0.25 each. Food markets have only recently been developed in Louisiana for larger soft-shelled crawfish weighing 10–30 g each. Wholesale prices in 1983 of $14.30 per kg ($6.50 per lb) for food-size soft-shelled crawfish compared favorably to maximum wholesale prices of $2.20 per kg ($1.00 per lb) for hard-shelled crawfish. Soft-shelled crawfish are readily accepted as substitutes for soft-shelled blue crabs.

There is no precise definition of a soft-shelled crawfish. Certainly, an animal that has just completed molting is soft, but the calcification process begins immediately. The soft stage is usually completed in about 12 hours; the shell becomes progressively harder as calcium carbonate is deposited in the uncalcified premolt shell and new matrix is generated. However, crawfish remain very flexible during the postmolt stage that lasts about 36–48 hours after the soft stage ends. These so-called "paper shell" crawfish enter traps and can be sold as "soft shell" crawfish. Late premolt crawfish are often called "busters" because they are ready to "bust" free from the old shell. It is not unusual for such busters to die, and the shell may be physically removed as a salvage operation. While added labor is involved in removing shells, such animals are still being sold at the same prices as soft crawfish.

Rapidly growing crawfish have much thinner shells than reproductively active crawfish. In Louisiana, these are referred to as "green" crawfish because they have a brownish-green color. Intermolt, "hard," crawfish, less than 7.5 cm long (3 in.), can usually be fried whole and, although a bit crunchy, can be a less desirable "soft-shelled" crawfish substitute. In some areas of Texas, whole tails of larger, green crawfish are fried, shell and all.

Production

Soft-shelled crawfish can be produced in two ways. Hard-shelled crawfish can be caught, preferably in the premolt stage, and held until they molt. Alternatively, soft animals can be caught with active gear like a seine or electro-trawl. Both methods are discussed below.

It is very difficult to distinguish intermolt and early premolt crawfish by eye. Reliable techniques are available for observing the development of new setae, or hairs, at the margins of the tail fan (uropods and telson). Magnification of 40–100 × is required. Briefly, flesh separates from the edge of the uropods in early premolt and new setae gradually develop as tubes within tubes that resemble columns. When the tubes are completed, the crawfish has reached the mid-premolt stage. Breaking the tip of a claw will reveal a new uncalcified claw tip below. Stevenson (1972) and Huner and Barr (1981) provide useful guides for distinguishing various molt stages microscopically.

Calala (1976) discusses trough systems for holding shedding *Orconectes* spp. in Ohio. Essentially the same techniques were developed by Huner and Avault (1981) in Louisiana. Calala (1976) suggests that densities of 500 5.0-cm *Orconectes* spp./m^2 (46 2-in. animals/ft^2) can be maintained in shedding tanks. This seems too high in view of the growth studies of Goyert and Avault (1978, 1979). We suggest that, when mixed intermolt and early premolt crawfish are to be held, densities of 5.0-cm (2-in.) animals should not exceed 40/m^2 (4/ft^2) and that they not exceed 5–10/m^2 for animals that are 7.5–10.0 cm long (0.5–1.0/m^2 for 3- to 4-in. animals). Once the mid-late premolt stages are reached, densities can be increased to near the biological limits of the individual system, as molting will take place regardless of crowding. Various molt stages must be segregated to prevent cannibalism.

Huner and Avault (1981) recommended that deionized water be used in shedding tanks but not in tanks where crawfish are to be held until they enter the mid-late premolt stages. Crawfish can successfully molt in deionized water but cannot harden completely in the absence of dissolved, environmental calcium and/or food. The use of deionized water, thus, reduces the necessity for constant, late night vigilance of shedding tanks. Adequate circulation and aeration must be maintained to remove metabolic wastes where crawfish are being fed. They are not fed in shedding tanks. Algae blooms can develop in a green house, interfering with attempts to observe crawfish for signs of impending molt.

Crawfish held in molting facilities may be caught with passive devices, such as traps, or active devices, such as seines and trawls. Because crawfish cease to feed toward the middle of the premolt period and can stop molting activities in early premolt, traps are not the best devices for capturing crawfish. However, they may be the only practical device in typical weed-choked crawfish ponds. Recent developments suggest that seines weighted with relatively heavy chains catch crawfish effectively in ponds choked with soft-stemmed vegetation. The labor required to use such seines makes them somewhat impractical when hard-shelled crawfish are the sole target but may be justified if a molting operation is to be supplied. An alternative to a hand-drawn seine is a boat-mounted surface trawl equipped with an electrofishing device (Cain and Avault 1983). Though open water is required, it can be provided by mowing lanes before filling ponds. By March most pond vegetation has decomposed providing open water.

Whenever crawfish are caught in seines or trawls, the various molt stages must be hand-sorted. Soft and late premolt crawfish may suffer mechanical damage during capture, reducing their attractiveness to customers. If trap-caught crawfish are to be used in molting facilities, they should not be refrigerated. Incomplete molts are often experienced in such cases.

Molt Induction

If crawfish are reluctant to molt, either because environmental conditions are unfavorable or because they are reproductively active and not molting, they may be induced to molt by bilateral removal of both eyestalks at their bases (Huner and Avault 1977). Mortality will be minimal if the animals are quickly returned to the water, as this accelerates the natural clotting process. Eyestalkless crawfish eat incessantly and should be provided with adequate food (see Intensive Crawfish Culture section). Reproductively active crawfish are almost certain to die during the molt but dying and hand-peeled "busters" have commanded the same price as cleanly molted animals. Additional labor is necessary to handle these animals because the shell must be physically removed by hand; limbs are commonly lost at this time.

Ecdysones may be injected to stimulate molting activities in crawfish (Huner and Avault 1977), but they are expensive and do not seem to be any real improvement over eyestalk removal. Apparently beta-ecdysone initiates molting activities in such a way that they do not proceed normally, and most treated animals die during the molt. Longer periods are required for molting if alpha-ecdysone is used.

INTENSIVE CRAWFISH CULTURE

Crawfishes have been cultured in "closed" systems since the late 1800s by scientists using them in various aspects of basic research. There is nothing novel about this although some species fair better than others. Commercially important species that have responded well in intensive culture have included *P. clarkii* (Black and Huner 1977; Goyert and Avault 1978, 1979; Dendy 1979) and *O. virilis* (Leonard 1981). The astacid signal crawfish, *P. leniusculus,* can also be maintainef and spawned in closed systems (Goldman *et al.* 1975; Cabantous 1975; Arrignon 1981; Biggs 1980).

At this writing, there may be no more than two or three American commercial firms utilizing intensive, closed or semi-closed systems in some form of crawfish culture. Although it is technically feasible to produce young *P. clarkii* more or less on demand, the supply of crawfish from wild stocks and ponds has kept prices too low to justify intensive culture to produce crawfish for pond stocking, food, or fish bait. This situation may change if crawfish prices stabilize and crawfish, specifically *P. clarkii,* become a principal rather than a supplemental crop.

The following information about intensive culture is applicable to *P. clarkii* only. Brood stock should be healthy and intact. They may be stocked at 8–20 animals/m^2 (1–2/ft^2) with a sex ratio of 1:1. Provision should be made for rapid flushing of tanks, but normally water input need only replace evaporation and seepage losses. Feeding must be conservative to maintain water quality. Water depth need not exceed 15–20 cm (6–8 in.), but precautions must be taken to prevent crawfish from climbing out of the container on dangling air hoses, stand pipes, or rough, textured tank corners. *P. clarkii* mature in 2½–6 months and will copulate immediately. Photoperiod does not appear very critical in this species. Females may lay eggs within 2–8 weeks of copulation at temperatures around 22°C (72°F). Newly berried females should not be disturbed for several days lest eggs be lost before their attachment to the swimmerets becomes firm. The females should then be removed from the breeding tank and may be held individually in 4-liter (1-gal) containers with about 5 cm (2 in.) of water, which is changed every 3–5 days, or in communal tanks provided with hiding areas. The advantage of using individual containers is that one can keep track of each animal; however, much more labor is involved. Females held in communal tanks require less care.

Young crawfish can be stocked into tanks or ponds as soon as they swim freely from the female, about 2–2½ weeks after hatching. In close confinement, the young will continue to be attracted by brooding pheromones of the female for several more weeks. Free-swimming 3rd instar crawfish are 0.7–1.0 cm (less than 0.5 in.) long and are very vulnerable to predators in ponds; thus, it may be wise to hold them in tanks, at densities of about 100–200/m^2 (9–18/ft^2) for about a month, before stocking. Mortality will be low, 15% or less, and crawfish should reach 2.5–3.5 cm (1–1.5 in.) during this period. Using larger juveniles to stock open ponds reduces losses to various predators, primarily arthropods.

Goyert (1978) and Goyert and Avault (1978, 1979) discuss raising *P. clarkii* to maturity in closed systems. Goyert and Avault (1978) employed individually heated cylindiral fiberglass tanks each with a bottom area of approximately 2 m^2 (22 ft^2) and a capacity of 1600 liters (422 gal). Each was equipped with an undergravel biological filter operated by an air lift pump. The filter consisted of a 7-cm (2.8 in.) layer of small gravel over a perforated filter plate resting on 4-cm (1.6-in.) high supports above the bottom of each tank. Flow rate was about 160 liters (42 gal) per minute. Water temperatures in tanks varied from 25° to 27°C (77° to 81°F), and oxygen was near saturation. Growth and survival, at densities of 10 to 40 3.3-cm juvenile *P. clarkii*/m^2 of bottom area (1 to 4 1.3-in. animals/ft^2), were best in tanks filled with a number of large loops of 26-cm (10-in.) wide window screen layered horizontally, forming artificial floors. There were only negligible differences in the growth and survival of crawfish in tanks with no substrate and in those with tubes on the bottom for hiding places. When a screen substrate was used, final size and mortality were 7.4 cm (3.0 in.) and 20% at 10 animals/m^2 (1/ft^2) and were 8.8 cm (3.5 in.) and 30% at 40 animals/m^2 (4/ft^2). Final size ranged from 6.2 to 7.1 cm (2.5 to 2.8 in.), and mortality ranged from 33 to 73%, in the other configurations. It is not clear if increased surface area in the screened system promoted better growth because mortality was greatest in that configuration. Surface area per animal was not given for any configuration.

In a concurrent study, Goyert and Avault (1979) noted a positive relationship between surface area and growth of individual crawfish. When held individually in circular containers with a bottom area of 660 cm^2, at a density equivalent to 15 animals/m^2 (1.4/ft^2), juvenile *P. clarkii* (2.5–3.0 cm or about 1 in. in total length) reached a size of around 7.0 cm (nearly 3 in.) before growth ceased after 49 days. Those held at 40/m^2 (4/ft^2) exhibited reduced growth after 14 days. Also, growth was severly limited when 2.2-cm (0.9-in.) crawfish stocked at 100 individuals/m^2 (9.3/ft^2) reached 3.0 cm (1.2 in.).

Morrissy (1976) discusses breeding and early rearing of the Australian parastacid crawfish *Cherax teniumanus*. Those seriously considering crawfish hatchery programs should review this reference.

Before investing in intensive culture systems, whether the aim is to produce juveniles for stocking or to raise salable adults, an individual must look at specific costs and potential returns. One should not rule out the use of existing fish hatchery facilities to reduce costs. In areas located far from crawfish production centers in Texas and Louisiana, such systems may be justified.

Leonard (1981) provides detailed information on culture of *O. virilis,* an annual spawner, in closed systems for bioassays. Photoperiod and temperature changes are used to induce early egg production and hatching (Aiken 1969). Leonard's methods are summarized below:

1. Holding containers should have a flow of 200 ml/min. Loading density is 1 g of crawfish per ml of water flow.
2. During winter, photoperiod should be 8 hr of light and 16 hr of dark; during summer–fall, it should be 16 hr of light and 8 hr of dark.
3. Temperature should be 12°C for winter or 18°C for summer–fall.
4. To change seasonal regimes, raise or lower temperature 2°C per day, and then change photoperiod 2 hours per day.
5. During the mid-October to mid-February period, water flow should be 1 ml/g of crawfish; feeding should stop; and tanks should be covered to exclude light. Spawning should occur by February. Hatching takes place soon after temperatures and photoperiod are gradually readjusted to summer–fall conditions.
6. Female *O. virilis* require total darkness for normal ovarian development and egg laying.

Hatchery systems have been developed in Europe to produce young *P. leniusculus* for stocking natural waters (Abrahamsson 1973; Cabantous 1975; Arrignon 1981). This species normally lay eggs in the fall which hatch the following spring. Raising temperatures in January and February will lead to rapid development and hatching, up to 3 months earlier than in nature. General culture conditions are similar to those mentioned for *P. clarkii* and *O. virilis,* although photoperiod does not seem as important to *P. leniusculas* as to the other species; however, opaque shelters and subdued lighting should be employed for both *P. leniusculus* and *P. clarkii.*

Feeding is far more important in intensive culture systems than in open ponds since trace elements, growth factors, and vitamins need to be supplied. Water-stable diets developed for crustaceans are accepted by crawfish. Pertinent references include Meyers *et al.* (1970), Huner *et al.* (1975), and Huner and Meyers (1979). Water-stable feeds developed for lobsters (Conklin *et al.* 1980) and penaeid shrimps (Colvin and Brand 1977) are probably also suitable for crawfish as long as protein levels exceed about 28%. Fish pellets, including trout chow (Goyert and Avault 1978) and catfish pellets (Clark *et al.* 1975), are suitable but disintegrate very rapidly in the water necessitating close attention to feeding rates and wastes. The provision of green, leafy aquatic vegetation, such as alligator weed or elodea (*Anacharis* sp.), results in normal pigmentation and improved growth of crawfish (Black and Huner 1977; Huner and Meyers 1979, 1981). Fish or other flesh may be fed, but care must be taken to avoid fouling the water. Feeding rates of chows should be approximately 3% of body weight per day for growing juveniles and 1% of body weight per day for adults. Vegetation should be present at all times. Most researchers report that they use some form of dry or semi-moist pet food to maintain their crawfish prior to using them in basic research studies. Thus, these materials may be used, but their cost will usually be higher than that of fish feeds. Moreover, such pet feeds will also foul the culture water if care is not taken to avoid the problem.

Crawfish eggs may be stripped from females and hatched in modified trout/salmon egg incubators (Mason 1979; Arrignon 1981). There seems to be little benefit to this practice with warm-water species such as *P. clarkii* whose embryos develop rapidly in 2–3 weeks, but the practice may be of value with cool-water species such as *P. leniusculus,* especially in areas like Europe where young crayfish are very valuable and high prices can justify the extra effort. Before eggs are stripped from females, a period of development must take place or mortality will be high. *P. leniusculus* eggs laid in October and incubated by the female at ambient temperatures of around 10°C (50°F) must develop at least 3–4 months before they are stripped for rapid incubation at higher temperatures (Mason 1979; Arrignon 1981).

STATUS OF CRAWFISH CULTURE

Climatic Restrictions on Culture of Crawfish

Crayfish suitable for use as fish bait, research animals, and food are found in all states except Alaska (Huner 1978a). *P. clarkii* and *P. acutus* require frost-free growing seasons of roughly 120–150 days to reach edible sizes in excess of 7.5 cm (3 in.). Thus, many states can support the culture of these species. The various *Orconectes* spp. that have been "cultured" reach sizes suitable for fish bait before the end of the summer growing season in the Midwest and Northeast and may reach edible sizes. Virtually no data are available on the sizes and population structures of *O. nais, O. rusticus,* and *O. virilis* in culture situations; however, life history studies (see Life Cycles section) indicate that a significant number of young-of-the-year do reach sexual maturity at sizes of 5.0–7.0 cm (2.0–2.8 in.) in one growing season. It seems reasonable to assume that improved conditions in terms of food, space, and water quality would lead to production of larger animals within the same period by increasing growth increments per molt. This is clearly apparent in *P. clarkii* and *P. acutus* populations where animals mature at sizes ranging from 5.0–12.5 cm (2–5 in.) in one growing season (Avault *et al.* 1975; Huner 1975; Huner and Romaire 1979) depending on environmental conditions. One management practice that may warrant study is the hatchery production of *P. clarkii* for stocking during shortened growing seasons at northern latitudes. This would not be necessary, however, if the various *Orconectes* spp. are found to exhibit comparable growth rates.

Whether or not *P. leniusculus* can be grown to an edible size in 4–8 months remains to be seen. Biggs' (1980) review of the suitability of wild stocks for culture of *P. leniusculus* to edible sizes suggests that 12–18 months may be required in open-pond culture at more northerly latitudes. Certainly, this is an area worth investigating.

Current Acreage Under Culture

In Louisiana, crawfish are cultured by individual farmers and families, often as second crops. About 40,486 ha (100,000 Ac) were devoted to crawfish culture in 1983–1984 (B. Craft, Soil Conservation Service, Alexandria, LA, personal communication, 1983). Individual ponds range in size from less than 1 ha to 100 ha (2.47 to 247 Ac), but the average size is 8–16 ha (20–40 Ac) (Gary 1974). A 1980 Soil Conservation Service survey (Craft 1980) reported that there were 361 crawfish farmers in Louisiana, farming a total of 22,389 ha (55,300 Ac). He listed three categories of ponds: rice-crawfish ponds, 8611 ha (21,269 Ac); open-water ponds with native vegetation, 5281 ha (13,044 Ac); and wooded ponds with native vegetation 8479 ha (20,943 Ac). Many of the rice-crawfish ponds were double-cropped with crawfish. Three uses for the ponds were noted: commercial, 274 ponds; family, 134 ponds; and recreational fish-out (fee-fishing) operations, 6. Many farmers used ponds for both commercial and family use. As found in Gary's (1974) survey, most operations were concentrated in the southern part of the state within 80–100 km (50–65 mi) of Lafayette.

Table 1.2 shows the acreage of crawfish culture in Louisiana over the past three decades. Acreage grew rapidly until the early 1970s but then remained relatively constant for several years. Reduced profits for row crops made crawfish farming more attractive in the early 1980s and spurred further expansion of the industry.

Whether or not large companies can establish vertically integrated crawfish production systems in the near future is questionable. Rather consistent but unpredictable pond failures make vertical integration

TABLE 1.2. CRAWFISH CULTURE ACREAGE IN LOUISIANA—1949–1982

Year	Hectares	Acres
1949	16	40
1960	810	2,000
1966	2,429	6,000
1968	4,049	10,000
1969	4,858	12,000
1970	7,287	18,000
1971	9,717	24,000
1973	17,814	44,000
1976	18,219	45,000
1978	19,433	48,000
1980	22,267	55,000
1982	40,486	100,000

Source: Huner and Barr 1981; B. Craft, personal communication, 1983.

risky, whereas family ventures can tolerate losses better because crawfish often are a second crop. Another important point is that crawfish are usually available from late November or December through early June. Thus, there are 5–6 months with no crawfish production each year during which a vertically integrated business would have to involve itself with other products. Most Louisiana crawfish processors process crabs and/or finfish during the "off" season. A few Louisiana crawfish processor–dealers have vertically integrated operations in the sense that they either own or control crawfish ponds and restaurants featuring crawfish products. Ponds assure minimal supplies of whole crawfish and crawfish meat for their restaurants.

Crawfish are cultured in many states in the sense that they are harvested incidental to various finfish operations (Huner 1976, 1978a); however, in addition to Louisiana, crawfish, are intentionally cultivated only in Texas, Mississippi, South Carolina, and Arkansas. Best estimates of the current area devoted to crawfish culture in these states are 4000 ha (9880 Ac) in Texas (J. Davis, Texas A&M University, College Station, TX, personal communication, 1983), 810 ha (2000 Ac) in Mississippi (R. Callahan, Soil Conservation Service, Jackson, MI, personal communication, 1983), 220 ha (550 Ac) in South Carolina (W. Melvin, Soil Conservation Service, Columbia, SC, personal communication, 1983), and 200 ha (500 Ac) in Arkansas (P. Brady, Soil Conservation Service, Little Rock, AR, personal communication, 1983). Texas production is concentrated in rice-growing areas in southeastern Texas near Orange. Returns have been so good that rapid expansion to 8000 ha (19,760 Ac) is expected within 5 years. While great interest has been generated in northwestern Mississippi (The Delta), expansion there will depend on the success of individual farmers, as there has been substantial variation in yields among ponds. South Carolina and Arkansas farmers are very enthusiastic, but their efforts are modest compared to those of farmers in other states. South Carolina farmers do, however, have the added advantage of being insulated from the Louisiana–Texas–Mississippi supplies.

Expansion Possibilities

It is apparent that crawfish farming, at least culture of *P. clarkii* and *P. acutus,* is feasible anywhere rice is cultured and clay pans or lenses are thick enough to permit burrows of 1–2 m (39–78 in.) without pond drainage. Substantial areas of Texas (200,000 ha or 247,000 Ac), Mississippi (60,000 ha or 148,000 Ac), and Arkansas (340,000 ha or 840,000 Ac) are devoted to rice production and could be used for crawfish culture; there are altogether, 1.01–1.13 million ha (2.49–2.79 million Ac) of rice production in the United States, including substantial areas in California, so the potential for expansion is great (Joint Subcommittee on Aquaculture 1980). There is no available estimate for the marginal, poorly drained lands that could be converted to crawfish culture throughout the country, but the Joint Subcommittee on Aquaculture (1980) reported about 121,000 ha (300,000 Ac) in Texas alone. Expansion will depend on the economics of supply and demand.

As long as there is a demand for bait crawfish, production will continue in midwestern finfish operations. While there are indications that the bait market can stand more input (Huner 1978a), there has been no indication that farmers intend to expand production. It would not be surprising, then, if farmers in major crawfish production states eventually seek these outlets as they saturate their domestic markets. *P. clarkii* is so well established in Hawaii that its burrowing activities are considered a serious problem in irrigation systems (Huner and Barr 1981). Several dozen fish, crustacean, and mollusk species are referenced in various aquacultural development plans but *P. clarkii* is not even considered.

CRAWFISH PROCESSING

Although most crawfish are sold alive for boiling in seasoned water, as much as 30% of the Louisiana crop is processed. That is, the abdomens, or "tails," are peeled to obtain the large muscle therein. Processing will not be discussed in detail here; the reader is referred to Moody (1981), Huner and Barr (1981), and Hudson and Fontenot (1970, 1971) for details (Fig. 1.20).

In most processing plants, crawfish are killed by immersion in boiling water and then hand-peeled. Meat is packed with or without the hepatopancreas ("fat") for fresh sales, and without the hepatopancreas for freezing, storage, and later sales. A roller-type mechanical peeler has been in use at one of the 70+ processing plants in Louisiana. In this plant, crawfish are killed by freezing them in a brine freezer, then thawed, beheaded by hand, and machine-peeled. This meat is raw, not cooked as is the hand-peeled meat.

Processing yields average around 15-20% of live weight. Wastes can be used as fertilizers, lime substitutes, sources of crustacean meal, and sources of carotenoid pigments, especially for salmonid feeds. A major waste processing plant is scheduled to begin operation in mid-1984. This should improve the profitability of peeling crayfish for meat. There is, for all practical purposes, no crawfish processing capacity outside the state of Louisiana. Some restaurants in Texas and Mississippi do peel crawfish for preparation of entrees but this is a limited activity. However, continued expansion of production in other states will necessitate development of processing capacity.

FIG. 1.20 Modern Louisiana crawfish plant selling live and processed crawfish. *Courtesy of J. Huner.*

ECONOMICS AND MARKETS

As Roberts (1980) noted, "There is a feverish scientific effort to make aquaculture of crustaceans something other than interesting reading. The question of commercial feasibility, yet to be answered for most crustaceans, is not an issue with crawfish." These comments still hold true today as do several generalizations that Roberts (1980) made to explain the economic success of crawfish in Louisiana:

1. Most ponds are built with minimum investment. The farmer routinely uses existing farm labor and equipment and frequently attaches new levees to existing levees previously constructed for drainage purposes.
2. Farmers have usually developed surplus lands of marginal usefulness in producing traditional crops. Few depend on crawfish for their sole income and most typically grow sugarcane or rice as a principal crop.
3. Recent expansion of crawfish farms has occurred in rice-producing areas of south Louisiana. Expansion is not being limited to double cropping with rice on a single tract of land. Crawfish, in general, fit more smoothly into a rotation with rice or a rice-soybean rotation.
4. Most farmers (91% of those surveyed) do not plant a forage crop for crawfish.*
5. Most farmers (82% of those surveyed) do not stock crawfish after the initial stocking.
6. Farmers experience widely differing average prices for the year's production. Price depends not only on time of year but also on the type of market the farmer enters. The highest average prices for the year are experienced by farmers harvesting a significant portion of their crop in winter and marketing to wholesalers or restaurants.
7. Crawfish culture offers competitive returns for skilled farmers, and returns should increase as more cost-effective harvesting methods are developed.

Roberts (1982) has developed a budget for a crawfish operation based on a single 40.5-ha (100-Ac) crawfish pond. This should provide a useful guide to those interested in pursuing crawfish culture. Numerous

*As of 1983 as many as 50% of Louisiana crawfish farmers were planting forage crops (virtually all Texas crawfish farmers plant forages).

qualifications may be made, but certain costs must be taken into account before initiating a crawfish operation (Table 1.3). Fixed costs were $6603 in years 1 and 2. Variable costs were $59,671 in year 1 and $38,167 in year 2. The reduction in variable costs the second year resulted from lower pumping and forage crop costs, absence of restocking, and lower interest charges.

The Texas Cooperative Extension Service (1981) prepared a cost analysis for Texas crawfish production in 1981. It is similar to that prepared by Roberts (1982) but is based on a 16-ha (40-Ac) operation with eight 2-ha (5-Ac) ponds; several items such as trap type and number, water costs, etc., are specific to Texas. Capital costs also were presented: these amounted to $9157 per ha ($3707 per Ac) including

TABLE 1.3. BUDGET SUMMARY FOR A 40.5-HA, SINGLE-POND CRAWFISH FARM IN LOUISIANA

	Year One		Year Two	
Fixed Costs				
Depreciation:				
Pump engine	$945		$945	
Harvesting boats & motors	911		911	
Truck (½ of cost charged)	817		817	
Pump and pipes	203		203	
		$2,876		$2,876
Interest on investment		1,927		1,927
Amortized pond construction		1,800		1,800
		6,603		6,603
Cash Costs				
Establishing rice crop as forage[a]	$18,922		$ 8,500	
Harvest labor[b]	11,500		11,500	
Pump engine fuel[c]	6,167		4,367	
Bait[d]	5,033		5,033	
Stocking[e]	4,000		-0-	
Harvest fuel	1,648		1,648	
Repairs	1,100		1,100	
Supplies	900		900	
Interest on operating capital	4,626		2,627	
		$59,671		$38,167

Source: Roberts 1982.

[a] A cheaper forage like millet ($210 per ha or $85 per Ac) can be planted; if no forage is planted, no cost other than minimal (recommended) fertilization would be incurred.

[b] Many arrangements are made to cover harvesting (share systems, rentals, etc.). No "cost" is incurred if family labor harvests the crawfish.

[c] The pond must be filled twice in year 1 and only once in year 2.

[d] This is a conservative estimate based on 27 traps/ha (11/Ac). Most farmers use twice as many traps which doubles bait usage and cost.

[e] Most ponds are not restocked after the first year; stocking rates can be halved in areas with good cover and/or native crawfish populations.

(Notes are based on Roberts (1982) and authors' knowledge of the industry.)

land costs of $2500 per ha ($1000 per Ac) and pond construction costs of $611 per ha ($247 per Ac). Many other capital items included (service facilities, tractors, pumps, mowers, grader blade, truck, and shop equipment) at a cost of $3300 per ha ($1296 per Ac) are usually on hand but, as pointed out by Roberts (1980), should be accounted for in terms of fixed costs in annual operating budgets.

Gross and net revenues are determined by supply and demand. Crawfish are invariably scarce and expensive early in the season regardless of location. As the season progresses young-of-the-year reach harvestable size roughly 4–6 months after ponds are flooded and increased supplies depress prices. This is not especially noticeable unless this period coincides with substantial harvests of wild crawfish, something that usually occurs 2 years out of every 5 in Louisiana (Huner and Barr 1981). This has not been the case in other states as they lack substantial supplies of wild crawfishes.

In Louisiana, crawfish farmers generally sell crawfish live to one of three markets: wholesalers (dealers), processors, or restaurants. The highest prices are usually paid by restaurants and the lowest by processors (Roberts 1980). Many wholesalers also process crawfish (peel them for meat as a salvage operation when live crawfish cannot be sold). Few processors deal only in crawfish meat. Therefore, it is sometimes difficult to distinguish between these two categories of markets.

Carrol (1975) and Blades (1975) surveyed crawfish processing and marketing in Louisiana in the early 1970s, and their findings are still applicable today. They found that the most important distribution channel for Louisiana crawfish is wholesaler/processor to seafood market then to consumer; the second most important channel is wholesaler/processor to restaurant to consumer; and the least important distribution channel is wholesaler/processor to regular food market to consumer. Live crawfish account for 60 to 70% of a wholesaler/processor's sales. Approximately three-fourths of the retail crawfish sales in seafood markets is live crawfish. Most restaurant sales (80%) consist of peeled meat in the form of entree components, although boiled (whole alive) crawfish is the most popular individual restaurant entree. The eight most popular items in restaurants are boiled, crawfish etoufee, crawfish bisque, fried crawfish tails, crawfish gumbo, crawfish dinner (a combination of entrees), crawfish jambalaya, and crawfish stew.

In Louisiana, prices for live crawfish were so good in the late 1970s and early 1980s that little meat was processed. The price of meat reached over $20 per kg ($9 per lb), a price level that led to substantial resistance by individual consumers and institutions. As a result, most "processors" cultivated the more lucrative live market as much as possible; they produced peeled meat only on a limited scale until late in the season when supplies rose and prices fell to more desirable levels. It is ironic that a crawfish-peeling machine was developed in the early 1980s (Huner and Barr 1981) at a time that wholesale prices for live crawfish made it a marginal system.

Wholesalers and processors often supply crawfish bait to pond owners and crawfish fishermen. This is a means to insure supplies of crawfish, as few farmers or fishermen have freezers to store bait for extended periods. Most agreements between farmers and buyers are informal, and few contracts are made because of the volatile nature of the supply and demand for crawfish. Few crawfish, whole or as peeled meat, leave Louisiana because suppliers cannot assure the volume demanded by institutional markets, even at fairly high prices. Then, too, shipping crawfish, even processed crawfish, to other states involves more trouble than most Louisiana wholesalers/processors have been willing to undertake when acceptable profits can be made in the state. However, this situation should change as production acreage increases and the crawfish promotion board in the Louisiana Department of Agriculture becomes active.

The overseas market in Europe appears quite lucrative (Huner 1978), but it is now largely ignored by Louisiana wholesalers because many have lost money in trying to cater to that market. Large supplies of Spanish crawfish, resulting from successful introductions of *P. clarkii* in 1973, have entered French markets lately, and this may eventually depress prices there (Huner and Barr 1981).

Texas crawfish producers prefer to sell their crawfish live at the farm (Dash 1982). They have been reluctant to sell to jobbers or brokers because of the lower prices offered. Sales have been expanded to include out-of-state buyers as far away as Colorado and Utah. Several of the larger Texas producers have banded together to insure bait supplies and markets. Virtually all of Mississippi's crawfish are sold within the state.

SUMMARY

Commercial pond culture of two species of cambarid crawfishes, *Procambarus clarkii* and *Procambarus acutus acutus,* is well established in the lower Mississippi River valley. Expanded markets should lead to development elsewhere in the Continental United States and, perhaps, Hawaii. Pond culture of cambarids of the genus *Orconectes* is practiced on a much less sophisticated scale in the Midwest and

Northeast. Most *Orconectes* spp. are used for fish bait, although several reach edible sizes. Culture of the astacid *Pacifastacus leniusculus* remains in the experimental stage, and existing fisheries have yet to be fully exploited.

Crawfish culture currently involves the establishment of self-sustaining populations in relatively low cost earthen ponds that are periodically dewatered. Hatcheries are not needed. Water depths need not exceed 0.5 m (19 in.). Expensive feeds are not required. Food is supplied by growing cover crops of water-tolerant grains or volunteer grasses, sedges, and forbs. Microbially enriched detritus, formed when ponds are filled, provides feed for young crawfish that hatch in burrows during the dry period. Detritus creates a marked biological oxygen demand necessitating substantial amounts of water circulation during warm months (water temperatures greater than 20°C or 68°F) until the initial period of rapid decomposition has been completed. Annual crop yields range from 500 to 1500 kg/ha (445 to 1335 lb/Ac). Harvesting is labor intensive, requiring use of boats or amphibious vehicles, wire mesh traps, and bait (usually cut, frozen rough fish). Most crawfish are sold alive to be boiled in seasoned water. The remainder are hand-peeled to obtain the meat, with recovery being about 15–20% of live weight, making prices quite high. Peeled meat usually sells for 10 times the price of the live product. Soft-shelled crawfish have long been a highly sought after fish bait and are now being produced on a limited scale for food at excellent prices. Diseases and parasites currently do not appear to be a problem.

Intensive culture, egg to adult or egg to stocking-size juveniles for open ponds, is not now practiced commercially. The economics are just not favorable even though the technology has existed for some time. Such systems could open areas in more northern latitudes to food crawfish production. They could also be used for selection of genetically superior crawfishes.

Culture of *Procambarus* spp. integrates well with rice culture especially in the deep South where the two crops can be cultivated in the same 12-month period—rice in the late spring and summer, crawfish in the fall–winter–spring. Rotation of rice, crawfish, and soybeans over 2- and 3-year intervals creates fewer management problems and is probably more suitable in the central South than double cropping. Polyculture of fish, especially fingerlings and minnows, and crawfish (both *Procambarus* spp. and *Orconectes* spp.) is feasible throughout the country, but crawfish can reduce production by destroying fish spawn and spawning mats and damaging fish in seines.

The area devoted to culture of *Procambarus* spp. is in excess of 45,000 ha (110,000 Ac) with about 90% in Louisiana, 8% in Texas, and 2%

in Arkansas, Mississippi, and South Carolina combined. Acreage will probably increase three- to fourfold in Texas, Mississippi, and South Carolina within the next 5 years because current production does not meet the demand. Further expansion in Louisiana will depend on development of both in-state and out-of-state markets. Crawfish culture will probably remain a "second" crop in the forseeable future and extensive vertical integration of culture, processing, and marketing systems will probably have to wait for more predictable culture systems than earthen ponds with established crawfish populations.

LITERATURE CITED

ABRAHAMSSON, S. 1973. The crayfish *Astacus astacus* in Sweden and the introduction of the American crayfish *Pacifastacus laniusculus*. Int. Symp. Freshwater Crayfish **1,** 27–40.

ABRAHAMSSON, S., and GOLDMAN, C. R. 1970. Distribution, density, and production of the crayfish *Pacifastacus leniusculus* Dana in Lake Tahoe, California-Nevada. *Oikos* **21,** 83–91.

AIKEN, D. E. 1969. Ovarian maturation and egg laying in the crayfish *Orconectes virilis:* Influence of temperature and photoperiod. Can. J. Zool. **47,** 931–935.

AIKEN, D. E. 1968. Further extension of the known range of the crayfish *Orconectes virilis* (Hagen). Natural Museum Can. Bull. **223,** 43–48.

AIKEN, D. E. 1980. Molting and growth. The Biology and Management of Lobsters, Vol. 1, J. S. Cobb and B. F. Phillips (editors), Academic Press, New York.

ALON, N. C., and DEAN, J. M. 1980. Production guidelines for crawfish farming in South Carolina. Publication USC-BI-80-1. South Carolina Sea Grant Consortium, Columbia, South Carolina.

AMBORSKI, R. L. 1980. Diseases of crayfish. Proc. 1st Nat. Crawfish Culture Workshop, D. Gooch and J. Huner (editors), Res. Ser. No. 50. University of Southwestern Louisiana, Lafayette, Louisiana.

AMBORSKI, R. L., GLORIOSO, J. C., and AMBORSKI, G. F. 1975a. Common potential pathogens of crayfish, frogs, and fish. Int. Symp. Freshwater Crayfish **2,** 317–326.

AMBORSKI, R. L., LOPICCOLO, G., AMBORSKI, G. F., and HUNER, J. V. 1975b. A disease affecting the shell and soft tissues of Louisiana crayfish, *Procambarus clarkii*. Int. Symp. Freshwater Crayfish **2,** 299–316.

ARRIGNON, J. 1981. L'ecrevisse et son 'elevage. Gauthier-Villars. Bordas, Paris.

AVAULT, J. W., JR. 1980. Management of aquatic species. Animal Agriculture, H. H. Cole and W. N. Garett (editors), W. H. Freeman and Co. San Francisco, California.

AVAULT, J. W., JR., DE LA BRETONNE, JR., L. W., and JASPERS, E. 1969. Culture of Crawfish, Louisiana's crustacean king. Am. Fish Farmer **1**(10), 18–14, 27.

AVAULT, J. W., JR., DE LA BRETONNE, JR., L. W., and HUNER, J. V. 1975. Two major problems in culturing crayfish in ponds: Oxygen depletion and overcrowding. Int. Symp. Freshwater Crayfish **2,** 139–144.

AVAULT, J. W., JR., HERNANDEZ, R., and GIAMALVA, M. 1978. Sugar cane waste products and chicken manure as supplemental feed for crawfish. Sugar Azucar **73**(6), 42.

BEAN, R. A. and HUNER, J. V. 1978. Comparison of the growth and survival of red swamp crawfish (*Procambarus clarkii* (Girard)) and white river crawfish (*Procambarus acutus acutus* (Girard)) (Decapoda: Cambaridae). Southeastern Assoc. Biol. Bull. **25,** 71 (Abstract).

BIGGS, E. D. 1980. A review of crayfish biology, culture, and potential future of the industry in the Pacific Northwest. Master's Project, Oregon State University, Corvallis, Oregon.

BLACK, J. B. and HUNER, J. V. 1979. Producing your own crayfish stock. Carol. Tips **42**(4), 1–3.

BLACK, J. B. and HUNER, J. V. 1980. Genetics of the red swamp crawfish, *Procambarus clarkii* (Girard): State-of-the-art. Proc. World Maric. Soc. **11,** 535–543.

BOETTCHER, P. A. 1979. Trematodes of crayfishes collected in the Tangipahoa River, Louisiana drainage system. M.S. Thesis, Louisiana State University, Baton Rouge, Louisiana.

BOVBJERG, R. V. 1970. Ecological isolation and competitive exclusion in two crayfish *(Orconectes virilis* and *Orconectes immunis).* Ecology **51,** 225–234.

BOYD, C. 1979. Water quality in warmwater fish ponds. Auburn University Agricultural Experiment Station, Auburn, Alabama.

BROWN, P. L. 1955. The biology of the crayfishes of central and southeastern Illinois. Ph.D. Dissertation, University of Illinois, Urbana, Illinois.

BROWN, R. T. and AVAULT, JR., J. W. 1974. Toxicity of antimycin to crayfish. Int. Symp. Freshwater Crayfish **2,** 351–370.

CABANTOUS, M. A. 1975. Introduction and rearing of *Pacifastacus* at the research center of Les Clouizious 18450 Brinon S/Sauldre France. Int. Symp. Freshwater Crayfish **2,** 49–56.

CAIN, C. D., JR., and AVAULT, JR., J. W. 1983. Evaluation of a boat-mounted electro-trawl as a commercial harvesting system for crawfish. Aquaculture Eng. **2**(2), 135–152.

CALALA, L. 1976. Crawfish. Keeping crawfish for bait. Making softshell crawfish and their care. Calala's Water Haven, Inc., New London, Ohio.

CANGE, S., M. MILTNER, and J. W. AVAULT, JR. 1982. Range pellets as supplemental crayfish feed. Prog. Fish-Cult. **44**(1), 23–24.

CARROL, J. C. and BLADES, JR., H. C. 1974. A quantitive analysis of the amounts of south Louisiana crawfish that move through selected channels of distribution. Res. Ser. (Marketing) No. 35. University of Southwestern Louisiana, Lafayette, Louisiana.

CHEAH, M., AVAULT, JR., J. W. and GRAVES, J. B. 1979. Some effects of rice pesticides on crawfish. Int. Symp. Freshwater Crayfish **4,** 349–362.

CHEAH, M., AVAULT, JR., J. W. and GRAVES, J. B. 1979–80. Some effects of rice pesticides on crawfish. Louisiana Agric. **23**(2), 8–9, 11.

CHEAH, M. L., AVAULT, JR., J. W. and GRAVES, J. B. 1980. Acute toxicity of selected rice pesticides to crayfish *Procambarus clarkii.* Prog. Fish-Cult. **42**(3), 169–171.

CHIEN, Y., and AVAULT, JR., J. W. 1979. Double cropping rice, *Oryza sativa,* and red swamp crawfish, *Procambarus clarkii.* Int. Symp. Freshwater Crayfish **4,** 263–276.

CHIEN, Y., and AVAULT, JR., J. W. 1980a. Production of crayfish in rice fields. Prog. Fish-Cult. **42**(2), 67–70.

CHIEN, Y., and AVAULT, JR., J. W. 1980b. Effects of flooding dates and type disposal of rice, *Oryza sativa,* straw on the crawfish *Procambarus clarkii* (Girard) cultured in rice fields. Abstr. Fish Culture Sect. Am. Fish. Soc. New Orleans, **1980,** 14.

CLARK, D. F., AVAULT, JR., and MEYERS, S. P. 1975. Effects of feeding, fertilization, and vegetation on production of red swamp crayfish, *Procambarus clarkii*. Int. Symp. Freshwater Crayfish **2,** 125–138.

COLLAZO, J. A. B. 1981. Development and evaluation of several artificial experimental baits for trapping crawfish. M.S. Thesis. Louisiana State University, Baton Rouge, Louisiana.

COLVIN, L. B., and BRAND, C. W. 1977. The protein requirements of penaeid shrimp at various life-cycle stages in controlled environment systems. Proc. World Maric. Soc. **8,** 821–840.

CONKLIN, D. E., D'AMBRAMO, L. R., BORDNER, C. E., and BAUM, N.A. 1980. A successful purified diet for the culture of juvenile lobsters: The effects of lecithin. Aquaculture **21,** 243–249.

CRAFT, B. 1980. Louisiana inventory of crawfish farmers 1980. Soil Conservation Service, Alexandria, Louisiana.

CROOK, G. 1980. Marron and marron farming. Department of Fisheries and Wildlife, Perth, Western Australia, Australia.

DASH, S. 1982. Aquaculture. Outlook and situation. Economic Research Service AS-3, U.S. Department of Agriculture, Washington, D.C.

DAY, C. H. 1983. Crawfish *(Procambarus clarkii)* production in ponds receiving varying amounts of soybean *(Glycine max)* stubble or rice *(Oryza sativa)* straw as forage. M.S. Thesis, Louisiana State University, Baton Rouge, Louisiana.

DE LA BRETONNE, L., JR., and AVAULT, JR., J. W. 1971. Liming increases crawfish production. Louisiana Agric. **15**(1), 10.

DE LA BRETONNE, L., JR. and J. W. AVAULT, JR. 1977. Egg development and management of *Procambarus clarkii* (Girard) in a south Louisiana crawfish pond. Int. Symp. Freshwater Crayfish **3,** 133–140.

DE LA BRETONNE, L. W., JR., AVAULT, JR., J. W., and SMITHERMAN, R. O. 1969. Effects of soil and water hardness on survival and growth of the red swamp crawfish, *Procambarus clarkii,* in plastic pools. Proc. 23rd Annu. Conf. Southeastern Assoc. Game Fish Commissioners, pp. 626–633.

FORNEY, J. L. 1968 (revised). Raising bait fish and crayfish in New York ponds, Extension Bull. 986. Cornell University, Ithaca, New York.

FROST, J. V. 1975. Australian crayfish. Int. Symp. Freshwater Crayfish **2,** 87–96.

GARY, D. L. 1974. The commercial crawfish industry in south Louisiana. Center for Wetland Resources, Sea Grant Publ. No. LSU-SG-74-01. Louisiana State University, Baton Rouge, Louisiana.

GOLDMAN, C. R., RUNDQVIST, J. C., and FLINT, R.W. 1975. Ecological studies of the California crayfish, *Pacifastacus leniusculus,* with emphasis on their growth from recycling waste products. Int. Symp. Freshwater Crayfish **2,** 481–488.

GOOCH, D., and HUNER, J. (editors). 1980. Proceedings of the First National Crawfish Culture Workshop, Res. Ser. No. 50. University of Southwestern Louisiana, Lafayette, Louisiana.

GOYERT, J. C. 1978. The intensive culture of crayfish, *Procambarus clarkii* (Girard), in a recirculating water system. Ph.D. Dissertation, Louisiana State University, Baton Rouge, Louisiana.

GOYERT, J. C., and AVAULT, JR., J. W. 1977. Agricultural by-products as supplemental feed for crayfish, *Procambarus clarkii.* Trans. Am. Fish. Soc. **106,** 629–633.

GOYERT, J. C., and AVAULT, JR., J. W. 1978. Effects of stocking density and substrate on growth and survival of crawfish *(Procambarus clarkii)* grown in a recirculating system. Proc. World Maric. Soc. **9,** 731–736.

GOYERT, J. C., and AVAULT, JR., J. W. 1979. Effects of container size on growth of crawfish *(Procambarus clarkii)* in a recirculating system. Int. Symp. Freshwater Crayfish **4,** 277–286.

GOYERT, J., AVAULT, JR., J. W., RUTLEDGE, J. E., and HERNANDEZ, T. E. 1975–76. Agricultural by-products as supplemental feed for crawfish. Louisiana Agric. **19**(2), 10–11.

HAMAJIMA, F., FUJINO, T., and KOGA, M. 1976. Studies on the host-parasite relationship of *Paragonimus westermani* (Kerbert 1878). IV. Predatory habits of some freshwater crabs and crayfish on the snail *Semisulcospira libertina* (Gould 1859). Annot. Zool. Jpn. **49,** 274–278.

HILL, L., and CANCIENNE, W. A. 1966 (revised). Grow crawfish in rice fields. Publ. No. 1346. Louisiana Cooperative Extension Service. Baton Rouge, Louisiana.

HOBBS, H. H., JR. 1972. Crayfishes (Astacidae) of North and Middle America. *In* Biota of Freshwater Ecosystems. Ident. Manu. No. 9, U.S. Government Printing Office, Washington, D.C.

HOBBS, H. H., JR. 1974a. Synopsis of the families and genera of crayfishes (Crustacea:Decapods). Smithson. Contrib. Zool. **164,** 1–32.

HOBBS, H. H., JR. 1974b. A checklist of the North and Middle American crayfishes Decapoda:Astacidae and Cambaridae). Smithson. Contrib. Zool. **166,** 1–161.

HOBBS, H. H., JR. 1981. The crayfishes of Georgia. Smithson. Contrib. Zool. **318,** 1–549.

HOLT, P. C. 1975. The branchiobdellid (Annelida:Clitellata) associates of astocoidean crayfishes. Int. Symp. Freshwater Crayfish **2,** 337–346.

HUDSON, J. F., and FONTENOT, W. J. 1970. Profitability of crawfish peeling plants in Louisiana. Res. Rep. No. 408. Dept. of Agric. Economics and Agribusiness, Louisiana State University, Baton Rouge, Louisiana.

HUDSON, J. F., and FONTENOT, W. J. 1971. The crawfish peeling industry in Louisiana. Res. Rep. No. 421. Department of Agricultural Economics and Agribusiness, Louisiana State University, Baton Rouge, Louisiana.

HUNER, J. V. 1975. Observations on the life histories of recreationally important crawfishes in temporary habitats. Proc. Louisiana Acad. Sci. **38,** 20–24.

HUNER, J. V. 1976. Raising crawfish for fish bait and food: A new polyculture crop with fish. Fish. Bull. **1**(2), 7–9.

HUNER, J.V. 1978a. Exploitation of freshwater crayfishes in North America. Fish. Bull. **6**(3), 2–5, 16–19.

HUNER, J. V. 1978b. Crawfish population dynamics as they affect production in several small, open crawfish ponds in Louisiana. Proc. World Maric. Soc. **9,** 619–640.

HUNER, J. V. 1982. Thermal tolerance and oxygen requirements of *Procambarus clarkii* and *Procambarus acutus acutus.* Spec. Symp. Crayfish Distributions. Annu. Meet. Crustacean Soc., Louisville, Kentucky (Abstr.).

HUNER, J. V., and ABRAHAM, G. R. 1981. Observations of wading bird activity and feeding habits in and around crawfish ponds in South Louisiana with management recommendations. Unnumbered Spec. Publ. Division of Special Services, U.S. Fish & Wildlife Service, Baton Rouge, Louisiana, 5 pages (Mimeo.).

HUNER, J. V., and AVAULT, JR., J. W. 1976. The molt cycle of subadult red swamp crawfish, *Procambarus clarkii* (Girard). Proc. World Maric. Soc. **7,** 267–273.

HUNER, J. V., and AVAULT, JR., J. W. 1977. Investigations of methods to shorten the intermolt period in a crawfish. Proc. World Maric. Soc. **8,** 883–893.

HUNER, J. V., and AVAULT, JR., J. W. 1981 (revised). Producing crawfish for fish bait. Sea Grant Publ. No. LSU-T1-76001. Center for Wetland Resources, Louisiana State University, Baton Rouge, Louisiana.

HUNER. J. V., and BARR, J. E. 1981. Red swamp crawfish. Biology and exploitation. Sea Grant Publ. No. LSU-T-80-001. Center for Wetland Resources, Louisiana State University, Baton Rouge, Louisiana.

HUNER, J. V., and MEYERS, S. P. 1979. Dietary protein requirements of the red swamp crawfish, *Procambarus clarkii* (Girard) (Decapoda:Cambaridae), grown in a closed system. Proc. World Maric. Soc. **10,** 751–760.

HUNER, J .V., and MEYERS, S. P. 1981. Growth responses of juvenile male crawfish, *Procambarus clarkii,* fed artificial diet supplemental with elodea, a vascular aquatic plant. Southwest Associa. Naturalists News (Abstract).

HUNER, J. V., and ROMAIRE, R. P. 1979. Size at maturity as a means of comparing populations of *Procambarus clarkii* (Girard) (Crustacea:Decapoda) from different habitats. Int. Symp. Freshwater Crayfish **4,** 53–64.

HUNER, J. V., KOWALCZUK, J. G., and AVAULT, JR., J. W. 1978. Postmolt calcification in subadult red swamp crayfish, *Procambarus clarkii* (Girard) (Decapoda: Cambaridae). Crustaceana **34,** 275–280.

JASPERS, E., and AVAULT, JR., J. W. 1969. Environmental conditions in burrows and ponds of the red swamp crawfish, *Procambarus clarkii* (Girard) near Baton Rouge, Louisiana. Proc. 23rd. Annu. Conf. Southeastern Assoc. Game Fish Commiss. pp. 634–647.

JOHNSON, S. K. 1977a. Crayfish and freshwater shrimp diseases. Publ. No. TAMU-SG-77-605. Department of Marine Resources Information, Texas A&M University, College Station, Texas.

JOHNSON, S. K. 1977b. Some disease problems in crawfish and freshwater shrimp culture. Publ. No. FDDL-S11. Fish Disease Diagnostic Laboratory, Texas A&M University, College Station, Texas.

JOHNSON, W., and AVAULT, JR., J. W. 1980. Some effects of poultry manure supplementation to rice/crawfish experimental earthen ponds. Abstr. Fish Culture Sect. Am. Fish. Soc. New Orleans, **1980,** 15.

JOINT SUBCOMMITTEE ON AQUACULTURE. 1980. Draft. National Aquaculture Plan. U.S. Department of Commerce, Washington, D.C.

KONIKOFF, M. 1977. Study of the life history and ecology of the red swamp crawfish, *Procambarus clarkii,* in the lower Atchafalaya Basin Floodway. Final Rep. for U.S. Fish & Wildlife Service, Department of Biology, University of Southwestern Louisiana, Lafayette, Louisiana.

LACAZE, C. G. 1981 (revised). Crawfish farming. Fish. Bull. No. 7. Louisiana Wildlife & Fisheries Commission (now Louisiana Department of Wildlife & Fisheries), Baton Rouge, Louisiana.

LANGLOIS, T. H. 1935. Notes on the habits of the crayfish, *Cambarus rusticus,* in fish ponds in Ohio. Trans. Am. Fish. Soc. **65,** 189–192.

LANTZ, K. E. 1973. Acanthocephalan occurrence in cultured red crawfish. Proc. 27th Annu. Conf. Southeastern Assoc. Game Fish Commiss. **27,** 735–738.

LASHER, C. V., JR. 1975. Epizooites of crayfish. I. Ectocommensals and parasites of crayfish of Brazos County, Texas. Int. Symp. Freshwater Crayfish **2,** 277–286.

LEONARD, S. L. 1981. *Orconectes virilis* (Hagen). *In* Manual for the Culture of Selected Freshwater Invertebrates, S. C. Lawrence (editor), pp. 95–108. Canadian Spec. Publ. Fisheries and Aquatic Sciences, Department of Fisheries and Oceans, Winnipeg, Manitoba, Canada.

LINDQVIST, O. V., and MIKKOLA, H. 1979. On the etiology of the muscle wasting disease in *Procambarus clarkii* in Kenya. Int. Symp. Freshwater Crayfish **4,** 363–372.

LORMAN, J. G., and MAGNUSON, J. J. 1978. The role of crayfishes in aquatic ecosystems. Fish. Bull. **6**(3), 8–10, 16–19.

LOYACANO, H. 1967. Some effects of salinity on two populations of red swamp crawfish, *Procambarus clarkii*. Proc. 21st Annu. Conf. Southeastern Assoc. Fish Game Commiss. **21,** 423–434.

LUTZ, C. G. 1983. Population dynamics of red swamp crawfish (*Procambarus clarkii*) and white river crawfish (*Procambarus acutus acutus*) in two commercial ponds. M.S. Thesis, Louisiana State University, Baton Rouge, Louisiana.

MASON, J. C. 1975. Crayfish production in a small woodland stream. Int. Symp. Freshwater Crayfish **2,** 449–481.

MASON, J. C. 1979. Significance of egg size in the freshwater crayfish, *Pacifastacus leniusculus* (Dana). Int. Symp. Freshwater Crayfish **4,** 83–92.

MELANCON, E., JR., and AVAULT, JR., J. W. 1977. Oxygen tolerance of juvenile red swamp crayfish, *Procambarus clarkii* (Girard). Int. Symp. Freshwater Crayfish **3,** 371–380.

MILTNER, M., and AVAULT, JR., J. W. 1980. An evaluation of rice (*Oryza sativa*) and Japanese millet (*Echinocloa frumentacea*) as forage for red swamp crawfish (*Procambarus clarkii*) (abstract only). Abstr. Fish Culture Sect. Am. Fish. Soc. New Orleans, **1980,** 15.

MILTNER, J., and AVAULT, JR., J. W. 1981. Rice millet as forages for crawfish. Louisiana Agric. **24**(3), 8–10.

MOMOT, W. T., and GOWING, H. 1977. Production and dynamics of *Orconectes virilis* in three Michigan lakes. J. Fish. Res. Board Can. **34,** 2056–2066.

MOMOT, W. T., and ROMAIRE, R. P. 1981. Use of a seine to detect stunted crawfish populations in ponds. A preliminary report. J. World Maric. Soc. **12**(2),384–390.

MOMOT, W. T., GOWING, H., and JONES, P. D. 1978. The dynamics of crayfish and their role in ecosystems. Am. Midl. Nat. **99,** 10–35.

MOODY, M. W. 1981. Louisiana seafood delight—The crawfish. Sea Grant Publ. LSU-T1-80-002. Louisiana Cooperative Extension Service, Baton Rouge, Louisiana.

MORRISSY, N. M. 1976. Aquaculture of marron, *Cherax tenuimanus* (Smith). Part 2. Breeding and early rearing. Fish. Res. Bull. No. 17. Department of Fisheries and Wildlife, Perth, Western Australia.

NELSON, R. G., and DENDY, J. S. 1979. Conditions for holding and propagating crawfish brood stock (*Procambarus clarkii*.) Proc. World Maric. Soc. **10,** 503–512.

PARK, T. 1945. A further report on toleration experiments by ecology classes. Ecology **26,** 305–308.

PAYNE, J. F. 1972. The life history of *Procambarus hayi*. Am. Midl. Nat. **87,** 25–35.

PENN, G. H., JR. 1943. A study of the life history of the Louisiana red crawfish, *Cambarus clarkii* Girard. Ecology **24,** 1–18.

PENN, G. H. 1956. The genus *Procambarus* in Louisiana (Decapoda Astacidae). Am. Midl. Nat. **56,** 406–422.

PERRY, W. G., JR. 1972. Marsh pond construction. Proc. World Maric. Soc. **3,** 149–165.

PFISTER, V. A. 1982. Catchability of ten commercial crawfish traps and an evaluation of selected crawfish harvesting strategies. M.S. Thesis, Louisiana State University, Baton Rouge, Louisiana.

POLLOCK, B. A. 1982. Development and evaluation of artificial baits for trapping crawfish. M.S. Thesis, Louisiana State University, Baton Rouge, Louisiana.

PRINS, R. 1968. Comparative ecology of the crayfishes *Orconectes rusticus* and *Cambarus tenebrosus* in Doe Run, Meade County, Kentucky. Int. Rev. Gesamten Hydrobiol. **53,** 667–714.

RICKETT, J. D. 1974. Trophic relationships involving crayfish of the genus *Orconectes* in experimental ponds. Prog. Fish Cult. **36,** 207–211.

RIVAS, R., ROMAIRE, R., AVAULT, JR., J. W., and GIAMALVA, M. 1979. Agricultural forages and by-products as feed for crawfish, *Procambarus clarkii*. Int. Symp. Freshwater Crayfish **4**, 337–342.

ROBERTS, K. J. 1980. Louisiana crawfish farming: An economic view. Proc. 1st Nat. Crawfish Culture Workshop, D. Gooch and J. Huner (editors). Res. Ser. No. 50. University of Southwestern Louisiana, Lafayette, Louisiana.

ROBERTS, K. J. 1982. Crawfish production cost. Louisiana Coop. Ext. Serv., Baton Rouge, Louisiana.

SHARFSTEIN, B. A., and CHAFIN, C. 1979. Red swamp crayfish: Short-term effects of salinity on survival and growth. Prog. Fish Cult. **41**, 156–157.

SMITH, R. I. 1940. Studies on the effects of eyestalk removal upon young crayfish *Cambarus clarkii* (Girard). Biol. Bull. **79**, 145–152.

SMITHERMAN, R., AVAULT, JR., J. W., DE LA BRETONNE, JR., L. J., and LOYACANO, H. A. 1967. Effects of supplemental feed and fertilizer on production of red swamp crawfish, *Procambarus clarkii,* in pools and ponds. Proc. Annu. Conf. Southeastern Assoc. Game Fish Commiss. **21**, 452–458.

SOGANDARES-BERNAL, F. 1965. Parasites from Louisiana crayfishes. Tulane Stud. Zool. **12**, 79–85.

STEVENSON, J. R. 1972. Changing activities of the crustacean epidermis during the molting cycle. Am. Zool. **12**, 373–380.

SUKO, T. 1956. Studies on the development of the crayfish. IV. Development of winter eggs. Sci. Rep. Saitama Univ. (Japan) **2B**, 213–219.

TACK, P. T. 1941. The life history and ecology of the crayfish *Cambarus immunis* (Hagen). Am. Midl. Nat. **25**, 420–466.

TEXAS COOPERATIVE EXTENSION SERVICE. 1981. Cost analysis for Texas crawfish production. Texas Coop. Ext. Serv., College Station, Texas.

THREINEN, C. W. 1958. A summary of observations on the commercial harvest of crayfish in northwestern Wisconsin, with notes on the life history of *Orconectes virilis*. Misc. Rep. No. 2, Wisconsin Conservation Department, Fisheries Management Division, Madison, Wisconsin.

TRAVIS, D. F. 1965. The deposition of skeletal structures in the Crustacea. 5. The histomorphological and histochemical changes associated with the development and calcification of the branchial exoskeleton of the crayfish *Orconectes virilis* Hagen. J. Ultrastruct. Res. **9**, 285–301.

TURNER, H. M., and CORKUM, K. C. 1977. *Allocorrigia filiformis* gen. et sp. a (Trematoda:Dicrocoeliidae) from the crayfish *Procambarus clarkii* (Girard, 1852). Proc. Helminthol. Soc. Wash. **44**, 65–67.

TUTEN, J. S., and AVAULT, JR., J. W. 1981. Growing red swamp crayfish *Procambarus clarkii* and several North American fish species together. Prog. Fish Cult. **43**,97–99.

UNESTAM, T. 1969. Resistance to the crayfish plague in some American, Japanese and European crayfishes. Rep. Inst. Freshwater Res., Drottingholm (Sweden) **49**, 202–209.

UNESTAM, T. 1973. Significance of disease on freshwater crayfish. Int. Symp. Freshwater Crayfish **1**, 135–150.

VISOCA, P., JR. 1966. Crawfish farming. Educ. Bull. No. 2. Louisiana Wildlife and Fisheries Commission (now Louisiana Department of Wildlife and Fisheries), Baton Rouge, Louisiana.

WILLIAMS, J. W., and AVAULT, JR., J. W. 1977. Acute toxicity of acriflavin, formalin, and potassium permanganate to juvenile red swamp crayfish, *Procambarus clarkii* (Girard). Int. Symp. Freshwater Crayfish **3**, 397–404.

WILLIAMS, E., CRAIG III, F. S., and AVAULT, JR., J. W. 1975. Some legal aspects

of catfish and crawfish farming in Louisiana: a case study. Bull. No. 689. Louisiana State University Agric. Exp. Stat., Baton Rouge, Louisiana.
WISHART, M. A., and LOYACANO, H. A. 1974. A survey of edible crawfish from the coastal plain of South Carolina. Completion Rep. for Coastal Plain Regional Commission, by Dept. of Entomology and Economic Zoology, Clemson University, Clemson, South Carolina.
WITZIG, J. F., AVAULT, JR., J. W., and HUNER, J. V. 1980. Insect dynamics in a crawfish pond with emphasis on predaceous insects (abstract only). Abstr. Fish Culture Sect. Am. Fish. Soc., New Orleans, **1980,** 14.

2

Freshwater Prawns

Paul A. Sandifer
Theodore I. J. Smith

Introduction
Basic Biology
  Taxonomy and Species of Interest
  Distribution and Life Cycle
  Reproduction
  Molting, Growth, and Behavior
  Environmental Requirements
  Feeds and Nutrition
Culture Techniques
  Hatchery Phase
  Nursery Phase
  Grow-Out Systems
  Processing
Diseases, Parasites, and Predators
Constraints on Development of Prawn Farming
Status of Culture in the United States
Economic Overview and Outlook
Literature Cited

INTRODUCTION

Historically, freshwater prawns of the genus *Macrobrachium* have been esteemed as food in many parts of the world. Of this group, the giant prawn of the Indo-Pacific, *M. rosenbergii,* has become so popular that demand exceeds the supply from natural production. Attempts to increase production by stocking wild-caught juveniles in ponds was initiated in Thailand during the 1950s, but an insufficient supply of young and low yields limited development of this production method (Dugan *et al.* 1975).

Modern aquaculture of freshwater prawns may be dated from July 1959 when Dr. Shao-wen Ling began his research with *M. rosenbergii* in Penang, Malaysia. By June 1961 Ling had discovered that the larvae of *M. rosenbergii* require brackish water for survival and development, and he produced the first postlarval prawns via culture. Ling's subsequent work on the biology, larval rearing, and pond culture of *M. rosenbergii* resulted in two landmark papers, "The General Biology and Development of *Macrobrachium rosenbergii* (de Man)" and "Methods of Rearing and Culturing *Macrobrachium rosenbergii* (de Man)," which were presented at the FAO World Conference on the Biology and Culture of Shrimps and Prawns in Mexico City in 1967. These papers were published two years later (Ling 1969a,b) and serve as basic references for prawn culturists.

American aquaculture of freshwater prawns began in Hawaii in the mid-1960s. Takuji Fujimura of the Hawaii Department of Land and Natural Resources learned of Ling's work with *M. rosenbergii* and in 1964 arranged to import brood stock to Hawaii (Fujimura 1966). Over the next several years, Fujimura and his associates developed a practical large-scale hatchery technique for mass production of postlarvae and initiated pond-rearing trials in Hawaii (Fujimura 1966, 1974; Fujimura and Okamoto 1972). Fujimura's success led to the initiation of a small prawn aquaculture industry in Hawaii, the establishment of the Anuenue Fisheries Research Center as the world leader for research and development activities with prawns, and the rapid spread of private and public interest in prawn aquaculture.

Interest in prawn culture spread to the continental United States in the late 1960s, where culture efforts began in Florida. Subsequently, major research activities developed at Florida's Marine Research Laboratory and South Carolina's Marine Resources Research Institute, and smaller programs evolved in several other states. Prawns have been or are being cultured experimentally in at least a dozen states besides Hawaii, and one large commercial farm is currently in operation in the southern United States.

M. rosenbergii has also been introduced into Africa, the Caribbean, Central and South America, Israel, Japan, Mauritius, Tahiti, Taiwan, and the United Kingdom for aquaculture studies. Commercial prawn farms exist or are being developed in several of these areas, as well as in countries within the species' natural range—e.g., Malaysia, India, Thailand—(Ling and Costello 1979; Sandifer, in press).

BASIC BIOLOGY

Taxonomy and Species of Interest

Macrobrachium is a large, cosmopolitan genus of caridean shrimp belonging to the family Palaemonidae. Caridean shrimp are differentiated from penaeid shrimp by having the pleura of the second abdominal segment overlapping those of the first and third segments and by the lack of chelae (claws) on the third pair of pereiopods, or walking legs (Williams 1965). The family Palaemonidae is distinguished by the following characteristics (Williams 1965): chelae on the first and second pairs of pereiopods, with the carpus of the second pair not subdivided; a rostrum that is usually toothed and immovable; and mandibles that usually have an incisor process. The genus *Macrobrachium* Bate 1868 is defined as follows (Holthuis 1952, p. 10):

> Palaemonid shrimps with rostrum well developed, compressed and toothed. Carapace armed with antennal and hepatic spines; branchiostegal groove present. Telson with 2 pairs of dorsal and 2 pairs of posterior spines. Mandible with a three-jointed palp. Exopods on all maxillipeds. Pleurobranchs on the third maxilliped and all pereiopods. Last three legs with the dactylus simple. Propodus of fifth leg with numerous transverse rows of setae on the posterior margin. First pleopod of male without appendix interna.

Most of the approximately 125 species of *Macrobrachium* inhabit fresh- and brackish-water environments in tropical and subtropical areas around the world (Holthuis 1950, 1952, 1980). Six species are known to occur in the continental United States (Holthuis and Provenzano 1970), one of which, *M. ohione,* supports small fisheries in a few southern states (Huner 1977). In addition, *M. lar* has been introduced into Hawaii and *M. rosenbergii* into Hawaii and several other states (Atkinson 1977; Sandifer 1980). Because of its temperature requirements, it is unlikely that *M. rosenbergii* will become established in the wild in any continental state.

On a worldwide basis, 49 species of *Macrobrachium* are of interest to fisheries (Holthuis 1980), and at least 15 species are being considered for aquaculture (Ling and Costello 1979). By far the most important species for culture is *M. rosenbergii,* and the remainder of this chapter is limited to treatment of this species.

Distribution and Life Cycle

Macrobrachium rosenbergii is a tropical species widely distributed in the Indo-Pacific region (Fig. 2.1) where it ranges from Australia to

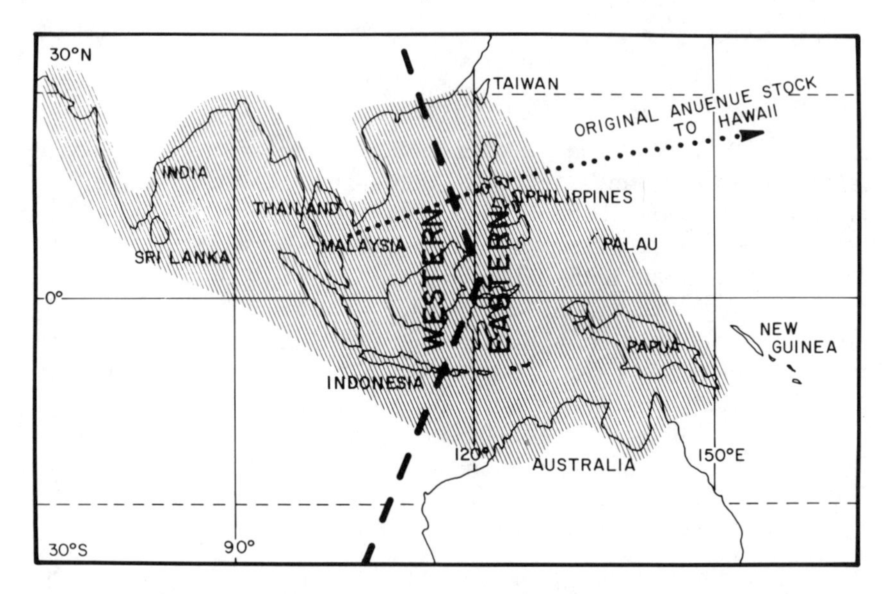

FIG. 2.1 Native distribution of *Macrobrachium rosenbergii* (shaded land areas, not including oceanic waters), showing geographic origin of the Hawaiian (Anuenue stock) and division of stocks into two major groups, Eastern and Western. *Adapted from Malecha 1980a.*

New Guinea to the Indus River delta (Johnson 1960; George 1969). Populations have also been reported from the Palau Islands (McVey 1975). This species occurs at temperatures of approximately 25°C (77°F) in Malaya (Johnson 1967) and 27°–34°C (81°–93°F) in India (John 1957; Rao 1967).

Adult *M. rosenbergii* usually inhabit freshwater reaches of coastal rivers and lakes, although juveniles and adults have been captured at salinities as high as 18 ‰ (George 1969). During the breeding season, the mature female prawn mates following a prespawning molt. The eggs are spawned, usually within 24 hours following mating, and they become attached to special setae on the abdominal appendages (pleopods). Here the eggs are carried for an embryonic-development period of approximately 3 weeks. Ovigerous females migrate downstream into estuarine areas where the eggs hatch as free-swimming larvae. The larvae are then swept downstream to develop in salinities of 5–20 ‰ (George 1969). Such brackish water conditions are required for developement; the larvae die within a few days in freshwater or high salinities.

The larvae pass through a series of about 11 stages during a planktonic period of about 20–50 days. They then metamorphose into post-

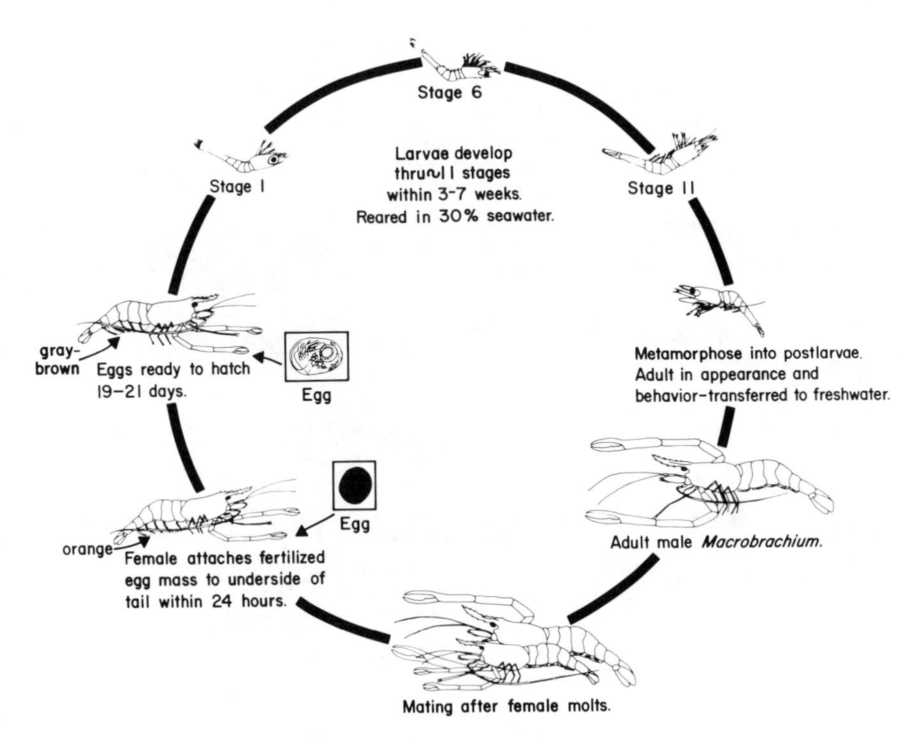

FIG. 2.2 Life cycle of *Macrobrachium rosenbergii* in culture.
After Sandifer and Smith (1978) by Institute of Food Technology.

larvae and assume a more benthic life style. The postlarvae are nearly transparent, but otherwise generally resemble the adult prawn. Within a few weeks, the postlarvae begin to migrate upstream into freshwater environments where they grow to maturity (John 1957; Raman 1967; George 1969; Ling 1969a). Prawns are capable of long-distance migrations, and adults have been found over 200 km (125 mi) inland from their natal brackish-water environment (Ling 1969a). The life cycle of *M. rosenbergii* in culture reflects these major events (Fig. 2.2).

Reproduction

Sexual Characteristics. The sexes are separate in *M. rosenbergii,* and external sexually dimorphic characteristics allow easy differentiation of mature male and female prawns (Figs. 2.3 and 2.4).

Male *M. rosenbergii* generally grow much larger than females, regularly reaching 200 g (7 oz) or more in the wild. The largest specimen documented to date weighed 654 g (23.4 oz), and there are uncon-

FIG. 2.3 An adult male *Macrobrachium rosenbergii*.

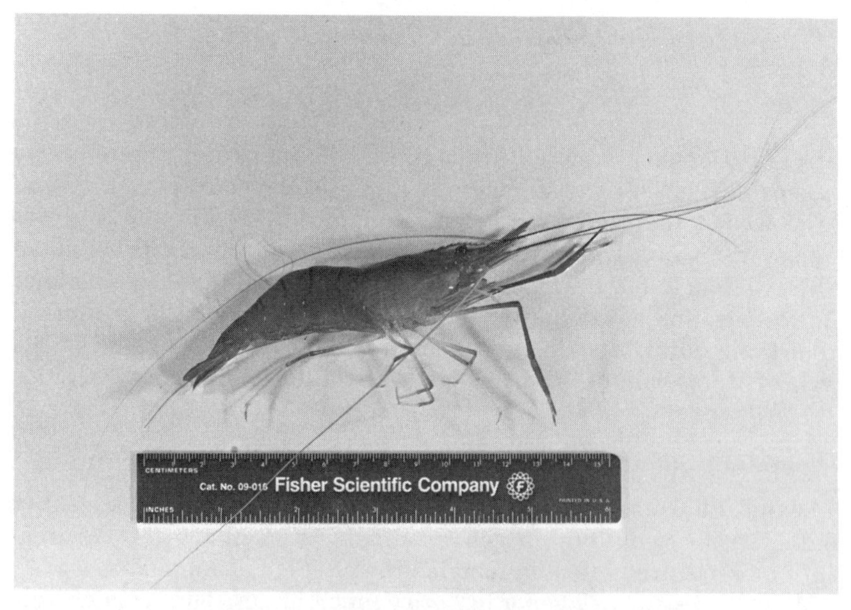

FIG. 2.4 An adult female *Macrobrachium rosenbergii*.

firmed reports of even larger prawns weighing 1000g or more (Malecha 1980b). Males have proportionately larger heads than females, and the second pereiopods occur as very long, robust, spinous claws, or chelipeds (Ling 1969a; Smith *et al.* 1980). The chelipeds of an adult male are typically dark blue, but some have shorter claws that are more golden in color (Sandifer and Smith 1977a). Males with the short-claw growth form are believed to be less aggressive and less sexually active than those with long blue claws. Differentiation of the mature male cheliped form may be apparent in prawns as small as 28 mm (1 in.) in carapace length (CL), but differences are often not obvious until the prawns are considerably larger (Nagamine and Knight 1980). Male prawns also may be recognized by the presence of a flap-covered gonopore on the coxa of each fifth pereiopod and by the appendix masculina, a spinous process adjacent to the appendix interna on the endopod of the second pleopod (Fig. 2.5). Male gonopores have been observed, with difficulty, in very small prawns (down to 5.9 mm (¼ in.) CL). However, the appendices masculinae do not begin to appear until later when prawns have reached about 10 mm CL or 30 mm total length (0.4 in. or 1¼ in., respectively) and are not fully formed until a size of about 70 mm (about 3 in.) TL is attained (Tombes and Foster 1979; Nagamine and Knight 1980). The pereiopods of the male also are set close together in nearly parallel lines, with little open space between the lines.

Female *M. rosenbergii* have proportionately smaller heads and much smaller, slenderer claws than males (Fig. 2.4). Their gonopores are located on the coxae of the third pereiopods, and there is a more or less triangularly shaped, unspecialized sperm receptacle area of the thoracic sterna between the last three pairs of pereiopods. In addition, the first, second, and third abdominal pleura are elongated and broadened to form an egg-brooding chamber, there are special reproductive setae on the thorax and pleopods, and the female's second pleopod lacks an appendix masculina (Fig. 2.5). Broadening of the abdominal pleura to form a brood chamber begins at about 20 mm (¾ in.) CL and, once formed, it is an obvious permanent characteristic (Nagamine and Knight 1980). The reproductive setae that appear on mature females are of two types, ovipositing and ovigerous. The ovipositing setae occur on the coxae of the last three pairs of pereiopods, the posterior margin of the sperm receptacle area, and the pleopods and serve to help guide and propel the eggs during spawning (Sandifer and Smith 1979a; Nagamine and Knight 1980). Although generally permanent, ovipositing setae are much more pronounced following a prespawning molt. Ovigerous setae, on the other hand, occur only fol-

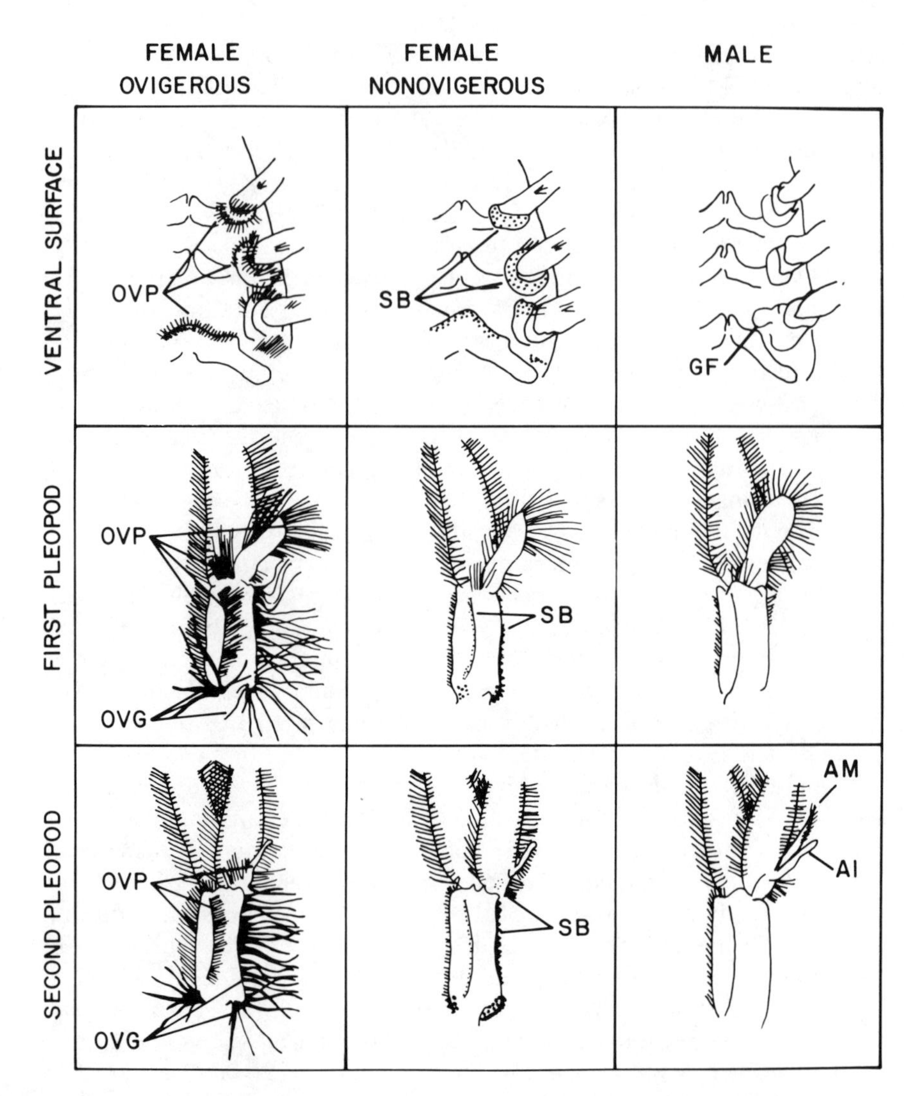

FIG. 2.5 Sexual dimorphism of the thorax ventral surface and the first and second pleopods of sexually mature male and female *Macrobrachium rosenbergii* and differences between ovigerous and nonovigerous females. Tips of pleopods have been deleted. AI = appendix interna; AM = appendix masculina; GF = gonopore flaps; OVG = ovigerous setae; OVP = ovipositing setae; SB = setal buds.
Redrawn from Nagamine and Knight (1980) with permission of the authors.

lowing a prespawning molt, and they serve to anchor the eggs to the pleopods for brooding.

The primary internal reproductive structures of males include paired testes, vasa deferentia, and terminal ampullae (Fig. 2.6). The fused testes lie mid-dorsally in the head (cephalothorax) and give rise to vasa deferentia, which are highly coiled just anterior to the heart and then extend as more or less straight tubes down the posterolateral margin of the head, ending in enlarged terminal ampullae (Sandifer and Smith 1979a). The ampullae open at the gonopores, and each contains one half of the spermatophore. At ejaculation, the musculature surrounding the ampullae contracts, extruding the two spermatophore halves, which fuse along their medial margins to form a complete spermatophore (Sandifer and Lynn 1980). Each half of the spermatophore contains a sperm mass in which the nonmotile, thumbtack-shaped sperm cells are embedded in a dense fibrous matrix (Fig. 2.7).

In females, the ovaries are located dorsal to the stomach and hepatopancreas in the head. When fully developed, the ovaries extend from just behind the eyes and beneath the rostral crest to the first

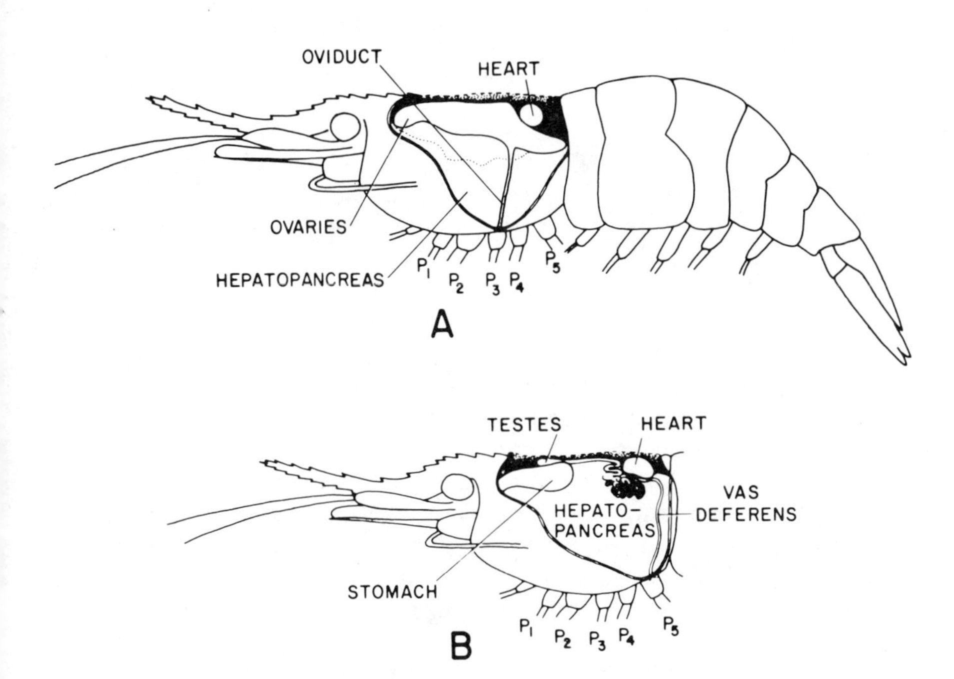

FIG. 2.6 Internal reproductive structures of (a) female and (b) male *Macrobrachium rosenbergii*.
Reprinted by permission of the publisher from Sandifer and Lynn (1980), in Advances in Invertebrate Reproduction, *1980 by Elsevier Science Publishing Co.*

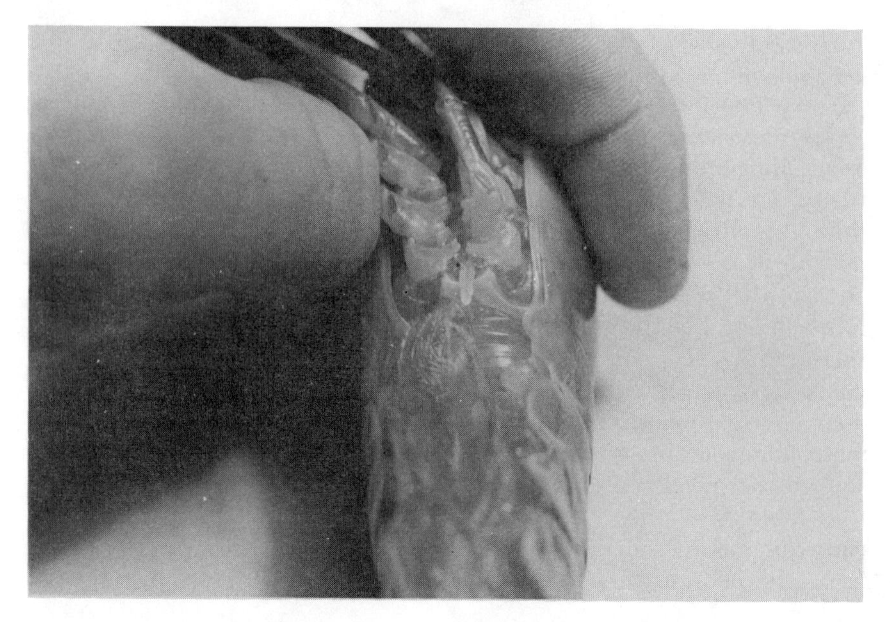

FIG. 2.7 Complete spermatophore extruded from gonopores of male *Macrobrachium rosenbergii.*

abdominal segment (Fig. 2.6). An oviduct arises laterally from each ovary just anterior to the heart and extends ventrolaterally to the gonopore on the coxa of the third pereopod. The color of the ripe ovary in *M. rosenbergii* is bright orange, and this color is generally visible through the carapace.

Mating Behavior. Prior to mating, a ripe female (i.e., one with fully developed ovaries) undergoes a prespawning molt. This molt usually occurs at night and is preceded by 2–3 days of fasting and reduced activity (Rao 1965; Ling 1969a). Within a few hours following the molt, the female becomes receptive to mating. Occasionally copulation may proceed rapidly, with little or no precopulatory behavior, but generally premating activities require 20–35 minutes (Ling 1969a) and may not be initiated until several hours after the female's molt. Chow *et al.* (1982) reported that the interval between molting and mating for 20 females ranged from 1.2 to 21.8 hours, with an average duration of 9.1 hours. Precopulatory behavior may be characterized by male displays and attempts by the male to contact the female with his claws and antennae. Once firm contact is established, the male may turn the female over on her dorsal surface and clean her thoracic sterna with his pereiopods (Rao 1965; Ling 1969a; Lynn 1981). Then copulation takes place, with the female generally on her back and the male above her at an angle (Rao 1965; Ling 1969a; Chow *et al.* 1982). This position brings the male gonopores close to the female's sternum (Fig. 2.8). Actual transfer of the spermatophore takes only a few seconds, and it is probably accomplished by the male's first and second pleopods (see Bauer 1976 for description of probable transfer mecha-

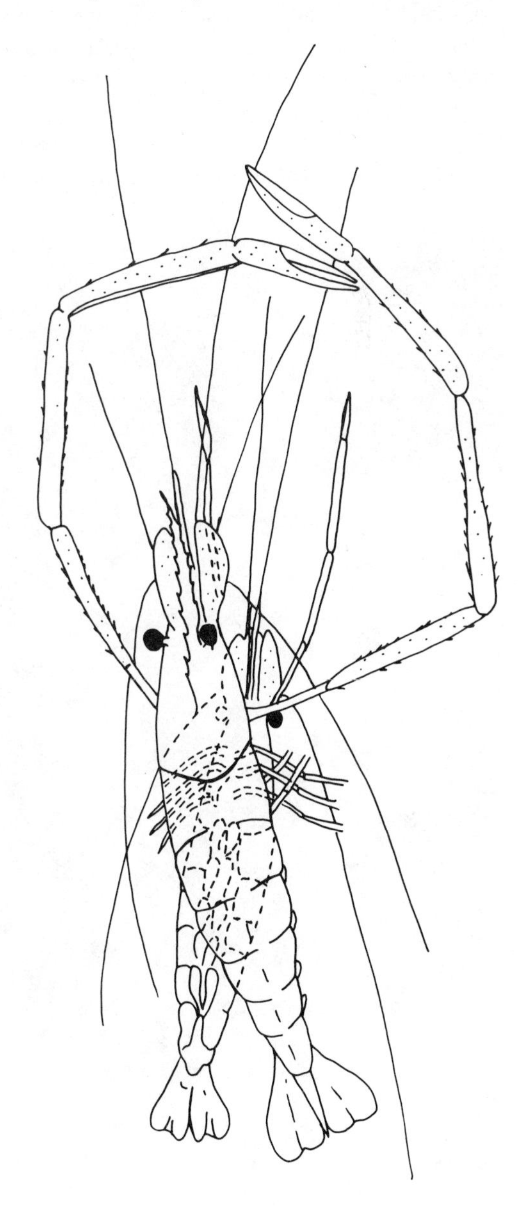

FIG. 2.8 Copulation in *Macrobrachium rosenbergii:* female below, male above.
Redrawn from Rao (1965).

nism). Because of the glutinous nature of the attachment matrix, the spermatophore adheres to the female's exoskeleton, and is placed in the sperm receptacle area between the last three pairs of pereiopods (Fig. 2.9).

Sexually active male prawns may build breeding depressions or nests in the bottoms of culture ponds (Smith and Sandifer 1979a). At night several ripe females may be observed in and around such depressions, each of which is usually occupied by one male. Similarly, in laboratory tanks equipped with artificial habitat units, mature males are often found on the topmost, least preferred layers, surrounded by a "harem" of up to six ripe females. Presumably, the male provides protection for the females during their period of vulnerability immedi-

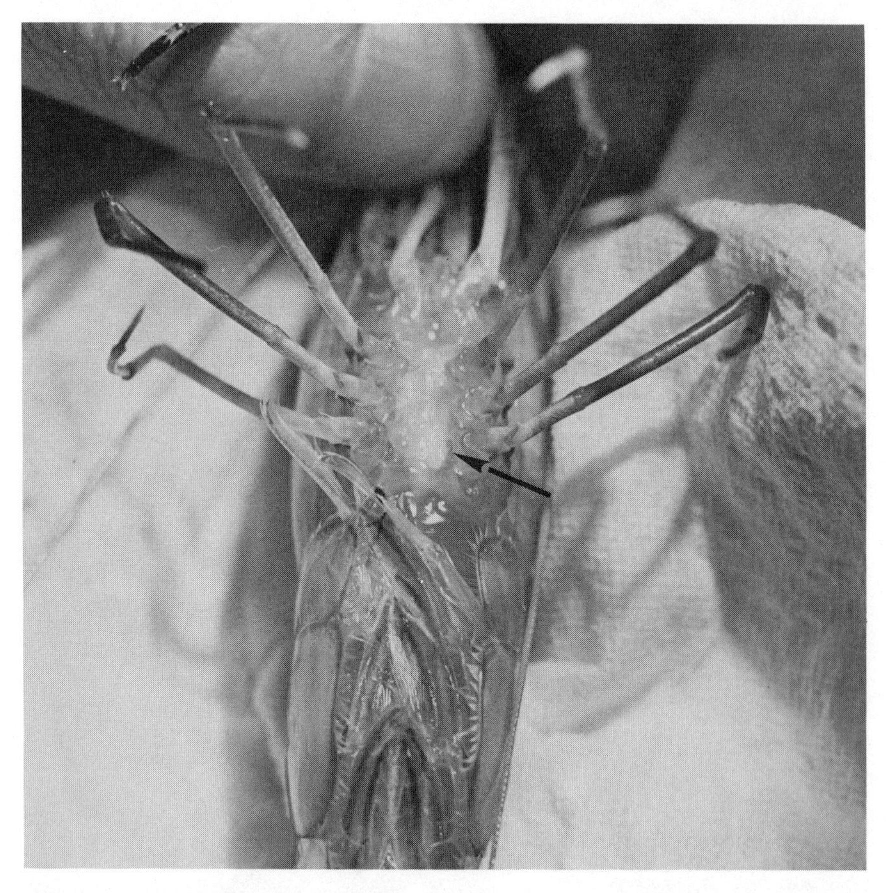

FIG. 2.9 Adult female *Macrobrachium rosenbergii* with spermatophore in place (arrow) following copulation.
Reprinted from Sandifer and Smith (1979).

ately preceding and following the molt, and then copulates with each female as she becomes receptive.

After copulation, female *M. rosenbergii* may spend up to several hours in prespawning preening behavior. In this activity, the small chelae of the first pereiopods are apparently used to arrange the ovipositing and ovigerous setae and the spermatophore prior to spawning.

Typically, spawning in *M. rosenbergii* occurs several hours after copulation and generally within 24 hours after the premating molt (Rao 1965; Ling 1969a; Sandifer and Smith 1979a; Chow *et al.* 1982). During spawning the female's abdomen is tightly flexed and the pleopods are extended so that the first two pairs overlap the sternum, providing a "floor" for the egg passageway. The eggs exit the gonopores in steady streams and are guided over the spermatophore and into the abdominal brood chamber by the ovipositing setae. The spermatophore is typically oriented so that a sperm mass is aligned with each gonopore to facilitate fertilization. The mechanism by which the sperm cells are released from the dense spermatophore matrix to come into contact with passing eggs is not known, but it is likely that an enzymatic reaction is involved (King 1948; Sandifer and Lynn 1980). Details of the fertilization sequence (interaction of sperm and egg) are described by Lynn (1981) and Lynn and Clark (1983a, b).

The act of spawning has been reported to require as few as 10 and as many as 60 minutes (Rao 1965; Lynn 1981), but it is usually completed within 15–25 minutes (Ling 1969a; Sandifer and Smith 1979a; Chow *et al.* 1982). After they are spawned, the eggs become attached to each other and to the ovigerous setae of the first four pairs of pleopods by a cementing substance. Several minutes are required for this substance to harden into an attachement membrane. If the female is disturbed before the cementing substance hardens, the eggs will be lost from the brood chamber.

After the cementing substance hardens, the female uses her first pereiopods to examine and clean the egg mass, apparently removing debris and damaged eggs. Such inspection and cleaning occurs intermittently throughout embryonic development. Regular beating of the pleopods, which serves to aerate the egg mass, also occurs throughout the development period.

Molting, Growth, and Behavior

In order to grow, *M. rosenbergii,* like other crustaceans, must periodically shed its exoskeleton. The frequency of molting is related to size, since small, rapidly growing prawns must molt more frequently

TABLE 2.1. EFFECTS OF SIZE ON MOLT FREQUENCY IN *MACROBRACHIUM ROSENBERGII* MAINTAINED AT 28°C

Live Weight (g)	Days Between Molts
2– 5.8	9
6–10	13.5
11–15	17
16–20	18.5
21–25	20
26–35	22
35–60	22–42

Source: Wickins 1976a.

than larger, more slowly growing animals (Table 2.1). Molt frequency is also affected by temperature, quantity and quality of food, water chemistry, sex, physiological condition of the animal, and other factors.

The molt cycle may be divided into four major stages as follows:

Premolt—Calcium and other materials are resorbed from the "old" exoskeleton to soften it; a "new" exoskeleton is formed beneath the "old" one. Stage D.

Molt—The "old" exoskeleton splits dorsally between the carapace and the intercalary sclerite of the abdomen; the cephalothorax and anterior appendages, including setae, are withdrawn, followed by body flexure to free the abdomen (Ling 1969a; Wickins 1976a). The act of molting requires only about 10 minutes and is accompanied by rapid uptake of water through the gills and consequent increase in size. Stage E.

Postmolt—This is the period during which the "new" exoskeleton hardens as the result of deposition of calcium and other materials; it may last a few hours to days, depending on the size of the prawn. Stages A and B.

Intermolt—During this relatively long period between molts, the water absorbed at molting is gradually replaced by tissue growth, and organic and mineral reserves are accumulated. Stage C.

The recognizable morphological characteristics of the various molt stages are summarized in Table 2.2.

During and immediately following molting, the prawns are extremely vulnerable to predation. Thus, it is not surprising that prawns generally seek shelter prior to molting and that most molts occur at night.

The best growth rates yet documented for *M. rosenbergii* in culture

TABLE 2.2. MOLT STAGES OF *MACROBRACHIUM ROSENBERGII*

Molt Stage		Distinguishing Morphological Characteristics	Approximate Duration (days)[a]
A.	Immediate Postmolt	Exoskeleton soft; rostrum deflectable; animal unable to raise chelae if held out of water	1
B.	Late Postmolt	Exoskeleton pliable; rostrum hard; no pigment retraction from margins of rostrum or abdominal pleura	3–5
C.	Intermolt	Exoskeleton hard; pigment retraction starts in margins of dorsal surface of abdominal segments	29–79
D_0	Early Premolt	Pigment retraction complete in lateral margins of abdominal segments; pigment retraction evident at pleopod tips	3–5
D_1	Early Premolt	New setae develop in pleopods	3–5
D_2	Early Premolt	Extensive pigment retraction on both dorsal and lateral margins of abdominal segments; no new setal development within antennal scale	3–5
D_3'	Late Premolt	Setal development in antennal scale	2–3
D_3''	Late Premolt	Pigment retraction from margin of dorsal surface of rostrum	2–3
D_3'''	Late Premolt	Exoskeleton becomes flexible	1–2

Source: Adapted from Peebles 1977.
[a] Based on healthy adult animals.

were recorded by Ling (1969a,b). Under good conditions, postlarvae reared in small nursery ponds grew to about 2 g in mean weight within 60 days. These juveniles were then stocked into grow-out ponds at densities of 6,000–15,000/ha or 2430–6075/Ac (0.6–1.5 prawns/m^2 or 0.06–0.15/ft^2), depending on the productivity of the ponds, and grew to a mean weight of about 100 g (3.5 oz) in 5–6 months.*

With one exception, other culturists have not reported such rapid average growth rates. In part, this probably reflects the culture densities used by other prawn growers. Growth in *M. rosenbergii* varies inversely with population density (Sandifer and Smith 1975; Willis and Berrigan 1977; Smith *et al.* 1978, 1981), and most U.S. growers tend to use stocking densities 10–15 times those recommended by Ling (Shang 1981; Smith *et al.* 1981). However, even at relatively moderate culture densities (2–6.5/m^2), mean sizes attained after 5–6 months have ranged from less than 20 to only about 45 g, depending in part on the size of the prawns at stocking (Smith *et al.* 1976a, 1978, 1981; Willis and Berrigan 1977; Tarver and Perry 1981). The one exception

*454 g = 1 lb and there are 10.76 ft^2 in each m^2.

involved a small group of juvenile prawns (mean weight of 0.7 g, stocking density of $0.5/m^2$) that were stocked into a pond along with several thousand postlarvae (stocking density of $6.3/m^2$). After 5 months the mean size of the juveniles had increased to 75 g, with the males in this group attaining 100 g on average (Smith *et al.* 1976a). Depending on stocking density and other factors, most prawns are likely to require 9–12 months to grow from postlarvae to a mean weight of 40–50 g.

Growth in *M. rosenbergii* is characterized by sexually dimorphic patterns and highly variable rates (Fig. 2.10). Male and female prawns grow at similar rates until the female begins to channel much of her energy intake into ovarian development and less into growth. According to Ling (1969a), slowing of female growth occurs at a size of about 60 g. However, we have found that, while males and females grow at the same rate up to a mean size of at least 17 g, slowing of the female growth rate becomes apparent at sizes above 25 g or so.

Typically, the size distribution for females is normal and encompasses a moderately broad range, whereas that for males is markedly skewed to the right, usually multi-modal, and extends over a very wide range (Figs. 2.11 and 2.12) (Fujimura and Okamoto 1972; Smith *et al.* 1978). The skewness of the male population has been

FIG. 2.10 Example of variable growth in *Macrobrachium rosenbergii.* These prawns are the same age and were reared in the same pond for 5 months.

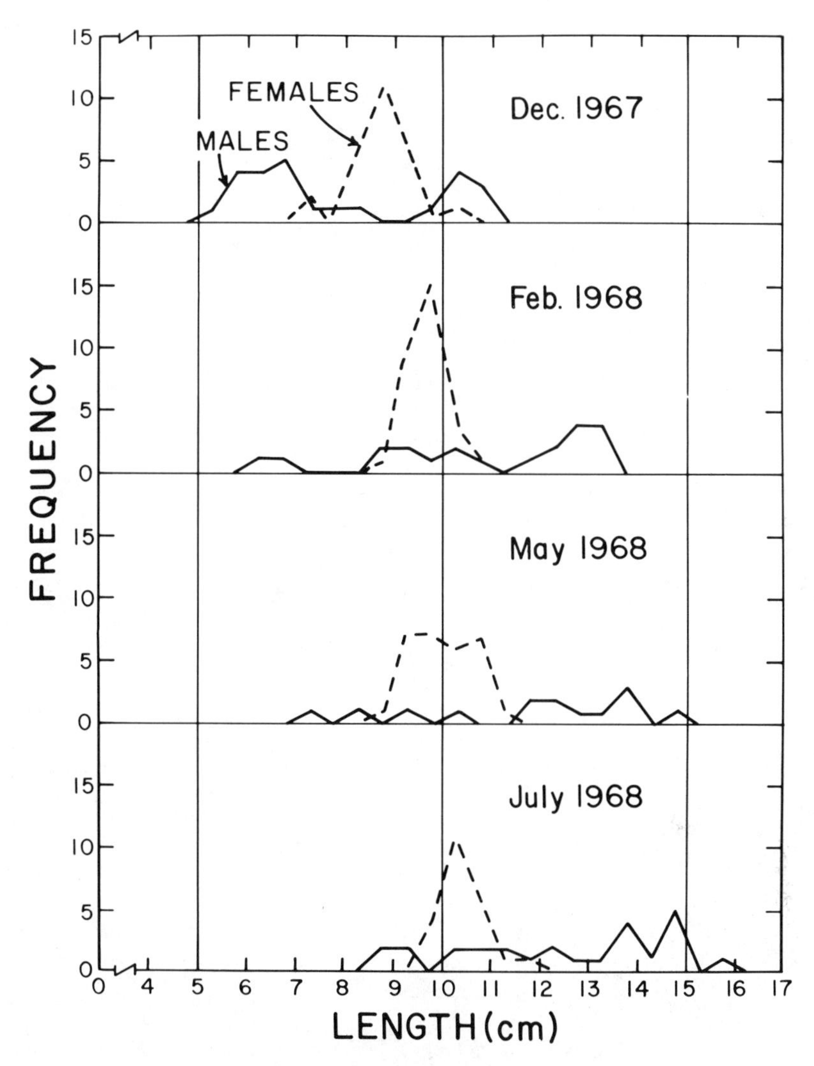

FIG. 2.11 Changes in size distributions for male and female *Macrobrachium rosenbergii* reared in a pond in Hawaii.
Redrawn from Fujimura and Okamoto (1972) with permission of the authors.

attributed to a social interaction whereby large "bull" males inhibit growth of small "bachelors" (Fujimura and Okamoto 1972). The wide variation in male growth rates begins during larval development when some larvae metamorphose to postlarvae much earlier than others. However, early metamorphosis apparently confers no growth advantage over late metamorphosis (Sandifer and Smith 1979b). As juve-

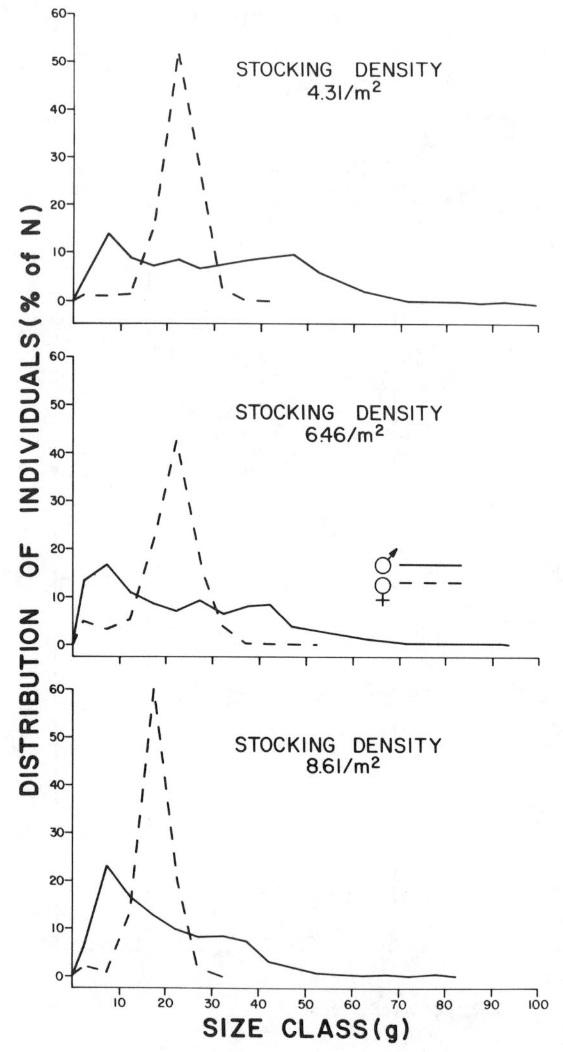

FIG. 2.12 Harvest size distributions of male and female *Macrobrachium rosenbergii* reared in ponds in South Carolina at different population densities.

niles, prawns continue to exhibit more variable growth rates than most other shrimp (Forster and Beard 1974), and the size variance increases considerably as the animals grow (Sandifer and Smith 1975; Malecha 1980b, Cohen *et al.* 1981; Malecha *et al.* 1981).

Because of the prawn's highly variable growth characteristics, Fujimura and Okamoto (1972) recommended selective harvesting of the larger, faster-growing animals with a large mesh seine. Then, as the remaining prawns grow into the larger size classes, they too are periodically seined out. The selective harvesting of large animals is believed to reduce competition for food, oxygen, and territory and to result in increased growth rates among the remaining prawns.

Macrobrachium rosenbergii is considered to be a rather aggressive, territorial, and cannibalistic species when reared in captivity (Wickins 1972; Forster and Beard 1974; Segal and Roe 1975). Increasing population density leads to increased incidence of agonistic encounters with the result that survival decreases. As density increases, acute mortalities occur among progressively smaller prawns (Sandifer and Smith 1975). Experimental evidence suggests that these deaths may result from aggressive competition for shelter (microhabitats) and food among animals of different molt stages and sizes (Peebles 1978, 1979, 1980). Prawn culturists have long noted that late premolt and early postmolt animals are frequently in the shallow, vegetated areas along the margins of culture ponds, whereas intermolt prawns occur in the deeper, more protected areas. In the laboratory, Peebles (1980) found that isolated premolt and postmolt animals chose the same deep habitat as did intermolt prawns. However, when a pre- or postmolt prawn was paired with an intermolt, the intermolt took possession of the deeper laboratory habitat and displaced the other to a shallower, less preferred area. As expected, pre- or postmolt prawns were much more susceptible to aggression-related deaths than were intermolt animals, and smaller prawns were more likely to suffer in encounters with larger ones than with those of comparable size (Peebles 1978).

Ling (1962, 1969b) first suggested that aquatic plants, branches, gravel, or shells be placed in nursery and grow-out ponds to provide refuges for molting prawns. Later, Smith and Sandifer (1975) demonstrated that addition of artificial habitats to laboratory tanks significantly improved the growth and survival of juvenile prawns reared at high densities. These authors then evaluated a series of habitat configurations and noticed that prawns exhibited a pronounced preference for the edges of habitat materials (Smith and Sandifer 1979b). Because of this "edge effect," they constructed artificial habitats to maximize edge area available to the prawns (e.g., see Fig. 2.13). This work on artifical habitats has made possible the development of intensive, high-density nursery systems for juvenile prawns (Sandifer and Smith 1977b, 1978; Smith and Sandifer 1979c).

Environmental Requirements

Temperature. Optimum temperatures for culture of all stages of *M. rosenbergii* are believed to be approximately 28°–30°C (82°–86°F), with a range of 26°–31°C (79°–88°F) considered satisfactory (Table

FIG. 2.13 Example of artificial habitat developed for use in tank culture of *Macrobrachium rosenbergii.*
Reprinted with permission from Handbook of Mariculture, Vol. I, Crustacean Aquaculture, CRC Press, Inc., Boca Raton, Florida; © 1983.

2.3). Lethal temperature extremes for acclimated juvenile and adult prawns are approximately 13° and 38°C (55° and 100°F), respectively, but mortality increases rapidly at sustained temperatures of less than about 18°C (64°F) or greater than 33°C (91°F) (Uno *et al.* 1975; Armstrong 1978; Farmanfarmaian and Moore 1978; Smith and Sandifer, unpublished data). Although prawns can withstand a broad temperature range, reductions in growth, activity, and survival are generally noted at temperatures outside the range of 22°–33°C (72°–91°F) (Uno *et al.* 1975). However, while low temperatures may slow or stop prawn growth, they do not seem to permanently stunt the animals. For example, Silverthorn and Reese (1978) reported that postlarvae that were held at approximately 16°C (61°F) for 3 weeks grew as rapidly when returned to 27°C (81°F) as those animals that had been maintained continuously at 27°C.

Salinity. *M. rosenbergii* larvae require saline water for development, and they are usually reared at salinities in the range of 8–17

TABLE 2.3. EFFECTS OF TEMPERATURE ON GROWTH AND SURVIVAL OF JUVENILE PRAWNS

Temperature (°C)	Initial Mean Weight (mg)	Final Mean Weight (mg)	Mean Growth Rate (mg/day)	Daily Growth Rate (% of rate at 29°C)	Survival (%)
17[a]	140	—	—	—	—
21	140	330	3.02	24.7	58.0
25	140	680	8.57	70.1	88.0
29	140	910	12.22	100.0	83.0
33	140	790	10.32	84.5	67.0

Source: Smith and Sandifer, unpublished data.

[a] This treatment was terminated on day 26 when mean weight was 160 mg and survival was 9.1%. All other treatments were continued for 63 days at which time final weights and % survival were determined.

‰, although salinities as low as 6 ‰ have been used for some stages (Sick and Beaty 1974). A salinity range of 12–16 ‰ is generally preferred for hatchery operations (Sandifer *et al.* 1977).

Typically, juvenile prawns are reared in fresh water, but they can survive and grow in a variety of saline environments. Sandifer *et al.* (1975) showed that postlarval *M. rosenbergii* exposed to high salinities began to die when exposed to a salinity of 25 ‰, and mortality increased rapidly at salinities of 30 ‰ and higher. The prawns were able to maintain a nearly constant blood osmotic concentration in waters of 0 to about 27 ‰ by hypoosmotic regulation up to 17–18 ‰ and hyperosmotic regulation at higher salinities. At salinities above 30 ‰ the osmoregulatory mechanism broke down completely. Castille and Lawrence (1981) suggested that the postlarva's ability to hypoosmoregulate at salinities above 17 ‰ is probably lost as the animal matures, and eventually juvenile prawns become osmoconformers at salinities above 14–15 ‰ (Armstrong 1978; Castille and Lawrence 1981). Further, the higher the temperature, the lower the salinity at which stress manifests itself in depressed metabolism—e.g., 28 ‰ at 20°C, 20 ‰ at 27°C, and any increase above 0 ‰ at 34°C (Nelson *et al.* 1977).

In slightly saline environments (2–5 ‰), small *M. rosenbergii* have been observed to grow a little faster than in fresh water and much more rapidly than at 15 ‰ salinity (Sandifer and Smith 1974; Perdue and Nakamura 1976). Ling and Costello (1979) suggested that salinities up to about 10 ‰ were suitable for pond culture of *M. rosenbergii,* and Smith *et al.* (1982) presented field data to support their recommendation.

Dissolved Oxygen. Sharp (1976) determined that there is a critical, temperature-dependent oxygen concentration above which the prawn's respiratory rate is independent of oxygen concentration but below which respiration is limited by the amount of oxygen present. For a 0.2-g (dry weight) prawn, the concentration at which oxygen becomes limiting is 2.1 ppm at 23°C (73°F), 2.9 ppm at 28°C (82°F) and 4.7 ppm at 33°C (91°F). Although prawns can withstand lower dissolved oxygen levels (e.g., see Smith *et al.* 1976a, 1978, 1981), stress can be minimized by maintaining oxygen levels above the critical minima.

Larger prawns require more oxygen than smaller ones, and thus the larger animals are more susceptible to low oxygen concentrations. This means that failure to manage oxygen levels properly in a culture system may lead to the loss of the larger, more valuable animals. For example, Smith *et al.* (1982) reported that the mean size of prawns killed by overnight oxygen depletion in a pond was twice that of survivors (10.8 vs. 5.3 g).

Nitrogenous Wastes. The principal form of nitrogen excreted by crustaceans is ammonia (Parry 1960), which may be highly toxic. Through a bacterially mediated process of nitrification, the execreted ammonia is converted to nitrite (NO_2^-), which is also toxic, and then to relatively nontoxic nitrate (NO_3^-). The toxicity of nitrogenous wastes, then, depends in part on the concentrations of ammonia, nitrite, and nitrate in solution. Further pH, and to a much lesser extent temperature, determine the proportion of the total ammonia present as the highly toxic un-ionized (NH_3) and less toxic ionized (NH_4^+) forms. The proportion of ammonia present as NH_3 increases with increasing pH (especially) and temperature (Emerson *et al.* 1975).

For *M. rosenbergii* larvae, the effects of pH on ammonia toxicity are clear. Values for 24-hr and 144-hr LC_{50}'s show that as pH increases, the tolerance of larvae to total ammonia levels decreases, primarily as the result of increasing concentrations of un-ionized ammonia (Table 2.4). A sublethal effect, reduced growth, was observed at a total ammonia level of 32 mg/liter at pH 6.83 and 7.60 (Armstrong *et al.* 1978).

For nitrite, LC_{50}'s for 10- to 14-day-old larvae in static bioassays were 130 mg/liter at 24 hr, 8.6 mg/liter at 96 hr, and 4.5 mg/liter at 192 hr (Armstrong *et al.* 1976). The highest nitrite concentrations at which no mortality occurred were 9.7 and 1.4 mg/liter over 24- and 168-hr periods, respectively. Reduced growth was observed among larvae exposed to a sublethal concentration of 1.8 mg/liter.

Juvenile *M. rosenbergii* and penaeid shrimp exhibit similar tolerances to ammonia, but the prawns are generally less tolerant of nitrite and nitrate (Wickins 1976b). Acute tolerance to ammonia is generally similar to that observed for larvae (Table 2.5). However, the

TABLE 2.4. TOXICITY[a] OF AMMONIA TO *MACROBRACHIUM ROSENBERGII* LARVAE (TOTAL AMMONIA = $[NH_3] + [NH_4^+]$) AT DIFFERENT PH LEVELS

pH	Total Ammonia	NH_3	NH_4^+	Percent Un-ionized
		24-hr LC_{50} (mg/liter)		
6.83	200	0.66	199.34	0.33
7.60	115	2.10	112.90	1.83
8.34	37	3.58	33.42	9.68
		144-hr LC_{50} (mg/liter)		
6.83	80	0.26	79.74	0.33
7.60	44	0.80	43.20	1.82
8.34	14	1.35	12.65	9.64

Source: Adapted from Armstrong *et al.* 1978.
[a] Expressed as LC_{50}, the concentration that is lethal to 50% of the experimental animals.

growth rate of juveniles exposed to chronic ammonia levels of only 0.16 mg/liter or greater for 6 weeks was approximately one-third less than that of controls. For nitrites and nitrates, LC_{50}'s in 3- to 4-week bioassays were 15.4 and 160 mg/liter, respectively.

pH. No specific studies relating water pH to growth and survival of prawns have been conducted. In South Carolina, prawns have been reared in ponds where the pH has ranged from 6.0 to 10.5 with no apparent adverse effects. However, chronic high pH may on occasion be associated with the formation of calcium carbonate precipitates on prawns (Cripps and Nakamura 1979; Smith and Sandifer, unpublished data). Such precipitates can cause death by occlusion of the gill surfaces. Generally, pH levels reported for prawn ponds range from 7.0 to about 8.5, and the pH should not be allowed to remain below 6.5 or above 9.0 for long periods.

Hardness and Alkalinity. Prawns require minerals, especially calcium and magnesium, for new exoskeleton formation and other bio-

TABLE 2.5. ACUTE TOXICITY[a] OF AMMONIA TO JUVENILE *MACROBRACHIUM ROSENBERGII*

Total Ammonia (mg/liter)	Un-ionized Ammonia (mg/liter)	LT_{50} (hours)
192	1.43	28.33
298	2.21	23.33
378	2.81	9.33

Source: Adapted from Wickins 1976b.
[a] Expressed as LT_{50}, the exposure time associated with 50% mortality of the experimental animals.

logical processes. However, relatively little is known concerning the effects of the mineral content of water supplies on prawns.

Only one study has examined the effect of hardness on *M. rosenbergii* larvae. Sick and Beaty (1974) reported that early larvae of *M. rosenbergii* experienced mass mortality when reared in an artificial seawater medium prepared with well water which had total hardness ($CaCO_3$) levels of 50–100 mg/liter. In the same study, normal survival and development were observed among larvae reared in a medium made with distilled water. No other investigators have reported problems with the hardness of the freshwater supplies used to prepare hatchery culture media.

Juvenile prawns have been reared successfully over a broad range of hardness levels. Good growth and survival were obtained for prawns reared at chronically low hardness levels (5–7 mg/liter) in South Carolina (Smith *et al.* 1976a), but there was a noticeable softening of the animals' exoskeletons at these low calcium levels. No such effect was apparent at slightly higher concentrations of $CaCO_3$ (15–26 mg/liter). In the laboratory, Heinen (1977) noted no differences in survival and growth of postlarvae reared at hardness levels of 10–310 mg/liter for 28 days, but Cripps and Nakamura (1979) reported decreasing growth rate with increasing hardness over the range 65–500 mg/liter. Prawns maintained at the lowest hardness level gained two to five times as much weight as those at the highest in the Cripps and Nakamura (1979) study, and the intermolt duration was significantly increased among prawns in the 500-mg/liter medium. Retarded growth rates, high incidence of encrustations of *Epistylis,* and precipitation of calcium carbonate on prawns are believed to be associated with hardness levels greater than 300 mg/liter in Hawaii (Fujimura, in Goodwin and Hanson 1975; Cripps and Nakamura 1979). For culture operations, water with hardness levels of 300 mg/liter or greater should be avoided, and considerably lower hardness levels are preferred.

Below 180 mg/liter, total $CaCO_3$ alkalinity appears to have little if any effect on prawn growth in ponds. However, alkalinity levels above 180 mg/liter occasionally have been associated with mass mortality of prawns in South Carolina. These deaths were related to $CaCO_3$ precipitation, which occurred at elevated pH levels associated with phytoplankton blooms. The calcium precipitate apparently occluded the gills, resulting in suffocation.

Toxic Materials. Although little is known about the sensitivity of *M. rosenbergii* to pollutants, it is assumed that they are susceptible to the pesticides, heavy metals, and other materials known to be toxic to other crustaceans. Thus, culturists should avoid sites having possible toxic residues of pesticides and other materials from previous agricultural or industrial uses or receiving drainage from areas that receive pesticide applications or effluents.

Feeds and Nutrition

Reviews of shrimp and prawn nutrition studies by New (1976) and Biddle (1977) indicate that knowledge of the specific dietary requirements of *Macrobrachium* is extremely limited. Recent research has provided additional information concerning protein requirements, protein–energy relationships, digestive enzymes, acceptable levels of dietary fiber, and other matters (Clifford and Brick 1978, 1979; Stahl and Ahearn 1978; Farmanfarmaian and Lauterio 1979, 1980; Fair *et al.* 1980; Lee *et al.* 1980; Millikin *et al.* 1980), but sufficient data are not yet available for the formulation of nutritionally complete prepared diets. Thus, prawn culturists have had to rely primarily on natural products or feeds prepared for other animals.

In its natural environment, *M. rosenbergii* is an omnivore, feeding on zooplankton (as larvae), worms, insects and insect larvae, small mollusks and crustaceans, flesh and offal of fish and other animals, grain, seeds, nuts, fruits, algae and other aquatic plants (Ling 1969a). However, since such natural foods are generally not available in sufficient quantities in culture tanks and ponds to support the desired prawn population densities, the amount of food must be increased, either by direct feeding, fertilization, or both.

Larval prawns were first reared to metamorphosis on live zooplankton (rotifers, copepods, and insect larvae) supplemented with finely chopped fish, mollusks, and steamed hen's egg (Ling 1962). Live *Artemia* nauplii have since been used as the basic diet in larval culture, supplemented with the foods mentioned above and others, including frozen adult *Artemia,* fish eggs, ground beef heart, ground euphausiids, freeze-dried oyster flesh, freeze-dried catfish flesh, ground vegetable matter, trout fry feed, tropical fish food, and a number of other prepared diets and natural products (Fujimura 1966; Sick and Beaty 1974, 1975; Sick 1975; Smith *et al.* 1976b; Hanson and Goodwin 1977; Murai and Andrews 1978; Sandifer and Smith 1978; Manzi and Maddox 1980). Algae are also used in some prawn hatcheries with beneficial results (Fujimura 1966; Manzi *et al.* 1977), but excellent production rates have been obtained by other culturists without addition of algae to the culture water (Smith *et al.* 1976b; Aquacop 1977a,b; Sandifer *et al.* 1977).

Although some progress has been made in the development of artificial formulated diets for prawn larvae (Sick 1975; Sick and Beaty

1975), no prepared ration for larval *M. rosenbergii* is yet available. Production hatcheries must still rely on natural food items; *Artemia* nauplii supplemented with minced fish flesh, fish eggs, or hen's egg are the generally preferred feeds.

A variety of animal feeds—including chicken broiler starter, gamebird feed, catfish and trout chows, pig feed, "prawn pellets" (a pelleted formulation similar in composition to broiler starter), Purina Experimental Marine Rations (penaeid shrimp feed), and others—have been used to rear prawns in ponds. However, none of these rations is nutritionally complete, and the prawns must obtain substantial (but unquantified) nutriment from the natural biota of the ponds. That the natural pond biota plays an extremely important role in prawn nutrition is indicated by Willis and Berrigan's (1978) report that prawn yields from fed ponds were only about twice those from unfed ponds.

Another important limitation of most of the foods used in prawn culture is their lack of water stability. In water, these feeds rapidly break down into fine particles. Since prawns are slow, continuous feeders, much of the feed may not be directly available to them and may benefit them only indirectly through a pond-fertilization effect. This question was examined experimentally by Fair and Fortner (1981) who compared the growth responses of prawns fed a water-stable pelleted diet with those of animals fed pulverized pellets or provided with fertilization-enhanced natural biota. They found that prawns fed the pulverized diet grew significantly more rapidly than those in the fertilization (manure) treatment, suggesting that even rations that rapidly decompose into small particles may provide considerable direct nutritive benefit. However, the prawns fed the water-stable pellets grew twice as fast as those given the pulverized diet, demonstrating that a water-stable preparation is a much more efficient means of delivering nutrients to prawns than are formulations that rapidly decompose in water. Despite their lack of water stability, broiler starter and, more recently, prawn pellets have been used extensively as prawn food in Hawaii and elsewhere. Food conversions (kg of dry food needed to produce 1 kg of live prawns) with these diets average around 3.3 in Hawaii (Gibson and Wang 1977).

Although not formulated specifically for prawns, the Purina Marine Rations were designed for penaeid shrimp and retain their integrity in water for several hours. These rations, especially the 25% protein formulation, have been used widely in culture trials in the continental United States. Food conversions of 2 or less have been obtained routinely by experienced culturists, and levels of 1 or less have been reported for a number of ponds (Willis and Berrigan 1977; Smith *et al.* 1981). Despite the Purina ration's apparent water stability, it suffers considerable breakdown and leaching during its first 2 hours in water and loses 60% of its initial weight in 24 hours (Farmanfarmaian *et al.* 1982). Repelletization of the ration with an algin binder significantly improved its stability in water, and this greater stability was reflected in improved growth rates and feed conversions. Augmentation of the levels of five amino acids in the ration (lysine, arginine, tryptophan, leucine, and isoleucine) also improved its performance in laboratory tests (Farmanfarmaian and Lauterio 1979, 1980).

CULTURE TECHNIQUES

The production cycle for prawns can be divided into three major phases: hatchery, nursery, and grow-out. During the hatchery phase prawn larvae are reared to postlarvae in saltwater tanks. In the nursery phase the small postlarval prawns are placed in freshwater tanks or small ponds and reared to juveniles at relatively high population densities. The grow-out or production phase involves culture of postlarvae or nursery-reared juveniles to marketable size. Some operators are involved in only one phase (e.g., hatchery or grow-out), whereas other farmers incorporate different combinations of the various phases in their operations. In general, the larger the prawn farm, the more integrated the production system.

Hatchery Phase

In the United States, the hatchery phase is generally conducted indoors where proper temperatures can be maintained. However, in tropical environments outdoor hatcheries function quite well. In both situations, careful monitoring and control over the rearing environment are essential.

There are two basic types of *Macrobrachium* hatchery systems: the Anuenue type and the Aquacop type. The Anuenue system was developed at the Anuenue Fisheries Center in Hawaii by Fujimura and his associates, following earlier work by Ling in Malaysia. The Aquacop-type design resulted from the work of the Aquacop team in Tahiti (1977a,b) and it is generally similar to the Gavleston hatchery method for penaeid shrimp and the experimental recirculating hatchery system developed for prawns by Smith *et al.* (1976b) in South Carolina.

The Anuenue system, which may or may not involve the use of "green water" (algal blooms, batch-cultured outdoors), is the most widely used type of hatchery for production of *Macrobrachium* postlarvae (Figs. 2.14–2.16). It utilizes large rectangular tanks (on the

FIG. 2.14 Anuenue-type hatchery, Anuenue Fisheries Research Center, Hawaii.
Courtesy of Division of Aquatic Resources, Hawaii Department of Land and Natural Resources.

FIG. 2.15 An outdoor Anuenue-type prawn hatchery at the Chacheongsao Fisheries Station, Thailand.

FIG. 2.16 A small "backyard" hatchery of the Anuenue type in Thailand. Note that each tank is divided at the end to provide space for a biological filter. This hatchery is operated as a recirculating system.

order of 10–30 m^3 each) made of concrete or fiberglass and is operated at a less intensive level than the Aquacop system; stocking rates with Anuenue systems typically range from 30 to 50 larvae/liter and yields average from 5 to 25 postlarvae/liter. Aquacop-type systems use relatively small cylindrical tanks (not larger than 2 m^3) with conical bottoms and require intensive management and often the routine application of bactericides (Figs. 2.17 and 2.18); stocking and production levels are on the order of 100–200 larvae/liter and 25–100 postlarvae/liter, respectively. The fundamental differences between the two systems are that the Aquacop system is more productive per unit of tank volume but requires very careful management, whereas the Anuenue system is less productive and needs less intensive care.

In both types of hatcheries, water management generally involves partial or complete exchanges of water on a daily basis, although in some instances water exchanges are made much less frequently. Both Anuenue- and Aquacop-type hatcheries have been operated as recirculating systems, with little water exchange, although flow-through operation is much more common. Use of a recirculating water system

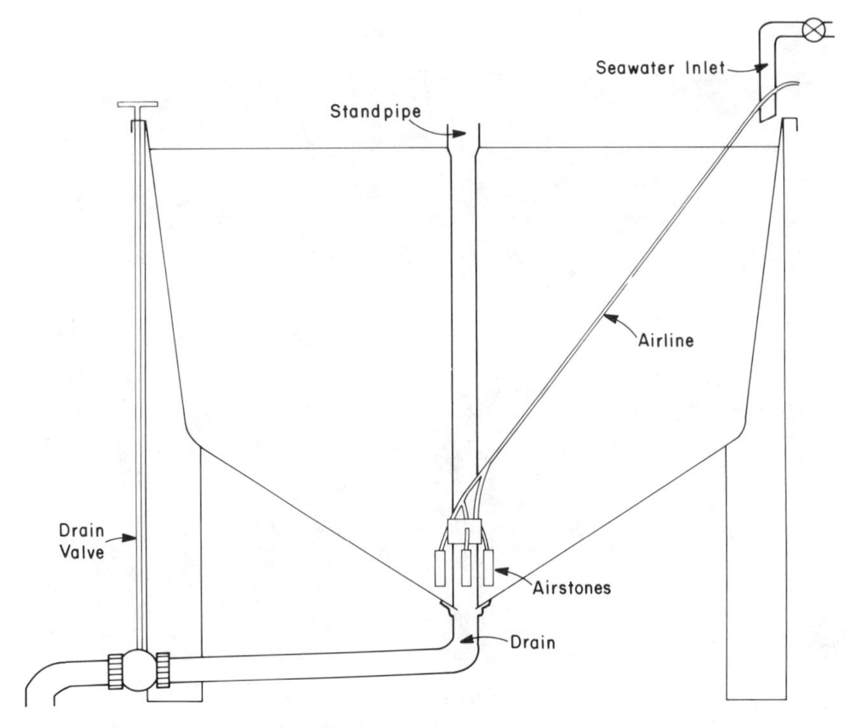

FIG. 2.17 Schematic of an Aquacop-type hatchery.
Redrawn from Aquacop (1977b) with permission of the authors.

FIG. 2.18 Example of an Aquacop-type hatchery.
Courtesy of Aquacop, Tahiti.

allows co-location of the prawn hatchery with the freshwater grow-out ponds and eliminates the need for a coastal hatchery site. Recirculating hatchery systems using rectangular and conical tanks and natural and synthetic seawaters have been implemented at both research and production scales (Sandifer *et al.* 1974; Manzi and Maddox 1976; Smith *et al.* 1976b; Cohen 1976; Menasveta 1980; Wulff 1982).

Larvae for stocking the hatchery tanks are obtained from cultured brood stock. In tropical climates the egg-bearing prawns are selected from production ponds or special brood stock ponds. In temperate areas, however, brood stock must be maintained in indoor facilities overwinter. Here, the egg-bearing females are obtained from breeding tanks that contain sexually mature males and females (Sandifer and Smith 1978).

Ovigerous females are selected from the ponds or breeding tanks and placed in hatching tanks until their eggs hatch. Depending on the stage of egg development and water temperature, females may remain in the hatching tanks for up to 20 days. Typically, only females carrying late-stage eggs are placed in hatching tanks. After their larvae hatch, the spent females are returned to breeding tanks, ponds, or processed for market. The newly hatched larvae are positively phototrophic and can be concentrated with a light and then removed with a fine mesh net or siphon hose. If the larvae hatch in fresh water, they should be placed in salt water (12–16 ‰) as soon as possible, and preferably within 24 hours. Alternatively, the ovigerous females can be maintained in brackish water during the egg incubation period so the larvae hatch directly into a saline environment.

Stocking density, feeding schedule, water treatment and management, etc., vary depending on specific hatchery design and the manager's discretion. As mentioned previously, depending on the system and management approach, larvae may be stocked initially at low (50/liter or less) or high (100/liter or more) densities. Alternatively, larvae may be stocked at high densities for the first 7–10 days of rearing (up to stage IV or V) after which the larvae are distributed among several tanks at a substantially lower density for development to postlarvae. Under normal hatchery conditions and temperatures of 27°–30°C (81°–86°F), the larval cycle should be completed within 25–40 days. At time of metamorphosis, postlarval prawns weigh about 6 mg and are about 8 mm in orbital length (orbit of eye to tip of telson).

The primary food used in most *Macrobrachium* hatcheries is the freshly hatched nauplius larva of the brine shrimp *Artemia. Artemia* nauplii are presented to the larval prawns beginning on day 2 of the rearing cycle, and are usually fed throughout the larval–rearing cycle. Feeding rates range from about 1 to 10 nauplii/ml of rearing water, and normally the *Artemia* concentration is gradually decreased during the rearing period as the amount of nonliving supplemental foods (e.g., minced fish, fish eggs, hen's egg, etc.) is increased.

Maintenance of a suitable rearing environment is critical and directly related to population density, age of the larvae, type and amount of food. Because prawns are aggressive and cannibalistic, the provision of attractive, nutritionally complete and properly sized food is very important to maintain satisfactory survival rates. Care must be exercised in feeding to provide an adequate amount of food without overfeeding, as most larval diets cause rapid deterioration of water quality. The feeding schedule should be arranged to provide frequent feedings with small amounts of food rather than less frequent feedings with larger amounts of food. This is particularly important during the early life stages. Tank hygiene usually involves draining off the bottom water or siphoning out bottom debris daily, with regular sterilization of all tank facilities between larval production cycles.

Nursery Phase

During the nursery phase the newly metamorphosed postlarval prawns are reared to larger juveniles for stocking in various grow-out systems. In tropical areas the nursery phase is usually incorporated into the production cycle of larger farms to provide more efficient utilization of the grow-out facilities and to improve predictability of crop yields. However, smaller farms operating in tropical climates often bypass the nursery phase and stock postlarvae directly into production ponds. Although this simplifies their operations, production results are not likely to be as predictable as when a nursery step is employed. In the temperate zone, the stocking of nursed juveniles appears necessary to increase the production rate and average size of the prawns at harvest to economically attractive levels (Sandifer and Smith 1979c; Smith *et al.* 1981; Sandifer *et al.* 1982a). Prawn nurseries are typically established outdoors in the tropics but must be enclosed and heated in temperate-zone areas.

Nurseries in the Tropics. In general, nursery systems in tropical areas are small-scale versions of grow-out ponds. Typically, nursery ponds are 0.2 ha (0.5 Ac) or less in size. Stocking density in nursery ponds is generally about 70–170 postlarvae/m^2 (7–16/ft^2) of pond surface area, and the duration of the nursery phase is usually 2–4 months. Netting, tree branches, or other types of materials or structures are sometimes placed in the ponds to serve as prawn habitats (see section on Molting, Growth, and Behavior). By increasing the surface area available to the prawns, such materials increase holding capacity and improve growth and survival rates. However, habitat materials can become fouled and adversely affect water quality, especially dissolved oxygen concentrations. Juvenile prawns in nursery ponds are generally fed the same prepared diet used in grow-out ponds. Crude protein levels in these feeds usually does not exceed 25%. Recent research suggests that use of a higher-protein ration (e.g., 40%) during the nursery phase might be beneficial (Millikin *et al.* 1980). Supplemental food is also obtained from animals and plants growing in the nursery ponds. Management procedures for nursery ponds are essentially the same as for grow-out ponds, but special care must be taken to exclude predators and competitors. Depending on density, feed, and environmental conditions (especially temperature), postlarval prawns should grow to a mean weight of $1\text{–}3^{+}$ g in a 2- to 4-month nursery period.

Of critical importance in the design of nursery ponds is the means of harvesting. Harvesting must be rapid and efficient to minimize handling and transfer stress. The ponds should have a graded, sloped bottom to facilitate draining and be equipped with a water-control structure, an interior or exterior drainage basin (Fig. 2.19) or a drain sump in which the young prawns are collected for rapid removal. The juvenile prawns may be harvested from the drainage structure with

FIG. 2.19 Example of corrugated aluminum water-control structure (flat-board riser) and in-pond harvest basin.

nets or a fish pump and then transferred in a live-haul truck to grow-out ponds. If desired, the animals can be size-graded, either by hand or mechanically, before restocking them in the production ponds.

Nurseries in Temperate Zones. Nursery systems in areas of temperate climate differ considerably from those in tropical areas due to climatic considerations. These temperate-zone systems are still under development, but their basic characteristics have been documented (Sandifer and Smith 1978; Smith and Sandifer 1979c). To produce nursed juveniles for seasonal grow-out in the continental United States, nursery systems must be operated during winter and early spring, necessitating enclosure and heating. In general, the major costs are associated with water containment structures (i.e., tanks), heating, and the nursery enclosure. Tanks fabricated of concrete, fiberglass, plastic liners, or other materials are currently in use as are very small earthen ponds of 500 m^2 (5380 ft^2) or less. Relatively low-cost structures such as inflatable greenhouses, which also provide considerable solar heating, are generally preferred to enclosed nursery tanks or ponds (see Fig. 2.20). Recirculating water management systems employing biological filters may be used in these indoor nursery systems to minimize heating costs (Sandifer and Smith 1977b). Since optimal

FIG. 2.20 Example of greenhouse-covered nursery pond. Note that plastic roof has been removed for summer use.
Photo used by permission of Commercial Shrimp Culture International, Inc.

rearing temperatures for this tropical species are 28°–30°C (82°–86°F), alternate energy sources such as heated effluents and geothermal resources are being evaluated to reduce heating costs. Nursery tanks or ponds are generally filled with fresh water, although if available low-salinity (1–5‰) water may offer a slight initial growth advantage (see section on Environmental Requirements).

Because of the higher capital and operating costs associated with enclosed nursery systems, postlarvae are stocked at much higher densities than those typically employed in outdoor nursery ponds. In the indoor systems, initial densities typically fall within the range 750–3000 postlarvae/m^2 (70–279/ft^2) of tank surface area. At the higher densities, populations are generally thinned by about 50% after the first month of rearing. Typical densities for our indoor nursery tank systems are on the order of 1000–1500 prawns/m^2 (93–140/ft^2) of tank floor area. The higher population densities are made possible through intensive management of water quality and the concentrated use of artificial habitats. These habitats may consist of fabricated cylindrical or rectangular units with numerous layers of plastic webbing or simply of plastic webbing hung vertically or horizontally in the nursery container (see Figs. 2.13 and 2.21). To be highly effective, artificial habitats must maximize the amount of edge available to the prawns

FIG. 2.21 Large concrete tank used for intensive culture experiments in South Carolina. Note hanging mesh habitat material.

and allow rapid prawn mobility to and from all habitat levels (Smith and Sandifer 1979b).

Preferably the juvenile prawns should be fed several times daily. A prepared ration (e.g., "Prawn Pellets," Purina Marine Ration 25, commercial fish feeds) is usually used, supplemented with a variety of natural foods including chopped fish, fish eggs, cooked hen's egg, various plant materials, etc. In some cases large plants such as water hyacinths and elodea are placed in nursery tanks to provide additional habitat area, supplemental feed, and water purification.

The juvenile prawns can be observed and their growth and survival rates monitored relatively easily in nursery tanks. Growth is usually slower at the higher densities used in most indoor systems compared with that in outdoor ponds, and the prawns generally attain mean weights of $0.5–1^{+}$ g after 2–3 months of rearing. At the end of the nursery period the prawns are harvested by tank drainage or by use of nets and transferred to grow-out units. In carefully managed, intensive nursery systems, survival should be 75% or higher.

Grow-Out Systems

Grow-out systems can be divided into extensive and intensive systems, both of which may be operated in a continuous production cycle or in a discontinuous, "batch-harvest" cycle. The rearing of prawns in earthen ponds in tropical areas has been shown to be economically viable (Shang and Fujimura 1977; Shang 1981), whereas similar culture in temperate climates and grow-out in intensive production systems are still under development (Sandifer and Smith 1978; Sandifer *et al.* 1982c; Smith *et al.* 1982).

Pond Culture in the Tropics. The general production system in tropical areas consists of stocking postlarvae or juveniles into earthen ponds and rearing them until an optimal market size is attained, about 20–25 whole prawns/kg (9–11/lb). Ponds are typically rectangular and most have a surface area of about 0.4–1 ha (1–2.5 Ac), although both larger and smaller ponds are in commercial production (Fig. 2.22). Dikes are steeply sloped (usually 3:1) to minimize predation losses due to wading birds. Pond bottoms usually have a slight slope to allow drainage. Preferred pond soils are impervious to water (e.g., clay, clay loam), and pond water depth averages about 1 m. Pond dikes are seeded with soil-stabilizing vegetation to limit erosion and to provide cover and feed for prawns. Moderate algal blooms should be maintained in grow-out ponds to shade out aquatic macrophytes that might interfere with water management, predator control, and harvesting.

FIG. 2.22 Amorient, the largest freshwater prawn farm in the United States, occupies a 48-ha site at Kahuku on the north shore of Oahu, Hawaii.
Courtesy of Aquaculture Development Program, Hawaii Department of Land and Natural Resources.

However, dense phytoplankton blooms should be avoided because of the potentially detrimental effects such blooms may have on water quality, especially the dissolved oxygen concentration. In some cases, various materials—webbing, nets, branches, etc.—have been placed in grow-out ponds to enchance production rates. Such materials may be effective in increasing crop yields, but documentation of their cost effectiveness is lacking. For convenience sake, such materials should not be placed in ponds where harvesting is done with seines.

In warmer climates, ponds are stocked initially at rates of about 65,000–163,000 prawns/ha (26,300–66,000/Ac), depending on harvesting and marketing strategy and the availability of seed stock. After an initial growing period of 7–9 months, selective harvesting of large prawns is begun using a long, large-mesh (4–5 cm, 1.5 in.) seine equipped with a bag. The seine may be pulled by hand or mechanically (e.g., with tractors). The prawns are dip-netted from the bag, culled (usually by hand), and the small animals (less than 35–40 g) returned to the pond for further growing. This harvest method takes advantage of the typically skewed growth performance of the male prawns (see section on Molting, Growth, and Behavior and Figs. 2.11 and 2.12). Although seine-harvesting is only about 50–75% efficient (Malecha 1980b), it yields prawns of a relatively uniform, large size.

A given pond is selectively seined at 2- to 4-week intervals, depending on conditions and production. The ponds are restocked with postlarvae or nursed juveniles one or more times during the year. Once selective seining is begun, production is more or less continuous, although it fluctuates considerably. Overall, yields from continuous production ponds average about 2000 kg/ha/year (1780 lb/Ac/year), and much higher yields have been obtained.

Continuous production ponds are usually drained only every several years as required for pond maintenance. Thus, a sluice gate may suffice as a water-control structure, although many growers prefer a flat-board riser-type drain structure (Fig. 2.19). Some growers in tropical areas are experimenting with more frequent drain-downs (e.g., every 2 years or so) to improve predictability of yields.

Major considerations in pond management involve the maintenance of proper water quality (especially dissolved oxygen concentrations), water depth, feeding, control of predators and competitors, and harvesting. Predators and competitors (especially fishes) can become serious problems, in some cases requiring draining of the pond for eradication. However, depletion of dissolved oxygen in culture ponds has probably been responsible for the catastrophic loss of more prawns than any other factor.

The prawns are fed once daily, usually in late afternoon, with a commercially prepared ration. The feed is broadcast over as much of the pond surface area as possible, either by hand or by a truck- or tractor-mounted blower (Fig. 2.23). The feeding rate is based on the estimated biomass in the pond. High feed levels (10–15% of estimated biomass per day) are used at the beginning of the culture period when the prawns are small, but these rates are decreased to 1–3% or less after 5–6 months.

Pond Culture in Temperate Zones. Ponds used for prawn culture in the continental United States and other climatically restricted areas are generally similar to those used in the Tropics except that provisions must be made to harvest the prawns by complete pond drainage (Fig. 2.24). Typically, the water management structure consists of a flat-board riser fitted with a bottom drain pipe. This unit is installed at the deepest area of the pond and the pond bottom is sloped to the riser to allow total drainage. The drain pipe should be sized to allow complete drainage in no more than 24 hours and placed slightly below the bottom of the pond so that all water will be removed. The face of the riser should contain two tracks for the installation of boards and screens that can be manipulated to flush out either bottom or surface waters during normal operations. For freshwater installations

FIG. 2.23 Feeding a prawn pond using a blower mounted on a pickup truck.
Photo used by permission of Commercial Shrimp Culture International, Inc.

FIG. 2.24 Removing prawns from harvest basin in a drain-harvested pond in South Carolina.

a semicylindrical corrugated aluminum riser is sufficient, while a concrete riser is recommended for use in brackish water. Drain pipes should be of PVC (preferred), concrete, or aluminum.

Profit margins are less in temperate areas than in tropical areas, so harvesting should be as efficient and complete as possible. Timing of the harvest is critical to maximize the growing season while avoiding potentially lethal cold water temperatures. Under normal conditions, prawns will move out of the pond with the outflow water and can be collected in a harvest basin or other collection device. The collection unit preferably should be placed at the outside end of the outflow pipe and may consist of a large three-sided trough having tracks for removable boards and screens at the outflow end or a simple net bag attached to the end of the drain pipe. Alternatively, an in-pond harvest kettle, such as is often found in ponds used for production of fingerling fish, may be employed (Fig. 2.19). The in-pond kettle is not preferred because of the amount of mud that accumulates in it during pond drainage. The outside trough arrangement allows easy washing and handling of the animals. Such structures can be designed to receive the effluent/harvest from two or more ponds, but they are certainly more expensive than a net bag.

To maximize growth and production during the limited growing season in temperate regions, nursed juvenile prawns of about 0.25–1.0 g in mean weight are stocked into ponds at relatively low densities, typically 43,000–65,000/ha (17,400–26,300/Ac). In cooler climates, the periodic selective harvest of large animals does not appear to be economically attractive because such prawns are generally not available in quantity until just before the final harvest. At harvest the ponds are drained and all prawns removed or "batch-harvested." Because prawns exhibit highly variable growth rates (see section on Molting, Growth, and Behavior), the harvested crop consists of a broad range of prawn sizes (Smith *et al.* 1978). All of these must be marketed to make prawn farming commercially viable; thus, size grading of the harvest may be necessary. Overall production levels of about 1100 kg of whole prawns/ha (980 lb/Ac) are readily obtainable in a 5- to 6-month growing season (Smith *et al.* 1981). Mean size of the harvested crop should be about 20–25 g (40–50 prawns/kg or 18–23 prawns/lb), with about 80% of the crop biomass made up of 18-g or larger prawns (55 prawns/kg or 25 prawns/lb).

Intensive Grow-Out Systems. Intensive production systems are designed to provide much higher crop yields per unit of area than those attained in ponds. These systems are characterized by higher investment costs, greater management requirements, and higher risks compared with pond systems. Intensive culture operations can be established in both tropical and temperate climates; however, heating costs in cooler climates will be substantial unless inexpensive heat sources can be incorporated into the operation. At present, these systems are under development (see Sandifer and Smith 1978 for details).

In many respects, intensive culture systems resemble nursery systems. Culture containers may be large tanks or small earthen ponds, and in temperate areas they must be enclosed for year-round operation. A large 185-m^2 (1990-ft^2) tank used for intensive culture experiments in South Carolina is shown in Fig. 2.21. The use of artificial habitats is necessary to increase growth and survival of prawns, but unfortunately little information is available on the relationships between habitat design and production. Generally, fabricated layered habitat units or draped netting are used (see Figs. 2.13 and 2.21); both are relatively easy to remove for harvesting.

Initial stocking density in intensive production systems should be at least five to ten times that employed in extensive grow-out ponds. However, growth and survival rates of prawns are inversely related to prawn density and size (Sandifer and Smith 1975). To moderate these interactive effects, the production phase may be subdivided into several subphases during which the population density is decreased in step-wise fashion as prawn size and biomass increase. This may be accomplished in several ways: subdividing the population into additional rearing units; stocking the entire population into progressively larger rearing units; and routine selective harvesting of marketable prawns or eliminating the slower-growing animals or "runts."

Good control of water quality and feed management are especially important in intensive production systems. Water quality must be closely monitored and controlled to provide as optimal a rearing environment as possible. Water may be recirculated, regularly exchanged, or used in a continual flow-through system. Feed should be provided several times daily and consist of a well-balanced diet. At present, a nutritionally complete diet for intensive culture is not available commercially, and so various prepared diets are supplemented with a variety of animal and plant materials. In some cases, macrophytes are also stocked in the production tanks to provide additional feed, shelter, and water purification. Production data from private intensive operations are generally lacking, but results of three seasonal pilot-scale trials in South Carolina have been reported. These trials, conducted in the $37 \times 5 \times 1.3$ m ($120 \times 16 \times 4$ ft) tank shown in Fig. 2.21, achieved production rates equivalent to 3800–4700 kg/ha (3385–4190 lb/Ac) during 3.5–4.5 months of culture; production rates on the order of 10,000 kg/ha/year (8900lb/Ac/year) were projected under conditions of continuous operation (Sandifer *et al.* 1982b; Smith and Sandifer, unpublished data). However, growth rates during these short-term trials were slower than those typically observed in ponds where prawn density is much lower and there is a greater diversity of natural foods.

Intensive systems offer the potential for increased production rates, reduced land and water requirements, and improved product quality and marketing control. In the long term, such systems may become more commercially attractive in the United States, especially if low-cost energy sources can be utilized to heat the systems.

Processing

Proper handling of the prawns at harvest is critical to the production of a high-quality product. Depending on the production system and market, prawn processing can range from simple to complex. In tropical areas selectively seined prawns are often maintained in large floating baskets in the ponds while being size-graded. The live prawns may then be transferred to tanks for sale to wholesalers or individual customers. Prawns that will not be sold live should be immediately chill-killed upon removal from the pond and then processed as soon as feasible. In general, whole prawns and tails can be maintained on ice for up to 8 days before sale, but their quality should be closely monitored (Waters and Hale 1981). Prawns can be frozen whole or as tails with an expected shelf life of about 7 months for whole prawns and about 10 months for tails (Hale and Waters 1981). Pretreatment (e.g., blanching) of fresh prawns before freezing has not been found to improve the keeping quality of prawns if they are handled properly. In fact, Hale and Waters (1981) found that whole prawns frozen raw have a longer shelf life than those frozen after cooking. For freezer storage, the prawns should be glazed, wrapped to prevent dessication, and maintained at $-20°$ to $-40°C$ ($-8°$ to $-40°F$).

DISEASES, PARASITES, AND PREDATORS

Disease and parasite problems can at times cause severe economic hardship in commercial aquaculture operations. However, at present such problems are not considered to present major obstacles to successful prawn farming.

Production rates are far more variable in the hatchery phase than in the other culture phases, in part reflecting the greater apparent susceptibility of larvae to stress and disease. Often larval populations suffer high attrition rates without apparent cause, and most commercial hatcheries are poorly equipped with the facilities and personnel needed to conduct the tedious diagnostic procedures used to identify microbial problems. In most cases, such affected cultures simply are discarded and the tanks sterilized before being restocked with new larvae; the source of the problem may never be identified.

Larval prawns are reared at high population densities and fed several times daily. As a consequence, there is a heavy loading of the rearing water with particulate and dissolved organic matter. Unless the cultures are carefully managed, this organic loading can cause poor water quality and provide excellent substrates for the proliferation of primary and secondary pathogens. For example, filamentous bacteria such as *Leucothrix* sp. and a number of protozoans including *Epistylis* sp., *Vorticella* sp., and *Cothurnia* sp. may bloom under such conditions. They attach to the gills and exoskeletons of larvae, restricting respiration and mobility (Hall 1977; Johnson 1977). Heavy infestations can ultimately result in mass mortality as the larvae become unable to feed, respire, or molt. Primary bacterial diseases also occur; Aquacop (1977c) reported a bacterial necrosis in *Macrobrachium* larvae that resulted in mortality rates to 100% among larvae of stages 4 and 5 within 48 hours. Additional bacterial problems likely will be identified as more diagnostic work is undertaken.

Fortunately, most disease problems encountered during the hatchery phase can be prevented or controlled. Careful maintenance of water quality and tank hygiene coupled with utilization of proper feeds and feeding techniques will greatly reduce the incidence of disease problems. In addition, certain antibiotics and therapeutic drugs have been successfully employed to control disease (Delves-Broughton 1974; Aquacop 1977a,b).

Disease problems are not significant once prawns attain the postlarval stage. Mass mortalities among pond-reared prawns are routinely associated with limiting environmental conditions (especially low dissolved oxygen levels) rather than with the presence of pathogenic organisms. In general, ectocommensal protozoans and algae and "black spot" disease are the most common problems encountered with juvenile and adult prawns. Epibionts observed on freshwater prawns are often similar to those reported on freshwater crawfish. Common genera include *Epistylis, Zoothamnium, Lagenophrys, Cothurnia,* and *Acineta;* less common genera include *Vorticella, Vaginicola,* and *Opercularia* (Johnson 1977). Prawns exhibiting dense infestations of these protozoans, especially *Zoothamnium* and *Vorticella,* are often referred to as "moss-backs." A similar moss-back condition is sometimes caused by dense growths of algae (Smith *et al.* 1979). Prawns covered with dense growths of protozoans or algae (Fig. 2.25) are occasionally seen in grow-out ponds, but the incidence of occurrence is usually low, and thus such infestations are generally of little economic importance.

The occurrence of darkly pigmented areas on broken appendages and various parts of the exoskeleton is quite common among cultured prawns. This so-called "black spot" disease (Fig. 2.26) results from invasion of chitinoclastic bacteria and fungi into physically damaged prawn tissues (Johnson 1978; Burns *et al.* 1979). Incidence of this infection is related to handling stress and frequency of prawn interactions. Generally, the more intense production systems have a greater incidence of animals displaying black spot disease. This condition appears to cause little harm to healthy prawns, which typically molt out of the infected condition.

Although diseases and ectocommensals do occur among cultured prawns and can cause high mortality rates especially during the hatchery phase, no known pathological condition currently poses a major threat to successful prawn farming. Unfortunately, a similar statement cannot be made concerning predators.

Predators can be responsible for major losses at all stages of production. Competitive organisms, which do not prey on prawns but in-

FIG. 2.25 An adult male prawn covered with algal epibionts.

FIG. 2.26 An example of black spot disease on the dorsal abdomen of an adult prawn.

stead utilize their feed and other pond resources, can also cause production losses. For example, in hatcheries small hydrozoan jellyfish may become major predators on prawn larvae or competitors for their food. Such infestations can decimate larval cultures and are extremely difficult to eradicate (Sandifer *et al.* 1974b). In nursery systems and grow-out ponds major predators include many species of fish, wading and diving birds, snakes, turtles, alligators, bullfrogs, otters, raccoons, and man. Special care should be taken to eradicate fish populations in ponds before stocking them with prawns, and any inflowing surface water should be screened to prevent fish from entering the ponds.

CONSTRAINTS ON DEVELOPMENT OF PRAWN FARMING

In the continental United States, climate is probably the major obstacle inhibiting development and expansion of prawn farming into a well-defined aquaculture industry. Even so, there is considerable interest in growing freshwater prawns commercially throughout much of the southeastern United States (Smith *et al.* 1976a, 1982; Sandifer and Smith 1979c; Willis and Berrigan 1977; Liao *et al.* 1981). Here, with a minimum 5- to 6-month growing season, production rates of 1100 kg/ha (980 lb/Ac) can be obtained (Smith *et al.* 1981). However, to attain this yield, nursery-reared juvenile prawns are recommended for stocking, and these animals will be more expensive than newly metamorphosed postlarvae. Some increase in production is likely if prawn farms are established in more southerly locations in the United States (e.g., the southern parts of Florida, Texas, California), but even in these areas, climate will play a major role in determining profitability.

Proper timing of harvests is critical in areas where water temperatures decrease to 16°C (61°F and below. In such areas, the batch harvest should be scheduled to maximize the growing season while minimizing the risk of unexpected lethal water temperatures. If the harvest is poorly timed, sublethal low temperatures may immobilize the prawns so that they cannot be drain-harvested, and manual harvesting of the entire pond area would then be necessary. Manual harvesting is likely to be expensive, and product quality could be affected adversely.

Marketing prawns in climatically limited areas requires a more diversified strategy than is necessary in tropical areas (Sandifer *et al.* 1982a). Instead of the regularly available, fresh and consistently sized product found in Hawaii, the product from continental U.S. farms will be available only seasonally and in variable sizes and volumes. Small farmers can generally sell their entire crop direct to consumers at the time of harvest, but larger producers will have to grade their product and sell the prawns in a variety of forms, including fresh and frozen, whole and tails only. These growers may also sell direct to consumers and to wholesalers, retail seafood markets, and restaurants (Liao and Smith 1980, 1981, 1982; Liao *et al.* 1981).

Construction of facilities for seasonal production of prawns may be economically unacceptable as a new venture. Loan officers and private investors may be reluctant to invest in construction programs where facilities will be productive for only a portion of the year (i.e., fixed investment costs for pond construction, well installation, etc., are the same regardless of the level of output). It may also be difficult to attract the necessary skilled labor for seasonal operations. The solution to these problems will be to diversify farm operations to include hatchery and nursery, as well as grow-out operations and to perhaps include other species that could be grown in the various facilities when they are not being used for prawns.

In summary, climate is a major impediment to the development of prawn farming in the continental United States. Even in Hawaii, prawn farms are subject to lower winter temperatures that reduce production rates and profits. Such climatic restrictions are difficult to overcome, although use of low-cost energy sources (e.g., power plant and industrial effluents, geothermal resources, solar energy), may help extend the growing season. Additionally, research to identify or develop a more temperature-tolerant prawn for culture in cooler climates may eventually help (Sandifer and Smith 1979a; Malecha 1980a,b; Sandifer and Lynn 1980).

Another important constraint to the development of prawn aquaculture, especially in the continental United States, is the limited availability of seed stock. As pointed out by Sandifer *et al.* (1982c), small private farms cannot diversify into prawn culture until supplies of low-cost seed stock are regularly available. On the other hand, private businessmen cannot afford to set up commercial hatcheries and nurseries until there is a sufficient demand for the product to warrant their investment. Thus, prawn farming is constrained by a supply–demand block on seed-stock availability. At present, only about four private hatcheries in the United States sell postlarvae, and all but one of these are located in Hawaii. One solution to this problem would be for government agencies to establish hatcheries and provide postlarvae to farmers at a low cost until there was sufficient demand to support private hatcheries. A similar approach was used successfully in Hawaii, and it is doubtful if a prawn aquaculture industry would have developed there had it not been for the government hatchery and extension program.

STATUS OF CULTURE IN THE UNITED STATES

Due to climatic conditions, the prawn-farming industry in the United States has developed primarily in Hawaii (Table 2.6). Between 1972 and 1981, the cultivated pond area there increased from 0.6 ha (about 1 Ac) to more than 100 ha (250 Ac). Only 10% of the farms are 40 ha (100 Ac) or larger, and 60% of all Hawaiian prawn farms have less than 4 ha (10 Ac) of production ponds (Shang 1981). Most of the farms are operated in conjunction with other agricultural or business activities. Along with its successes, prawn aquaculture in Hawaii has also had some failures. One large venture ceased operations several years ago and another major corporate pilot project was cancelled recently. Despite these failures, further expansion of the industry in Hawaii is highly likely (Shang 1981).

Production rates in Hawaii vary, but an average annual crop yield

TABLE 2.6. GROWTH OF THE FRESHWATER PRAWN AQUACULTURE INDUSTRY IN HAWAII—1972–1981

Year	Production (metric tons)	Number of Firms	Area Cultivated (ha)	Wholesale Value ($)
1972	1.95	1	0.61	15,000
1973	2.00	1	0.61	15,300
1974	4.99	7	2.02	38,500
1975	18.28	11	10.52	140,900
1976	19.64	13	10.52	151,600
1977	24.90	15	13.36	206,000
1978	49.99	20	43.30	420,000
1979	95.26	21	111.29	753,000
1980	136.08	24	125.46	1,200,000
1981	119.75	21	116.15	1,100,000

Source: Sandifer et al. 1982.

of about 2250 kg/ha (2000 lb/Ac) of whole prawns should be obtained after the ponds have been in production for 2 to 3 years (Shang 1981). Prawns are typically sold locally as a live or fresh iced product by the small producers. The largest producer currently exports a substantial portion of his crop to the mainland. The small producers receive about $9.37 per kg ($4.26 per lb) for whole prawns, while an overall weighted price for the industry is approximately $8.82 per kg, or $4.00 per lb (Shang 1981).

Within the United States, Puerto Rico probably has the best climate for prawn aquaculture. Despite this advantage, commercial prawn farming has been unsuccessful there to date, but the failures appear to be largely the results of poor site selection, inadequate capital resources, management difficulties, and an unusual frequency of hurricanes. A new commercial project is now underway in Puerto Rico, and it appears to have excellent potential for success. In addition, at least one other large prawn farm is being considered for Puerto Rico.

Development of prawn farming in the continental United States has been slower than expected due to the limited availability and high cost of seed stock and climatic constraints. In some areas of temperate climate, as in most other areas, small-scale prawn farming appears economically feasible only if it can be incorporated into existing agricultural operations or related businesses (Smith *et al.* 1981). Trials on small private farms have been conducted in a number of states, including Arkansas, California, Florida, Georgia, Louisiana, Mississippi, South Carolina, and Texas, and a number of small aquafarms in California are currently involved in some prawn culture (Boeing 1982). However, as yet few, if any, small farms have become commercially viable. During the next few years, small-scale trials by private farmers are expected to continue and expand, but total production levels will remain low for several years (e.g., private production in South Carolina was 1.1 metric tons in 1981 and 1982). The economic feasibility of establishing intensive production systems has also been explored in several states, but as yet no true commercial development has occurred.

Of particular note in the continental United States is the development of a large prawn farm in southern Texas. This commercial venture has constructed its own hatchery and nursery systems and 34 ha (84 Ac) of ponds for seasonal production of *Macrobrachium* (Fig. 2.27). In addition to supplying its own needs, this farm may enhance the development and feasibility of small prawn farms throughout the southeastern United States.

ECONOMIC OVERVIEW AND OUTLOOK

The number and size of freshwater prawn farms in the United States are expected to increase rather slowly but steadily. In Hawaii, recent economic analysis suggests that small farms are likely to be successful only when operated in conjunction with other business activities

FIG. 2.27 The largest freshwater prawn farm in the continental United States is Commercial Shrimp Culture International's facility in Texas.
Photo used by permission of Commercial Shrimp Culture International, Inc.

and often as family ventures (Shang 1981). Shang's analysis further indicates that the smallest truly commercial unit would be approximately 8 ha (20 acres). Summary tables from Shang's study (Tables 2.7 and 2.8) show that investment and operating costs even for small-scale prawn farming are substantial. However, his analysis (Table 2.9) also shows excellent potential for profit at the larger farm sizes of 8–40 ha (20–100 Ac) when production reaches 2240 kg/ha/year (1995 lb/Ac/year) or more on average and market price is $8.82 per kg ($4.00 per lb) or higher. Unfortunately, not all prawn farmers have been able to attain production levels greater than 2000 kg/ha/year (1781 lb/Ac/year) routinely.

Although lacking Hawaii's tropical climate, the continental United States offers a number of advantages for establishing prawn farms. These include ready availability of low-cost land for purchase or lease; abundant and relatively inexpensive water resources; less expensive feed and labor; and greater marketing opportunities. Although the climatic constraints of mainland sites offset these advantages to a large degree, the development of inexpensive sources of heat and development of a more temperature-tolerant race or hybrid prawn may ameliorate the effect of climate.

Recent economic studies suggest that seasonal prawn farming in the continental states will be similar to that in Hawaii—that is, primarily small family ventures operated concurrently with related agricultural enterprises and, for the most part, using existing facilities and equipment (Roberts and Bauer 1978; Smith *et al.* 1981; Sandifer *et al.* 1982a; Bauer *et al.*, 1983). Tables 2.10 and 2.11 show estimates of the investment and annual operating costs for a 4-ha (10-Ac) prawn farm in South Carolina for comparison with the figures of Shang (1981) for Hawaii. Analysis of revenue estimates shows that seasonal prawn farming can be profitable but only under certain conditions (Table 2.12). For example, if the harvest is to be sold as shrimp tails, profit potential exists only if the initial seed stock can be acquired for $10 per thousand or less and if the basic facilities needed for prawn farming are already available. In contrast, if the prawns can be sold as a heads-on specialty item at an average price of $6.60 per kg ($3.00 per lb) of whole prawns, then profit potential exists at all levels of seed cost (up to $50 per thousand) for both existing and new facilities.

The cost estimates presented here indicate that the largest expenses involved in prawn farming are for initial capital investment, interest, labor, feed, and seed stock. As pointed out previously, a major constraint to the further development of private prawn aquaculture, especially in the continental states, is the limited availability of seed stock. If this problem can be solved, either through private en-

TABLE 2.7. ESTIMATED AVERAGE CONSTRUCTION AND EQUIPMENT COSTS FOR PRAWN FARMS IN HAWAII BY FARM SIZE

Cost Item	Costs by Farm Size				
	0.4 Hectare (1 Acre)	4 Hectares (10 Acres)	8 Hectares (20 Acres)	20 Hectares (50 Acres)	40 Hectares (100 Acres)
Construction					
Pond	$ 5,724	$45,790	$ 85,860	$200,350	$372,100
Pipe	295	10,180	31,740	79,350	158,700
Sluice gate	800	8,000	16,000	40,000	80,000
Storage	150	1,000	2,000	15,000	30,000
Subtotal	6,969	64,970	135,600	334,700	640,800
$/0.4-ha pond	6,969	6,497	6,780	6,694	6,408
Equipment					
Seine & nets	869	1,740	1,640	3,500	7,000
Holding & transport tanks	400	1,000	1,000	2,000	3,000
Portable pump	300	600	600	1,200	1,800
Mowing equipment	300	10,000	10,000	10,000	10,000
Truck	7,000	7,000	7,000	15,000	22,000
Freezer	—	—	—	10,000	10,000
Water pump	—	1,500	3,000	6,000	9,000
Oxygen meter	700	700	700	1,400	1,400
pH meter	150	150	150	300	300
Ice machine	—	3,000	3,000	3,000	3,000
Miscellaneous[a]	486	1,285	1,504	2,620	3,875
Subtotal	10,205	26,975	28,594	55,020	71,375
$/0.4-ha pond	10,205	2,698	1,430	1,100	714
TOTAL	17,174	91,945	164,194	389,200	712,175
$/0.4-ha pond	17,174	9,195	8,210	7,794	7,122

Source: Modified from Shang 1981.
[a] 5% of total equipment cost.

TABLE 2.8. ESTIMATED AVERAGE ANNUAL OPERATING COSTS PER 0.4-HECTARE PRAWN POND BY FARM SIZE

Cost Item	Operating Costs by Farm Size									
	0.4 Hectare (1 Acre)		4 Hectares (10 Acres)		8 Hectares (20 Acres)		20 Hectares (50 Acres)		40 Hectares (100 Acres)	
	($)	(%)	($)	(%)	($)	(%)	($)	(%)	($)	(%)
Labor	1,184	12.4	3,144	33.7	2,094	26.7	1,542	21.8	1,325	20.0
Postlarvae[a]	520	5.4	520	5.6	520	6.6	520	7.4	520	7.9
Feed	1,660	17.4	1,660	17.8	1,660	21.2	1,660	23.5	1,660	25.1
Electricity	—	—	207	2.2	165	2.1	205	2.9	153	2.3
Land lease	882	9.2	708	7.6	708	9.0	708	10.0	708	10.7
Gasoline and oil	163	1.7	81	0.9	81	1.0	36	0.5	27	0.4
Maintenance	382	4.0	305	3.2	286	3.7	267	3.8	248	3.7
Interest	2,605	27.3	1,406	15.0	1,281	16.4	1,228	17.4	1,146	17.3
Depreciation	1,652	17.3	493	5.3	389	5.0	357	5.0	322	4.9
Tax	40	0.4	40	0.4	40	0.5	40	0.5	40	0.6
Insurance	—	—	334	3.6	232	3.0	172	2.4	149	2.3
Miscellaneous[b]	454	4.8	445	4.8	373	4.8	337	4.8	315	4.8
TOTAL	9,542	99.9	9,343	100.1	7,829	100.0	7,072	100.0	6,613	100.0

Source: Shang 1981.

[a] Cost used for postlarvae was $8.00 per 1000 animals. This is considerably lower than selling prices advertised by private hatcheries.

[b] 5% of other operating costs.

TABLE 2.9. ESTIMATED PROFIT PER 0.4-HECTARE PRAWN POND BY PRODUCTION LEVEL, FARM PRICE, AND FARM SIZE IN HAWAII

Production Per 0.4-ha Pond kg (lb)	Farm Price[a]		Profit by Farm Size				
	($/kg)	($/lb)	0.4 Hectare (1 Acre)	4 Hectares (10 Acres)	8 Hectares (20 Acres)	20 Hectares (50 Acres)	40 Hectares (100 Acres)
680	$7.72	$3.50	$ −4,292	$ −4,093	$ −2,579	$ −1,822	$ −1,363
(1,500)	8.27	3.75	−3,917	−3,718	−2,204	−1,477	−988
	8.82	4.00	−3,542	−3,343	−1,829	−1,072	−613
	9.37	4.25	−3,167	−2,968	−1,454	−697	−238
907	7.72	3.50	−2,542	−2,343	−829	−72	387
(2,000)	8.27	3.75	−2,042	−1,843	−329	428	887
	8.82	4.00	−1,542	−1,343	171	928	1,387
	9.37	4.25	−1,042	−843	671	1,428	1,887
1,134	7.72	3.50	−792	−593	921	1,678	2,137
(2,500)	8.27	3.75	−167	32	1,546	2,303	2,762
	8.82	4.00	458	657	2,171	2,928	3,387
	9.37	4.25	1,083	1,282	2,796	3,553	4,012
1,361	7.72	3.50	958	1,157	2,671	3,428	3,887
(3,000)	8.27	3.75	1,708	1,907	3,421	4,178	4,637
	8.82	4.00	2,458	2,657	4,171	4,928	5,387
	9.37	4.25	3,208	3,407	4,921	5,678	6,137

Source: Shang 1981.
[a] Prices for whole prawns.

TABLE 2.10. ESTIMATED INITIAL INVESTMENT AND ANNUAL FIXED COST PER POND TO ESTABLISH A NEW 4-HECTARE PRAWN FARM OR TO ADAPT EXISTING POND FACILITIES FOR PRAWN PRODUCTION IN SOUTH CAROLINA[a]

Cost Item	New Facilities		Existing Facilities	
	Total Investment	Annual Fixed Cost/Pond	Total Investment	Annual Fixed Cost/Pond
Ponds				
Pond construction	$26,000	$218.09	$ —	$ —
Harvest basin and drain	3,500	47.33	—	—
Levee stabilization (gravel and grass)	1,080	16.20	1,080	16.20
Well and Equipment				
Well	25,000	250.00	—	—
Pump	5,000	66.67	5,000	66.67
Pipe	1,840	24.53	1,840	24.53
Seines	396	9.90	396	9.90
Instruments	500	12.50	500	12.50
Feeder	1,300	19.50	1,300	19.50
Feed Storage	1,900	28.50	1,900	28.50
Aerators	2,800	42.00	2,800	42.00
Land Rental	—	38.75	—	38.75
TOTAL	$69,316	$773.97	$14,816	$258.55

Source: Adapted from Bauer *et al.* 1983.

[a] Based on hypothetical farm consisting of ten 0.4-ha ponds.

TABLE 2.11. ESTIMATED ANNUAL OPERATING COSTS PER 0.4-HECTARE PRAWN POND ON A 4-HECTARE FARM IN SOUTH CAROLINA

Cost Item	$/0.4-ha Pond
Labor	269.50
Juveniles[a]	780.00
Feed	237.12
Fertilizer	6.50
Lime	30.00
Electricity	62.40
Ice	5.00
Transportation	15.00
Tractor[b]	147.25
Truck[b]	85.47
Repair and maintenance	138.63
Miscellaneous	50.00
Interest on operating capital	21.96
TOTAL	1,848.83

Source: Adapted from Bauer *et al.* 1983.

[a] Estimated cost for 50:50 mixture of nursed juveniles and post-larvae used is $30/1,000 animals. Stocking density is assumed to be 65,000/ha.

[b] It is assumed that the prawn farm is part of a larger agricultural operation and that a tractor and truck are available. Costs for these vehicles are based on the estimated time they are needed for prawn-farming activities.

TABLE 2.12. ESTIMATED NET REVENUES FOR PRAWN AQUACULTURE IN SOUTH CAROLINA AS A FUNCTION OF SEED COST, MARKETING STRATEGY, AND FACILITY INVESTMENT[a]

Seed Stock Price ($/1,000 Prawns)	Market Strategy	Net Revenue ($/0.4-ha Pond) New Facilities	Existing Facilities
0	Tails Only	$ 190.88	$ 324.54
	Whole + Tails	606.60	1,122.02
	Whole Only	1,457.20	1,972.62
10	Tails Only	−449.28	66.14
	Whole + Tails	348.20	863.62
	Whole Only	1,197.20	1,712.62
20	Tails Only	−1,283.95	−192.26
	Whole + Tails	89.80	605.22
	Whole Only	937.20	1,452.62
30	Tails Only	−1,369.95	−450.66
	Whole + Tails	−168.60	346.82
	Whole Only	677.20	1,192.62
40	Tails Only	−1,455.95	−709.06
	Whole + Tails	−427.00	88.42
	Whole Only	417.20	932.62
50	Tails Only	−1,541.95	−967.46
	Whole + Tails	−685.40	−169.98
	Whole Only	157.20	672.62

Source: Adapted from Bauer *et al.* 1983.

[a] Estimates are based on the following assumptions: (1) the prawn farm consists of ten 0.4-ha (1-acre) ponds in both investment cases; (2) the ponds are stocked with a 50:50 mixture of postlarvae and nursed juveniles at a density of 65,000/ha; (3) average production is approximately 500 kg of whole prawns/0.4-ha pond; and (4) the prawns are marketed simply as shrimp tails at local ex-vessel prices for marine shrimp (tails only), or the large prawns, 30 g or larger, are sold whole at an average price of $8.82/kg and smaller prawns are sold as shrimp tails (whole + tails), or all prawns are marketed whole at an average price of $6.60/kg (whole only).

terprise or government intervention, and nursery-reared juveniles made available to prawn farmers at competitive prices, then prawn farming is likely to continue to expand in the continental United States. If research is continued to improve yields, cut costs, and develop markets, prawn farming should have a bright future. At present, prawn farming is still in the pioneering stage, with all the attendant risks and dangers. Over the next decade, prawn aquaculture should progress to become an established aquabusiness activity in the United States, and prawns should become commonly available in fine restaurants.

LITERATURE CITED

AQUACOP. 1977a. *Macrobrachium rosenbergii* (de Man) culture in Polynesia: progress in developing a mass intensive larval rearing technique in clear water. Proc. World Maric. Soc. **8,** 311–326.

AQUACOP. 1977b. Production de masse de poste-larves de *Macrobrachium rosenbergii* (de Man) en milieu tropical: unité pilote. Third Meeting of the International Council for the Exploration of the Seas, Working Group on Mariculture, Brest, France, May 1977.

AQUACOP. 1977c. Observations on diseases of crutacean cultures in Polynesia. Proc. World Maric. Soc. **8,** 685–703.

ARMSTRONG, D. A. 1978. Responses of *Macrobrachium rosenbergii* to extremes of temperature and salinity. *In* Studies on Bioenergetics, Osmoregulation, Development, Behavior and Survival of Several Aquaculture Organisms, A. W. Knight *et al.* (editors), pp. 18–51. Water Sci. Eng. Papers No. 4507. University of California, Davis, California.

ARMSTRONG, D. A., STEPHENSON, M. J., and KNIGHT, A. W. 1976 Acute toxicity of nitrite to larvae of the giant Malaysian prawn, *Macrobrachium rosenbergii*. Aquaculture **9,** 39–46.

ARMSTRONG, D. A., CHIPPENDALE, D., KNIGHT, A. W., and COLT, J. E. 1978. Interaction of ionized and unionized ammonia on short-term survival and growth of prawn larvae, *Macrobrachium rosenbergii*. Biol. Bull. **154,** 15–31.

ATKINSON, J. M. 1977. Larval development of a freshwater prawn, *Macrobrachium lar* (Decapoda, Palaemonidae), reared in the laboratory. Crustaceana **33**(2),119–132.

BAUER, L. L., SANDIFER, P. A., SMITH, T. I. J., and JENKINS, W. E. 1983. Economic feasibility of *Macrobrachium* production in South Carolina. J. Aquacult. Eng. **2**(3),181–201.

BAUER, R. T. 1976. Mating behaviour and spermatophore transfer in the shrimp *Heptacarpus pictus* (Stimpson) (Decapoda:Caridea:Hippolytidae). J. Nat. Hist. **10,** 415–440.

BIDDLE, G. N. 1977. The nutrition of *Macrobrachium* species. *In* Shrimp and Prawn Farming in the Western Hemisphere, J. A. Hanson and H. L. Goodwin (editors), Dowden Hutchinson & Ross, Stroudsburg, Pennsylvania.

BOEING, P. 1982. Desert aquaculture in Southern California. Aquacult. Dig. **7**(11),1–2.

BURNS, C. D., BERRIGAN, M. E., and HENDERSON, G. E. 1979. *Fusarium* sp. infection in the freshwater prawn *Macrobrachium rosenbergii* (de Man). Aquaculture **16,** 193–198.

CASTILLE, F. L, Jr., and LAWRENCE, A. L. 1981. The effect of salinity on the osmotic, sodium, and chloride concentrations in the hemolymph of the freshwater shrimp, *Macrobrachium ohione* Smith and *Macrobrachium rosenbergii* De Man. Comp. Biochem. Physiol. **70A,** 47–52.

CHOW, S., OGASAWARA, Y., and TAKI, Y. 1982. Male reproductive system and fertilization of the palaemonid shrimp *Macrobrachium rosenbergii*. Bull. Jpn. Soc. Sci. Fish. **48**(2),177–183.

CLIFFORD, H. C., III, and BRICK, R. W. 1978. Protein utilization in the freshwater shrimp *Macrobrachium rosenbergii*. Proc. World Maric. Soc. **9,** 195–208.

CLIFFORD, H. C., III, and BRICK, R. W. 1979. A physiological approach to the study of growth and bioenergetics in the freshwater shrimp *Macrobrachium rosenbergii*. Proc. World Maric. Soc. **10,** 701–719.

COHEN, D. 1976. The introduction of the fresh-water prawn *Macrobrachium rosenbergii* into Israel's fish ponds: a. On the role of algae in *Macrobrachium* hatchery. b. The use of geothermic springs in heating *Macrobrachium* hatchery and nursery. 5th Conf. Marine Biological Laboratory, Eylat, Israel (mimeo).

COHEN, D., RAANAN, Z., and BRODY, T. 1981. Population development and morphotypic differentiation in the giant freshwater prawn *Macrobrachium rosenbergii* (de Man). J. World Maric. Soc. **12**(2),231–243.

CRIPPS, M. C., and NAKAMURA, R. M. 1979. Inhibition of growth of *Macrobrachium rosenbergii* by calcium carbonate water hardness. Proc. World Maric. Soc. **10,** 579–580.

DELVES-BROUGHTON, J. 1974. Preliminary investigations into the suitability of a new chemotherapeutic, furanace, for the treatment of infectious prawn diseases. Aquaculture **3,** 175–185.

DUGAN, C. C., HAGOOD, R. W., and FRAKES, T. A. 1975. Development of spawning and mass larval rearing techniques for brackish-freshwater shrimps of the genus *Macrobrachium* (Decapoda Palaemonidae). Florida Marine Res. Publ., No. 12, Department of Natural Resources, Marine Research Laboratory, St. Petersburg, Florida.

EMERSON, K., RUSSO, R. C., LUND, R. E., and THURSTON, R. V. 1975. Aqueous ammonia equilibrium calculations: effects of pH and temperature. J. Fish. Res. Board Can. **3,** 2379–2383.

FAIR, P. H. and FORTNER, A. R. 1981. The role of formula feeds and natural productivity in culture of the prawn, *Macrobrachium rosenbergii.* Aquaculture **24,** 233–243.

FAIR, P. H., FORTNER, A. R., MILLIKIN, M. R., and SICK, L. V. 1980. Effects of dietary fiber on growth, assimilation and cellulase activity of the prawn (*Macrobrachium rosenbergii*). Proc. World Maric. Soc. **11,** 369–381.

FARMANFARMAIAN, A., and LAUTERIO, T. 1979. Amino acid supplementation of feed pellets of the giant shrimp (*Macrobrachium rosenbergii*). Proc. World Maric. Soc. **10,** 674–688.

FARMANFARMAIAN, A., and LAUTERIO, T. 1980. Amino acid composition of the tail muscle of *Macrobrachium rosenbergii*—comparison to amino acid patterns of supplemented commercial feed pellets. Proc. World Maric. Soc. **11,** 454–462.

FARMANFARMAIAN, A., and MOORE, R. 1978. Diseasonal thermal aquaculture—1. Effects of temperature and dissolved oxygen on the survival and growth of *Macrobrachium rosenbergii.* Proc. World Maric. Soc. **9,** 55–66.

FARMANFARMAIAN, A., LAUTERIO, T., and IBE, M. 1982. Improvement of the stability of commercial feed pellets for the giant shrimp (*Macrobrachium rosenbergii*). Aquaculture **27,** 29–41.

FORSTER, J. R. M., and BEARD, T. W. 1974. Experiments to assess the suitability of nine species of prawns for intensive cultivation. Aquaculture **3,** 355–368.

FUJIMURA, T. 1966. Notes on the development of a practical mass culturing technique of the giant prawn *Macrobrachium rosenbergii.* Rep. No. IPFC/C66/WP 47, Indo-Pacific Fisheries Council, 12th Session, Honolulu, Hawaii.

FUJIMURA, T. 1974. Development of a prawn culture industry in Hawaii. Job Completion Report to National Marine Fisheries Service, U.S. Department of Commerce, National Oceanic and Atmospheric Administration, (mimeo.). Washington, D.C.

FUJIMURA, T., and OKAMOTO, H. 1972. Notes on progress made in developing a mass culturing technique for *Macrobrachium rosenbergii* in Hawaii, *In* Coastal Aquaculture in the Indo-Pacific Region, T. V. R. Pillay (editor), Fishing News (Books) Ltd., London.

GEORGE, M. J. 1969. Genus *Macrobrachium* Bate 1868. *In* Prawn Fisheries of India, Bull. Central Mar. Fish. Res. Inst. (Mandapam Camp, India) **14,** 178–216.

GIBSON, R. T., and WANG, J.-K. 1977. An alternative prawn production systems design in Hawaii. Sea Grant Tech. Rep., UNIHI-SEAGRANT-TR-77-05, University of Hawaii, Honolulu, Hawaii.

GOODWIN, H. L., and HANSON, J. A. 1975. The aquaculture of freshwater prawns (*Macrobrachium* species). Augmented summary of Proc., Workshop on Culture of Freshwater Prawns, St. Petersburg, FL. The Oceanic Institute, Waimanalo, Hawaii.

HALE, M. B. and WATERS, M. E. 1981. Frozen storage stability of whole and headless freshwater prawns, *Macrobrachium rosenbergii*. Marine Fish. Rev. **43,** 18–21.

HALL, T. J. 1977. Ectocommensals of the freshwater shrimp, *Macrobrachium rosenbergii,* in culture facilities at Homestead, Florida. TAMU-SG-79-114: Texas A&M University, College Station, Texas.

HANSON, J. A. and GOODWIN, H. L. (editors). 1977. Shrimp and Prawn Farming in the Western Hemisphere. Dowden, Hutchinson & Ross, Inc., Stroudsburg, Pennsylvania.

HEINEN, J. M. 1977. Influence of dissolved calcium and magnesium on postlarval growth of the freshwater shrimp *Macrobrachium rosenbergii.* M.S. Thesis, Florida Atlantic University, Boca Raton, Florida.

HOLTHUIS, L. B. 1950. Subfamily Palaemoninae. The Palaemonidae collected by the Siboga and Snellius Expeditions with remarks on other species. I. The Decapoda of the Siboga Expedition. Part 10. Siboga Expedition Monogr. **39**(a)9,1–268.

HOLTHUIS, L. B. 1952. A general revision of the Palaemonidae (Crustacea Decapoda Natantia) of the Americas. II. The subfamily Palaemoninae. Occasional Paper 12, pp. 1–396. Allan Hancock Foundation Publications, Los Angeles, California.

HOLTHUIS, L. B. 1980. FAO Species Catalogue. Vol. 1. Shrimps and Prawns of the World. FAO Fish. Synop. (125)1,1–261.

HOLTHUIS, L. B. and A. J. PROVENZANO, JR. 1970. New distribution records for species of *Macrobrachium* with notes on the distribution of the genus in Florida (Decapoda Palaemonidae). Crustaceana **19**(2),211–213.

HUNER, J. V. 1977. Observations on the biology of the river shrimp from a commercial bait fishery near Port Allen, Louisiana. Proc. Annu. Conf. Southeastern Assoc. Fish Wildl. Agencies **31,** 380–386.

JOHN, M. C. 1957. Bionomics and life history of *Macrobrachium rosenbergii* (de Man). Bull. Central Res. Inst. University of Kerala (India), Ser. C **15,** 93–102.

JOHNSON, D. S. 1960. Some aspects of the distribution of freshwater organisms in the Indo-Pacific area and their relevance to the validity of the concept of an oriental region in zoogeography. Proc. Cent. Bicent. Congr. Biol., Singapore, 1958, pp. 170–181.

JOHNSON, D. S. 1967. Some factors influencing the distribution of freshwater prawns in Malaya. Proc. Symp. Crustacea, Mar. Biol. Assoc. India **1,** 418–433.

JOHNSON, S. K. 1977. Crawfish and freshwater shrimp diseases. Rep. No. TAMU-SG-77-605: Texas A&M University, College Station, Texas.

JOHNSON, S. K. 1978. Some disease problems in crawfish and freshwater shrimp culture. Rep. No. FDDL-S11, Fish Disease Diagnostic Laboratory, Texas A&M University, College Station, Texas.

KING, J. E. 1948. A study of the reproductive organs of the common marine shrimp, *Penaeus setiferus* (Linnaeus). Biol. Bull. **94,** 244–262.

LEE, P. G., BLAKE, N. J., and RODRICK, G. E. 1980. A quantitative analysis of digestive enzymes for the freshwater prawn *Macrobrachium rosenbergii*. Proc. World Maric. Soc. **11,** 392–402.

LIAO, D. S., and SMITH, T. I. J. 1980. The marketing opportunity for freshwater shrimp in South Carolina: a preliminary survey. Proc. Trop. Subtrop. Fish. Technol. Conf. Am. **5,** 67–69.

LIAO, D. S., and SMITH, T. I. J. 1981. Test marketing of freshwater shrimp, *Macrobrachium rosenbergii,* in South Carolina. Aquaculture **23,** 373–379.

LIAO, D. S., and SMITH, T. I. J. 1982. Marketing of cultured prawns, *Macrobrachium rosenbergii,* in South Carolina. J. World Maric. Soc. **13,** 56–62.

LIAO, D. S., SMITH, T. I. J., and TAYLOR, F. S. 1981. The marketability of prawns *(Macrobrachium rosenbergii)* in restaurants in South Carolina: a preliminary analysis. Proc. Trop. Subtrop. Fish. Technol. Conf. Am. **6,** 38–41.

LING, S. W. 1962. Studies on the rearing of larvae and juveniles and culturing of adults of *Macrobrachium rosenbergii* (de Man). Current Affairs Bull. 35. Indo-Pacific Fisheries Commission.

LING, S. W. 1969a. The general biology and development of *Macrobrachium rosenbergii* (de Man). FAO UN Fish. Rep. (57)3, 589–606.

LING, S. W. 1969b. Methods of rearing and culturing *Macrobrachium rosenbergii* (de Man). FAO UN Fish. Rep. (57)3, 607–619.

LING, S. W. and COSTELLO, T. J. 1979. The culture of freshwater prawns: a review. *In* Advances in Aquaculture, T. V. R. Pillay and W. A. Dill (editors), Fishing News Books Ltd., Farnham, England.

LYNN, J. W. 1981. The reproductive biology and gamete interaction in the freshwater prawn *Macrobrachium rosenbergii.* Ph.D. Dissertation, University of California, Davis, California.

LYNN, J. W., and CLARK, W. H., JR. 1983a. A morphological examination of sperm-egg interaction in the freshwater prawn, *Macrobrachium rosenbergii.* Biol. Bull. **164,** 446–458.

LYNN, J. W., and CLARK, W. H., JR. 1983b. The fine structure of the mature sperm of the freshwater prawn, *Macrobrachium rosenbergii.* Biol. Bull. **164,** 459–470.

MALECHA, S. R. 1980a. Development and general characterization of genetic stocks of *Macrobrachium rosenbergii* and their hybrids for domestication (University of Hawii Sea Grant College Program). Sea Grant Q. **2**(4), 1–6.

MALECHA, S. R. 1980b. Research and development in freshwater prawn *(Macrobrachium rosenbergii)* culture in the U.S.: current status and biological constraints with emphasis on breeding and domestication. Ninth Joint Meet., U.S. Japan Aquacult. Panel, Kyoto, Japan, May 1980.

MALECHA, S. R., MASUNO, S., BIGGER, D., BRAND, T., WEBER, G., LEVITT, A., and CHOW, D. 1981. Genetic and environmental sources of growth pattern variation in the cultured freshwater prawn, *Macrobrachium rosenbergii.* Unpublished manuscript.

MANZI, J. J. and MADDOX, M. B. 1976. Algal supplement enhancement of static and recirculating system culture of *Macrobrachium rosenbergii* (de Man) larvae. Helgol. Wiss. Meeresunters. **28,** 447–455.

MANZI, J. J., and MADDOX, M. B. 1980. Requirements for *Artemia* nauplii in *Macrobrachium rosenbergii* larviculture. *In* The Brine Shrimp Artemia, Vol. 3, Ecology, Culturing, Use in Aquaculture, G. Persoone, O. Roels, and E. Jaspers (editors), Universa Press, Wetteren, Belgium.

MANZI, J. J., MADDOX, M. B., and SANDIFER, P. A. 1977. Algal supplement enhancement of *Macrobrachium rosenbergii* (de Man) larviculture. Proc. World Maric. Soc. **8,** 207–223.

MCVEY, J. P. 1975. New record of *Macrobrachium rosenbergii* (de Man) in the Palau Islands (Decapoda, Palaemonidae). Crustaceana **29,** 31–32.

MENASVETA, P. 1980. Effect of ozone treatment on the survival of prawn larvae (*Macrobrachium rosenbergii* de Man) reared in a closed-recirculating water system. Proc. World Maric. Soc. **11,** 73–78.

MILLIKIN, M. R., FORTNER, A. R., FAIR, P. H., and SICK, L. V. 1980. Influence of dietary protein concentration on growth, feed conversion and general metabolism of juvenile prawn *(Macrobrachium rosenbergii).* Proc. World Maric. Soc. **11,** 382–391.

MURAI, T., and ANDREWS, J. W. 1978. Comparison of feeds for larval stages of the giant prawn *(Macrobrachium rosenbergii).* Proc. World Maric. Soc. **9,** 189–193.

NAGAMINE, C. M. and KNIGHT, A. W. 1980. Development, maturation, and function of some sexually dimorphic structures of the Malaysian prawn, *Macrobrachium rosenbergii* (de Man) (Decapoda, Palaemonidae). Crustaceana **39**(2), 141–152.

NELSON, S. G., ARMSTRONG, D., KNIGHT, A. W., and LI, H. W. 1977. The effects of temperature and salinity on the metabolic rate of juvenile *Macrobrachium rosenbergii* (Crustacea:Palaemonidae). Comp. Biochem. Physiol. **56A,** 533–537.

NEW, M. B. 1976. A review of dietary studies with shrimp and prawns. Aquaculture **9,** 101–144.

PARRY, G. 1960. Excretion, *In:* The Physiology of Crustacea, Vol. 1, Metabolism and Growth. T. H. Waterman (editor). Chapter 10, pp. 341–366. Academic Press, New York.

PEEBLES, J. B. 1977. A rapid technique for molt staging in live *Macrobrachium rosenbergii.* Aquaculrure **12,** 173–180.

PEEBLES, J. B. 1978. Molting and mortality in *Macrobrachium rosenbergii.* Proc. World Maric. Soc. **9,** 39–46.

PEEBLES, J. B. 1979. The roles of prior residence and relative size in competition for shelter by the Malaysian prawn, *Macrobrachium rosenbergii.* Fish. Bull. **76**(4), 905–911.

PEEBLES, J. B. 1980. Competition and habitat partitioning by the giant freshwater prawn *Macrobrachium rosenbergii* (de Man) (Decapoda, Palaemonidae). Crustaceana **38**(1), 49–54.

PERDUE, J. A., and NAKAMURA, R. 1976. The effect of salinity on the growth of *Macrobrachium rosenbergii.* Proc. World Maric. Soc. **7,** 647–654.

PERRY, W. G., and TARVER, J. W. 1981. Malaysian prawn culture in brackish water ponds in Louisiana. J. World Maric. Soc. **12**(2), 214–222.

RAMAN, K. 1967. Observations on the fishery and biology of the giant freshwater prawn *Macrobrachium rosenbergii* (de Man). Proc. Symp. Crustacea, Mar. Biol. Assoc. India, Part 2, pp. 649–669.

RAO, R. M. 1965. Breeding behaviour in *Macrobrachium rosenbergii* (de Man). Fish. Tech. **2**(1), 19–25.

RAO, R. M. 1967. Studies on the biology of *Macrobrachium rosenbergii* (de Man) of the Hooghly estuary with notes on its fishery. Proc. Nat. Inst. Sci. India, Part B. **33,** 252–279.

ROBERTS, K. J., and BAUER, L. L. 1978. Costs and returns for *Macrobrachium* grow-out in South Carolina, U.S.A. Aquaculture **15,** 383–390.

SANDIFER, P. A. 1980. Freshwater prawn species plan, National Aquaculture Plan (Draft), pp. 227–252. U.S. Department of Commerce, National Oceanic and Atmospheric Administration, May 1980.

SANDIFER, P. A. (1984). Some recent advances in the culture of crustaceans. Proc. Eur. Maric. Soc./World Maric. Soc. World Conf. Aquacult., Venice, Sept. 1981, in press.

SANDIFER, P. A., and LYNN, J. W. 1980. Artificial insemination of caridean shrimp. *In* Advances in Invertebrate Reproduction, W. H. Clark, Jr. and T. S. Adams (editors). Elsevier North Holland, Inc., Amsterdam.

SANDIFER, P. A., and SMITH, T. I. J. 1974. Development of a crustacean mariculture program at South Carolina's Marine Resources Research Institute. Proc. World Maric. Soc. **5,** 431–439.

SANDIFER, P. A., and SMITH, T. I. J. 1975. Effects of population density on growth and survival of *Macrobrachium rosenbergii* reared in recirculating water management systems. Proc. World Maric. Soc. **6,** 43–53.

SANDIFER, P. A., and SMITH, T. I. J. 1977a. Preliminary observations on a short-claw growth form of the Malaysian prawn. *Macrobrachium rosenbergii* (de Man). Proc. Nat. Shellfish. Assoc. **67,** 123–124.

SANDIFER, P. A., and SMITH, T. I. J. 1977b. Intensive rearing of postlarval Malaysian prawns, *Macrobrachium rosenbergii,* in a closed cycle nursery system. Proc. World Maric. Soc. **8,** 225–235.

SANDIFER, P. A., and SMITH, T. I. J. 1978. Aquaculture of Malaysian prawns in controlled environments. Food Technol. **32**(7), 36–38, 40–42, 44–45, 83.

SANDIFER, P. A., and SMITH, T. I. J. 1979a. A method for artificial insemination of *Macrobrachium* prawns and its potential use in inheritance and hybridization studies. Proc. World Maric. Soc. **10,** 403–418.

SANDIFER, P. A., and SMITH, T. I. J. 1979b. Possible significance of variation in the larval development of palaemonid shrimp. J. Exp. Mar. Biol. Ecol. **39,** 55–64.

SANDIFER, P. A., and SMITH, T. I. J. 1979c. Experimental aquaculture of the Malaysian prawn, *Macrobrachium rosenbergii* (de Man), in South Carolina, U.S.A. *In* Advances in Aquaculture, T. V. R. Pillay and W. A. Dill (editors). Fishing News Books, Ltd., Farnham, England.

SANDIFER, P. A., ZIELINSKI, P. B., and CASTRO, W. E. 1974a. A simple airlift-operated tank for closed system culture of decapod crustacean larvae and other small aquatic animals. Helgol. Wiss. Meeresunters. **26**(1), 82–87.

SANDIFER, P. A., SMITH, T. I. J., and CALDER, D. R. 1974b. Hydrozoans as pests in closed-system culture of larval decapod crustaceans. Aquaculture **4,** 55–59.

SANDIFER, P. A., HOPKINS, J. S. and SMITH, T. I. J. 1975. Observations on salinity tolerance and osmoregulation in laboratory-reared *Macrobrachium rosenbergii* postlarvae (Crustacea:Caridea). Aquaculture **6,** 103–114.

SANDIFER, P. A., HOPKINS, J. S., and SMITH, T. I. J. 1977. Status of *Macrobrachium* hatcheries, 1976. *In* Shrimp and Prawn Farming in the Western Hemisphere, J. A. Hanson and H. L. Goodwin (editors), Dowden, Hutchinson, & Ross, Inc., Stroudsburg, Pennsylvania.

SANDIFER, P. A., SMITH, T. I. J., and BAUER, L. L. 1982a. Economic comparisons of stocking and marketing strategies for aquaculture of prawns, *Macrobrachium rosenbergii* (de Man), in South Carolina, U.S.A. Proc. Symp. Coastal Aquacult. Cochin, India 1980 **1,** 88–97.

SANDIFER, P. A., SMITH, T. I. J., STOKES, A. D., and JENKINS, W. E. 1982b. Semi-intensive grow-out of prawns *(Macrobrachium rosenbergii):* preliminary results and prospects, *In* Giant Prawn Farming, M. B. New (editor), Developments in Aquaculture and Fisheries Science Vol. 10, Elsevier Scientific Publishing Co., Amsterdam.

SANDIFER, P. A. *et. al.* 1982c. Prawn species plan, *In* Sea Grant Aquaculture Plan, 1983–1987, B. Crowder (editor). Texas A&M University Sea Grant Program, TAMU-SG-82-114, College Station, Texas.

SEGAL, E., and ROE, A. 1975. Growth and behavior of post juvenile *Macrobrachium rosenbergii* (de Man) in close confinement. Proc. World Maric. Soc. **6,** 67–88.

SHANG, Y. C. 1981. Freshwater prawn *(Macrobrachium rosenbergii)* production in Hawaii: practices and economics. Sea Grant Misc. Rep., UNIHI-SEAGRANT-MR-81-07, University of Hawaii, Honolulu, Hawaii.

SHANG, Y. C., and FUJIMURA, T. 1977. The production economics of freshwater prawn *(Macrobrachium rosenbergii)* farming in Hawaii. Aquaculture **11,** 99–110.

SHARP, J. 1976. Effects of dissolved oxygen, temperature, and weight on respiration of *Macrobrachium rosenbergii.* Laboratory Studies on Selected Nutritional, Physical and Chemical Factors Affecting the Growth, Survival, Respiration, and Bioenergetics of the Giant Prawn, *Macrobrachium rosenbergii,* A. W. Knight *et al.* (editor). Water Science and Engineering Pap. 4501. University of California, Davis, California.

SICK, L. V. 1975. Selected studies of protein and amino acid requirements for *Macrobrachium rosenbergii* larvae fed neutral density diets, Proc. 1st Int. Conf. Aquacult. Nutrition, pp. 215–228. University of Delaware, Newark, Delaware.

SICK, L. V., and BEATY, H. 1974. Culture techniques and nutrition studies for larval stages of the giant prawn, *Macrobrachium rosenbergii.* Georgia Marine Science Center, Tech. Rep. 74–5, University of Georgia, Skidaway Island, Georgia.

SICK, L. V., and BEATY, H. 1975. Development of formula foods designed for *Macrobrachium rosenbergii* larval and juvenile shrimp. Proc. World Maric. Soc. **6,** 89–102.

SILVERTHORN, S. U., and REESE, A. M. 1978. Cold tolerance at three salinities in post-larval prawns, *Macrobrachium rosenbergii* (de Man). Aquaculture **15,** 249–255.

SMITH, T. I. J., and SANDIFER, P. A. 1975. Increased production of tank-reared *Marcobrachium rosenbergii* through use of artificial substrates. Proc. World Maric. Soc. **6,** 55–66.

SMITH, T. I. J., and SANDIFER, P. A. 1979a. Breeding depressions in culture ponds for Malaysian prawns. Aquaculture **18,** 51–57.

SMITH, T. I. J., and SANDIFER, P. A. 1979b. Observations on the behavior of the Malaysian prawn, *Macrobrachium rosenbergii* (de Man), to artificial habitats. Mar. Behav. Physiol. **6,** 131–146.

SMITH, T. I. J., and SANDIFER, P. A. 1979c. Development and potential of nursery systems in the farming of Malaysian prawns, *Macrobrachium rosenbergii* (de Man). Proc. World Maric. Soc. **10,** 369–384.

SMITH, T. I. J., SANDIFER, P. A., and TRIMBLE, W. C. 1976a. Pond culture of the Malaysian prawn, *Macrobrachium rosenbergii* (de Man), in South Carolina, 1974–1975. Proc. World Maric. Soc. **7,** 625–645.

SMITH, T. I. J., SANDIFER, P. A., and TRIMBLE, W. C. 1976b. Progress in developing a recirculating synthetic seawater hatchery for rearing larvae of *Macrobrachium rosenbergii,* H. H. Webber and G. D. Ruggieri (eds.), Food-Drugs from the Sea, Proc. 1974, Puerto Rico. Marine Technology Society, Washington, D.C.

SMITH, T. I. J., SANDIFER, P. A., and SMITH, M. H. 1978. Population structure of Malaysian prawns, *Macrobrachium rosenbergii* (de Man), reared in earthen ponds in South Carolina, 1974–1976. Proc. World Maric. Soc. **9,** 21–38.

SMITH, T. I. J., SANDIFER, P. A., and MANZI, J. J. 1979. Epibionts of pond-reared adult Malaysian prawns, *Macrobrachium rosenbergii* (de Man), in South Carolina. Aquaculture 16, 299–308.

SMITH, T. I. J., WALTZ, W., and SANDIFER, P. A. 1980. Processing yields for Malaysian prawns and their implications. Proc. World Maric. Soc. **11,** 557–569.

SMITH, T. I. J., SANDIFER, P. A., JENKINS, W. E., and STOKES, A. D. 1981. Effects of population structure and density at stocking on production and commercial feasibility of prawn *(Macrobrachium rosenbergii)* farming in temperate climates. J. World Maric. Soc. **12**(1), 233–250.

SMITH, T. I. J., SANDIFER, P. A., and JENKINS, W. E. 1982. Growth and survival of prawns, *Macrobrachium rosenbergii,* pond-reared at different salinities. *In* Giant Prawn Farming, M. B. New (editor), Developments in Aquaculture and Fisheries Science Vol. 10, Elsevier Scientific Publishing Co., Amsterdam.

STAHL, M. E., and AHEARN, G. A. 1978. Amino acid studies with juvenile *Macrobrachium rosenbergii.* Proc. World Maric. Soc. **9,** 209–216.

TOMBES, A. S., and FOSTER, M. W. 1979. Growth of appendix masculina and appendix interna in juvenile *Macrobrachium rosenbergii* (de Man) (Decapoda, Caridea). Crustaceana, Suppl. **5,** 179–184.

UNO, Y., BEJIE, A. B., and SGARASHI, Y. 1975. Effects of temperature on the activity of *Macrobrachium rosenbergii.* La Mer **13**(3), 150–154.

WATERS, M. E., and HALE, M. B. 1981. Quality changes during iced storage of whole freshwater prawns *(Macrobrachium rosenbergii).* Proc. Trop. Subtrop. Fish. Technol. Conf. Am. **6,** 116–127.

WICKINS, J. F. 1972. Experiments on the culture of the spot prawn *Pandalus platyceros* Brandt and the giant freshwater prawn *Macrobrachium rosenbergii* (de Man). Fishery Invest. Minist. Agric. Fish. Food (GB), Ser. II, Salmon Freshwater Fish. **27**(5), 1–23.

WICKINS, J. F. 1976a. Prawn biology and culture. Oceanogr. Mar. Biol. Annu. Rev. **14,** 435–507.

WICKINS, J. F. 1976b. The tolerance of warm-water prawns to recirculated water. Aquaculture **9,** 19–37.

WILLIAMS, A. B. 1965. Marine decapod crustaceans of the Carolinas. Fish. Bull. **65**(1), 1–298.

WILLIS, S. A., and BERRIGAN, M. E. 1977. Effects of stocking size and density on growth and survival of *Macrobrachium rosenbergii* (de Man) in ponds. Proc. World Maric. Soc. **8,** 251–264.

WILLIS, S. A., and BERRIGAN, M. E. 1978. Effects of fertilization and selective harvest on pond culture of *Macrobrachium rosenbergii* in Central Florida. National Marine Fish. Serv., PL88-309, No. 2-298-R-1 Job 3B. Completion Report for U.S. Department of Commerce, National Oceanic and Atmospheric Administration. Florida Department of Natural Resources, Marine Research Laboratory.

WULFF, R. E. 1982. The experience of a freshwater prawn farm in Honduras, Central America. *In* Giant Prawn Farming, M. B. New (editor), Developments in Aquaculture and Fisheries Science Vol. 10, Elsevier Scientific Publishing Company, Amsterdam.

3

Penaeid Shrimp Culture

A. L. Lawrence,
J. P. McVey,
J. V. Huner

Introduction
Species of Importance
Basic Biology and Culture Requirements
Culture Techniques
Shrimp Maturation and Spawning
Hatchery Procedures
Grow-out Procedures
Diseases and Parasites
Status of Shrimp Farming in United States
Climatic and Other Constraints
Current and Future Shrimp Culture by U.S. Companies
Economic Overview and Outlook
Literature Cited

INTRODUCTION

The development of technology for the culture of penaeid shrimp for commercial purposes has been slow with most of the initial work being done in Japan. The earliest documented attempt to spawn and hatch a penaeid shrimp *(Penaeus japonicus)* was by Hudinaga in 1933 (Fujinaga 1969b). Dr. Motosaku Fujinaga did most of the pioneering work of penaeid larviculture (Hudinaga 1942; Hudinaga and Miyamura 1962), though it was not until 1964 that Hudinaga and Kittaka (1966, 1967) developed a method for the mass culture of marine shrimp larvae for the purpose of releasing postlarvae into the Seto Inland Sea of Japan. During the 1960s and 1970s several research groups throughout the world expanded on the work done by the Japanese by working on other species *(Metapenaeus monoceros, M. joyneri, Penaeus aztecus, P. duorarum, P. monodon, P. semisulcatus, P. setiferus, P. teraoi)* under different culture conditions (Cook and Murply 1966, 1969; Cook 1969; Liao *et al.* 1969a,b; Mock and Murphy 1970; Mock 1971; Liao and Huang 1972; Tabb *et al.* 1972). The nauplii (first larval stage) used for these early studies were obtained from mated mature females captured from the ocean and spawned in captivity. This method is still being used to support commercial hatcheries in Japan (Kittaka 1981), Taiwan, the Philippines, Panama, and Ecuador.

However, during the last decade a number of species *(Metapenaeus ensis, Penaeus aztecus, P. indicus, P. japonicus, P. merguiensis, P. monodon, P. orientalis, P. plebejus, P. semisulcatus, P. setiferus, P. stylirostris,* and *P. vannamei)* have been matured, mated, and spawned in captivity producing viable nauplii (Aquacop 1975, 1977a,b, 1979, 1980; Arnstein and Beard 1975; Beard and Wickins 1980; Beard *et al.* 1977; Brown *et al.* 1980; Caubere *et al.* 1979; Chamberlain and Lawrence 1981a,b; Emerson 1980; Kelemec and Smith 1980; Laubier-Bonichon 1975, 1978; Laubier-Bonichon and Laubier 1979; Lawrence *et al.* 1980; Primavera 1978; Primavera and Borlongan 1978; Primavera *et al.* 1978, 1979; Sandifer 1981; Santiago 1977). Though the technology for the maturation and reproduction of penaeid shrimp in captivity has greatly increased, this is still the limiting phase for the development of shrimp farming on a commercial basis not only in the United States but worldwide. In addition, technology for artificial insemination has been developing rapidly, offering new opportunities for genetic selection and improved spawning efficiency (Persyn 1977; Bray *et al.* 1982, 1983; Lawrence *et al.* 1983; Sandifer *et al.* 1983). Using this newly developed technology, the first successful interspecific cross between two commercial penaeid species *(P. setiferus* and *P. stylirostris)* has been accomplished, producing the first hybrid marine shrimp (Lawrence *et al.* 1984).

Two basic approaches are being used for larviculture of penaeid shrimp (hatchery phase): intensive, small-scale production—the Galveston method—and extensive, large-scale production—the Japanese method. Much of the early work on the intensive method was initially done by Cook and Murphy (1969) with later modifications by Mock and Murphy (1970), Mock (1971), Mock and Neal (1974), and Salser and Mock (1974). Other investigators doing initial work on the intensive method were Kittaka (1971) and Tabb *et al.* (1972). The initial studies on the development of the extensive method were reported by Hudinaga and Kittaka (1966, 1967) and Fujinaga (1969a,b). Additional information concerning this method, including comparisons with the intensive method, have been published (Shigueno 1968; Yang 1975; Heinen 1976; Kurata and Shigueno 1976; Kittaka 1976, 1981). Simon (1981) presents a description of a large tank system that incorporates many of the "good" aspects of the Japanese and Galveston systems.

Pond culture of marine shrimp in the United States was initially investigated by Lunz using *Penaeus aztecus, P. duorarum,* and *P. setiferus* (Lunz 1951, 1956, 1958, 1967; Lunz and Bearden 1963). There have been numerous studies since then on pond production of marine shrimp in the United States (Beynon *et al.* 1981; Broom 1968, 1970, 1971; Caillouet *et al.* 1974; Chamberlain *et al.* 1981; Conte 1975; Latapie *et al.* 1972; Lawrence *et al.* 1983; Liao and Huang 1972; Ojeda *et al.* 1980; Parker and Holcomb 1973; Parker *et al.* 1974; Rubright *et al.* 1981; Tabb *et al.* 1972; Tatum and Trimble 1978; Trimble 1980; Wheeler 1967, 1968; and others). Extensive culture in tidal impoundments ("ponds") is discussed by de la Bretonne and Avault (1970), Rose *et al.* (1975), and Huner (1979). At the present time, pond and/or impoundment culture of shrimp has been very successful only in Ecuador (Hirono 1983), making shrimp the most valuable renewable resource in that country (Anon. 1982).

Grow out of shrimp in intensive culture in tanks and raceway systems has attracted considerable attention because sites for pond culture (semi-intensive or extensive) are limited in the continental United States and Hawaii. Shigueno (1975) describes successful intensive culture systems in Japan. Mahler *et al.* (1974), Salser *et al.* (1978), and NOAA (1980) describe the intensive system developed jointly through the Universities of Arizona and Sonora (Mexico) and Coca-Cola Company and F. H. Prince Company. As of 1983, this sytem was not yet in commercial use, but it will probably be put into commercial use in Hawaii within the near future.

The technology available to shrimp culturists in the early 1980s is either close to or at the level needed for successful shrimp culture. Commercial success is still marginal within the United States (Lawrence *et al.* 1983a), but shrimp farms in tropical countries (e.g., Ecuador) have been successful (Hirono 1983). There are many reasons for the general lack of commercial success but the usual reason is the inability to reproduce marine shrimp in captivity with the predictability necessary to support a commercial operation. Complete control over the life history of several shrimp species must be obtained if commercial shrimp culture is to become successful within the United States.

It is impossible to cover all the existing knowledge concerned with marine shrimp in this chapter. This chapter attempts to provide a summary of the present state of the art in penaeid shrimp culture and will discuss the various problems and restrictions that need to be overcome in order for a shrimp culture industry to develop in the United States.

SPECIES OF IMPORTANCE

More than 20 species of shrimp have been investigated for their suitability for commercial culture. Each of these species has attributes that may make it more suitable than others for culture in specific areas. Culturists generally agree that *Penaeus vannamei* may be the species of choice for most culture situations. However, *P. stylirostris, P. monodon,* and *P. japonicus* may be highly appropriate in certain situations. Unfortunately, at this time, it is not simple to choose any of the species for a particular culture situation because all of the species have problems that inhibit commercial success.

Penaeus vannamei (Fig. 3.1A), a native of the Pacific west coast from northern Peru to Sonora, Mexico (Holthuis 1980), grows very well in pond culture. It is the primary culture species being used in Ecuador at the present time (Hirono 1983). It can reach a 20-g market size from 5- to 15-day-old postlarvae in approximately 4–6 months at average densities of 50,000–75,000/ha (20,000–30,000/Ac). Survival rates in ponds are fairly dependable compared with other species, and 60–80% survival is common. Most commercial companies would like to have a steady supply of seed stock of *P. vannamei* because of its excellent grow-out characteristics. However, captive reproduction is very hard to obtain. Mating and spawning with this species occur less frequently than with *P. stylirostris, P. japonicus,* or *P. monodon. Penaeus vannamei* tolerates a wide range of salinities and temperatures and grows well in lower salinities.

Penaeus stylirostris (Fig. 3.1B), a native of the Pacific west coast from Peru to Baja California, Mexico, (Holthuis 1980), is easier to reproduce under captive conditions than *P. vannamei.* Large numbers of nauplii can be obtained but survival in the grow-out phases is not as predictable as that of *P. vannamei,* and occasional unexplainable losses occur. The time to a 20-g market size from 5- to 20-day-old postlarvae is approximately 4–6 months at densities of 25,000–50,000/ha (10,000–20,000/Ac). Because seed stock can be obtained without too much difficulty, *P. stylirostris* is being seriously considered for the development of the shrimp culture industry in the United States. The unexplained deaths (probably due to disease or intolerance to some environmental condition) that sometimes occur in pond culture of this species make economic projections very difficult and have occasionally led to failure of commercial operations.

Penaeus monodon (Fig. 3.1C), a native of east and southeast Africa, Pakistan to Japan, the Malay Archipelago, and northern Australia (Holthuis 1980), is one of the largest shrimps being cultured and is used for extensive culture wherever it is found. It takes approximately 4–6 months for 5- to 25-day-old postlarvae to reach a 20- to 40-g market size at densities of 5,000–250,000/ha (2,000–100,000/Ac). Captive reproduction has been accomplished but needs further improvement to support a commercial operation. Survival of *P. monodon* in the hatchery phase is low compared with that of *P. japonicus* and *P. vannamei*. *Penaeus monodon* is being cultured under high-density culture conditions by the Taiwanese and is being stocked at low densities in extensive pond culture and in polyculture with fish in the Philippines and Indonesia. This species does well in low-salinity conditions and can be cultured in salinities up to 25 ppt with good growth.

Penaeus japonicus (Fig. 3.1D), is a native of the Red Sea, east and southeast Africa to Korea, Japan, and the Malay Archipelago and has recently entered the eastern Mediterranean through the Suez Canal (Holthuis 1980). This species is one of the most valuable of the species presently under culture. The Japanese market will absorb large quantities of this shrimp at very high prices. It takes approximately 5–6 months to reach a 20-g market size at densities up to 25,000–37,000/ha (10,000–15,000/Ac). *Penaeus japonicus* prefers a sandy bottom and a higher-protein diet than the other species considered for aquaculture. Methods for captive reproduction and intensive larviculture have been established (Shigueno 1968; Aquacop 1975; Laubier-Bonichon 1975). A major problem for the successful culture of this species is the cost of the high-protein feeds in intensive culture techniques.

Several additional exotic species are promising for commercial culture but few are being cultured routinely. *Penaeus semisulcatus* is frequently mixed in with *P. monodon* in Southeast Asian ponds. *Penaeus orientalis* and *P. kerathurus* may be good species because they can tolerate lower temperatures. *Penaeus merguiensis* and *P. indicus* are easy to reproduce and grow well in high densities but are a relatively small species and have not been shown to produce well in ponds.

It is important to emphasize that while exotic species may be superior to native species in cultural situations, they are still exotics. Most states have strict regulations on exotic species because of the potential adverse environmental problems that could be associated with an accidental introduction.

Initial shrimp culture studies in the southern United States in-

(A)

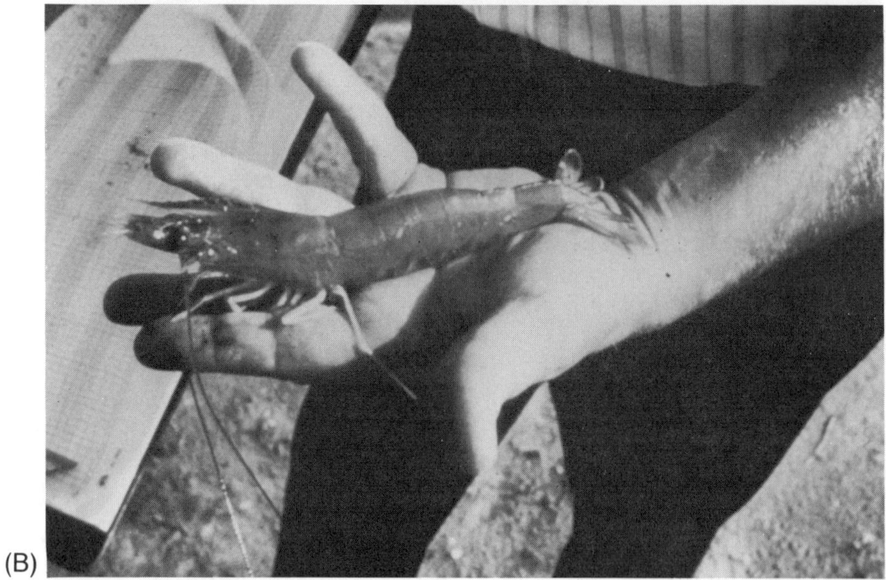

(B)

FIG. 3.1 (A) *Penaeus vannamei*, approximately 18 g wet body weight and 4.5 months old; (B) *Penaeus stylirostris*, approximately 24 g wet body weight and 5.0 months old.

(C)

(D)

FIG. 3.1 *(Continued)*
(C) *Penaeus monodon*, approximately 30 g wet body weight and 5.0 months old;
(D) *Penaeus japonicus:* approximately 25 g wet body weight and 5.0 months old.
All shrimps were reared in ponds within the Texas A&M University system.
Courtesy of Texas A&M University Shrimp Mariculture Program.

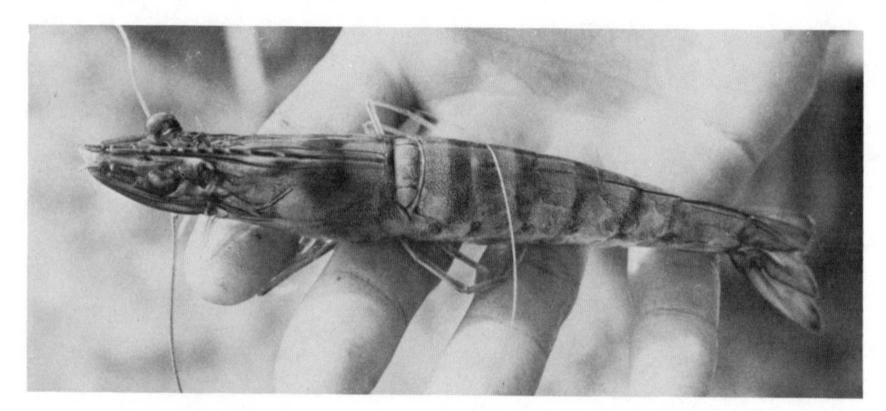

FIG. 3.2 *Penaeus aztecus,* a native, grooved Gulf Coast brown shrimp. Note the grooves extending across the middorsal surface of the carapace.
Courtesy of J. Huner.

volved the three native species of *Penaeus: P. aztecus,* the Gulf Coast brown shrip; *P. duorarum,* the pink shrimp; and *P. setiferus,* the Gulf Coast white shrimp (Figs. 3.2 and 3.3). This is not surprising because the only source of shrimp seed was wild-caught, mated females. It became obvious, however, that of the three species, *P. setiferus* was the superior species. Two of the three native species *(P. setiferus* and *P. aztecus)* have now been matured and spawned in captivity, but most interest continues to be directed toward *P. setiferus* for commer-

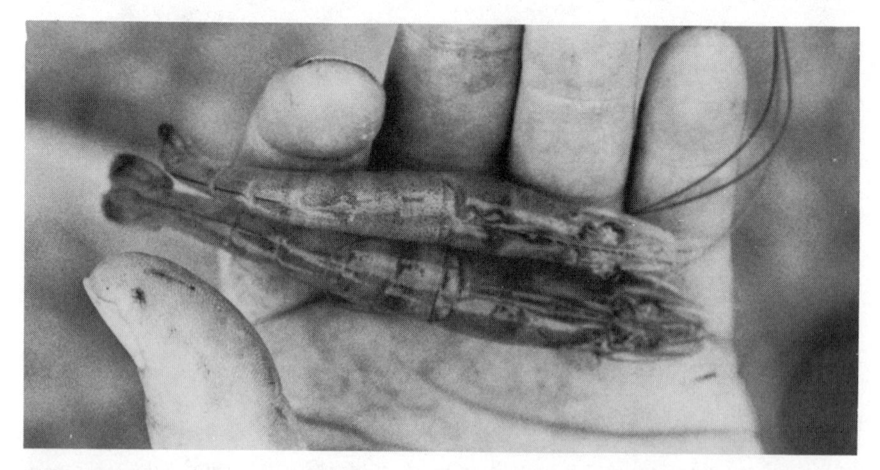

FIG. 3.3 Comparison of the native, grooved Gulf Coast brown shrimp, *Penaeus aztecus* (bottom) and the native, nongrooved Gulf Coast white shrimp, *Penaeus setiferus* (top).
Courtesy of J. Huner.

cial food production, particularly with the recent improved success of this species in ponds (Lawrence *et al.* 1983a).

In the Gulf Coast and South Atlantic states, *P. aztecus* and *P. setiferus* are the two species encountered in extensive impoundment shrimp culture where postlarve enter impounded areas on an incoming tide and are subsequently trapped by screened water-control structures. Life cycles are such that *P. aztecus* enters the estuaries in late winter–early spring and leaves in late spring–early summer when *P. setiferus* enters. *Penaeus setiferus* subsequently leaves in late summer through mid-fall (de la Bretonne and Avault 1971).

BASIC BIOLOGY AND CULTURE REQUIREMENTS

Basic Biology

Penaeid shrimps are decapod crustaceans and are very similar in morphology and physiology to freshwater prawns, crawfishes, and lobsters. Readers are referred to the chapters dealing with those species for information about basic decapod biology. An excellent summary of the biology and culture of many shrimps and prawns is given in a review edited by Mistakidis (1968). It should be noted, however, that there are two general taxa of penaeid shrimps: the brown and white shrimps, with closed thelyca; and the nongrooved white shrimps, with open thelyca. Grooved shrimps have parallel grooves beginning on either side of the rostrum and continuing posteriorly to the back of the carapace (Fig. 3.3). Molting is just as important in shrimps as it is in the other crustaceans. The decapod molt cycle has been described by several authors but readers may wish to refer to Huner and Colvin (1979) and Huner *et al.* (1979a,b) for details and additional references. Individuals interested in penaeid nutrition should consult Sandifer (1981) and New (1976) as well as references to decapod nutrition in other chapters of this volume.

Shrimp reproduce sexually, and males and females differ only slightly in morphology, although females may attain a larger size than males. Mating and spawning usually occur in offshore waters where the eggs are released into the open ocean. The eggs usually hatch within 18 to 24 hr at 28°C (82.4°F), and the larval shrimp go through five naupliar stages, three protozoea stages, and three mysis stages before reaching the postlarval stage. It can take from 10 to 18 days or longer to reach the postlarval stage depending on temperature and the quality and quantity of food.

In natural conditions postlarval shrimp migrate from offshore waters into estuary systems that serve as nursery grounds. In temperate climates shrimp migrate out of the estuaries during the cold winter season and reach sexual maturity offshore and the whole cycle starts again. In tropical climates shrimp may be associated with estuaries on a year-round basis, only moving offshore to mature and spawn.

Culture Requirements

The environmental and dietary conditions encountered during the shrimp's life cycle must be approximated by the culturist in the shrimp maturation and hatchery facility in order to produce commercial numbers of healthy shrimp. There are two main steps necessary for the production of postlarvae: (1) the mating and spawning of brood stock with the production of viable offspring (maturation/reproduction) and (2) rearing of larvae through the approximately 11 larval stages. These two steps require different facilities and procedures, though both require oceanic-quality water. Once postlarvae are obtained, they can be placed in either extensive, semi-intensive, or intensive culture systems for grow out to marketable size.

The general culture requirements for most species of shrimp have been defined. Subtropical and tropical shrimp grow best under the following conditions: water temperatures between 23°C and 32°C (73.4° and 89.6°F); dissolved oxygen above 3 ppm; pH near 8.0; and salinity between 10 and 30 ppt (some shrimp are more tolerant of higher and lower salinities). For best growth, adequate food must be provided for high-density cultures or the shrimp must be kept at lower densities in extensive culture systems using less or no supplemental feeds. Flow-through water systems are advantageous to good growth in that they eliminate metabolic waste products produced by the shrimp and provide some temperature control during hot periods. The soil should be clay–loam or sand, depending upon the species to be cultured, and should hold water without leaking. It is important to construct ponds that are completely drainable so that predators can be eliminated and organic materials oxidized.

The most successful shrimp farm operations are in the tropics where the natural climatic conditions are adequate for shrimp growth year-round. Extensive culture operations have been the most successful, as the financial inputs are less and the yields—though lower than with intensive systems—have been sufficient to meet expenses. Addition of organic and inorganic fertilizers has increased yields and profits in most areas, and more recently, supplemental diets have been used to increase yields and economic return.

In recent years there have been several attempts to develop more-intensive culture systems using raceways and tanks. Intensive culture systems require the same environmental optimums as pond culture, but the factors such as oxygen, salinity, temperature, and pH must be more rigidly controlled. This requires special aerating, heating, and filter systems to keep these parameters in the optimal range. Intensive culture also requires a more complete diet rather than the supplemental diets that are provided in pond culture. The cost of environmental controls and special diets are high, and even though yields have been increased in intensive culture operations, very few have met with economic success. The one exception to this is the culture of *P. japonicus* in Japan where high prices can sustain the high costs of production.

In the United States culture conditions for extensive pond culture are only suitable in the southern states from about April through June to September–November. This has usually meant that only one crop could be produced each season. Recent advances in controlled reproduction and improved intensive culture systems using recirculated water have made it possible to begin shrimp culture indoors before the normal growing season. Consequently, two crops per year have been produced at Texas A&M University (Lawrence *et al.* 1983b). There is continued interest in developing raceway systems with better temperature control for high-density indoor culture. This would permit shrimp culture in more northern areas, but the high energy and food costs associated with this method are presently prohibitive.

Low-level management systems using naturally stocked tidal impoundments were studied in South Carolina and Louisiana. A major problem with this technique is the question of ownership, public or private, of the naturally produced seed. However, investigators are currently examining the feasibility of culturing shrimp in abandoned rice fields in estuarine areas of South Carolina. In this system, hatchery-reared shrimp seed is often stocked, eliminating ownership questions, but tidal flushing is used to circulate water.

CULTURE TECHNIQUES

Culture techniques for penaeid shrimp vary from one locality to another. The literature generated by various research projects has been reviewed in the introduction to this chapter and the reader should refer to specific papers to obtain the various details of the culture sequence. However, this section will summarize some of the more basic techniques.

Shrimp Maturation and Spawning

Up until the mid 1970s there had been very little success with spawning penaeid shrimp in captivity. All spawners had to be obtained from wild stock that were in advanced reproductive stages and mated so that they would spawn on the same day as captured. During the next decade, however, several species of penaeids were induced to spawn in captivity. The general procedure involves bringing adult shrimp, of a size known to be reproductive in nature, into the laboratory and placing them in large tanks. The usual sex ratio is 1:1, although some increase the number of females relative to males. The minimum size for circular tanks is usually 3.05 m (10 ft) in diameter and 0.8 m (2.6 ft) in depth. Square or rectangular tanks, 10–25 m^3 (350–883 ft^3) and 1.5–2.0 m (4.9–6.6 ft) deep, are also used at some locations, especially for *P. monodon* which seems to require deeper tanks. Care must be taken to avoid placing obstacles around the perimeter of the tank so that chasing associated with mating is not interrupted. Air stones are, therefore, placed in the center of the tank. Most tanks are free of sand and gravel substrates, but sand is a substrate necessary for *P. japonicus*.

If tanks are placed indoors, they are equipped with fluorescent lighting providing approximately 100–200 lux of light intensity. Outdoor tanks are usually shaded to provide 10–20% of ambient light intensities. Photoperiod is very important and optimums vary with the species. A light period of 14 hours followed by 10 hours of dark has proven successful for *P. stylirostris*. The French have reversed the normal cycle, so that darkness occurs during regular daylight work periods. This facilitates the work of staff, as most shrimp mating and spawning activities occur during periods of darkness. Also, there is probably a light spectral requirement for shrimp reproduction in captivity though very little work has been done in this area. More work needs to be done to understand light and photoperiod requirements for all commerically valuable species. There are also temperature and salinity requirements for shrimp reproduction. Changing temperature regimes have been used for reproduction of *P. japonicus* in captivity (Laubier-Bonichon 1978, 1979). Temperatures of 24°C–31°C (75.2–87.8°F) and salinities of 26–35 ppt are generally used for shrimp reproduction in captivity.

Diet is also an important aspect of the maturation regime. Lawrence *et al.* (1980) have indicated that long-chain fatty acids are needed for proper egg development in female shrimp, which cannot synthesize them from shorter, less complex, food components. Long-chain fatty acids are usually destroyed in the extrusion of prepared diets, so it is necessary to provide fresh, natural foods to the shrimp brood stock. The best success has thus far been obtained with marine polychaete worms (*Glycera* spp.), squid, clams, and mussels. These may or may not be augmented with dried pelleted foods. Marine worms have been found especially effective and should be used whenever available. A typical feeding regime for brood stock is approximately 5% of the body weight per day of dry food distributed in four feedings: early morning, noon, mid-afternoon, and evening. Diet components can be alternated in the feeding plan and worms may be given in the early morning and evening. Presently, shrimp maturation and reproduction facilities still use frozen foods. An adequate dry formulated feed is needed to lower cost and to provide better water quality.

The system of water exchange used in maturation tanks depends on the particular facility, but it is important to keep oceanic water-quality conditions. An exchange two to three times per day is considered effective. There is some belief that a little turbidity is beneficial to mating, but this has not been verified.

In general, unilateral eyestalk ablation is practiced on all females used in maturation facilities (Fig. 3.4). Although spawning success is definitely increased through this procedure, there is some question as to the overall quality of seed stock produced from ablated females. Hatchery and pond survival are slightly less for seed stock obtained from ablated females than for stock from wild spawners. Ultimately,

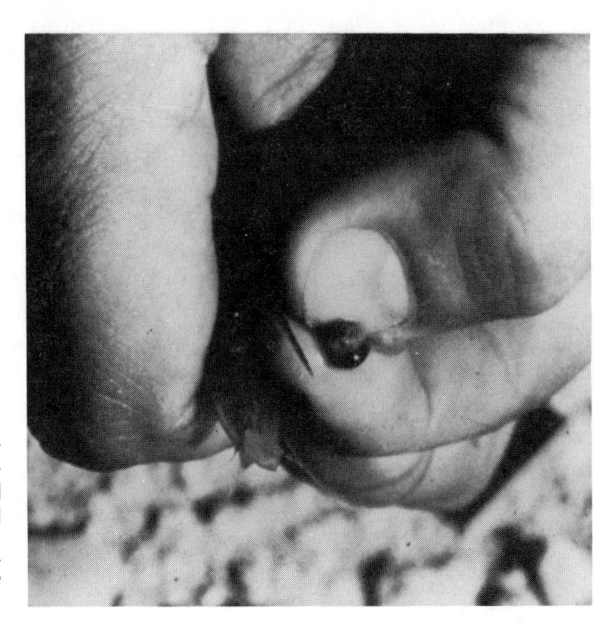

FIG. 3.4 Eyestalk removal is used to facilitate maturation and spawning in several penaeid shrimps.
Courtesy of Texas A&M University Shrimp Mariculture Program.

it may be important to obtain spawning by more natural means than unilateral eyestalk ablation. This will probably be accomplished through manipulation of photoperiod, light intensity, temperature, diet, and holding systems.

Penaeids may be divided into three groups: those that mature, mate, and spawn freely in response to controlled environmental conditions *(P. merguiensis* and *P. japonicus);* those that require unilateral eyestalk ablation for maturation *(P. monodon* and, perhaps, *P. aztecus);* and those that will spawn without ablation but which usually respond better to a combination of controlled environmental conditions and ablation *(P. setiferus, P. stylirostris,* and *P. vannamei).*

Once the maturation tanks are set up and the shrimp are ablated and in place, the shrimp should be observed daily for sexual behavior (chasing and mating) and maturation. Most of this activity will occur in the late afternoon and early evening if natural photoperiods are followed. It is helpful to have an egg collector on the outfall of the maturation tank (mesh size 200 micron) to observe whether any eggs have been released. Spawning may go unnoticed and an egg collector is a good method of double checking spawning activities. Once eggs are observed in the egg collectors, or mating is observed, a close watch is kept on all females for spermatophore attachment or for stage development, depending on the species (*P. monodon* and *P. japonicus* are mated during the previous molt and stage development is the best indicator, whereas *P. stylirostris* and *P. setiferus* are mated a few hours before spawning, without a molt, and an attached spermatophore is the best indicator). Mated or advanced-stage females are usually isolated into special spawning tanks between 7 and 10 P.M., and spawning usually occurs in the late evening or early morning hours (11 P.M. to 4 A.M.).

Penaeids with closed thelyca (brown and white shrimps) mate shortly after the molt preceding spawning, and ovarian development is the key factor in determining the timing of spawning. However, those with open thelyca (nongrooved white shrimps) carry spermatophores that may be lost or may never be successfully afixed before spawning. For this reason, techniques have been developed for mechanically removing spermatophores from males, isolating the sperm masses, and placing them in the thelycum area of a gravid female. The results from this technique have been encouraging; it has been used to produce interspecific hybrids and to increase the effectiveness of sourcing (capture of wild, gravid female shrimp) by increasing the total number of nauplii produced.

Fertilized eggs develop rapidly and nauplii hatch within 24 hr at

28°–30°C (82.4°–86°F). The hatching tanks are aerated lightly and nauplii are moved to the hatchery tanks within 24 hr of hatching.

Hatchery Procedures

There are two basic culture systems used in shrimp hatcheries today: the Japanese method, which uses very large tanks of 100–250 MT (Fig. 3.5) and the Galveston method, which uses smaller (2-MT) more controllable tanks (Fig. 3.6). There are many gradations between the two systems, and individual hatchery operators have adopted various components from both.

In the Japanese method, water is prefiltered to exclude zooplankton and the water is fertilized with inorganic fertilizers to obtain a bloom of phytoplankton. Shrimp larvae are obtained from spawners placed in the tank, so the original density is not predetermined. Most operations run at 30–60 nauplii/liter as a starting density. The nauplii do not have a complete digestive tract and do not eat. New fertilizer is added daily to encourage phytoplankton growth so that adequate food will be available when nauplii metamorphose into protozoea. Other foods such as copepods, soybean meal, or yeast may be used as supplemental food, but the majority of the food for protozoea comes from the algal community and associated fauna in the enriched tank water. *Artemia* (brine shrimp) are added for late protozoea and mysis stages

FIG. 3.5 This large-scale, extensive larval shrimp rearing unit is an example of the Japanese method. *Courtesy of J. Huner.*

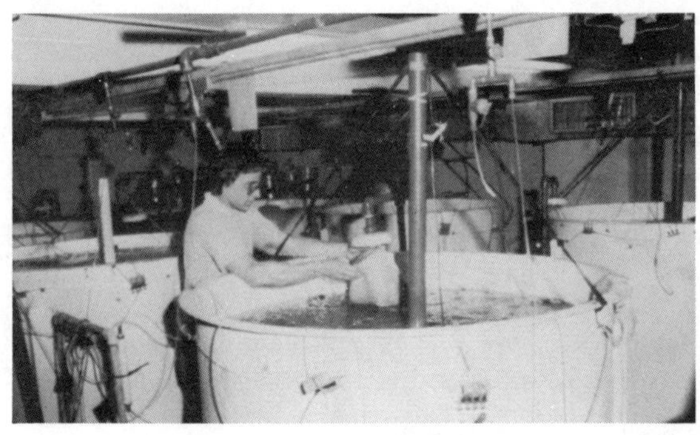

FIG. 3.6 These small-scale, intensive larval shrimp rearing units are an example of the Galveston method.
Courtesy of Texas A&M University Shrimp Mariculture Program.

to promote survival and metamorphosis. Survival usually ranges from 20 to 40% to the stage of 10- to 15-day-old postlarvae. Occasional failures due to blooms of blue green algae, collapse of the algal community or disease occur.

In the Galveston method, or adaptations of the method, all environmental and dietary parameters are rigidly controlled. Water is prefiltered to exclude anything larger than 1 micron. Temperature is maintained at 28°–30°C ± 1°C (82.4°–86°F). Salinity is kept constant at 28 to 35 ppt. Oxygen is kept high by aeration. Food is produced in separate tanks and usually consists of monocultures of selected algae (diatoms and phytoflagellates such as *Isochrysis, Nitzschia, Phaedactylum, Chaetocerus, Skeletonema, Tetraselmis,* and *Thalassiosira*) and *Artemia* nauplii. However, yeast, rotifers, nematodes, and other foods of appropriate size can be utilized. All foods are introduced so as to maintain a certain density of food, rather than on the basis of a certain amount per larvae, because larval shrimp must encounter food within certain time limits in order to survive. Algae fed to protozoea are maintained at densities around 100,000 cells/ml. *Artemia* nauplii or other animal plankton can be introduced as early as protozoea 2 but usually are introduced during the mysis stages. *Artemia* densities must be maintained at 2–4 individuals/ml. Larval density is higher (approximately 100/liter) than with the Japanese method, and survival to 5-day-old postlarvae may be as high as 80–90% with an average of about 50%. Because of the controlled conditions and close observations used in the Galveston method, the shrimp can be monitored for stage development, health, survival, behavior, and general appearance on a regular basis. This is very helpful in the improvement of production. However, this method is labor and capital intensive and technically more difficult than the Japanese method. The cost per 1000 larvae is somewhat higher than with the Japanese method.

Grow-Out Procedures

There are almost as many grow-out procedures as there are shrimp ponds. Shrimp culture may generally be classified into intensive, semi-intensive and extensive techniques. Intensive culture of *P. japonicus* occurs in Japan and Taiwan; the stocking density can be as high as 160 juveniles/m^2 (640,000/Ac) and production of 4.5–24 MT/ha (4000–21,900 lb/Ac) can be obtained. *P. monodon* culture in Taiwan, with stocking densities of 20–40 juveniles/m^2 (80,000–160,000/Ac) and yields of 1.4–9.6 MT/ha (1200–8500 lb/Ac), is classified as semi-intensive. Typical extensive culture occurs in most areas where shrimp culture is practiced (i.e., in the Philippines, Indonesia, Southeast Asia, and Central and South America). Less intensive culture operations stock at low densities of 2.5–5 juveniles/m^2 (4,000–20,000/Ac). With extensive culture conditions shocking less than 2.5 juveniles/m^2, very little if any supplemental feed is provided when stocking densities are less than 1–2 junveiles/m^2. Organic and inorganic fertilizers are sometimes used to increase productivity in extensive systems. Intertidal systems that are self-stocked are common in some areas (Fig. 3.7).

There is presently considerable research on improving techniques of semi-intensive and intensive pond culture. Further, much research is being done on recirculating and flow-through raceway systems designed to get the most production from as little space as possible. Additional research is needed on the dietary requirements and disease problems associated with these systems, and such programs are nec-

FIG. 3.7 Screened floodgate facilitates water circulation in extensive and semi-intensive shrimp culture ponds. Screen can be removed to allow wild postlarval shrimp to enter ponds on flood tides.
Courtesy of J. Huner

essarily multidisciplinary in approach and require substantial sums of risk capital. In general, there have not been any real economic success stories with intensive culture systems because of their high operational costs. Perhaps Japan, with its very high-priced market is the exception to this generalization, but many of the high-technology farms in Japan are presently in economic difficulties because of the rising costs of energy and food.

A model intensive culture system was developed jointly by the University of Arizona and the University of Sonora (Mexico) at Puerto Penasco, Sonora (Salser *et al.* 1981). Grow-out units are called "aquacells" (Fig. 3.8). Each contains two long raceways, 3.35 m wide × 61 m long × 0.3 m deep (11 ft × 200 ft × 1 ft), which are covered with air-inflated plastic domes. Sea water from wells is delivered through side manifolds and is turned over about seven times a day (Fig. 3.9). All shrimp are visible. The hatchery phase of this system takes roughly 20 days, the posthatchery nursery phase some 18 weeks, and the final grow-out phase about 21 weeks. After this 220- to 230-day period, shrimp are harvested at a large size of 21 g. Final carrying capacity is 5.6 kg/m^2 (1.1 lb/ft^2). Based on 1980 data, 18 grow-out units could produce 3632 kg (8000 lb) of shrimp per month.

All extensive pond culture systems respond well to fertilization with

FIG. 3.8 This experimental, intensive shrimp culture unit in Hawaii was developed by the University of Arizona's Environmental Research Laboratory. Raceways are covered by opaque, air-inflated plastic covers.
Courtesy of University of Arizona, Environmental Research Laboratory, Tucson, AZ.

FIG. 3.9 Raceways in intensive shrimp culture units require large volumes of clean salt water to support economical and profitable shrimp production.
Courtesy of J. Huner.

both inorganic and organic (including treated sewage) materials, as shrimp obtain a substantial amount of their daily ration from natural biota in a pond. However, shrimp are usually fed to achieve production in excess of the 500 kg/ha (445 lb/Ac) needed per crop to make shrimp culture profitable in the southern United States. Protein, an expensive component of any formulated feed, must equal or exceed 30–35% of the feed for intensive tank/raceway culture systems, but excellent results have been noted in pond cultures fed rations containing 20–25% protein (Lawrence *et al.* 1985; Rubright *et al.* 1981). Predators and competitors including wading birds, gulls, ducks, and many varieties of fish can seriously affect production and must be controlled if a pond facility is to operate at a profit (Fig. 3.10) (Huner 1979). Polyculture of different shrimp species (*P. stylirostris* and *P. vannamei,* for example) and of fish (such as the very valuable pompano) and shrimps has shown much promise in research situations in Texas and Alabama, respectively (Chamberlain *et al.* 1981; Tatum and Trimble 1978; Trimble 1980). Staggered stocking of shrimp of different ages has generated better survival and higher yields than those expected from one-time-stocked ponds at the same density (Ojeda *et al.* 1980).

DISEASES AND PARASITES

As shrimp culture has developed from extensive culture techniques into semi-intensive and intensive techniques, the culturist has encountered more and more problems with diseases and parasites that affect productivity. The papers cited in the introduction of this chap-

FIG. 3.10 Hardy, prolific killifishes such as these *Cyprinodon variegatus* can be serious competitors in shrimp ponds. Other species are voracious predators even though their size rarely exceeds 10–15 cm.
Courtesy of J. Huner.

ter provide an excellent synopsis of many of these diseases. However, Johnson (1975) provides a well-illustrated summary of shrimp diseases.

Shrimp diseases can be divided into infectious diseases, noninfectious diseases, and nutritional, environmental and toxic diseases. Infectious diseases include those caused by viruses (including Rickettsia and Chlamydia) bacteria, and fungi. Noninfectious diseases include those caused by epicommensals (*Leucothrix* disease and ciliate gill disease) and parasitic protozoa (microsporidians and gregarines). Nutritional, environmental, and toxic diseases include a broad range of problems from vitamin deficiencies to blooms of toxic algae. Any disease in any of the above categories can cause catastrophic mortality in shrimp cultures, especially during the delicate larval stages. The best cure is proper management and prevention, as very few drugs or chemical treatments can control a disease once it begins. Aquacop, the French research facility in Tahiti, uses a variety of antibiotics routinely to inhibit bacterial diseases. Most hatcheries depend on maintaining optimum water quality and diets to safeguard the health of their larval shrimp. Work is continuing on the control of diseases and some emphasis is now being placed on identifying and controlling virus diseases that have been a limiting factor in production of *P. monodon, P. vannamei* and *P. stylirostris,* the species used in intensive culture operations. Recently, a vaccine has been developed for postlarval shrimp (Lewis and Lawrence 1983) and is currently being tested on the larval stages.

It is presently impractical to treat large ponds in the commercial production of shrimp. Therefore, almost all disease research concerns disease problems in hatcheries, where the volume of water is more manageable and where most of the diseases are expressed because of the high density of larvae. If a severe disease problem develops in pond culture, the present practice is to drain the pond, dry the bottom, and start over again. Research projects involving immunization of shrimp seed prior to pond stocking have, however, been initiated at Texas A&M University.

STATUS OF SHRIMP FARMING IN THE UNITED STATES

Climatic and Other Constraints

Warm-water shrimp require temperatures of 23°–32°C, optimum 28°C (73.4°–89.6°F, optimum 82.4°F) and salinities of 10–35 ppt depending on the species. The required temperature conditions are found only in the tropics on a year-round basis. The southern United States, specifically south Florida, Hawaii, and Texas, have climatic conditions that permit the production of one or two crops per year depending on the stock management regime. Because of this, many commercial ventures by U.S. companies have been initiated outside the United States. Recent advances in control of shrimp reproduction in captivity and improvements in systems and diets will make shrimp farming a reality within the United States.

Production technology and environmental limitations are not the only stumbling blocks to commercial shrimp culture in the United States. The National Aquaculture Plan (NOAA 1980), recently completed by a consortium of government and private concerns, identifies the competition for estuary sites, laws concerning waste water discharge, the difficulties in obtaining permits, and unavailability of loan money (banks are hesitant to finance unproven ventures) as just some of the major problems facing the prospective shrimp aquaculturist. It may be necessary to get as many as 15 to 20 agencies to agree and to give permits before a shrimp farm can be built. Because federal laws prohibit modification of marshlands, the shrimp farmer must place

FIG. 3.11 Shrimp culture in intertidal areas such as salt marshes cannot be practiced today in the United States because of laws prohibiting modification of marshlands.
Courtesy of J. Huner.

ponds on land adjacent to marshes and estuaries (Fig. 3.11). After surmounting all these problems, the farmer still has to cope with unwanted predators, poor soil conditions, competition from other aquatic organisms, logistics, and daily operational problems.

Even though penaeid larval rearing is fairly routine in research laboratories and commercial hatcheries exist, it is obvious that much more work must be done on increasing predictability and survival. Poor timing of hatchery production to meet the narrow window for production in ponds has frequently led to failure of shrimp farms.

Another major obstacle is the inability to obtain enough seed stock of the best shrimp for grow out under culture conditions. *Penaeus vannamei* appears to be the best species for semi-intensive pond culture but it does not mate or produce eggs well in captivity. As a result many U.S. companies are locating facilities outside the United States close to natural populations of *P. vannamei* so they can obtain wild spawners to provide seed stock.

Current and Future Shrimp Culture by U.S. Companies

There are presently no financially successful shrimp farms within the United States. Several companies have been started and have failed; others are being formed and several are producing crops at this time. The main geographic areas where shrimp farms are being tried are south Texas, South Carolina, and Hawaii. These states are supportive of the concept of shrimp farming and permits have been obtained for several farms. Those farms planned for south Texas are designed for semi-intensive pond culture of shrimp based on work at Texas A&M University. Those farms planned for Hawaii include both semi-intensive pond culture systems and intensive raceway systems. Those in South Carolina use very extensive pond management techniques. Production predictions for the intensive raceway system range from 56,250 to 113,500 kg/ha/yr (50,000 to 100,000 lbs/Ac/yr). If successful this system, or one like it, will revolutionize shrimp culture, but first the prototype must be brought to a commercial scale and problems with diet and diseases overcome.

Most likely, investment in shrimp aquaculture by U.S. companies will first occur in those countries in Central and South America and Southeast Asia that provide the right combinations of climate, spawn and seed-stock supplies, labor, and economic incentives. Indeed, several U.S. companies already are involved in shrimp farming in Ecuador, Panama, Costa Rica, and other Central and South American countries. For example, a Ralston Purina Company facility in Panama appears to be successful and presently has close to 600 ha (1500 Ac) of shrimp ponds.

Commercial technology developed in foreign countries and in U.S. research facilities will be used for semi-intensive commercial shrimp culture in the United States. Intensive culture systems also have potential in this country. The present consensus is that any shrimp farming that develops in the United States will involve semi-intensive pond or intensive raceway culture for production of food (20-g shrimp) and/or intensive raceway culture for production of fish bait (1- to 3-g shrimp). This type of technology will require several more years to perfect; estimates of the time needed to develop successful systems for the United States range from 1 to 10 years depending on the source. In order for intensive culture to be successful, substantial investments will have to be made in developing complete diets, disease control procedures, water systems, and techniques for controlling reproduction in captivity to produce healthy and viable seed stock of the most suitable species.

This technological development must be accompanied by simplification of legal and permitting procedures to enable those interested in shrimp farming to obtain clear-cut guidelines on how and where they can place their production facilities. At present, each state may have different permit procedures and some of these procedures can take several months or even years to negotiate. Legislation is now being proposed in several states that will facilitate the process of obtaining permits.

ECONOMIC OVERVIEW AND OUTLOOK

Shrimp culture within the United States is being pursued most vigorously in Texas, South Carolina, and Hawaii. Most of the information about the economics of shrimp culture has been developed by researchers at Texas A&M University and relates to south Texas. Preliminary economic analyses of pond culture in areas such as Texas with a growing season of less than 1 year have indicated that production per crop must be greater than 600 kg/ha (534 lb/Ac) to support a commercial operation (Adams *et al.* 1980; Griffin *et al.* 1981; Pardy *et al.* 1983).

A typical, integrated semi-intensive shrimp culture system in south Texas includes a maturation/reproduction unit, a hatchery, and grow-out units (ponds). Johns *et al.* (1983) discuss pond production strategies and budget analyses. Johns *et al.* (1981a) present a budget analysis for a shrimp maturation/reproduction unit, while Johns *et al.* (1981b) present complimentary data in the form of a budget analysis for a shrimp hatchery facility. Lawrence *et al.* (1981) summarize the costs and revenues associated with an integrated, semi-intensive shrimp culture system. A 557-m^2 (6000-ft^2) metal frame maturation/reproduction unit would cost about $400,000 (1981) and could produce about 1.5 million shrimp larvae (nauplii) per day. Nauplii would be transfered to a hatchery/nursery facility housed in a 650-m^2 (7000-ft^2) metal frame building costing about $700,000 (1981). Some 750,000 6- to 10-day old postlarvae could be produced from seedstock at a cost of about $4 per 1000. Postlarvae would then be cultured in 2- to 8-ha (5- to 20-Ac) grow-out ponds and raised to a size of 10–18 g (25–45/lb), heads-on, in 3 to 4 months. Although not stated, the grow-out system would be roughtly 100 ha (240 Ac) in total size (Johns *et al.* 1983). Shrimp production would vary from 785 to 2243 kg/ha (700 to 2000 lb/Ac) with a value of $3700–10,574 per ha ($1500–4000 per Ac); operating costs would be about $2964–6916 per ha ($1200–2800 per Ac) of water.

Griffin *et al.* (1981) employed a bioeconomic model to determine the sensitivity of profit to various stochastic elements in the pond culture of shrimp based on southern Texas experience. They concluded that specific biological and environmental parameters are critical to profits. The length of the growing season has an important effect, suggesting that starting shrimp early in small, heated nursery ponds may be economically advantageous. Although salinity data were limited, profits were sensitive to decreased salinity and methods to control it should be pursued. Profits are very sensitive to minimum oxygen levels. In the model only water flow was used to control oxygen levels, but other methods should be studied.

Specific economic analyses are not generally available for intensive culture systems but Salser *et al.* (1978) projected a production of 84,100 kg/ha/yr (75,000 lb/Ac/yr) for the intensive culture system developed by the universities of Arizona and Sonora (Mexico) and Coca-Cola Company and F. H. Prince Company. This level of production should lead to profitable operation of intensive raceway systems, which would be suitable for other sites in the United States. Such a system will possibly be in operation by 1985 in Hawaii.

Marketing cannot be ignored in any discussion of the economics of shrimp culture. In the Sea Grant Aquaculture Plan, Jennings *et al.* (1982) state that while demand for shrimp is increasing, domestic production from capture fisheries is at or near the maximum sustainable yield. The shortfall is being made up with imports, which include substantial volumes of cultured Ecuadorian shrimp. Shrimp value varies according to size with the largest shrimp being the most valuable. Size is determined by counting the number of shrimp per unit of weight. Care must be taken in extrapolating values quoted by the seafood industry, as these "counts" may mean whole ("heads-on") or cleaned ("headed") with the cephalothorax removed and only the abdomen ("tail") remaining. Whole 12- to 22-g shrimp sold for $8–13 per kg ($3.60–5.90 per lb) in 1983.

While most emphasis is placed on raising shrimp for food, there is a substantial demand for live bait shrimp in the Gulf of Mexico coastal region. Once general production procedures are well established for food shrimps, cultivation of smaller bait shrimps may represent a profitable alternative to culture of food shrimps; however, only native species could be raised for live sales.

So far, commercial production of shrimp has, for the most part, only been economically successful in third world countries where conditions (available native shrimp from natural sources, climate, labor, and legal constraints) are favorable to shrimp farming. Low-technology systems, certain semi-intensive systems in Panama and Ecuador, and intensive systems in Japan and Taiwan have all shown a profit, indicating a progression in technology and understanding of the biological requirements of shrimp that will lead to the eventual success of shrimp culture in the United States.

The recent breakthroughs in captive reproduction of prime shrimp species, *P. vannamei* and *P. stylirostris,* once perfected, should provide a constant and reliable supply of seed stock, thus overcoming an important limiting factor for United States shrimp aquaculture. Also, continued improvement of production technology for the native shrimp *(P. setiferus)* in ponds and increased larvae production through captive reproduction and sourcing will be of significance. Assuming that further advances will occur in the development of shrimp diets and disease control methods, we optimistically expect profitable shrimp culture operations in the United States within the very near future.

LITERATURE CITED

ADAMS, C. M., GRIFFIN, W. L., NICHOLS, J. P., and BRICK, R. W. 1980. Bioengineering-economic model for shrimp mariculture systems, 1979. Sea Grant College Program Tech. Rep. TAMU-SG-80-203, Texas A&M University, College Station Texas.

ANON. 1982. Fishing exports now second largest. Weekly Analysis of Ecuadorean Issues, No. 18, pp. 148–156.

AQUACOP. 1975. Maturation and spawning in captivity of penaeid shrimp: *Penaeus merguiensis* de Man, *Penaeus japonicus* Bates, *Penaeus aztecus* Ives, *Metapenaeus ensis* de Hann, and *Penaeus semisulcatus* de Hann. Proc. World Maric. Soc. **6,** 123–132.

AQUACOP. 1977a. Observations on the maturation and reproduction of penaeid shrimp in captivity in a tropical medium. Third Meeting I.C.E.S. Working Group on Mariculture, Brest, France, May 1977.

AQUACOP. 1977b. Reproduction in captivity and growth of *Penaeus monodon,* Fabricius in Polynesia. Proc. World Maric. Soc. **8,** 927–945.

AQUACOP. 1979. Penaeid reared broodstock: closing the cycle of *P. monodon, P. stylirostris* and *P. vannamei.* Proc. World Maric. Soc. **10,** 445–452.

AQUACOP. 1980. Penaeid reared broodstock: closing the cycle of *P. monodon, P. stylirostris* and *P. vannamei.* Centre Oceanologique Pacifique, CNEXO-COP, Tahiti.

ARNSTEIN, D. R., and BEARD, T. W. 1975. Induced maturation of the prawn *Penaeus orientalis* Kishinouye in the laboratory by means of eyestalk removal. Aquaculture **5,** 411–412.

BEARD, T. W., and WICKINS, J. F. 1980. Breeding of *Penaeus monodon* Fabricius in laboratory recirculation systems. Aquaculture **20,** 79–89.

BEARD, T. W., WICKINS, J. F., and ARNSTEIN, D. R. 1977. The breeding and growth of *Penaeus merguiensis* de Man in laboratory recirculating systems. Aquaculture **10,** 275–289.

BEYNON, J. L., HUTCHINS, D. L., RUBINO, A. J., LAWRENCE, A. L., and CHAPMAN, B. R. 1981. Nocturnal activity of birds on mariculture ponds. J. World Maric. Soc. **12**(2), 63–72.

BRAY, W. A., CHAMBERLAIN, G. W., and LAWRENCE, A. L. 1982. Increased larval production of *Penaeus setiferus* by artificial insemination during sourcing cruises. J. World Maric. Soc. **13,** in press 123–133.

BRAY, W. A., CHAMBERLAIN, G. W., and LAWRENCE, A. L. 1985. Observations on natural and artificial insemination of *Penaeus setiferus.* Proc. 1st Int. Conf. Warm Water Aquacult.—Crustacea 1, 1982, to be published.

BROOM, J. G. 1968. Pond culture of shrimp on Grand Terre Island, Louisiana, 1962–1968. Gulf Carib. Fish. Inst. Univ. Miami Proc. **21,** 137–151.

BROOM, J. G. 1970. Shrimp culture. Proc. World Maric. Soc. **1,** 63–68.

BROOM, J. G. 1972. Shrimp studies in Honduras 1969–1971. Proc. World Maric. Soc. **3,** 193–204.

BROWN, A., JR., MCVEY, J. P., SCOTT, B. M., WILLIAMS, T. D., MIDDLEDITCH, B. S., and LAWRENCE, A. L. 1980. The maturation and spawning of *Penaeus stylirostris* under controlled laboratory conditions. Proc. World Maric. Soc. **11,** 488–499.

CAILLOUET, C. W., NORRIS, J. P., HERALD, E. J., and TABB, D. C. 1974. Growth and yield of pink shrimp (*Penaeus duorarum* Burkenroad) in a feeding experiment in ponds. Proc. World Maric. Soc. **5,** 125–135.

CAUBERE, J. L., LAFON, R., RENE, F., and SALES, C. 1979. Etude de la maturation et la ponte chez *Penaeus japonicus* en captivite. *In* Advances in Aquaculture, T. V. R. Pillay and W. A. Dill (editors), Fishing News Books Ltd., Surrey.

CHAMBERLAIN, G. W., and LAWRENCE, A. L. 1981a. Maturation, reproduction and growth of *Penaeus vannamei* and *P. stylirostris* fed natural diets. J. World Maric. Soc. **12**(1), 209–224.

CHAMBERLAIN, G. W., and LAWRENCE, A. L. 1981b. Effect of light intensity and male eyestalk ablation on reproduction of *Penaeus stylirostris* and *Penaeus vannamei*. J. World Maric. Soc. **12**(2), 357–372.

CHAMBERLAIN, G. W., HUTCHINS, D. L., and LAWRENCE, A. L. 1981. Mono- and polyculture of *Penaeus vannamei* and *Penaeus stylirostris*. J. World Maric. Soc. **12**(1), 251–270.

CONTE, F. S. 1975. Penaeid shrimp culture and the utilization of waste heat effluent. Proc. Waste Heat Aquacult. Workshop **1,** 27–47.

COOK, H. L. 1969. A method of rearing penaeid shrimp for experimental studies. FAO UN Fish. Rep. **57, 3,** 709–715.

COOK, H. L., and MURPHY, M. A. 1966. Rearing penaeid shrimp from eggs to postlarvae. Proc. Annu. Conf. Southeastern Assoc. Game Fish Commiss. **19,** 283–288.

COOK, H. L., and MURPHY, M. A. 1969. The culture of larval penaeid shrimp Trans. Am. Fish. Soc. **98,** 751–754.

DE LA BRETONNE, L. W., JR., and AVAULT, JR., J. W. 1970. Shrimp mariculture methods tested. Am. Fish Farmer **1**(12), 8–11, 27.

DE LA BRETONNE, L. W., JR., and AVAULT, JR., J. W. 1971. Movements of brown shrimp, *Penaeus aztecus,* and white shrimp, *Penaeus setiferus,* over weirs in marshes in South Louisiana. Proc. Annu. Conf. Southeastern Assoc. Game Fish Commiss. **25,** 651–654.

EMMERSON, W. D. 1980. Induced maturation of prawn *Penaeus indicus*. Mar. Ecol. Prog. Ser. **2,** 121–131.

FUJINAGA, M. 1969a. Kurama shrimp *(Penaeus japonicus)* cultivation in Japan. FAO UN Fish. Rep. 57, **3,** 811–832.

FUJINAGA, M. 1969b. Method of cultivation of penaeid shrimp. United States Patent no. 3,477,406. November 11, 1969.

GRIFFIN, W. L., HANSON, J. S., BRICK, R. W., and JOHNS, M. A. 1981. Bioeconomic modeling with stochastic elements in shrimp culture. J. World Maric. Soc. **12**(1), 94–103.

HEINEN, J. M. 1976. An introduction to culture methods for larval and postlarval shrimp. Proc. World Maric. Soc. **7,** 333–344.

HIRONO, Y. 1983. Preliminary report on shrimp culture activities in Ecuador. J. World Maric. Soc. **14,** 451–457.

HOLTHUIS, L. B. 1980. FAO Species catalogue. Vol. 1. Shrimps and prawns of the world. FAO Fish. Synop. No. 125, Vol. 1.

HUDINAGA, M. 1942. Reproduction, development and rearing of *Penaeus japonicus* Bate. Jpn. J. Zool. **10,** 305–393.

HUDINAGA, M., and KITTAKA, J. 1966. Studies on food and growth of larval stage of a prawn, *Penaeus japonicus,* with reference to the application to practical mass culture (in Japanese). Inf. Bull. Planktol. Jpn. **13,** 83–94.

HUDINAGA, M., and KITTAKA, J. 1967. The large scale production of the young kurama prawn, *Penaeus japonicus* Bate. Inf. Bull. Planktol. Jpn. Commem., pp. 35–46.

HUDINAGA, M., and MIYAMURA, M. 1962. Breeding of the kurama prawn (*Penaeus japonicus* Bate) (in Japanese). J. Oceanogr. Soc. Jpn. 20th Anniv. Vol. **20,** 694–706.

HUNER, J. V. 1979. Aquaculture in Louisiana's coastal zone. Proc. Third Coastal Marsh Estuary Management Symp. Baton Rouge, Louisiana.

HUNER, J. V., and COLVIN, L. B. 1979. Observations on the molt cycles of two species of juvenile shrimp, *Penaeus californiensis* and *Penaeus stylirostris* (Decapoda: Crustacea). Proc. Nat. Shellfish. Assoc. **69,** 77–84.

HUNER, J. V., COLVIN, L. B., and REID, B. L. 1979a. Postmolt mineralization of the exoskeleton of juvenile California brown shrimp, *Penaeus californiensis* (Decapoda: Penaeidae). J. Comp. Biochem. Physiol. **62A,** 889–893.

HUNER, J. V., COLVIN, L. B., and REID, B. L. 1979b. Whole-body calcium, magnesium, and phosphorous levels of the California brown shrimp, *Penaeus californiensis* (Decapoda: Penaeidae) as functions of molt stage. J. Comp. Biochem. Physiol. **64A,** 33–36.

JENNINGS, F. D., BEETON, A. M., DAVIDSON, J. R., GAITHER, W. S., and GILMARTIN, M. 1982. Sea Grant Aquaculture Plan 1983–1987. Sea Grant College Program Tech. Rep. TAMU-SG-82-114. Texas A&M University College Station, Texas.

JOHNS, M. A., GRIFFIN, W. L., LAWRENCE, A. L., and HUTCHINS, D. 1981a. Budget analysis of shrimp maturation facility. J. World Maric. Soc. **12**(1), 104–112.

JOHNS, M. A., GRIFFIN, W. L., LAWRENCE, A. L., and FOX, J. 1981b. Budget analysis of penaeid shrimp hatchery facilities. J. World Maric. Soc. **12**(2), 305–321.

JOHNS, M. A., GRIFFIN, W. L., PARDY, C., and LAWRENCE, A. L. 1985. Pond production strategies and budget analysis for penaeid shrimp. Proc. 1st Int. Biennial Conf. Warm Water Aquaculture—Crustacea 1, to be published.

JOHNSON, S. K. 1975. Handbook of shrimp diseases. Sea Grant College Program Tech. Rep. TAMU-SG-75-603. Texas A&M University College Station, Texas.

KELEMEC, J. A., and SMITH, I. R. 1980. Induced ovarian development and spawning of *Penaeus plebejus* in a recirculating laboratory tank after unilateral eyestalk enucleation. Aquaculture **21,** 55–62.

KITTAKA, J. 1971. Kuruma shrimp culture techniques (in Japanese). *In* Through Culture in the Shallow Sea, T. Imai (editor). Kohseisha Koheikaku, Tokyo.

KITTAKA, J. 1976. Food and growth of penaeid shrimp. Proc. 1st Int. Conf. Aquaculture Nutrition, University of Delaware, Newark.

KITTAKA, J. 1981. Large scale production of shrimp for releasing in Japan and in the United States and the results of the releasing programme at Panama City, Florida. Kuwait Bull. Mar. Sci. **2,** 149–163.

KURATA, H., and K. SHIGUENO. 1976. Recent progress in the farming of Kuruma shrimp *(Penaeus japonicus). In* Advances in Aquaculture. T. V. R. Pillay and W. A. Dill (editors). Fishing News Books Ltd., Surrey, England.

LATAPIE, W. R., JR., BROOM, J. G., and NEAL, R. A. 1972. Growth rates of *Penaeus aztecus* and *P. setiferus* in artificial ponds under varying conditions. Proc. World Maric. Soc. **3,** 241–254.

LAUBIER-BONICHON, A. 1975. Induction de la maturation sexuelle et ponte chez la crevette *Penaeus japonicus* Bate en milieu controle. C.R. Acad. Sci. Paris (Ser. D) 281, 2013–2016.

LAUBIER-BONICHON, A. 1978. Ecophysiologie de la reproduction chez la crevette *Penaeus japonicus* trois annees d'experience en milieu controlle. Oceanologica Acta **1,** 135–150.

LAUBIER-BONICHON, A., and LAUBIER, L. 1979. Reproduction controlee chez la crevette *Penaeus japonicus*. *In* Advances in Aquaculture. T. V. R. Pillay and W. A. Dill (editors). Fishing News Books Ltd., Surrey, England.

LAWRENCE, A. L., AKAMINE, Y., MIDDLEDITCH, B. S., CHAMBERLAIN, G. W., and HUTCHINS, D. L. 1980. Maturation and reproduction of *Penaeus setiferus* in captivity. Proc. World Maric. Soc. **11,** 481–487.

LAWRENCE, A. L., CHAMBERLAIN, G. W., and HUTCHINS, D. L. 1981. Shrimp mariculture, an overview. Sea Grant College Program Tech. Rep. TAMU-SG-82-503. Texas A&M University College Station, Texas.

LAWRENCE, A. L., JOHNS, M. A., and GRIFFIN, W. L. 1983. Shrimp mariculture: positive aspects and state of the art. *In* Aquaculture in Dredged Material Containment Areas, J. Homziak and J. Lunz (editors). U.S. Army Corps of Engineer, Washington, D.C. 216 pp.

LAWRENCE, A. L., BRAY, W. A., and LESTER, L. J. 1984. Successful interspecific cross of two species of marine shrimp *Penaeus setiferus* and *P. stylirostris*. Science (in review).

LAWRENCE, A. L., MCVEY, J. P., JOHNS, M. A., and GRIFFIN, W. L. 1985. Production of two shrimp crops per year in south Texas using a single phase pond production system. Proc. 1st Int. Biennial Conf. Warm Water Aquaculture—Crustacea. 1, to be published.

LEWIS, D. H., and LAWRENCE, A. L. 1985. Immunoprophylaxis to *Vibrio* sp. in pond reared shrimp. Proc. 1st Int. Biennial Conf. Warm Water Aquaculture—Crustacea. 1, to be published.

LIAO, I. C., and HUANG, T. L. 1972. Experiments on the propagation and culture of prawns in Taiwan. *In* Coastal Aquaculture in the Indo-Pacific Region. T. V. R. Pillay (editor), Fishing News Books Ltd., Surrey, England.

LIAO, I. C., HUANG, T. L., and KATSUTANI, K. 1969a. A preliminary report on artificial propagation of *Penaeus monodon* Fabricius. Chinese-American Joint Council on Rural Reconstruction (JCRR) Fish. Ser. **8,** 67–71.

LIAO, I. C., TING, Y. Y., and KATSUTANI, K. 1969b. A preliminary report on artificial propagation of *Metapenaeus monoceros* (Fabricius). JCRR Fish. Ser. **8,** 72–76.

LUNZ, G. R. 1951. A saltwater fish pond. Contributions **12,** 1–12. Bears Bluff Laboratories, South Carolina.

LUNZ, G. R. 1956. Harvest from an experimental one-acre salt-water pond at Bears Bluff Laboratories, South Carolina. Prog. Fish Cult. **19,** 92–94.

LUNZ, G. R. 1958. Pond cultivation of shrimp in South Carolina. Gulf Carib. Fish. Inst. Univ. Miami Proc. **10,** 44–48.

LUNZ, G. R. 1967. Farming the salt marshes. Proc. Marsh Estuary Management Symp. **1,** 172–177.

LUNZ, G. R., and BEARDEN, C. M. 1963. Control of predacious fishes in shrimp farming in South Carolina. Contrib. No. 36. Bears Bluff Laboratory, South Carolina.

MAHLER, L. E., GROH, J. E., and HODGES, C. N. 1974. Controlled-environment aquaculture. Proc. World Maric. Soc. **5,** 379–384.

MISTAKIDIS, M. E. (editor). 1968. Proc. World Sci. Conf. Biol. Cult. Shrimps Prawns. FAO UN Fish. Rep. No., 57, **2,** 5.

MOCK, C. R. 1971. Larval culture of penaeid shrimp at the Galveston Biological Laboratory. Contrib. No. 344, National Marine Fisheries Service, Galveston Laboratory, Galveston.

MOCK, C. R., and MURPHY, M. A. 1970. Techniques for raising penaeid shrimp from egg to postlarvae. Proc. World Maric. Soc. **1,** 143–156.

MOCK, C. R., and NEAL, R. A. 1974. Penaeid shrimp hatchery systems. Proc. FAO CARPAS Symp. Aquacult., Montevideo, Uraguay, 1974.

NEW, M. 1976. A review of dietary studies with shrimp and prawns. Aquaculture **9,** 101–144.

NOAA. 1980. National Aquaculture Plan (Draft). National Oceanic and Atmospheric Administration, U.S. Department of Commerce, Washington, D.C.

OJEDA, J. W., ALDRICH, D. V., and STRAWN, K. 1980. Staggered stocking of *Penaeus stylirostris* in brackish water ponds receiving power plant cooling water. Proc. World Maric. Soc. **11,** 23–29.

PARDY, C. R., GRIFFIN, W. L., JOHNS, M. A., and LAWRENCE, A. L. 1983. Preliminary economic analysis of stocking strategies for penaeid shrimp culture. J. World Maric. Soc. **14,** 49–63.

PARKER, J. C., and HOLCOMB, JR., H. W. 1973. Growth and production of brown and white shrimp *(Penaeus aztecus* and *P. setiferus)* from experimental ponds in Brazoria and Orange counties, Texas. Proc. World Maric. Soc. **4,** 215–234.

PARKER, J. C., CONTE, F. S., MACGRATH, W. S., and MILLER, B. W. 1974. An intensive culture system for penaeid shrimp. Proc. World Maric. Soc. **5,** 65–79.

PERSYN, H. O. 1977. Artificial insemination of shrimp. United States Patent No. 4,031,855. June 28, 1977.

PRIMAVERA, J. H. 1978. Induced maturation and spawning in five-month-old *Penaeus monodon* Fabricius by eyestalk ablation. Aquaculture **13,** 355–359.

PRIMAVERA, J. H., and BORLONGAN, E. 1978. Ovarian rematuration of ablated sugpo prawn *Penaeus monodon* Fabricius. Ann. Biol. Anim. Biochim. Biophys. **18,** 1067–1072.

PRIMAVERA, J. H., BORLONGAN, E., and POSADAS, R. A. 1978. Mass production in concrete tanks of sugpo *Penaeus monodon* Fabricius spawners by eyestalk ablation. Fish. Res. J. Philipp. **3,** 1–12.

PRIMAVERA, J. H., LIM, C., and BORLONGAN, E. 1979. Feeding regimes in relation to reproduction and survival of ablated *Penaeus monodon.* Philipp. J. Biol. **8,** 227–235.

ROSE, C. D., HARRIS, A. H., and WILSON, B. 1975. Extensive culture of penaeid shrimp in Louisiana salt-marsh impoundments. Trans. Am. Fish. Soc. **104,** 246–307.

RUBRIGHT, J. S., HARRELL, J. L., HOLCOMB, H. W., and PARKER, J. C. 1981. Responses of planktonic and benthic communities to fertilizer and feed applications in shrimp mariculture ponds. J. World Maric. Soc. **12,** 281–299.

SALSER, B. R., and MOCK, C. R. 1974. Equipment used for the culture of larval penaeid shrimp at the National Marine Fisheries Service, Galveston Laboratory. Proc. 5th Congr. Nac. Ocenogr., Guaymas, Mexico.

SALSER, B., MAHLER, L., LIGHTNER, D., URE, J., DANALD, D., BRAND, C., STAMP, N., MOORE, D., and COLVIN, B. 1978. Controlled environment aquaculture of penaeids. *In* Drugs and Food from the Sea, P. N. Kaul and C. J. Sindermann (editors) University of Oklahoma, Norman, Oklahoma.

SANDIFER, P.A. 1981. Recent advances in the culture of crustaceans. Spec. Publ. European Maric. Soc. Elsevier, New York.

SANDIFER, P. A., LAWRENCE, A. L., HARRIS, S. G., CHAMBERLAIN, G. W., STOKES, A. D., and BRAY, W. A. 1984. Electrical stimulation of spermatophore expulsion in marine shrimp, *Penaeus* spp. Aquaculture (to be published).

SANTIAGO, JR., A. C. 1977. Successful spawning of cultured *Penaeus monodon* Fabricius after eyestalk ablation. Aquaculture **11,** 185–196.

SHIGUENO, K. 1968. Problems on shrimp culture (in Japanese). Suisan Zoshoku Sosho, **19.**

SHIGUENO, K. 1975. Shrimp culture in Japan. Japan Publication Trading Co., P.O. Box 787, White Plains, New York.

SIMON, C. M. 1981. Design and operation of a large-scale, commercial penaeid shrimp hatchery. J. World Maric. Soc. **12**(2), 322–334.

TABB, D. C., YANG, W. T., HIRONO, Y., and HEINEN, J. 1972. A manual for culture of pink shrimp *Penaeus duorarum,* from eggs to postlarvae suitable for stocking. Sea Grant Spec. Bull. No. 7, University of Miami, Miami, Florida.

TATUM, W. M., and TRIMBLE, W. C. 1978. Monoculture and polyculture pond studies with pompano *(Trachinotus carolinus)* and penaeid shrimp *(Penaeus aztecus, P. duorarum,* and *P. setiferus).* Proc. World Maric. Soc. **9,** 433–446.

TRIMBLE, W. C. 1980. Production trials for monoculture and polyculture of white shrimp *(Penaeus vannamei)* or blue shrimp *(P. stylirostris)* with Florida pompano *(Trachinotus carolinus)* in Alabama, 1978–79. Proc. World Maric. Soc. **11,** 44–59.

WHEELER, R. S. 1967. Experimental rearing of postlarvae brown shrimp to a marketable size in ponds. Commer. fish. Rev. **29,** 49–52.

WHEELER, R. S. 1968. Culture of penaeid shrimp in brackish-water ponds, 1966–67. Proc. 23rd Annu. Conf. Southeastern Assoc. Game Fish Commiss.

YANG, W. T. 1975. A manual for large-tank culture of penaeid shrimp to the postlarval stages. Sea Grant Tech. Bull. No. 31. University of Miami, Coral Gables, Florida.

4

Lobster Aquaculture

Louis R. D'Abramo,
Douglas E. Conklin

Introduction
General Biology—Selected Aspects
  Morphology and Anatomy
  Molting and Growth
  Feeding, Aggressive, and Mating Behavior
Culture
  Broodstock
  Hatching
  Larval Rearing
  Grow Out of Juveniles
Diets for Larger Juveniles
  Physical Characteristics
  Food Consumption and Digestion
  Nutritional Requirements
  Other Aspects
Disease
Site Selection and Design of Facilities
Marketing and Associated Economics
Future Prospects
Literature Cited

INTRODUCTION

Given the large number of lobster species (163 according to Phillips *et al.* 1980), it may seem surprising that only the two closely related species of *Homarus, H. americanus* and *H. gammarus,* are considered to be serious candidates for aquaculture. The explanation, however, begins to become apparent upon examining differences in the life cycle of the various lobster families. One of the most critical problems facing lobster culture is the protracted larval development phase of most species. Most female lobsters carry their eggs for only limited periods and the larvae undergo a lengthy (4–22 months) and complex series of metamorphic molts (life stages) before settling to the bottom as benthic juveniles. The larval, or phyllosoma, stages of the spiny lobsters (family Palinuridae), slipper lobsters (family Scyllaridae), and coral lobsters (family Synaxidae) are all characteristically delicate, dorsoventrally flattened, transparent, and have numerous hairs, or setae, on the appendages (Phillips and Sastry 1980). Although these larvae are well suited for a pelagic existence in the open ocean, they present difficult and as yet insurmountable problems of time and labor for the culturist interested in producing large number of postlarvae. An additional culture burden is the apparent requirement for several different varieties of food in order to satisfy the size and textural requirements of the phyllosoma larvae. Phyllosoma larva from a number of species have been cultured all the way through the larval period; however these painstaking labors have been limited to investigators whose interests resided in taxonomic identification. Rudloe (1983) has successfully cultured some adults and juveniles of the slipper lobster *Scyllarides nodifer* through two molt cycles, and proposed particular conditions for commercial culture to be practical. Larval culture of these species has been unsuccessful, thereby causing an undesirable dependence upon the natural environment for postlarvae.

The family Nephropidae (recognized by the large claws, or chelae, on the most anterior set of legs) is unique among lobsters because females carry their eggs for relatively long periods. In addition, the portion of time spent as pelagic larvae, completing the transition to a benthic existence and to the morphology typical of the adult, is comparatively short. Larval homarid lobsters, which readily consume easily available feeds, take less than 2 weeks to complete the pelagic portion of the life cycle when reared at 20°C (68°F). Biologists taking advantage of this ease of larval rearing have carried out limited culture of homarid lobsters for approximately 100 years. These efforts, undertaken for a variety of purposes, have provided an extensive background of information that further focuses the attention of modern aquaculturists on the two species of *Homarus*.

In addition to the two species of *Homarus,* the family Nephropidae includes the Norway lobster, *Nephrops norvegicus,* as well as some lesser known genera. Unfortunately, information on the biology of these other species is quite limited, and their potential can not be presently evaluated. Some preliminary studies with *Nephrops norvegicus* (Figueiredo and Vileda 1972) were disappointing in that growth rates appeared slow; however, the culture conditions in these studies may have been less than ideal.

Given the paucity of information available about species of the family Nephropidae other than *Homarus* and the seemingly complex problems encountered in the culture of phyllosoma larvae, only the two species of *Homarus* can presently be considered as candidates for commercial culture. Consequently, the rest of this chapter deals exclusively with the aquaculture of *H. americanus* and *H. gammarus* (Fig. 4.1). For a comprehensive review of lobsters the reader is encouraged to read the recent two-volume work *The Biology and Management of Lobsters* (Cobb and Phillips 1980).

While the idea of rearing *Homarus* for profit is not new, a concerted effort to establish commercial lobster aquaculture was not initiated until the early 1970s. The research focus on the culture requirements of the American lobster *H. americanus* and its European counterpart *H. gammarus* was initially stimulated by the claim that the time required for *H. americanus* to grow to market size (1 lb) could be reduced approximately 60% (from 6 to 2 yr) by maintaining the culture temperature at 20°C or (68°F) (Hughes *et al.* 1972). A further impetus was the growing realization that the clawed lobster fishery could decline dramatically as a consequence of intense harvesting efforts in

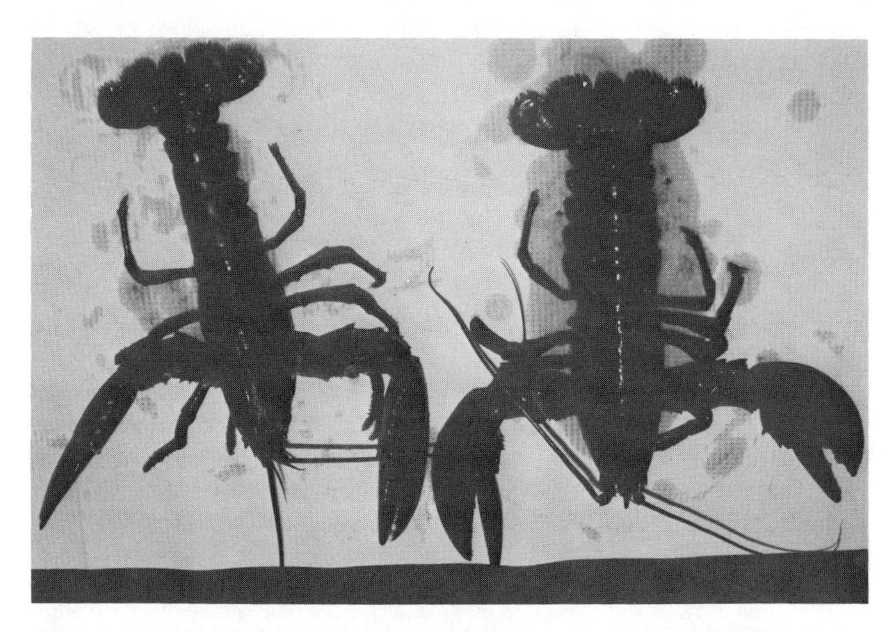

FIG. 4.1 Adult *Homarus americanus* (right) and *H. gammarus* (left). Note specifically the tapered abdomen and smaller claws that are characteristic of the European *H. gammarus*. Photo is very dark.

response to an unabating consumer demand (Dow 1980). Early studies to define culture parameters and to develop economically viable rearing techniques were conducted primarily by researchers in the United States with the support of the National Sea Grant Program,* as well as by several Canadian investigators. These loosely coordinated research efforts collected preliminary experimental data that were used, together with various assumptions, to develop a complex mathematical model to examine the impact of various parameters on the economics of lobster culture (Allen and Johnston 1976; see Johnston and Botsford 1980 for a recent updating to account for inflation.) Surprisingly, even though the assumed interrelationships among the various parameters of the model have never been thoroughly examined, the conclusion that economic success requires extensive production (on the order of 1 million marketable lobsters per year) and highly mechanized facilities is tacitly accepted. Confidence in these conclusions will only arise when a more exhaustive data base becomes available.

While intensive lobster acquaculture will undoubtedly be highly capital intensive, it should be remembered that no commercial production facilities exist and the actual costs involved in culturing significant numbers of homarid lobsters remain unknown. Nevertheless, techniques for rearing homarid lobsters in the laboratory through their complete life cycle with relative ease are now available (McVey 1983). Thus, even though the economic success of lobster aquaculture has yet to be demonstrated, interest in commercial enterprises remains high. Efforts to move from the research to the commercial phase will undoubtedly continue, and an economically viable lobster aquaculture industry based on existing and future technical information is an eventual certainty. In this chapter we will review the information available on the culture of these decapod crustaceans and speculate on possible adaptations and innovations that may be useful in the ultimate development of profitable commercial production units.

GENERAL BIOLOGY—SELECTED ASPECTS

If cultivation and ultimately domestication are to be successful, the physiological needs of the animal of interest must be understood either factually or intuitively. As most individuals are not very familiar with

*The National Sea Grant Program was established in the mid-1960s to accelerate the sound development of marine resources. This program can be compared with the Land Grant Program established in 1865 to stimulate agricultural development in the United States.

lobsters, a brief review of their biology is appropriate. In particular we will address those aspects of lobster biology that are vital to understanding the problems associated with culture.

Morphology and Anatomy

The general body configuration of *Homarus* (see Fig. 4.2) is well known. Characteristic of the phylum Arthropoda, its body and appendages are made of jointed and easily distinguished segments. The embryologically distinct body segments of the cephalic and thoracic region are fused together in the developed animal to form a cephalothorax, which is covered by a curved shieldlike carapace. Arising from most of the body segments are paried jointed appendages, which have been modified from front to rear to serve sensory, alimentary, ambulatory, sexual, and natatory functions.

At the anterior end of the cephalothorax are the first antennae and the long whiplike second antennae. The movable and stalked eyes are not true appendages, as they have a completely different embryological origin. Located around the mouth and not shown in Fig. 4.2 are the laterally opposed mandibles, or jaws, as well as the accessory mouth parts, the first and second maxillae (Fig. 4.3). From the original eight thoracic segments arise the first, second, and third maxillipeds, which are also located near the mouth and used in feeding, and the five pairs of pereiopods, or walking legs. In *Homarus* the first three pairs of pereiopods are chelate (clawed); the first pair is further modified into characteristically large claws. The last six segments and the terminal telson (not a true segment) comprise the abdomen and tail. The

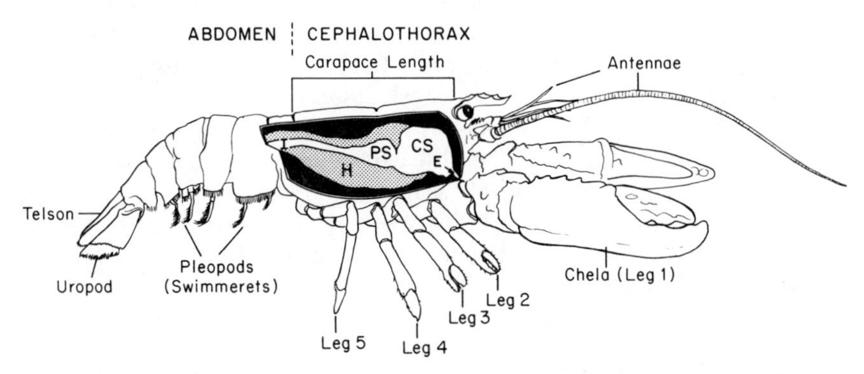

FIG. 4.2 Lateral view of a homarid lobster illustrating selected features of the internal and external anatomies. E = esophagus; CS = cardiac stomach; PS = pyloric stomach; H = hepatopancreas.

FIG. 4.3 Frontal view of *H. americanus.* Note various feeding appendages and mouth parts.

first five segments have paired pleopods (swimmerets). In the male the first pair of pleopods is structurally modified for transfer of sperm. Brooding females employ pleopod movement to maintain water quality around their eggs, which are ventrally attached on the abdomen.

The entire body of arthropods is covered by a chitinous shell (exoskeleton); in the case of the lobster, it is also heavily mineralized with calcium carbonate. This exoskeleton serves for the attachment of muscles and isolates the animal's internal tissues from direct exposure to the environment. Of course, the innate rigidity of the exoskeleton dictates that it must be periodically shed or molted for tissue growth to continue. When the old exoskeleton is molted, large amounts of water are engulfed by the lobster and upon being absorbed are used to expand internal volume. The new exoskeleton hardens around this increased volume and tissue growth commences. While the molting process creates the illusion that growth is of a discontinuous or incremental nature, in reality tissue growth occurs throughout much of the period between molts.

Molting and Growth

For reference, the molt cycle is divided into five stages that reflect different types of physiological activity. Following ecdysis (the actual shedding of the exoskeleton), lobsters are soft shelled and extensive water absorption and redistribution occur; this period is called stage A. Stage B begins after the initial hardening of the shell and is characterized by the thickening of the exoskeleton. The most immediate postmolt stages, A and B, generally represent 5% or less of the complete cycle. In stage C, or intermolt, the lobster is completely functional and actively forages for food. During intermolt, which comprises approximately 50% of the molt cycle, tissue growth occurs. The remaining part of the cycle before ecdysis (stage D) is termed premolt, and during this period all the complex metabolic activities necessary for ecdysis occur. Stage E, ecdysis, is quite rapid, lasting only 15–30 min out of the approximate 1-year period between molts in an *Homarus* adult.

The molt cycle is thought to be controlled by two opposing hormonal factors. Presently, there is uncertainty regarding whether each of these opposing factors is composed of one or several hormones. One factor, called the molt-inhibiting hormone (MIH, thought to be a peptide), is synthesized within a glandular complex located at the base of the eyestalk and is believed to suppress the secretion or activity of the molting hormone (presumably β-ecdysone, a derivative of cholesterol). Ecdysone is synthesized by the Y-gland located in the anterior region of the cephalothorax. Ecdysone is not stored but released into the hemolymph and hydroxylated to form 20-hydroxyecdysone (β-ecdysone) in the target tissues. Removal of the eyestalks causes molting to occur more frequently than normal.

The growth rate is determined by the increase in size at each molt and the time intrerval between molts. Linear size in lobsters is usually measured as carapace length (CL), which is shown in Fig. 4.2) As the lobster ages, both the frequency of molting and the percentage increase in size with each molt decrease. In the natural environment a variety of environmental factors can influence growth rate but the most significant is temperature.

Once lobsters have molted through the fourth stage (first postlarval stage) and assume the morphology of the adult, they then seek a benthic existence. An ideal substrate is apparently a sandy bottom with overlying rocks. The small juveniles engineer deep burrows, 40–70 cm (16–28 in.), into the substrate, and apparently stay in these burrows feeding on the encrusting invertebrates until they grow to a size of about 40 mm (1¾ in.) CL (Cooper and Uzmann 1980). Larger lobsters spend more time foraging for food outside of the burrows. Although the sand and rock substrate is a common habitat, homarid lobsters are highly adaptive and can be found in a variety of habitats that provide darkened shelter. As individuals grow larger, new shelters are developed or found; generally these quarters are appropriate to the lobster's size, so that physical contact can be maintained with the walls and roof. Although lobsters are generally solitary, multiple occupancy of shelters is observed during the winter and in areas with a scarcity of appropriate shelters.

Feeding, Aggressive, and Mating Behavior

The foraging for food by larger lobsters appears to be nocturnal, although this behavior is influenced by other factors, such as cloud cover, that reduce light intensity. Lobsters are most active in the summer months.

Lobsters in nature prey on a variety of bottom invertebrates such as crabs, polychaetes, mussels, periwinkles, sea urchins, and starfish. Other sessile invertebrates, fish, and seaweeds make up a relatively small portion of the diet. Consumption of fish flesh may be associated with its availability in lobster traps and ingestion of plant material may be incidental to the consumption of small crabs and other invertebrates. The frequency of various items in the diet generally reflects their abundance in the habitat, but lobsters may show preference for calcium-rich foods at the time of molting (Ennis 1973). Lobsters may exhibit other preferences, such as a 5 : 1 dietary preference of crabs over sea urchins (Evans and Mann 1977), but the relative abundance of prey items in the habitat seems to remain the dominant factor determining dietary intake (Hirtle and Mann 1978). The predominant prey of immature lobsters are sea urchins, mussels, rock crabs, polychaetes, and brittlestars (Carter and Steele 1982).

Food is torn apart by the joint action of the mandibles and maxillipeds and passed into the esophagus (see Fig. 4.2) with the aid of other mouth parts. Food is passed through the esophagus, with the aid of secreted mucopolysaccharides, into a bulbous cardiac stomach. At the posterior end of the cardiac stomach, the food in triturated by a gastric mill composed of three chitinous teeth, one positioned dorsally, the others laterally. The gastric mill activity is aided by a secretion of digestive enzymes that originate in the hepatopancreas (digestive gland). After this preparatory treatment, the food, now in small particles, is directed to the pyloric stomach. The pyloric stomach passes suitably sized particles to the hepatopancreas where the nutrients are absorbed. Oversized particles are rerouted to the gastric mill. Undigested material is passed through the mid gut and the hind gut to the anus.

A large number of studies in recent years have focused on both the feeding and the aggressive behavior of homarid lobsters. Although this behavior is intrinsically interesting and an important subject area in the general understanding of the fishery, there was also the early hope that behavior studies would provide insights enabling homarid lobsters to be communally cultivated. Communal culture at densities necessary for profitable commercial enterprises, however, appears impossible because cannibalism, a phenomenon unobserved at natural densities, is prevalent. Molting animals release chemicals that stimulate the feeding behavior of other lobsters. At the unnatural densities of intensive culture this invariably leads to cannibalism on the molted lobsters until densities are reduced.

At natural densities, the strong aggressive behavior used in maintaining possession of the burrow generally does not result in injury or death. According to Atema and Cobb (1980), the outcome of aggressive encounters to determine dominance and shelter possession between two lobsters is dependent on their relative size. If the size differential is not too great, other factors such as molt stage and experience may also be involved.

The normal solitary existence of homarid lobsters and the behavioral patterns that maintain it are altered for mating. A sexually mature female of 500 g (1.1 lb) will seek out the shelter of a large male about a week before mating. Presumably in response to a chemical secreted by the female (Atema and Engstrom 1971), the normal aggressive behavior of the male is tempered and pair formation and cohabitation are established. Soon after the female molts, mating takes place. Fortunately for the culturist, pair formation is not necessary for mating in the laboratory. Males of appropriate size, when placed in the same holding system, are generally willing to copulate with a freshly molted female.

CULTURE

Brood Stock

In the past, brood stock for lobster research was commonly obtained from wild "berried" (egg-bearing) females which are relatively common in the catch during the summer (Fig. 4.4). Because berried females are typically protected by fishing regulations, these lobsters were obtained under special collecting permits for the purposes of research. Such a source, with its intrinsic seasonality, would not support the needs of commercial interests even if existing regulations were changed to allow the harvest of berried females for commercial culture.

A previously used and alternative source of seed available to the

FIG. 4.4 Egg-bearing female homarid lobster.

commercial culturist is wild mated female lobsters that have yet to extrude their eggs. This source takes advantage of the phenomenon that mating and egg fertilization are not coincidental events in *Homarus*. Although mating and the transfer of a sperm packet, the spermatophore, occur immediately following the molt of a reproductively mature female, egg extrusion and fertilization of the eggs by sperm stored in the spermatophore take place months later. Thus, seed stock from the wild can be obtained not only from berried females but also from molted females, which can be identified by their somewhat leathery and clean, shiny (free of epibiotic growth) shell. Sixty to 80% of such females collected in August and September can be expected to extrude eggs over the following 7 months (Schuur *et al.* 1976). Nevertheless, use of wild-caught females as a seed supply is dangerous because of the potential introduction of disease into the brood-stock culture system (see Disease section).

Assurance of a continuous and predictable source of egg-bearing female lobsters for domestication requires complete control of the reproductive cycle. In nature, both *H. americanus* and *H. gammarus* exhibit a biennial reproductive cycle in which a barren year alternates with a fertile one. This biennial cycle is probably a consequence of temperature, light, growth, and reproductive interactions. In the laboratory, a number of lobsters, particularly *H. gammarus,* have been observed to regularly extrude eggs on an annual basis (D. Hedgecock, personal communication).

The reproductive maturity of a female lobster is often judged by the ratio of the maximum outside width of the second abdominal segment to the carapace length. A female is judged to be reproductively mature when this ratio exceeds a particular value that is often dependent upon geographical locale. A value of ≥ 0.60 has been generally acknowledged as a reliable indicator of reproductive maturity (see Aiken and Waddy 1980 for a detailed discussion).

In our laboratory, mature males and females are cultured in semi-recirculating systems equipped with UV treatment, sand filtration, and make-up flow rates approaching a turnover rate of twice per day. These systems are composed of tiered rack-mounted, injection-molded plastic troughs, $36 \times 152 \times 26$cm ($14 \times 61 \times 10$ in.). These troughs can be subdivided into as many as 10 compartments by perforated PVC partitions for the housing of lobsters (Figs. 4.5 and 4.6).

Controlled matings of lobsters in the laboratory are easy to perform and generally successful. Laboratory matings are commonly carried

FIG. 4.5 Tiered holding systems composed of PVC troughs for the culture of homarid lobsters.

FIG. 4.6 Trough with perforated PVC sections for the individual culture of late juvenile and adult homarid lobsters.

out by placing a reproductively mature male whose claws have been banded into a holding system occupied by a recently molted female and her shed exoskeleton. The holding system should be large enough to accommodate satisfactory movement. Typically, the animals are placed together in the morning and are then separated in the late afternoon. During the time they are in the same tank, the transfer of a spermatophore from the male to the female is assumed to take place. The opening of the seminal receptacle of the female can be examined for the presence of a sperm "plug"; however, this is not always an accurate indicator of whether insemination has occurred.

Recent work carried out by Dr. P. Talbot and coworkers at the University of California, Riverside, makes it probable that commercial facilities will utilize artificial insemination of females. These workers have found that spermatophore extrusion can be accomplished by electrically stimulating the male lobster (Kooda-Cisco and Talbot 1983) and that the resulting spermatophore can be used to artificially inseminate the female (P. Talbot personal communication). Some of the eggs extruded by artificially inseminated females have been carried to the eyespot stage, indicating that the eggs were normally fertilized. This technique has been successfully used with shrimp (Sandifer and Lynn 1981) and should be suitable for lobsters. Artificial insemination, if successful, would remove the occasional doubt as to the occurrence of spermatophore transfer and male fertility and may also lead to the development of lobster sperm banks to facilitate breeding programs.

After mating is accomplished, the wait for egg extrusion begins. Recently Hedgecock (1983) has suggested that the process of egg extrusion is controlled principally by photoperiod. A change in the photoperiod from a short-daylength to long-daylength cycle (16 hr of light, 8 hr of darkness) is necessary for egg extrusion. There is a high correlation between the date of the switch to long daylengths and the date of egg extrusion. Manipulation of photoperiod may permit the eventual control and timing of a reproduction cycle event in such a way that it will be predictable. Hedgecock has proposed that with proper manipulation of photoperiod and temperature, many females can be moved through their entire reproductive cycle on approximately an annual cycle. In its present simplified form, as diagrammed in Fig. 4.7A, the scheme consists of a sequence of three differing sets of environmental conditions. Females are placed on a

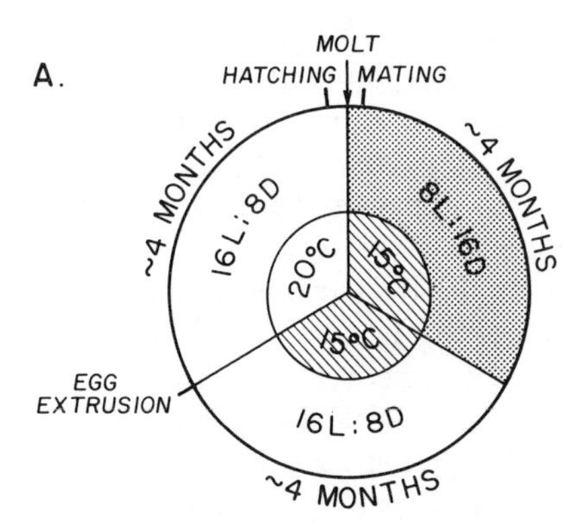

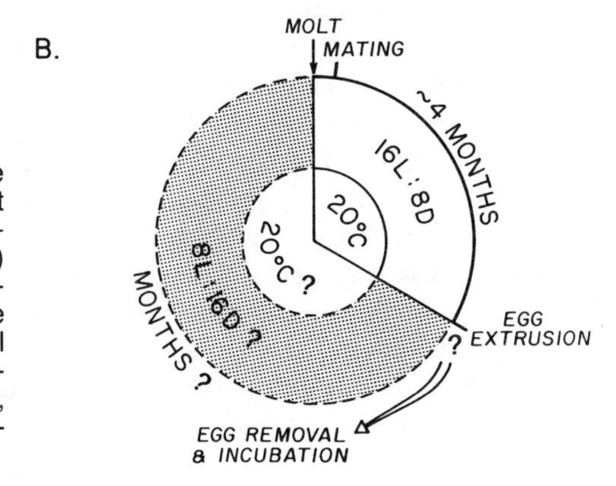

FIG. 4.7 (A) Simplified scheme of brood-stock management based upon an annual molt cycle. (B) Possible modifications (?) that would maximize the production of eggs by a single female lobster. Efficient management will ultimately depend upon a delicate regulation of temperature, molting, and photoperiod to insure induction of vitellogenesis.

short-day cycle in cold water at 10°–15°C (50°–59°F) after mating. There is some indication that after mating transfer to this short-day cycle should be accomplished before the onset of the C stage of the molt cycle (Aiken 1980). Without transfer, the probability of extrusion is significantly reduced. After 4 months, the photoperiod is then changed so the females are exposed to a long-day photoperiod (16 hr of light, 8 hr of dark) with the water temperature remaining at 10°–15°C (50°–59°F). At the end of the second 4-month period, the females extrude their eggs and are then moved into 20°C (68°F) water with long daylengths continued; the eggs hatch in this final part of the annual cycle.

Nelson *et al.* (1983a) have attempted to define further the subtle interrelationships existing between vitellogenesis and molting in *H. americanus* first described by Aiken and Waddy (1976). Active vitellogenesis is stimulated by the onset of long daylengths. Events associated with premolt (stage D) that occur beyond a critical period after the onset of long daylengths seem to interfere with active vitellogenesis, resulting in a delay beyond the expected date or even complete postponement of egg extrusion. Once extrusion occurs, the presence of eggs on the pleopods normally inhibits the molting process from proceeding to the premolt state (stage D). After a period of 120 days on long daylengths, molts are no longer followed by extrusions unless the animal experiences a period of short days to renew response to long daylengths.

Because of the various interactive factors that have recently been recognized, the proposed annual cycle method of brood-stock management would appear to be quite limited in its application. For a commercial facility with large numbers of reproductive adults, brood-stock management must be computerized. Because of the interactions among the processes of growth, ovarian maturation, extrusion, and egg brooding a significant amount of inter- and intrafemale variation over time will exist. Inevitably, a well-designed computer program will be required so that data can be readily entered and retrieved and proper management decisions can be duly executed.

Previous control of the reproductive cycle in lobsters was limited to general temperature manipulation. Cooler culture temperatures were used to retard egg development of selected berried females and thus spread out the cycle so that some egg production would take place throughout most of the year. Although this approach has been moderately successful, the time of egg extrusion by any one particular female remained unpredictable. Presumably much of the variation was due to the unrealized prolonged long-day photoperiod exposure, which was normally used in the laboratory. As discussed earlier, if exposure to long daylengths exceeds a 120-day period, exposure to short daylengths is required to reset the long-daylength responder that activates vitellogenesis. The more reliable control afforded through an understanding of the photoperiod effect and of the subtle antagonisms associated with growth and reproductive physiology will have a significant impact on the economics of commercial lobster aquaculture. This knowledge not only assists in freeing the industry from a continuous dependency on wild-caught females, but also permits the effective utilization of facilities for year-round production. The strict dependence of egg development on temperature adds an additional dimension to the control of seed production.

While a combination of the two techniques of photoperiod manipulation to control egg extrusion and temperature manipulation to control the rate of egg development offers the immediate promise of successful control and timing of the reproductive cycle in a commercial facility, further advances are anticipated. Since small adult females can complete a molt cycle in less than 12 months, the potential for speeding up the reproductive cycle is present. Further work with photoperiod manipulation may lead to the reduction of the length of the short-day period or use of the short-day photoperiod in the final third phase at 20°C (68°F) while egg development is occurring (Fig. 4.7B). There is some indication that a minimum period of approximately 3–4 months of short days is required before the change to the long-day photoperiod becomes effective in stimulating egg extrusion. Whether or not the period following egg extrusion but before the molt can be used for this short-day photoperiod time-setting factor is unknown. Moreover, culture of broodstock at 20°C exclusively may not be feasible if, for example, normal ovarian development does not proceed without a specific period of exposure to lower temperatures.

The reproductive cycle could also be compressed if eggs could be removed from the female so that, following egg extrusion, the female could rapidly proceed with the next molt cycle under the influence of increased water temperature (see Fig. 4.7B). A precedent for the successful removal of eggs exists. In the late 1800s, extensive hatcheries were developed, aimed at enhancing the lobster fisheries by planting fourth-stage animals. These hatcheries removed the eggs from the female (Galtsoff 1973), indicating that it is definitely possible to carry lobster eggs through to hatching removed from the female. However, the proper timing of egg removal and culture conditions approaching antisepsis are critical to good survival. The removal of eggs may be advantageous since the egg-stripped female could spend a portion of the short-day exposure period at 20°C (68°F). An adaption of the egg-removal procedure for commercial culture should be evaluated in the future.

The fertility of males used for insemination purposes also is integral to successful brood-stock culture. The spermatophore-extrusion technique was recently used to assess the fertility of males maintained at the Bodega Marine Laboratory and Aquaculture Enterprises at Oxnard and Monterey, California. Laboratory-reared males *(H. gammarus* and *H. americanus)* consistently extruded good to excellent spermatophores in the majority of the trials (Talbot *et al.* 1983). Males may exhibit some seasonality in their reproductive cycle, and this may explain observations that indicate some males to be consistently infertile.

Proper nutrition of brood-stock females will be important in assuring their fertility and fecundity in commercial culture. During vitellogenesis, significant amounts of macronutrients, particularly lipids, are transported to the eggs. A special brooding ration may have to be administered to females that have been inseminated and are to be exposed to the photoperiodic regime that induces vitellogenesis and eventual oviposition. It can be expected that further refinements in control of the female reproductive cycle will occur and other associated problems involving fertility and fecundity will be solved. Thus it is anticipated that steady year-round production of larvae will not be a problem in lobster aquaculture.

In the laboratory, growth rates of juveniles have been shown to be progeny dependent. As a result, once a predictable scheme for the control of reproduction and seed supply has been developed, proper efforts in hybridization and selection should eventually yield progeny with such desirable culture characteristics as large larval size, rapid growth, and enhanced disease resistance. Animals with these traits will probably be marked through some type of genetic fingerprinting based upon the presence or absence of particular alleles. At the Bodega Marine Laboratory of the University of California, hybrid progeny have recently been collected from crosses between *H. americanus* and *H. gammarus*. Backcrosses of progeny (F_1) to the pure species may have limited application; for example, hybrid males *(H. gammarus* × *H. americanus)* have extruded morphologically normal spermatophores containing no sperm, indicating infertility (Talbot *et al.* 1983). Additional sampling of male hybrids should be undertaken, as sample size is currently small and some individuals may successfully produce sperm (P. Talbot, personal communication). Female hybrids have been observed to extrude eggs, but information regarding hatching of eggs from hybrids is presently unavailable (D. Hedgecock, personal communication).

Hatching

The developmental stage of eggs of *H. americanus* can be determined by following the development of the eye spot. Perkins (1972) has suggested that an index based on the diameter of the eye spot is a reliable predictor of hatching time. A similar index has been suggested for *H. gammarus* (Richards and Wickins 1979). When the eyespot index suggests that actual hatching is approaching, the females are moved into a special larval-hatching container. This is a modified adult-rearing trough, described previously, to which has been added a small catch basket (Fig. 4.8). Water flow through this hatching trough is such that all water must pass through the catch basket before exiting, thus skimming the newly hatched larvae off the surface. Typically, lobsters hatch out their young in batches during the evening hours over a period of 3–4 days or longer. The larvae collected in the hatching basket are transferred to a specially designed "plankton kriesel" (see Larval Rearing section) where their subsequent larval development occurs.

Larval Rearing

As noted previously, the portion of time spent as pelagic larvae is short in *Homarus* compared with most other decapod crustaceans of

FIG. 4.8 Brooding trough for egg-bearing female. At left is catch basket for the collection of larvae after the eggs have hatched.

potential aquaculture importance. Homarid lobsters undergo three larval molts before taking on the typical morphology of the adult at the fourth stage. At 20°C (68°F), the most common rearing temperature, larval development is completed in less than 2 weeks. During this period, the larvae are generally reared in a kriesel-type system (Fig. 4.9) that is designed to maintain an homogeneous distribution of animals by means of a spiral upwelling flow pattern (Hughes *et al.* 1974); other simpler systems (Stewart and Castell 1979) have also been used with success. The critical factors are the provision of adequate food and a flow pattern of constant turbulence, which minimizes larval interactions and subsequent cannibalism.

A wide variety of diets (e.g., live or frozen zooplankton, chopped fish and molluscs, and artificial preparations) can be used as feed for lobster larvae. Live adult *Artemia* is a preferred choice of food, as it is both highly nutritious and relatively nonfouling. In the kreisel system, a ratio of at least four adult brine shrimp per lobster larva should be maintained (Carlberg and Van Olst 1976). When live brine shrimp

FIG. 4.9 Plankton kriesel for the culture of homarid lobster larvae.

are used with kriesel systems, an excess is typically fed once or twice a day so that a 4:1 ratio of shrimp:larva exists at the next feeding. In order to reduce the water fouling that is usually associated with nonliving diets, an automatic feeder (Serfling *et al.* 1974) should be used to deliver small amounts of feed at periodic intervals.

Larval lobsters are cannibalistic and will readily feed on each other if food is limiting or the culture systems are overcrowded. Large temperature fluctuations are deleterious to the general health of developing larvae and should be avoided. Larvae are particularly sensitive to ammonia, and ammonia concentrations should be kept below 0.2 mg/liter. Alkaline conditions will exacerbate ammonia toxicity (Delistraty *et al.* 1977). In a 40-liter (10.6-gal) kriesel system, maximum survival of 70% is achieved with an initial starting density of approximately 1500 larvae per kriesel.

Aiken *et al.* (1981) have suggested that short daylengths are optimal for the rearing of lobster larvae. Larvae exposed to 1 hr of light daily had significantly greater survival to fourth stage than those exposed to 12 or 23 hr of light daily. The time required to reach fourth stage is also protracted by long daylengths. Recent observations suggest that survival responses to different photoperiodic regimes vary by season. Larvae appear to be very sensitive to photoperiod and maintenance of the same photoperiod regime during both embryonic and larval development would appear to be advisable.

Grow Out of Juveniles

Water Quality. A successful commercial operation involving the culture of *Homarus* will be heavily dependent upon consistently good water quality. Disease problems can be minimized and growth maximized if proper attention is paid to such factors as salinity, ammonia concentration, and dissolved oxygen concentration. Salinity should be maintained at 30–33‰. Ammonia is the primary waste product of *Homarus* and concentrations in culture water should not exceed 1.5 mg/liter. Dissolved oxygen concentrations should be between 6 and 7 mg/liter. Water used for culture purposes contains small amounts of particulate organic material, which through bacterial activity contribute to the accumulation of nitrogenous wastes and the depletion of dissolved oxygen. Obviously, flow-through rather than closed systems would be less prone to the problems associated with ammonia toxicity and oxygen depletion. Precautions should also be taken to use water that is not susceptible to contamination activities, thereby creating the potential of transfer of human pathogens or toxicants to the lobsters.

Early Juvenile Culture. Fourth-stage larvae, although not exhibiting a truly benthic behavior until after molting into the fifth stage, are typically moved to grow-out systems that contain individual-rearing containers (Conklin *et al.* 1981). Communal-rearing systems containing high densities of juveniles have been tried (Carlberg *et al.* 1979), but use of these beyond one or two molts is rapidly limited because of cannabilism, which still occurs in spite of the provision of multiple shelters. Another problem is that lobsters cultivated in communal-rearing systems for any extensive time are extremely heterogeneous in size. As a result, the use of communal rearing in commercial operations is expected to be limited. However, the typical individual-rearing techniques used in experimental facilities are also considered inappropriate for use in a commercial setting. Provision of artificial diets of a ration size that would be entirely consumed by early juveniles on a daily basis is extremely difficult and extensive cleaning of the uneaten food is required. This situation would result in excessive labor and diet costs in commercial operations. These considerations point to the attractiveness of live brine shrimp as a diet for small juvenile lobsters.

Although chopped mollusks (mussels, squid) are often used for laboratory culture, live adult brine shrimp are a labor-minimizing, as well as nutritionally superior diet for lobsters up to 120 days old. Based upon carapace length (CL), growth rates of 0.1 mm/day have been observed in juvenile lobsters fed brine shrimp. Other appealing characteristics of brine shrimp include its motility and relative lack of fouling. Special rearing systems for small (<20-mm CL) lobster juveniles could be designed to take advantage of these positive characteristics of *Artemia*. Individual perforated containers that retain and separate the juvenile lobsters from each other should be used; perforations in the walls would allow the movement of brine shrimp throughout the entire unit. In such containers, juvenile lobsters could be cultured through the first 4 months of life with minimal labor input. The perforated containers could be suspended in some manner above the floor of the rearing unit so that dead, uneaten brine shrimp which settled to the bottom could be flushed out without disturbing the juveniles.

If, as suggested here, live adult brine shrimp are used extensively in commercial lobster aquaculture, then large-scale brine shrimp production units will undoubtedly be required (see Appendix). Important considerations in such production systems are the food costs and nutritional quality of *Artemia*. Although work done on the cultivation of brine shrimp using inexpensive nonliving food sources appears promising (Sorgeloos 1980), the nutritional value of these brine shrimp has yet to be adequately assessed. D'Agostino (1980) points out that nonliving particles serving as nucleii for extensive bacterial growth may not be as effective a food as algae for brine shrimp culture. However, unless toxic conditions invariably develop from these types of cultures, nutrient deficiencies could probably be overcome by providing a mixture of algal species with the nonliving foods toward the end of the culturing period.

Larger Juvenile Culture. As juvenile lobsters mature, live brine shrimp no longer support maximum growth rates, presumably because the relatively smaller size of the prey becomes limiting in terms of energy derived from food intake versus energy expended for capture. Larger juveniles can be fed on chopped natural food items such as molluscs, crustaceans, and fish. A number of problems associated with these foods—including the difficulty of procurement, extensive storage needs, seasonal and locational vagaries in chemical composition, disease introduction, and labor intensive preparation—makes them less attractive than artificial feeds for use in commercial facilities. Systems that presently exist for large-scale laboratory culture (Fig. 4.10) will require significant modification for commercial production (see Site Selection and Design of Facilities section).

Associated Growth Factors. To decrease the molt cycle interval and thereby significantly increase growth rates, eyestalk ablation of juvenile lobsters has been proposed (Castell *et al.* 1977). The success of such a procedure would reduce overall labor and energy costs required for culture to market size. Eyestalk ablation suppresses the effect of the molt-inhibiting hormone, so that normal growth rates are increased threefold. Subsequent to eyestalk ablation, rapid increases of hemolymph ecdysteroids in lobsters have been observed through the use of radioimmunoassay techniques (Chang and Bruce 1980). The increases resemble those that occur in normal animals that are approaching premolt.

Unfortunately, considerable mortality has been associated with individuals that have been ablated, usually after they have completed several molts. The abrupt physiological changes and accelerated growth elicited by eyestalk ablation may lead to unique qualitative and quantitative nutritional demands. Further research is needed to assess this possibility accurately. Certainly, as suggested by Castell *et al.* (1976), ablated animals would be excellent candidates for a rapid and exacting determination of the nutritive value of particular diets.

There have been several reports detailing the effects of limited space on the growth of *Homarus* (Shleser 1974; Stewart and Squires 1968).

FIG. 4.10 Perforated compartmentalized PVC trays for the culture of larger juvenile homarid lobsters.

To avoid significant growth inhibition and mortality, Van Olst and Carlberg (1978) and Aiken and Waddy (1978) suggest that an area equivalent to 75 times the carapace length squared (CL^2) is required for juvenile lobsters cultured individually. This hypothesis needs to be rigorously reexamined because previous results may have been confounded by a possibly short-lived, density-dependent, chemical growth-inhibiting factor that has recently been observed by Nelson *et al.* (1980, 1983b,c). Undoubtedly, as lobsters are cultured to market size, they need to be transferred to larger units to accommodate molting and some movement for normal feeding activities. However, current estimates of the optimal relationship between animal size and culture area appear questionable. The total amount of culture space required could be reduced if more attention were directed toward a design that would assure excellent water quality via optimal volume and flow rate characteristics.

DIETS FOR LARGER JUVENILES

Nutritional research with *Homarus* has focused on the definition of nutrient requirements and the development of a suitable artificial ration for the late grow-out phase. The chemical and physical aspects of an artificial diet suitable for commercial lobster aquaculture must optimize the availability, delivery, and metabolic use of nutrients at a reasonable cost.

Analysis of the chemical composition of the natural foods of *Homarus* can only provide baseline information regarding nutritional requirements. A purified diet (chemically defined) is necessary to precisely define both qualitative and quantitative nutrient requirements. Although the formulation of artificial diets for lobsters is still in its infancy, the development of a purified diet that supports good growth (0.05 mm/day) and high survival of lobsters (Conklin *et al.* 1980) represents significant progress (see Table 4.1 for a more recent modification of this diet). This information will serve as a foundation for the formulation of commercial diets that provide optimum nutrient levels and are composed of a mixture of inexpensive, readily available feedstuffs. These rations would be pelletized or adapted for automated feeding and ideally could be stored at room temperature.

Physical Characteristics

Physical characteristics of the diet such as size, shape, and moisture content and integrity are important considerations, which have

TABLE 4.1. CURRENT PURIFIED DIET BEING USED TO INVESTIGATE THE NUTRITIONAL REQUIREMENTS OF THE LOBSTER, *HOMARUS*

Ingredient	% Dry Weight	Source[b]
Casein	31	(2)
Corn Starch	24	(1)
Cellulose	7.1	(1)
Soy Lecithin, refined	10	(1)
Lipid Mix S	6	(3)
Gluten	10	(1)
Vitamin Mix BML-2[a]	4	(1)
Spray Dried Egg White	4	(2)
Mineral Mix BTm	3	(4)
Cholesterol	0.5	(1)
Vitamin E acetate 50% (500 IU/g)	0.2	(2)
Vitamin A acetate (500,000 IU/g)	0.1	(2)
Vitamin D_3 (400,000 IU/g)	0.1	(2)

[a] Vitamin mix BML-2 contains thiamin mononitrate 0.5%, riboflavin 0.8%, nicotinic acid 2.6%, Ca-pantothenate 1.5%, pyridoxine HCl 0.3%, cobalamine 0.1%, folic acid 0.5%, biotin 0.1%, inositol 18%, ascorbic acid 12.5%, PABA 3%, cellulose 60%, and BHA 0.1%.

[b] Key to sources:

(1) ICN Pharmaceuticals, Inc., Cleveland, OH 44128.

(2) Bio-serv, Inc., P.O. Box 100-B, Frenchtown, NJ 08825.

(3) Contains: cod liver oil 66% (Bio-serv), corn oil 33.8% (Bio-serv), ethoxyquin 0.2% (Monsanto).

(4) Mineral Mix Bernhart-Tomarelli; modified (ICN).

only recently been the subject of investigations. The effect of storage and processing on the stability of particularly labile dietary nutrients must be evaluated. Food stability is a primary concern and a diet for commercial purposes should employ an inexpensive binder that can be conveniently incorporated into the ration to assure a 24-hour stability in water. Both moist and dry diets have been used in laboratory research. Moist diets contain a gelling agent such as agar or gelatin; dry diets contain a binder such as gluten, starch, or carboxymethylcellulose. Dry pelleted crustacean diets appear to be less stable than moist ones (Heinen 1981), but no precise study has been directed toward the effect of moisture content or binder on diet palatability and lobster consumption behavior. The optimal physical characteristics of artificial diets may vary with lobster age.

An efficient delivery system for water-soluble nutrients is an obvious concern to commercial culturists. A partial solution may be microencapsulation of highly soluble nutrients that possibly would be supplemented to the diet in a pure form. However, microencapsulation technology is still at an early developmental stage, and its eventual use in diet formulation may be limited by labor and cost considerations. Another, perhaps more practical, way to minimize the problem of leaching would be the development of a highly attractive and palatable artificial ration that would be eagerly sought, easily found, and quickly ingested by lobsters.

Food Consumption and Digestion

Knowledge of the various factors affecting food consumption will be important in the development of a plan to optimize feed use and cost in commercial production facilities. As discussed earlier, lobsters, unlike fish, are slow, intermittent feeders. Using live *Artemia* as a food source, Bordner and Conklin (1981) determined that juvenile lobsters consume approximately 17% of a daily ration within the first 4 hr after feeding. How much consumption occurs, and when, during the next 20 hr is unknown.

Bordner and Conklin (1981) also found that certain environmental factors can affect food consumption. Consumption appears to be positively affected by a photoperiod regime of almost constant darkness and by increased temperature. Feeding twice daily does not significantly increase consumption. However, a 48-hr starvation period was followed by a significant increase in daily food consumption that subsided within 2 days. These consumption experiments raise several questions regarding the efficient administration of an artificial ration. Effective feeding regimes may eventually incorporate such procedures as skip-a-day feeding or administration of feed at a particular time of day. Nondaily feeding may be associated with an increase in food conversion efficiency (food acquisition by the lobster in nature is characteristically periodic). Observations with *H. gammarus* also indicate that an endogenous feeding pattern may be operative. All lobsters cease feeding for a period before and after molting; culture temperature affects the duration of this feeding cessation.

Generally juvenile and adult lobsters daily consume food at approximately 10% and 1% of their body weight, respectively. The quantity of food consumed may also depend upon its caloric content. Precise consumption studies involving artificial diets remain inconclusive because the lobster, when feeding, breaks up food into minute particles and complete retrieval of all uneaten food in large volumes of water is very difficult.

Knowledge of the digestive processes in the lobster is very limited. Barker and Gibson (1977) have provided an anatomical and histological review of digestion. Controversy exists regarding whether all digestion is exclusively extracellular. Studies of the hepatopancreas of lobsters indicate the existence of various classes of enzymes, including proteinases, lipases, and various carbohydrases. Barker and Gibson (1977) suggest that the complete digestive cycle in *H. gammarus* lasts approximately 12 hr. Lipid (triglyceride) digestion and absorption from the foregut was completed in 8–12 hr after ingestion in the Norwegian lobster *Nephrops norvegicus* (Dall 1981). Recent studies in our laboratory suggest that lobsters digest protein more efficiently than carbohydrate (Bordner *et al.* 1983). Whether this observation is related to the chemical composition of our lobster diet or is a general phenomenon is unknown. Additional research is necessary to characterize lobster digestive enzymes and their associated activities more completely. Ideally, artificial diets should contain the required nutrients in forms that are readily digested and do not inhibit enzymatic activity.

Nutritional Requirements

Protein and Energy. Information regarding the nutritional requirements of lobsters is currently limited and fragmentary. A great deal of interest has been devoted to protein levels, as protein is typically the most expensive component of an animal's diet. Protein quality and quantity in an artificial diet will be of major interest to the commercial culturist.

Although lobsters, like all other animals studied to date, are unable to synthesize the ten essential amino acids (Gallagher 1976) little is known about the quantitative requirements for these essential amino acids. Research on quantitative amino acid requirements of lobsters is currently not feasible given the design of artificial diets and the high solubilities of amino acids. Additionally, as pointed out by Friend and Dadd (1982), lobsters may have unique requirements for other essential amino acids. Also, some of the nonessential amino acids may be synthesized at a rate insufficient to promote maximum growth and thus dietary supplementation will be required for high production. Arginine and the three sulfur amino acids, methionine, cysteine, and taurine may require special attention since their concentration in lobster tissue exceeds the amounts available in proteins of terrestrial sources. The inclusion of a marine source of protein in artificial diets would appear to be most advantageous in eliciting maximum growth rates; such protein would assure a good representation of amino acids and probably be most efficiently digested.

An artificial diet should be so formulated that protein is efficiently channeled into growth functions and not utilized for energy needs. Work by Capuzzo and Lancaster (1979a) suggests that, as with diets for other animals, the inclusion of carbohydrates and lipids in lobster diets will effectively spare the amount of protein required. Laboratory experiments with lower protein levels in formulated diets would seem to support this contention (D'Abramo *et al.* 1981a). Protein levels can be reduced to 30% (dry weight) of the diet without any significant effects on growth and survival; further reductions may be possible if the proper complement of amino acids is made available. Critical to further understanding of the effect of protein sources and protein sparing will be information on the digestibility of various types of protein and carbohydrate sources.

The amount of protein and energy required in a lobster diet probably varies with the physiological state of the animal. For example, dietary concentrations of protein and energy for optimal growth may be age dependent. Young, rapidly growing juveniles may require more energy and protein (a higher dietary protein: energy ratio). Differences in metabolic activity between larval and early postlarval stages of *H. americanus* have already been shown (Capuzzo and Lancaster 1979b).

Lipids. The lipid composition of a commercial diet for the culture of lobster will certainly be of great importance. Lipids serve as excellent sources of energy and are known to be integrally associated with the transport of nutrients. Polyunsaturated fatty acids of the linolenic series have been shown to be required for growth of *H. americanus* (Castell and Covey 1976). Fish oil sources of polyunsaturated fatty acids, particularly docosahexanoic acid (22:6ω3)* and eicosapentaenoic acid (20:5ω3) have been implicated in the enhanced growth of lobsters (D'Abramo *et al.* 1980). These two fatty acids are characteristic of marine food chains. The ratio of ω6- to ω3-type fatty acids, already shown to exert effects on growth of fish and some crustacea, would seem to be an important area of study in the lobster.

Cholesterol is a required nutrient for the growth and survival of *H. americanus,* and a concentration of 0.5% (dry weight) in the diet had been considered to be optimal (Castell *et al.* 1975) but D'Abramo *et al.* (1984) found no difference in growth and survival of juvenile lobsters at dietary cholesterol levels of 0.12 and 0.5% (dry weight). For a commercial diet, ingredients such as marine-derived oils and meals could serve as the source of cholesterol. However, such ingredients

*A shorthand designation for identifying fatty acids. The first number is the number of carbon atoms in the molecule. The number following the colon is the number of double bonds and the number following the ω indicates the position of the first double bond counting from the methyl end of the molecule.

may not provide an adequate amount, thereby requiring supplementation. The ubiquity and low cost of plant-derived sterols (phytosterols) such as sitosterol and stigmasterol make them particularly attractive as possible substitutes for cholesterol in diets. However, recent work in our laboratory indicates that at least a portion and possibly all of the sterol requirement of juvenile lobsters is specific for cholesterol (D'Abramo *et al.* 1984). The ability to convert dietary phytosterols to cholesterol appears to be poor.

Lobsters and possibly crustaceans in general have unique requirements for dietary phospholipids. Preliminary research in our laboratory indicates that >60% of hemolymph lipid is phospholipid. Lobsters are known to have the ability to synthesize phospholipids (Shieh 1969), but apparently the rate of synthesis is limiting. Lack of lecithin, a phospholipid source in the purified diet developed by Conklin *et al.* (1980), leads to a characteristic deficiency syndrome that has been named "molt death" syndrome (Bowser and Rosemark 1980). Without the appropriate quantity and quality of phosphatidylcholine in this diet (D'Abramo *et al.* 1981b), the animal is unable to extricate itself from its molt. The exact mechanism involved in the development of molt death syndrome is not completely known at this time. One characteristic of this syndrome is apparent deposits of minerals on the shed exoskeleton of deficient animals.

The lack of phospholipids in the diet is correlated with a significant decrease of total cholesterol and total phospholipids in the serum (D'Abramo *et al.* 1982). Recent radiolabel experiments in our laboratory suggest that the reduction of cholesterol titers in the serum is associated with a reduction in the turnover rate of cholesterol from its main storage site, the hepatopancreas, to the hemolymph, and on to various target tissues. It is assumed that cholesterol is associated with high-density lipoprotein (HDL) molecules, which are known to be important in the transport of nutrients, particularly lipids, in other crustaceans. As phosphatidylcholine is the presumed major component of these high-density lipoproteins, the lack of a dietary supplement may cause a decrease in the total amount of lipoprotein molecules that can be involved in cholesterol transport. Cholesterol is used for membrane synthesis and also serves as a precursor for ecdysteroid synthesis. Preliminary research indicates that phospholipid-deficiency symptoms may be related to abnormalities in membrane synthesis during the premolt period. Currently the purified diet of Conklin *et al.* (1980) contains approximately 2.5% dietary phosphatidylcholine, which originates from the soy lecithin ingredient. This level is considered to be high and further reductions are anticipated by the possible use of a marine-derived phospholipid that is rich in polyunsaturated fatty acids, and by the use of a better quality protein. Recent experiments have demonstrated that replacement of casein or casein/albumin with purified protein from the rock crab *Cancer irroratus* will spare the phospholipid requirement (J. D. Castell, personal communication).

Carotenoids. Like other crustaceans, homarid lobsters are dependent upon an exogenous source of carotenoids for the production of their natural pigmentation. The primary pigment of the lobster exoskeleton is astaxanthin in free and esterified forms. The astaxanthin molecule is normally bonded to a protein molecule. Using pure dietary carotenoids incorporated into a purified diet, D'Abramo *et al.* (1983) have demonstrated that the efficacy of a particular carotenoid in producing pigmentation is directly related to its proximity to the astaxanthin end product in the metabolic scheme.

Juvenile lobsters do not appear to require carotenoids for growth and survival. However, since carotenoids constitute a large percentage of the lipid component of lobster eggs, they may be necessary for successful embryological development and hatching. In addition to satisfying particular physiological needs, a dietary carotenoid source may be necessary to produce lobsters that are acceptable to consumers. D'Abramo *et al.* (1983) have observed that oleoresin paprika oil and a carotenoid extract from crayfish waste are effective and relatively inexpensive sources of pigmentation when included at 12.5 and 3.0 mg/100 g of diet, respectively. More research is required to determine other potential carotenoid sources, optimal dietary concentrations, and the appropriate time of supplementation to the diet in preparation for marketing.

Vitamins and Minerals. A particular problem with the development of artificial rations for lobsters and other slowly-feeding aquatic invertebrates is the rapid leaching of water-soluble vitamins. Research efforts to determine optimal dietary levels of these nutrients have been plagued by excessive leaching and confounded by the potential contribution of endogenous gut bacteria.

Recent work in our laboratory confirms the observations of Canadian investigators (J. D. Castell, personal communication) indicating that the inclusion of lecithin in an artificial diet reduces the leaching rate of water-soluble nutrients. Leaching of B vitamins may prove to be a minor problem with commercial feeds based on complex feedstuffs in which the vitamins may be bound and less prone to rapid leaching than are the crystalline sources of vitamins used in defined research diets.

A critical question is the possible requirement for vitamin C. As ascorbic acid is very labile, it is seldom found in high levels in traditional feedstuffs and, if required, supplementation would be necessary. For various fish diets, ascorbic acid supplements are typically added to the diet after processing, often as a lipid-based sprayed-on coating to reduce leaching. However, this approach may not be effective for lobster aquaculture because lobsters feed slowly and continuously throughout the day, whereas fish feed quickly. If there is a requirement for vitamin C in lobsters, as has been shown in penaeid shrimp (Lightner *et al.* 1977), it is likely to represent a major stumbling block to the development of formulated feeds, at least until effective techniques to prevent leaching can be perfected. For example, highly soluble vitamins that must be supplemented into the diet could be lipid coated or microencapsulated.

Leaching is not a problem for the delivery of the lipid-soluble vitamins—A, D, E, and K. As yet no qualitative requirements for vitamin E or A have been demonstrated in the lobster. If needed, adequate levels of vitamins A, D and E could be easily provided by inclusion of vegetable and fish oils in commercial rations. Recent work in our laboratory indicates that there is no requirement for additional vitamin D_3 (cholecalciferol).

Little is currently known about the mineral requirements of lobsters. It is assumed that lobsters, unlike freshwater crustaceans, are able to obtain the calcium they need from sea water. Gallagher *et al.* (1978) have suggested that an optimal dietary calcium:phosphorus ratio would be 1:2.

Other Aspects

In addition to its chemical and physical characteristics, a particular diet should be assessed in terms of its conversion efficiency. An ideal diet would yield a conversion ratio of 2 to 1 (dry weight of food required to produce a corresponding wet weight of animal). More research is needed on the measurement of particular metabolic activities—such as respiration, digestion, assimilation, and excretion—and the relationship of these processes to the chemical and physical characteristics of the diet. Information about the dependence of ammonia excretion rates on diet, for example, would be useful not only in deciding which dietary nutrients to include but also in designing the flow characteristics of a culture system.

The nutritional value of particular diets has generally been assessed in terms of increases in wet weight over time. Other rapid and routine diagnostic measures of the lobster's physiological status are needed. Biochemical indices that incorporate the qualitative and quantitative measurement of particular nutrients circulating in the hemolymph would be advantageous. These indices would contribute to early recognition and possible alleviation of stressful conditions caused by the early stages of a pathogenic infection or the presence of an antinutritional factor that has been unintentionally incorporated into the diet.

More basic research—including studies on nutrient relationships, growth- and age-dependent requirements, optimal physical stability, feeding behavior, and digestive physiology—is required before an effective and inexpensive lobster ration can be composed. Nevertheless, the future looks promising. Artificial diets currently in use in our laboratory, though suboptimal and very expensive, yield growth rates at 20°C (68°F) that indicate traditional market-sized lobsters of 500 g (1.1 lb) can be attained in 2.5 yr (D'Abramo *et al.* 1981a).

DISEASES

Little is known regarding lobster diseases, their means of transmission or effective treatments. The larval stage appears to be the most critical period for lobster culture in terms of microbiological diseases. Epibiotic bacterial infestations (Nilson *et al.* 1975) and the fungal disease *Lagenidium* (Nilson *et al.* 1976) have been found to cause extensive larval mortality. Severe infestations of filamentous forms of bacterial epibionts can be visually identified by a fuzzy appearance of the lobster larvae, and nonfilamentous bacterial infestations can be diagnosed with the aid of a compound microscope. In contrast, dead larvae infected with *Lagenidium* appear white or opaque with individual fungal mycelium being detected with a dissecting microscope.

While some work has been done on evaluating antibiotic and fungitoxic compounds as treatments of larvae (Fisher 1977; Bland *et al.* 1976), none of the treatments appears consistently effective or convenient, and therefore are not particularly promising for use in commercial culture situations. A simple method to minimize the possibility of disease in larval culture is to maintain strict measures of cleanliness, including the use of ultraviolet-radiated water in the system. A system consisting of multiple isolated banks of kriesel systems will probably be essential to avoid disease problems. With this design, an entire unit can be conveniently and readily treated with chlorine bleach after the rearing of a particular batch of larvae.

Lobsters are particularly susceptible to a fatal disease called gaffkemia, which is caused by a nonmotile, tetrad-forming, encapsulated, gram-positive, coccoid bacterium, *Aerococcus viridans* variety *homari*. This pathogen and the disease gaffkemia have been recently reviewed by Stewart (1980). The hemolymph of infected lobsters often develops a pinkish or red coloration, which can be seen through the semitransparent ventral tail membranes; this condition gives rise to the common name of this disease, "red tail." Although injections of penicillin (Fisher *et al.* 1978) or vancomycin (Stewart and Arie 1974) have been suggested as treatments, the disease is generally fatal unless the antibiotic is given in the very early stages of infection. To prevent contamination of already existing broodstock, wild-caught females are generally kept in quarantine at 20°C (68°F) for 30 days. The infection in diseased animals is accelerated by this higher temperature and during the month-long quarantine period infected animals will develop gaffkemia symptoms and expire. Susceptibility to the disease seems to require a break in the lobster integument (Stewart *et al.* 1969), an observation that may explain the periodic outbreaks of the disease in lobster pounds where the claws are often pegged and additional damage can occur as the result of aggressive encounters. Epidemics of gaffkemia at culture facilities are not a particular worry since cultured lobsters are typically individually housed; however, it could take a significant toll on females brought in for the purpose of providing seed. Since the incidence of the disease is highly variable (0–40%) and details of its distribution and incidence from year to year are not available (Stewart 1980), its potential impact on commercial culture operations cannot be predicted.

Recently, a vibrio-like bacterium designated *Vibrio* sp. (BML 79-078) was reported as a pathogenic agent in the American lobster (Bowser *et al.* 1981). This bacterium, most probably a part of the natural marine flora, is characterized as a facultative pathogen. Susceptibility to infection by this bacterium would appear to be enhanced by the potentially stressful conditions inherent in high-density culture. Maintenance of proper nutrition and good water quality in culture facilities should significantly reduce the chances of infection.

Shell disease is another affliction that is characteristic of cultured juveniles and adults. This disease is associated with chitinolysis and a subsequent necrosis of the exoskeleton, which often takes on a pitted appearance. Causative agents are believed to include fungi and bacteria, although precise descriptions and identifications are lacking. Sawyer and Taylor (1949) believe that the chitinoclastic attack increases the probability of infection of living tissue by other pathogens. Shell disease has been associated with the abrasive destruction or injury of the first layer of the crustacean exoskeleton, the epicuticle. This condition leads to the microbial attack of the chitinous layers located below. It is therefore believed that conditions conducive to rapid epicuticle repair would diminish the possibility of a successful infection. Fisher *et al.* (1976) suggest that efficient epicuticle repair is dependent upon culture temperature and proper nutrition.

Observations in our laboratory suggest that shell disease sometimes afflicts animals that are devoid of cuticular injury. Infection in these cases may be related to nutritional deficiencies that create cellular abnormalities causing an unnatural susceptibility to attack. Shell disease is usually not fatal, and its symptoms are eliminated when the old exoskeleton is shedded. However, preventive measures would appear important since exoskeleton appearance may affect consumer acceptance.

SITE SELECTION AND DESIGN OF FACILITIES

The success of commercial lobster production facilities will depend upon an efficient, overall design and proper site location. Most onshore designs that have been proposed for juvenile culture incorporate a modular stacked system of compartmented culture units (trays), which would be conveniently accessible for automated checks and daily food distribution. Two-dimensional systems would be composed of multileveled units, each of which would have self-contained seawater. In this design, each level would require a holding tank of water. In a three-dimensional design, all units would be submerged in common holding tanks of sea water. The three-dimensional approach would yield considerable savings in terms of structural expense but would probably require increased flow rates due to low surface to volume ratios. A more complete discussion of the design of on-shore production units and associated advantages and disadvantages may be found in Van Olst *et al.* (1980). Brood-stock and grow-out areas should be isolated from one another. If early commercial-size operations are successful, later production facilities may specialize in either the juvenile grow-out phase or the broodstock and hatchery phases to optimize output.

Various sites along the eastern and western coasts of the United States are attractive for lobster aquaculture facilities. However, environmental constraints based upon conflicts with recreational land use, maintenance of federal and state effluent standards, proper salinity, and water quality could be severely limiting. An additional consideration in site selection concerns the average annual temperature of seawater in a location, and the temperature fluctuation during the year. Research indicates that lobsters grow optimally at 20°C (68°F); however, no areas in the continental United States have coastal waters of this temperature year-round. Heating ambient seawater with fossil fuels appears to be impractical given the recent unpredictable supplies and accompanying vacillating costs. Production facilities may be able to incorporate the use of heated seawater effluent from steam-electric power plants to reduce operating costs (Van Olst and Ford 1976). The use of such effluent to increase culture temperature may have deleterious effects due to heavy-metal contamination. However, studies have indicated that the concentrations of trace metals in lobsters growth in power-plant effluent do not significantly differ from those in animals taken from the natural habitat. A detailed study of the potential of this effluent causing metal contamination of lobster tissue is given in Dorband *et al.* (1976).

Alternatively, solar power may prove to be a complete or at least supplemental source of heated water, although the volumes of water required, the unpredictability of the number of sun days in many parts of the United States, and the state of existing technology currently make this approach suspect. Heated seawater could also be derived from geothermal energy or waste heat associated with industries other than fossil-fuel plants. Because of the many unsolved problems and constraints regarding the maintenance of optimal culture temperatures, the most likely initial sites for commercial culture of lobsters may be in semi-equatorial areas where water temperature is predictable and relatively uniform.

Culture facilities located inland could be a viable alternative in the future. Such facilities would use artificial seawater in closed systems in which potentially harmful metabolite accumulations, particularly of ammonia, would be minimized through the use of large natural populations of nitrifying bacteria. The bacterial populations would oxidize ammonia to nitrate and nitrite; the structural design would provide maximal surface area exposure of flowing seawater to the bacteria. A large-scale culture facility of this type would probably employ engineering similar to that in current secondary sewage treatment plants. Use of artificial seawater in such inland facilities should diminish the probability of exposure to various pathogens. The feasibility of such a closed-system culture unit has already been demonstrated on a small scale by researchers in Utah (R. Mickelsen and R. Infanger, personal communication).

Finally, the possible use of bays for the extensive culture of lobster has been proposed. A possible design consists of stacked individual compartments that would be incorporated into a platform anchored to the bottom. However, such an approach appears least promising. All feeding and harvesting would probably have to be done manually. Also, potential conflict with waterway access would restrict the size of operations and thereby reduce volume and profit potential. Finally, lack of control over such factors as water quality, diseases, and fouling would create a degree of unpredictability that minimizes the chances of success.

MARKETING AND ASSOCIATED ECONOMICS

The mathematical model developed by Allen and Johnston (1976) to describe the biological, physical, and economic aspects of a commercial lobster facility provides a means to evaluate the relative contribution of various parameters to the overall cost of production; this model will become more useful as better experimental data is generated for input. The model has indicated that the development of artificial foods yielding optimal growth and increased knowledge of diet-dependent ammonia excretion rates are two factors that can significantly affect cost. In addition, the model reveals that the use of heated seawater originating from power-plant effluent rather than fossil-fuel heating would reduce the cost of producing a market-sized lobster by approximately 45%.

Most of the inherent biological problems associated with the establishment of a profitable culture facility appear to be solved. Within the next 5 years, development of efficient engineering designs and improvement in feeds, brood-stock management, and disease treatment and control should assure success. The initial outlay of capital for start-up will be substantial, and no return on investment can be expected for at least 2–3 years.

Consumer demand for *H. americanus* remains high; in 1981 total supply of the American lobster in the United States amounted to 77.8 million lb (Morehead 1982). Marketing of a product that originates in a commercial facility would appear to be problem-free for a variety of reasons. Since 1974, approximately 48% of the lobster consumed by Americans has come from imports. Moreover, potentially large market areas in the United States have not even been partially saturated because established markets handle all the lobster that's available. Consumer acceptance also does not pose any concern since professional taste tests have revealed that cultured lobsters do not differ significantly from wild-caught animals in such characteristics as odor, texture and palatability (Stroud and Dalgarno 1982). In the future, consumers may accept a smaller, 340-g (0.75-lb), animal. Eventual acceptance of such a cultured product would translate into a higher turnover rate of stock, yielding possibly significant reductions in the cost of the product.

FUTURE PROSPECTS

Projecting when cultured lobsters will actually arrive on the market is not a simple proposition at present. Although it does not appear that any large conceptual void remains, a number of uncertainities exist that prevent accurate assessment of production costs. Those areas in which the lack of information or experience precludes reliable cost estimates are listed in Table 4.2. The assessments are based on a combination of our experience and available literature.

Extrapolating from available information, we believe that the reproductive cycle of *Homarus* is quite amenable to control and that present experience and possible future innovation should guarantee the year-round availability of stock. This control of the reproductive cycle also permits the conducting of specific breeding programs to develop strains with advantageous culture characteristics. In addition, nutritionally based fertility factors should be investigated. The limited larval period of *Homarus* also appears to represent little problem for the culturist.

Although early juveniles have been used extensively by biologists studying lobster culture requirements, more research is needed to develop large-scale culture techniques. Early grow-out techniques that are routinely employed in the laboratory would appear to be too labor intensive for use in commercial situations. As suggested, brine shrimp will probably be used extensively for both larval rearing and early juvenile growth in lobster aquaculture. Therefore, it is essential that techniques for the large-scale cultivation of nutritionally superior adult brine shrimp be adapted for on-site production at a reasonable cost.

Because public researchers have focused almost exclusively on the culture of early juveniles, there is concern that the actual cost involved in successfully growing out large numbers of larger juveniles to adults remains an unknown entity. The availability of a comparatively inexpensive yet growth-efficient diet is definitely the critical factor during this later phase. Certainly the space requirements of these larger animals needs to be reexamined, particularily in relation to water flow and perhaps other factors such as the provision of shelters and photoperiod. Photoperiod, which has been shown to be important to other aspects of the lobster's biology, may also have a strong impact on feeding and feed conversion. Continued development of di-

TABLE 4.2. STATUS OF LOBSTER CULTURE IN THE UNITED STATES

Culture Phase	Status[a]
BROOD-STOCK MANAGEMENT	
Control of reproduction	+ +
System design	+ + +
Female fertility	+
Male fertility	+
Breeding programs	+
LARVAL REARING	
Hatching	+ + +
System design	+ + +
Survival	+ +
EARLY GROW OUT	
System design	+
Brine shrimp culture	+ +
LATE GROW OUT	
System design	+
Diet:	
Chemical composition	+ +
Physical attributes	+
Conversion ratios	+
Growth-associated factors:	
Ablation	+
Space	+
Photoperiod	+
DISEASE	
Descriptions	+ + +
Therapeutics	+
ECONOMICS	
Pilot assessment	+
Site availability:	
Coastal	+ +
Inland	+
Marketing:	
Traditional	+ + +
Product variations	+

[a] + = further definition of parameters needed; + + = parameters defined but experience insufficient to establish accurate cost estimates; + + + = parameters well understood and costs can be accurately estimated based on existing experience. Assessments based on authors' experience and existing literature.

agnostic and therapeutic procedures for disease will obviously assist in preventing significant financial loss.

The availability of suitable sites is also difficult to judge at present. While the development of reliable systems for removing wastes from recycled inland facilities is possible, the cost and reliability of such systems are subjects of speculation. Site location will also determine to a great extent the type of heating system, if any, used to provide optimum culture temperatures. Coastal sites where large volumes of heated seawater are available, such as shoreline power stations, are limited but represent the most likely locations for the first plants. Future advances in solar-energy technology may eventually prove adaptable in obtaining heated seawater for production facilities.

The continuing development of appropriate techniques for the intensive culture of the lobster will also have impact in general on other types of crustacean culture. Techniques for brood-stock management, hatching, and feed formulation that have been developed, or will be, should be useful for other species. As an example, successful culture of the freshwater shrimp, *Macrobrachium rosenbergii,* in the continental United States appears, because of a limited growing season, to be dependent upon some type of intensive nursery culture system. Such a system must provide the necessary high juvenile survival and the larger size needed for stocking into managed ponds. Moreover, lobster diet research should serve as a foundation for the development of a ration that could be routinely used as a food supplement in extensive pond culture of crayfish and shrimp.

Future lobster aquaculture should be characterized by a burgeoning intensive culture industry that is complemented by continued research and development and novel approaches to marketing—all to the delight of the palates of consumers.

ACKNOWLEDGMENTS

The authors wish to thank Dr. Prudence Talbot, Dr. Keith Nelson, Dr. Cadet Hand, Mr. John Hughes, and Ms. Nancy Baum for their review of the manuscript and helpful comments. We also wish to thank Mrs. Nancy Heinzel for her perseverance during the preparation of this chapter.

LITERATURE CITED

AIKEN, D. E. 1980. Molting and growth. *In* The Biology and Management of Lobsters, J. S. Cobb and B. F. Phillips (editors). Vol. I, pp. 91–163. Academic Press, New York.

AIKEN, D. E., and WADDY, S. L. 1976. Controlling growth and reproduction in the American lobster. Proc. World Maric. Soc. **7,** 415–430.

AIKEN, D. E., and WADDY, S. L. 1978. Relationship between space, density, and growth of juvenile lobsters *(Homarus americanus)* in culture tanks. Proc. World Maric. Soc. **9,** 461–467.

AIKEN, D. E., and WADDY, S. L. 1980. Reproductive biology. *In* The Biology and

Management of Lobsters, J. S. Cobb and B. F. Phillips (editors). Vol. I, pp. 215–276. Academic Press, New York.

AIKEN, D. E., MARTIN, D. J., MEISNER, J. D., and SOCHASKY, J. B. 1981. Influence of photoperiod on survival and growth of larval American lobsters *(Homarus americanus)*. J. World Maric. Soc. **12,** 225–230.

ALLEN, P. G., and JOHNSTON, W. E. 1976. Research direction and economic feasibility: An example of systems analysis for lobster aquaculture. Agriculture **9,** 144–180.

ATEMA, J., and ENGSTROM, D. G. 1971. Sex pheromone in the lobster, *Hormarus americanus*. Nature (London) **232,** 261–263.

ATEMA, J., and COBB, J. S. 1980. Social behavior. *In* The Biology and Management of Lobsters, J. S. Cobb and B. F. Phillips (editors), pp. 409–450. Academic Press, New York.

BARKER, P. L., and GIBSON, R. 1977. Observations on the feeding mechanism, structure of the gut, and digestive physiology of the European lobster *Homarus gammarus* (L.) (Decapoda: Nephropidae). J. Exp. Mar. Biol. Ecol. **26,** 297–324.

BLAND, C. E., RUCH, D. G., SALSER, B. R., and LIGHTNER, D. V. 1976. Chemical control of *Lagenidium,* a fungal pathogen of marine crustacea. Sea Grant College Program Publ. UNC-SG-76-02. University of North Carolina Raleigh, North Carolina.

BORDNER, C. E., and CONKLIN, D. E. 1981. Food consumption and growth of juvenile lobsters. Aquaculture **24,** 285–300.

BORDNER, C. E., D'ABRAMO, L. R., and CONKLIN, D. E. 1983. Assimilation of nutrients by cultured hybrid lobsters (*Homarus* sp.) fed experimental diets. J. World Maric. Soc. **14,** 11–24.

BOWSER, P. R., and ROSEMARK, R. 1981. Mortalities of cultured lobsters, *Homarus,* associated with a molt death syndrome. Aquaculture **23,** 11–18.

BOWSER, P. R., ROSEMARK, R., and REINER, C. R. 1981. A preliminary report of vibriosis in cultured American lobsters, *Homarus americanus*. J. Invertebr. Pathol. **37,** 80–85.

CAPUZZO, J. M., and LANCASTER, B. A. 1979a. The effects of carbohydrate levels on protein utilization in the American lobster, *Homarus americanus*. Proc. World Maric. Soc. **10,** 689–700.

CAPUZZO, J. M., and LANCASTER, B. A. 1979b. Some physiological and biochemical considerations of larval development in the American lobster, *Homarus americanus*. Milne Edwards. J. Exp. Mar. Biol. Ecol. **40,** 53–62.

CARLBERG, J. M., and VAN OLST, J. C. 1976. Brine shrimp *(Artemia salina)* consumption by the larval stages of the American lobster *(Homarus americanus)* in relation to food density and water temperature. Proc. World Maric. Soc. **7,** 379–389.

CARLBERG, J. M., VAN OLST, J. C., and FORD, R. F. 1979. Potential for communal rearing of the nephropid lobsters (*Homarus* spp.). Proc. World Maric. Soc. **10,** 840–853.

CARTER, J. A., and STEELE, D. H. 1982. Stomach contents of immature lobsters *(Homarus americanus)* from Placentia Bay, Newfoundland. Can. J. Zool. **60,** 337–347.

CASTELL, J. D., and COVEY, J. F. 1976. Dietary lipid requirements of adult lobsters, *Homarus americanus* (M.E.). J. Nutr. **106,** 1159–1165.

CASTELL, J. D., MASON, E. G., and COVEY, J. F. 1975. Cholesterol requirements in the juvenile lobster *Homarus americanus*. J. Fish. Res. Board Can. **32,** 1431–1435.

CASTELL, J. D., MAUVIOT, J. C., and COVEY, J. F. 1976. The use of eyestalk ablation in nutritional studies with American lobsters *(Homarus americanus)*. Proc. World Maric. Soc. **7,** 431–441.

CASTELL, J. D., COVEY, J. F., AIKEN, D. E., and WADDY, S. L. 1977. The potential for ablation as a technique for accelerating growth of lobsters *(Homarus americanus)* for commercial culture. Proc. World Maric. Soc. **8,** 895–914.

CHANG, E. S., and BRUCE, M. J. 1980. Ecdysteroid titers of juvenile lobsters following molt induction. J. Exp. Zool. **214,** 157–160.

COBB, J. S., and PHILLIPS, B. J. (editors). 1980. The Biology and Management of Lobsters, Vols. I and II. Academic Press, New York.

CONKLIN, D. E., D'ABRAMO, L. R., BORDNER, C. E., and BAUM, N. A. 1980. A successful purified diet for the culture of juvenile lobsters: the effect of lecithin. Aquaculture **21,** 243–249.

CONKLIN, D. E., BORDNER, C. E., GARRETT, R. E., and COFFELT, R. J. 1981. Improved facilities for experimental culture of lobsters. Proc. World Maric. Soc. **12,** 59–63.

COOPER, R. A., and UZMANN, J. R. 1980. Ecology of juvenile and adult *Homarus*. *In* The Biology and Management of Lobsters, J. S. Cobb and B. F. Phillips (editors), Vol. II, pp. 97–142. Academic Press, New York.

D'ABRAMO, L. R., BORDNER, C. E., CONKLIN, D. E., DAGGETT, G. R., and BAUM, N. A. 1980. Relationships among dietary lipids, tissue lipids and growth in juvenile lobsters. Proc. World Maric. Soc. **11,** 335–345.

D'ABRAMO, L. R., CONKLIN, D. E., BORDNER, C. E., BAUM, N. A., and NORMAN-BOUDREAU, KAREN A. 1981a. Successful artificial diets for the culture of juvenile lobsters. J. World Maric. Soc. **12,** 325–332.

D'ABRAMO, L. R., BORDNER, C. E., CONKLIN, D. E., and BAUM, N. A. 1981b. Essentiality of dietary phosphatidylcholine for the survival of juvenile lobsters. J. Nutrition **111,** 425–431.

D'ABRAMO, L. R., BORDNER, C. E., and CONKLIN, D. E. 1982. Relationship between dietary phosphatidylcholine and serum cholesterol in the lobster *Homarus sp.* Mar. Biol. **67,** 231–235.

D'ABRAMO, L. R., BAUM, N. A., BORDNER, C. E., and CONKLIN, D. E. 1983. Carotenoids as a source of pigmentation in juvenile lobsters fed a purified diet. Can. J. Fish. Aquat. Sci. **40,** 699–704.

D'ABRAMO, L. R., BORDNER, C. E., CONKLIN, D. E., and BAUM, N. A. 1984. Sterol requirement of juvenile lobsters, *Homarus* sp. Aquaculture (in press).

D'AGOSTINO, A. 1980. The vital requirements of Artemia: Physiology and nutrition. *In* The Brine Shrimp *Artemia,* G. Persoone *et al.* (editors). Vol. 2, pp. 55–82. Universa Press, Wetteren, Belgium.

DALL, W. 1981. Lipid absorption and utilization in the Norwegian lobster *Nephrops norvegicus*. J. Exp. Mar. Biol. Ecol. **50,** 33–45.

DELISTRATY, D. A., CARLBERG, J. M., VAN OLST, J. C., and FORD, R. F. 1977. Ammonia toxicity in cultured larvae of the American lobster *(Homarus americanus)*. Proc. World Maric. Soc. **8,** 647–672.

DORBAND, W. R., VAN OLST, J. C., CARLBERG, J. M., and FORD, R. F. 1976. Effects of chemicals in thermal effluent on *Homarus americanus*. Proc. World Maric. Soc. **7,** 391–414.

DOW, R. L. 1980. The clawed lobster fisheries. *In* The Biology and Management of Lobsters, J. S. Cobb and B. F. Phillips (editors). Vol. II, pp. 265–316. Academic Press, New York.

ENNIS, G. P. 1973. Food, feeding, and condition of lobsters, *Homarus americanus,* throughout the seasonal cycle in Bonavista Bay, New Foundland. J. Fish. Res. Board Can. **30,** 1905–1909.

EVANS, P. D., and MANN, K. A. 1977. Selection of prey by American lobsters *(Homarus americanus)* when offered a choice between sea urchins and crabs. J. Fish. Res. Board Can. **34,** 2203–2207.

FIGUEIREDO, M. J., and VILELA, M. H. 1972. On the artificial culture of *Nephrops norvegicus* reared from the egg. Aquaculture **1,** 173–180.

FISHER, W. S. 1977. Microbial epibionts of lobsters. *In* Disease Diagnosis and Control in North American Marine Aquaculture. C. J. Sindermann (editor). Developments in Aquaculture and Fisheries Science, Vol. 6, pp. 163–167. Elsevier North Holland Publ. Co., Amsterdam.

FISHER, W. S., ROSEMARK, T. R., and NILSON, E. H. 1976. The susceptibility ot cultured American lobsters to a chitinolytic bacterium. Proc. World Maric. Soc. **7,** 511–520.

FISHER, W. S., MILSON, E. H., STEENBERGEN, J. F., and LIGHTNER, D. V. 1978. Microbial diseases of cultured lobsters: A review. Aquaculture **14,** 115–140.

FLEMING, L. C., and GIBSON, R. 1981. A new genus and species of monostiliferous hoplonemerteans, ectohabitant on lobsters. J. Exp. Mar. Biol. Ecol. **62,** 79–93.

FRIEND, W. G., and DADD, R. H. 1982. Insect nutrition: A perspective. Advan. Nutr. Res. **4,** 205–247.

GALLAGHER, M. L. 1976. The nutritional requirements of juvenile lobsters, *Homarus americanus.* Ph.D. Thesis, University of California, Davis, California.

GALLAGHER, M. L., BROWN, W. D., CONKLIN, D. E., and SIFRI, M. 1978. Effects of varying calcium/phosphorus ratios in diets fed to juvenile lobsters *(Homarus americanus).* Comp. Biochem. Physiol. **60A,** 467–471.

GALTSOFF, P. S. 1937. Hatching and rearing larvae of the American lobster, *Homarus americanus. In* Culture Methods for Invertebrate Animals, P. S. Galtsoff, F. E. Lutz, P. S. Welch and J. G. Needham (editors), pp. 233–236. Dover Publications, Inc., New York.

HEDGECOCK, D. 1983. Maturation and spawning of the American lobster. *In* CRC Handbook of Mariculture, J. P. McVey (editor), Vol. I. pp. 261–270. CRC Press, Boca Raton, Florida.

HEINEN, J. M. 1981. Evaluation of some binding agents for crustacean diets. Prog. Fish Cult. **43,** 142–145.

HIRTLE, R. W. M., and MANN, K. 1978. Distance chemoreception and vision in the selection of prey by American lobster *(Homarus americanus).* J. Fish. Res. Board Can. **35,** 1006–1008.

HUGHES, J. T., SULLIVAN, J. J., and SHLESER, R. A. 1972. Enhancement of lobster growth. Science **177,** 1110–1111.

HUGHES, J. T., SHLESER, R. A., and TCHOBANOGLOUS, G. 1974. A rearing tank for lobster larvae and other aquatic species. Prog. Fish Cult. **36,** 129–133.

JOHNSTON, W. E., and BOTSFORD, L. W. 1980. Systems analysis for lobster aquaculture. 11th Session FAO Eur. Inland Fish. Advisory Comm. Stavanger, Norway.

KOODA-CISCO, M. J., and TALBOT, P. 1983. A technique for electrically stimulating extrusion of spermatophores from the lobster, *Homarus americanus.* Aquaculture **30,** 221–227.

LIGHTNER, D. V., COLVIN, L. B., BRAND, C., and DANALD, D. A. 1977. "Black Death"—a disease syndrome of penaeid shrimp related to a dietary deficiency of ascorbic acid. Proc. World Maric. Soc. **8,** 611–624.

MCVEY, J. P. (editor). 1983. Handbook of Mariculture, Vol. I (Crustacean Aquaculture). CRC Press, Inc., Boca Raton, Florida.

MOREHEAD, B. C. 1982. Shellfish market review. Current economic analysis S-44, U.S. Department of Commerce, Washington, D.C.

NELSON, K., HEDGECOCK, D., BORGESON, W., JOHNSON, E., DAGGETT, R., and ARONSTEIN, D. 1980. Density-dependent growth inhibition in lobsters, *Homarus* (Decapoda, Nephropidae). Biol. Bull. **159,** 162–176.

NELSON, K., HEDGECOCK, D., and BORGESON, W. 1983a. Photoperiodic and ecdysial control of vitellogenesis in lobsters *(Homarus)* (Decapoda, Nephropidae). Can. J. Fish. Aquat. Sci. **40,** 940–947.

NELSON, K., and HEDGECOCK, D. 1983b. Size dependence of growth inhibition among juvenile lobsters *(Homarus).* J. Exp. Mar. Biol. Ecol. **66,** 125–134.

NELSON, K., HEDGECOCK, D., HEYER, B., and NUNN, T. 1983c. On the nature of short range growth inhibition in juvenile lobsters *(Homarus).* J. Exp. Mar. Biol. Ecol. **72,** 83–89.

NILSON, E. H., FISHER, W. S., and SHLESER, R. A. 1975. Filamentous infestations observed on eggs and larvae of cultured crustaceans. Proc. World Maric. Soc. **6,** 367–375.

NILSON, E. H., FISHER, F. S., SHLESER, R. A. 1976. A new mycosis of larval lobster *(Homarus americanus).* J. Invertebr. Pathol. **27,** 177–183.

PERKINS, H. C. 1972. Development rates at various temperatures of embryos of the northern lobster, *Homarus americanus* Milne-Edwards. Fish. Bull. **70,** 95–99.

PHILLIPS, B. F., and SASTRY, A. N. 1980. Larval ecology. *In* The Biology and Management of Lobsters, Vol. II, J. S. Cobb and B. F. Phillips (editors), pp. 11–57. Academic Press, New York.

PHILLIPS, B. F., COBB, J. S., and GEORGE, R. W. 1980. General biology. *In* The Biology and Management of Lobsters, Vol. I, J. S. Cobb, and B. F. Phillips (editors), pp. 1–82. Academic Press, New York.

RICHARDS, P. R., and WICKINS, J. F. 1979. Lobster culture research. Lab. Leafl. No. 47, Ministry of Agriculture, Fisheries and Food, Directorate of Fisheries Research, Lowesteft, United Kingdom.

RUDLOE, A. 1983. Preliminary studies of the mariculture potential of the slipper lobster, *Scyllarides nodifer.* Aquaculture. **34,** 165–169.

SANDIFER, P. A., and LYNN, J. W. 1981. Artificial insemination of caridean shrimp. *In* Advances in Invertebrate Reproduction, W. H. Clark, Jr. and T. S. Adams (editors), pp. 271–288. Elsevier North-Holland, Amsterdam.

SAWYER, W. H., JR., and TAYLOR, C. C. 1949. The effect of shell disease on the gills and chitin of the lobster *(Homarus americanus).* Maine Dept. Sea Shore Fish. Res. Bull. **1,** 1–10.

SCHUUR, A. M., FISHER, W. S., VAN OLST, J., CARLBERG, J., HUGHES, J. T., SHLESER, R. A., and FORD, R. A. 1976. Hatchery methods for the production of juvenile lobsters *(Homarus americanus).* Sea Grant Publ. No. 48. University of California La Jolla, California.

SERFLING, S. A., VAN OLST, J. C., and FORD, R. F. 1974. An automatic feeding device and the use of live and frozen *Artemia* for culturing larval stages of the American lobster, *Homarus americanus.* Aquaculture **3,** 311–314.

SHIEH, M. S. 1969. The biosynthesis of phospholipids in the lobster, *Homarus americanus.* Comp. Biochem. Physiol. **30,** 679–684.

SHLESER, R. A. 1974. Studies of the effects of feeding frequency and space on the growth of the American lobster, *Homarus americanus.* Proc. World Maric. Soc. **5,** 149–155.

SORGELOOS, P. 1980. The use of the brine shrimp *Artemia* in aquaculture. *In* The Brine Shrimp *Artemia,* Vol. III, G. Persoone *et al.* (editors). Universa Press, Wetteren, Belgium.

STEWART, J. E., and SQUIRES, H. J. 1968. Adverse conditions as inhibitors of ecdysis in the lobster, *Homarus americanus*. J. Fish. Res. Board Can. **25,** 1763–1774.

STEWART, J. E. 1980. Diseases. *In* The Biology and Management of Lobsters, J. S. Cobb and B. F. Phillips (editors), Vol. 1, pp. 301–342. Academic Press, New York.

STEWART, J. E., and CASTELL, J. D. 1979. Various aspects of culturing the American lobster, *Homarus americanus*. *In* Advances in Aquaculture, T. V. R. Pillay and W. A. Dill (editors), Fishing News (Books) Farnham, England.

STEWART, J. E., and ARIE, B. 1974. Effectiveness of vancomycin against gaffkemia, the bacterial disease of lobsters (genus *Homarus*). J. Fish. Res. Board Can. **31,** 1873–1879.

STEWART, J. E., DOCKRILL, A., and CORNICK, J. W. 1969. Effectiveness of the integument and gastric fluid as barriers against transmission of *Gaffkya homari* to the lobster *Homarus americanus*. J. Fish. Res. Board Can. **26,** 1–14.

STROUD, G. D., and DALGARNO, E. J. 1982. Wild and farmed lobsters *(Homarus gammarus),* a comparison of yield, proximate chemical composition and sensory properties. Aquaculture **29,** 147–154.

TALBOT, P., HEDGECOCK, D., BORGESON, W., WILSON, P., and THALER, C. 1983. Examination of spermatophore production by laboratory-maintained lobsters *(Homarus)*. J. World Maric. Soc. **14,** 271–278.

VAN OLST, J. C., and FORD, R. F. 1976. Use of thermal effluent in aquaculture. Sea Grant Annu. Rep. No. 57, pp. 39–44. University of California La Jolla, California.

VAN OLST, J. C., and CARLBERG, J. M. 1978. The effects of container size and transparency on growth and survival of lobsters cultured individually. Proc. World Maric. Soc. **9,** 469–479.

VAN OLST, J. C., CARLBERG, J. M., and HUGHES, J. T. 1980. Aquaculture. *In* The Biology and Management of Lobsters, Vol. II, J. S. Cobb and B. F. Phillips (editors), pp. 333–384. Academic Press, New York.

5

Other Crustacean Species *

Michael J. Oesterling,
Anthony J. Provenzano

Introduction
Blue Crab
Cancer Crabs
Spot Prawn
Spiny Lobsters
The Smaller Shrimps
Literature Cited

INTRODUCTION

Many crustaceans, because of their value in the marketplace or in the laboratory, have been considered as candidates for aquaculture in the United States. Chapters 1–4 have outlined the successes in culturing some of the more popular and well-known species. This chapter will address other crustaceans that are either currently being cultured or offer potential for culture.

Each region of the United States has its locally caught crab, lobster, or prawn that finds favor with the residents of that area. Many fisheries for wild-caught species exhibit dramatic fluctuations in stock abundance or harvest. This phenomenon, more than any other, has prompted investigation of the possibilities for culturing regionally available species. Some valuable crustaceans that have been considered for culture are spiny lobsters (*Panulirus* spp.), the spot prawn *(Pandalus platyceros)*, cancroid crabs *(Cancer magister, C. irroratus,*

*Virginia Institute of Marine Science Contribution Number 1078.

and *C. borealis)*, stone crabs (mainly *Menippe mercenaria*), the blue crab *(Callinectes sapidus)*, tanner or snow crabs *(Chionoecetes bairdi* and *C. opilio)*, and the king crab *(Paralithodes camtschatica)*. These species have been investigated for their use as human food. Other species have been studied for their applicability as laboratory animals for bioassay, as food sources for other animals (primarily fish) or as ornamentals; these species include the caridean shrimp *Palaemonetes*, the cladoceran *Daphnia*, and the mysids *Mysidopsis* and *Neomysis*.

The species discussed in detail in this chapter were chosen because of their potential for economic return or their current applications in commercial culturing operations. These species are the blue crab *(Callinectes sapidus)*, spiny lobsters *(Panulirus argus* and *P. interruptus)*, cancroid crabs *(Cancer magister, C. irroratus,* and *C. borealis)*, the spot prawn *(Pandalus platyceros)*, and carideans of the genus *Palaemonetes*. Brief remarks concerning other carideans and a few stenopidean shrimp are included. The techniques and procedures described in previous chapters on crustaceans may also be appropriate for the species discussed in this chapter.

BLUE CRAB

The blue crab, *Callinectes sapidus* Rathbun (Decapoda, Portunidae), supports a large commercial fishery along the eastern seaboard of the United States and Gulf of Mexico (Fig. 5.1). In actuality there are two blue crab fisheries—one for hard-shelled crabs and one for soft-shelled crabs. Soft-shelled crabs are not a separate species of crab, but are blue crabs that have shed (molted) their hard outer shells in preparation for growth. During 1980, over 73.9 million kg (162.6 million lb) of hard-shelled crabs were landed with a value of more than $35 million; soft-shelled crab landings totaled 861,825 kg (1,896,015 lb) valued at $2.4 million for the same period of time (U.S. Department of Commerce 1981; H. M. Perry, Gulf Coast Research Laboratory, Ocean Springs, MS, personal communication). Because of the blue crab's value to the eastern and southern United States, a great deal of information exists on all aspects of the fishery and biology of the species (Van Engel 1958, 1962; Tagatz and Hall 1971; Williams 1974).

The range of this economically valuable crustacean is from Nova Scotia to northern Argentina, including Bermuda and the Antiles. It has been reported from Øresund, Denmark; the Netherlands and adjacent North Sea; southwest France; Golfo di Genova; and the north-

FIG. 5.1 Dorsal view of a blue crab, *Callinectes sapidus*. Note the flattened pair of swimming legs used to determine closeness to molting.

ern Adriatic, Aegean, western Black, and eastern Mediterranean seas (Williams 1974). It is characterized as a coastal inhabitant ranging from the shoreline to approximately 90 m (295 ft) of water, but primarily inhabits shallow water up to 35 m (115 ft) in depth (Williams 1974). It has been taken from fresh water, such as Florida's Salt Springs, to hypersaline lagoons, such as Laguna Madre de Tamaulipas, Mexico (Williams 1974). Although the blue crab is considered a scavenger, its normal diet consists of a variety of materials, including fishes, benthic invertebrates, and plant material (Van Engel 1958; Darnell 1959; Tagatz 1968a; Williams 1974). It is more properly classified as an omnivore that prefers fresh to putrid flesh (Williams 1965). The blue crab life span is from 2 to 3 years (Jaworski 1972; Williams 1965, 1974). The adult stage is reached after 12 to 18 growing months (Churchill 1919; Van Engel 1958; Darnell 1959; Fischler 1965).

Female blue crabs mate only once, during the molt between the terminal juvenile instar and the adult (Churchill 1919; Van Engel 1958; Williams 1965). The males, however, are capable of mating more than once and at any time during their last three intermolts (Van

Engel 1958; Williams 1965). Mating takes place in the lower-salinity waters of estuaries (Futch 1965; Williams 1965). Spawning usually takes place 1 to 10 months after mating (Tagatz 1968a), depending on when mating took place, and occurs in the spring and summer months, thus assuring favorable environmental conditions for larval survival. Most females spawn only once or twice in their lifetime (Van Engel 1958; Tagatz 1968b; Williams 1974).

At the time of spawning, the female blue crab produces 700,000 to 2 million eggs (Churchill 1919; Van Engel 1958; Williams 1965), which she carries attached to the swimmerets under her abdomen. The eggs are carried 7–14 days at which time they hatch as zoeae (Sandoz and Rogers 1944; Van Engel 1958; Fischler 1965). The complete sequence of larval stages of the blue crab was first described by Costlow and Bookhout (1959). Working in the laboratory, they described seven (rarely eight) planktonic zoeal stages. These persisted for 31 to 49 days, depending upon the temperature and salinity in which they were reared. Sandoz and Rogers (1944) found the optimum ranges for development to be 19°–29°C (66°–84°F) and 23–28 ppt salinity. The last zoea stage metamorphoses into a single megalopal stage, which has both planktonic and benthic affinities (Williams 1971; Sulkin 1974). The megalopal stage lasts 6–20 days (depending upon environmental conditions); at the end of this period, the megalopa metamorphoses into the first crab, the stage in which the crab form is first seen (Costlow and Bookhout 1959).

In a later study, Costlow (1967) expanded upon the effects of salinity and temperature on megalopal duration. He examined the effects of all possible combinations of six different salinities (5, 10, 20, 30, and 40 ppt) and four different temperatures (15°, 20°, 25°, and 30°C or 59°, 68°, 77°, and 86°F) on survival, rate of development, and metamorphosis of the megalopa. Although some megalopae survived to the first crab stage in all temperature/salinity combinations (except 15°C/10 ppt and 20°C/5 ppt), better survival occurred at high temperatures and salinities. With salinities of 10–40 ppt, survival to the first crab ranged from 78.9 to 100% at 25° and 30°C and from 70 to 90% at 20°C. The time to metamorphosis varied greatly with different temperature/salinity combinations. Low temperatures combined with high salinity extended the duration of the megalopal stage. Length of megalopal life increased slightly in high salinities at 20°C, but at 25° and 30°C there was no significant difference in duration at 5 or 40 ppt salinity. Costlow (1967) postulated that survival and duration of the megalopal stage was directly related to the time of hatching, the time at which the megalopal stage was reached in relation to seasonal changes in water temperature, and the salinity of the water in which the final zoeal molt occurred.

Larval development takes place in more saline waters than the confines of the estuary (Gunter 1950; Daugherty 1952; Van Engel 1958; Tagatz 1965, 1968a; Dudley and Judy 1971; Sulkin 1974; Williams 1974). The young crabs, however, spend the majority of their growing life within the nursery grounds of estuaries (Williams 1965, 1974). During the megalopal and first few crab stages, there is a "directed" movement shoreward toward the nursery grounds (Van Engel 1958; Darnell 1959; Futch 1965; Tagatz 1968a; Sulkin 1974; Williams 1974). It has been suggested that the megalopal and first crab stages take advantage of incoming tidal currents by rising into the water column during flood tide, then settling and holding to the bottom during ebb tide, thus, eventually reaching the estuary (Futch 1965; Williams 1971; Sulkin 1974).

Although considerable information about the blue crab is available, total culture from egg to adult hard-shelled crab has not been practiced commercially. Initially high mortality rates, high labor demands to raise larvae, and a prolonged larval life are the main reasons why total culture from egg to market size has not been attempted. An additional factor mitigating against complete culture is the relatively low market value for hard crabs. The production of soft-shell crabs is an entirely different situation.

Soft-shelled crab production, in reality, should be considered short-term farming which consists of holding crabs for molting and not feeding or holding crabs for extended periods of time. Soft crab production relies entirely on the capture of premolt crabs (peelers). Before discussing the techniques and facilities for "shedding" crabs, we digress to provide a bit of background history on the soft-shelled crab industry.

Attempts to mass produce soft-shelled blue crabs began in the mid-1800s near Crisfield, Maryland (Warner 1976). "Controlled" shedding of crabs was first conducted in wire enclosures staked out in tidal zones. These crab "pounds" were filled with hard-shelled crabs which were fed and watched closely for molting. This method was difficult to manage and mortality was high due to cannibalism or to variations in water quality (Otwell *et al.* 1980). Later, crab pounds were equipped with floating boxes to house and protect those crabs near to molting (Otwell *et al.* 1980). These floating boxes were successful and were continually modified to suit specific requirements (e.g., size, depth, and location) of individual producers. In time, producers used more floating boxes, or live cars, and became less dependent on crab pounds in which the crabs required extra care and feed. Eventually production became entirely dependent on the selective harvest of peelers.

Little change occurred in the systems used to shed crabs until the 1950s when bank or shore floats were developed (Otwell *et al.* 1980). Shore floats were simply troughs or shallow tables used to hold running water pumped from an adjacent brackish-water supply. These open-flow systems were easier to manage than floating boxes, and soon evolved into enclosed shedding tables that were housed to provide shade and protection from rain and predators. Recently, attempts have been made to carry these facilities one step further with the development of closed, recirculating systems of shedding tables, which permit better control of water quality (Paparella 1979, 1982; Perry *et al.* 1982; Oesterling 1984).

All three types of facilities—floating boxes, open flow-through, and closed recirculating—are in use today. Any one of the three systems may be the most suitable depending upon location and water quality. The choice of facility will be determined by each producer's background, training, and financial situation.

All soft-shelled crab production systems, whether traditional floats or on-shore facilities, have a common requirement: an adequate supply of wild-caught peeler crabs (Haefner and Garten 1974; Otwell 1980). Peeler crabs are caught with a variety of devices, including crab traps (fykes), crab pots, "bush lines," and scrapes, as well as other gear. Descriptions of these are given by Otwell (1980), Otwell *et al.* (1980), Cupka and Van Engel (1982), Perry *et al.* (1982), and Oesterling (1984).

Peeler crabs can be distinguished from intermolt hard-shelled crabs by visual inspection. Although there are several indications that molting is approaching, the most reliable and widely used involves color changes associated with the formation of the new shell.

As the time for molting approaches, the new shell of the crab will begin to form and become visible underneath the old hard shell. These changes are most visible as a line along the edges of the last two flattened sections of the paddle fins, the last pair of pereiopods (Fig. 5.2). In the early stages, the line is white, indicating that the crab will molt within 2 weeks. Gradually, as molting time nears, this line goes through a series of color changes: a pink line means molting within 1 week, a red line indicates molting within 1 to 3 days. The last peeler stage is identified not by color but by the physical condition of the hard shell. A split develops under the lateral spines and along the posterior edge of the carapace. At this point the crab is termed a "buster" and has actually begun molting (Fig. 5.3), which may be completed in another 2–3 hours.

As mentioned previously, there are basically two methods for hold-

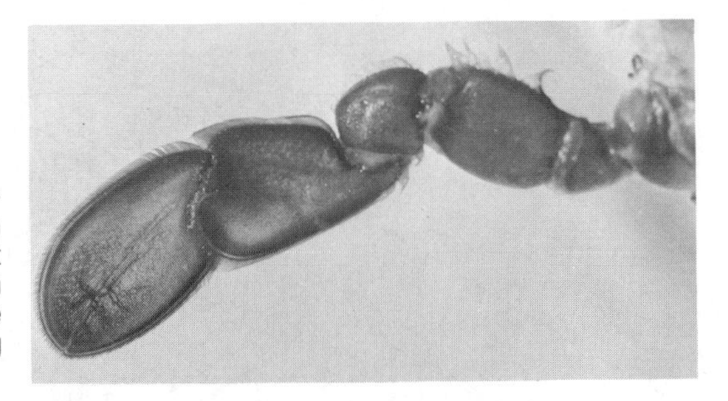

FIG. 5.2 Blue crab swimming leg showing "line" on the next to last segment used to identify peelers. In actuality this line is the edge of the new shell being formed under the old.

ing peelers for shedding purposes, either in a floating box or on-shore in tables with a flow-through or recirculating water supply.

Floating boxes (Fig. 5.4) are made entirely of wood. Although float sizes may vary, traditional dimensions have been 1.2 m (4 ft) wide, 3.6 m (12 ft) long and 38–45 cm (15–18 in.) deep (Haefner and Garten 1974; Otwell *et al.* 1980). The bottom of the float is made of closely fitting boards, while the sides are made of laths with 1.2-cm (0.5-in.) spaces between them. About midway up the sides is a wooden shelf 15.2–20.3 cm (6–8 in.) wide serving as a stabilizer and adjusting the depth at which the box floats.

Floats must be located where there is adequate water flow, either from tidal action or currents, in order to promote water movement

FIG. 5.3 Posterior view of a "buster" blue crab. The exoskeleton has split along a predetermined fracture line in advance of ecdysis. This is the last stage before a soft-shelled crab emerges.

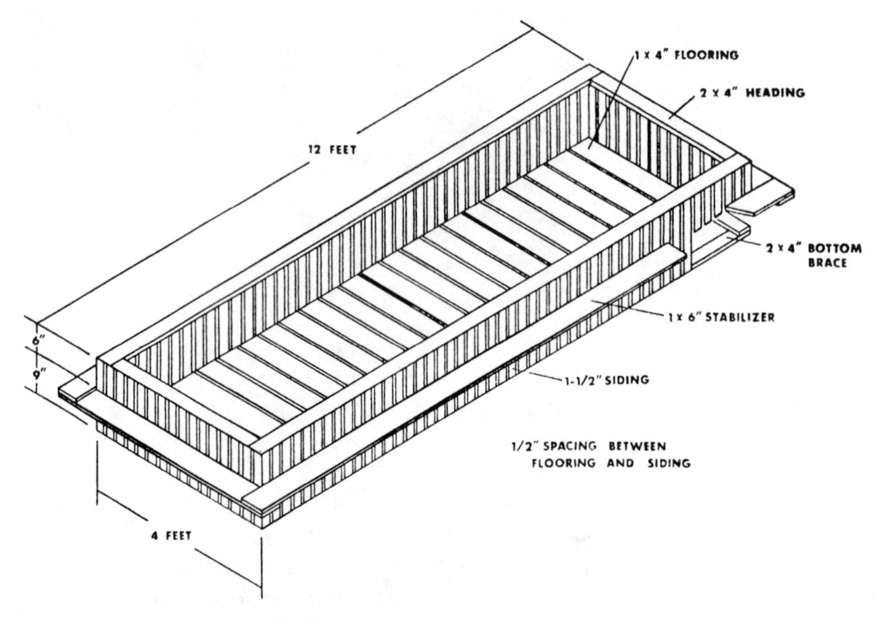

FIG. 5.4 Traditional blue crab shedding "float."
Courtesy of Otwell et al. *1980.*

through the box (Oesterling 1982, 1984). This will help remove any waste products as well as provide fresh, oxygenated water. Water depth should be sufficient to keep floats from resting on the bottom at ebb tide. In some cases, a breakwater may be required to keep high waves from rocking the floats and tearing them loose from their moorings. Floats should be removed from the water periodically, scrubbed to remove fouling plants and animals, and dried for several days to destroy rot and wood borers.

Although the least expensive method for shedding crabs, floats have several drawbacks (Oesterling 1982; Perry *et al.* 1982). The siting of many adjacent floats requires an area of good water quality and circulation. Since crabs are held in the upper few centimeters of water, they are subject to rapid temperature and salinity changes such as may occur during a rain storm. Being held in the "natural" environment leaves crabs open to predation by birds and fish, especially eels. Also there are the physical difficulties associated with tending a group of moored boxes. All of these factors have combined to cause more and more producers to turn to shore-based facilities.

Several reports include full descriptions on the construction of onshore shedding tables (Jachowski 1969; Haefner and Garten 1974; Bearden *et al.* 1979; Perry *et al.* 1982; Oesterling 1984). Only a general description of these facilities is given here; the papers cited may be consulted for detailed construction information.

The most common shedding table in use by the industry today is constructed of unprotected wood (Fig. 5.5); dimensions are 1.2 × 2.4 m (4 × 8 ft) and 25.4 cm (10 in.) deep. Tables can safely be constructed of fiberglass, concrete, or wood coated with a sealer/waterproofer (Fig. 5.6). Of course, no toxic paints or metals should be used in either the table tank or plumbing and pumps. The number of tables used will vary with the number of peelers available and the amount of effort the operator wishes to expend. However, for descriptive purposes, a system of 10–15 tables may be considered as an average commercial operation.

Just as floating boxes have certain requirements, so do on-shore tables. Foremost among these is the need for adequate water depth and circulation. Both of these are easily taken care of in the initial construction of the facility. Water depth is most commonly regulated by means of a standpipe drain. It has been found that a water depth of 10 cm (4 in.) is quite satisfactory for table systems (Haefner and Garten 1974; Oesterling 1982; Perry *et al.* 1982). Standpipes can either be a simple single-pipe or a double-pipe arrangement. In the double-pipe arrangement, the inner pipe is cut to the desired water depth, while the outer pipe is longer with notches (about 2.5 cm or 1 in.) cut at the bottom. The double-pipe causes water to be drawn from the

FIG. 5.5 A typical Virginia shore-based, wooden-table blue crab shedding system. The facility is located outdoors, with overhead water delivery, as well as lights. The drain sytem is simply a wooden V-trough that empties into a ditch.

FIG. 5.6 A Virginia blue crab shedding facility with tanks constructed of concrete blocks and enclosed in a block building. Water delivery is from opposite corners, facilitating water circulation.

bottom of the table, thus creating a self-cleaning system (Oesterling 1982; Perry *et al.* 1982). A 0.6-cm (0.25-in.) hole should be drilled approximately 2.5 cm (1 in.) from the bottom of the pipe that controls water depth. This will act as an emergency drain should water flow stop because of an electrical failure. Crabs completely submerged in still water can quickly deplete available oxygen and die. However, if held in only enough water to keep their gills moist, they will be able to "bubble" and receive atmospheric oxygen (Perry *et al.* 1982).

The easiest way to control water circulation is to control the introduction of water into the table. There are many different methods in use, as detailed by Jachowski (1969). These include over-head spray, nozzle injection, and simple splash. One method that has proven very successful involves below-surface or just above surface inflows at opposite corners of a table. This creates a circular motion within the table, eliminating dead spots. This type of arrangement should be coupled with a center, double-pipe drain, creating an efficient self-cleaning system (Fig. 5.7).

A table that is 1.2 × 2.4 m (4 × 8 ft) with 10 cm (4 in.) of water will contain approximately 303 liters (80 gal) of water. To maintain good water quality in a table of this size and volume, it is recommended that water within the table be replaced three to four times

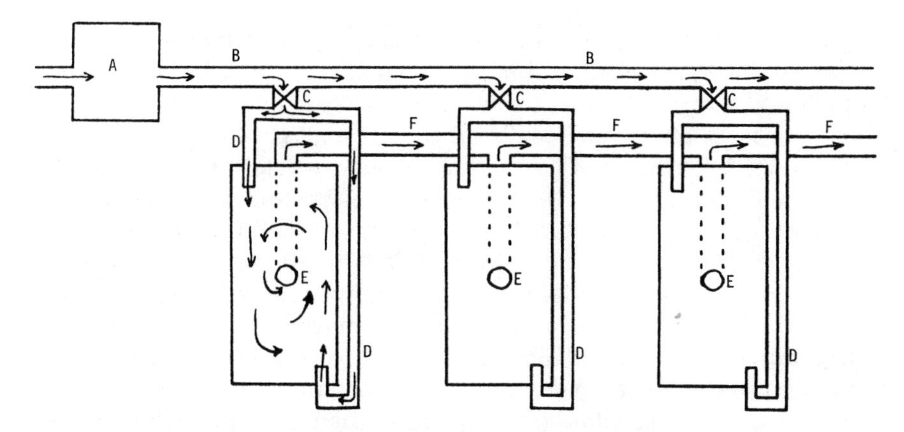

FIG. 5.7 Typical layout for a flow-through shedding system. This particular system employs dual water injection and a center drain for optimum water circulation. A = water pump; 2 hp is recommended for 10–15 tables. B = main water delivery line of 2-inch PVC. C = valve with reduction coupling (2 inch to 1 inch); either brass or PVC is recommended. D = table water delivery lines; 1-inch PVC is recommended. E = table drain (2-inch PVC). F = main drain line (6-inch PVC). Arrows indicate direction of water flow.

per hour (Haefner and Garten 1974; Oesterling 1982). Care should be taken during the design of the facility to provide for a pump with large enough capacity to handle the required flow rate.

This information on water depth and circulation applies to either a flow-through or closed, recirculating system. The only difference is that in a flow-through system drain water is returned to the original water source, whereas in a recirculating system drain water is passed through a filter (or series of filters) and then returned to the tables. Lately, interest has been expressed in the use of a recirculating system for crab shedding as a means of better controlling water quality (Paparella 1979, 1982; Perry *et al.* 1982; Oesterling 1984). The basic principles of closed, recirculating seawater systems discussed by Spotte (1979) apply to crab shedding as well as other culture systems. Operators have adapted these principles to the closed systems they currently are using. In any one system, a combination of biological filtration, foam fractionation, algal filtration, and mechanical filtration may be found. Again, the actual system selected will depend on individual preference.

Mortality rates in shedding systems are quite high, in some cases exceeding 50%. Oesterling (1982) has listed various causes of crab death and has offered possible solutions to these problems. Most deaths are caused either by bad handling practices or by adverse environmental conditions, not by diseases or parasites. Because of the relatively short period of time during which crabs are held within the system and the rapid water turnover (in the case of flow-through systems), diseases and parasites do not constitute a problem.

The actual operation of a shedding facility is relatively simple, albeit time-consuming. In a typical operation, peelers would be segregated by sign (i.e., color stage) into separate floats/tanks. Each 1.2 × 2.4 m (4 × 8 ft) float/tank would contain 200–300 or more crabs. At regular intervals (usually 4–6 hr), each tank would be "fished." Those crabs with changed color signs would be moved to the appropriate tank and those in the advanced buster stage would be placed in a separate tank for shedding. Soft-shelled crabs would be removed from the tank after being allowed to expand to their full size. When crabs are removed from the water, the hardening process of the new shell ceases; thus, the timing of removal is critical, for if crabs are left in the water too long the shell will begin to harden, producing an inferior product (Otwell *et al.* 1980). Depending on temperature and salinity, anywhere from 15 minutes to several hours will be required for expansion to full size.

Currently, soft crab production relies exclusively on wild-caught peeler crabs. However, in areas with low densities of shedding crabs, hard intermolt crabs may be held and fed until time for molting. Ogle *et al.* (1982) suggest a complete system for holding hard-shelled crabs in a closed recirculating facility. They discuss dietary requirements, and environmental and biological background information that should be considered in the design of facilities. They do concede that, at this time, the holding of intermolt crabs for use in a soft-shelled crab production system is not economically feasible.

It has been suggested that removal of eyestalks containing the molt inhibitory X-organ may hasten the advent of shedding (Otwell *et al.* 1980; Cupka and Van Engel 1982). Experience has shown that this method is not a reliable means of enhancing molting. In most cases, eyestalk removal in blue crabs causes death or other problems during the shedding process (Otwell *et al.* 1980; Cupka and Van Engel 1982). With improved technology and the possibility of rapid wound cauterization, eyestalk ablation may become feasible.

Soft-shelled crabs are much more valuable than hard-shelled crabs. During the 1970s the average annual value for soft crabs increased from \$1.08 to over \$2.42 per kg (\$0.42 to over \$1.10 per lb) (Otwell *et al.* 1980). In some sections of the U.S., their value was much higher, reaching \$5.57 per kg (\$2.53 per lb) in Louisiana in 1977 and an estimated \$8.80–13.20 per kg (\$4.00–6.00 per lb) in Florida during 1979 (Otwell *et al.* 1980). During this same time the price for hard-shelled crabs remained at $0.33–0.55 per kg ($0.15–0.25 per lb). Because of this price differential and the wide geographical range of the blue crab, more and more people are becoming involved in crab shedding.

The production of soft-shelled crabs is regulated by water temperature. Water temperatures of 21°C (70°F) or above are required for active shedding, although shedding begins at lower temperatures (Haefner and Garten 1974). Currently the Chesapeake Bay states of Virginia and Maryland produce the majority of soft-shelled crabs. In these states, the season runs from May through October. However, the areas to the south, including the Gulf of Mexico, with a longer period of warm temperatures have the potential for an extended season of production. Producers in these areas could enter the market earlier with a fresh product and continue longer, thus commanding premium prices in certain seasons.

Expansion of the soft-shelled crab industry will continue as more people become familiar with the relative ease with which crabs can be shed and with their market value. Marketing does not present a problem as demand consistently exceeds supply. The limiting factor in any operation will be a consistent supply of quality peeler crabs.

CANCER CRABS

Crabs of the genus *Cancer* (Decapoda, Cancridae) occur worldwide in temperate regions, with several species exploited commercially (MacKay 1942; Waldron 1958; Williams 1965; Sastry 1977; Bigford 1979). In the United States there are East and West Coast species that are harvested either in directed fisheries or as incidental catches in other fisheries. On the Pacific Coast, the Dungeness crab *(Cancer magister)* supports a large commercial fishery valued at over $36.6 million for 1983 (U.S. Department of Commerce 1984). Harvested to a lesser degree from the Atlantic Coast are the rock crab *Cancer irroratus* and the jonah crab *(Cancer borealis)*. Because of widely fluctuating landings for all three species, consideration has been given to their culture potential.

The life histories of the *Cancer* crabs are well known, from both natural populations and laboratory studies. As early as 1942, MacKay described the life cycle of Dungeness crabs *(C. magister)* found in British Columbia waters. He observed mating to occur from April through September, followed by an egg-bearing period lasting from October to June (MacKay 1942). Waldron (1958), working in Oregon waters, reported similar mating and egg-bearing periods. Larvae were observed from December through June with peaks during March

(MacKay 1942) and from April to July (Waldron 1958). MacKay (1942) identified five zoeal stages, lasting 89 days, from the wild population. This compares favorably with larvae reared in the laboratory by Poole (1966), who reported 80 days to complete all zoeal stages at a temperature of 10.5°C (51°F) and salinities ranging from 26 to 30 ppt. Reed (1969) succeeded in reducing zoeal development time from 90 days at 10.0°C (50°F) to only 65 days at 13.9°C (57°F), both in salinities ranging from 25 to 30 ppt. He also found that development and survival at 6.1°C (43°F) was poor, regardless of salinity, and that below 20 ppt all zoeae died rapidly (Reed 1969). This led him to hypothesize optimum zoeal development ranges of 10.0°–13.9°C (50°–57°F) and 25–30 ppt salinity (Reed 1969). Following the fifth zoeal larva is a single megalopal stage. In British Columbia waters, megalopae were collected by MacKay (1942) during July and August and were reported as present from May to September. In the laboratory the megalopal stage lasted for 31 days, after which time megalopae metamorphosed into the first crab instar (Poole 1966).

Similar development has been reported for the rock crab, *C. irroratus*. Bigford (1979) provided a very comprehensive synopsis of biological information on the rock crab, and Sastry (1971, 1977) described larval development under laboratory conditions. As with the Dungeness crab, rock crabs pass through five zoeal and one megalopal stages (Sastry 1971, 1977). Sastry (1971, 1977) observed maximum survival to the first crab stage in 37 to 58 days at a temperature of 15°C (59°F) and salinity of 30 ppt. At those conditions and with a photoperiod of 14 hrs light and 10 hrs dark, survival to the megalopal stage was 79% and to the first crab stage, 61% (Sastry 1971). Bigford (1979) reported that zoeae and megalopae were found to survive differentially in various temperature–salinity regimes, depending upon the stage of development. He suggested optimum developmental ranges of 15°–20°C (59°–68°F) and 20–35 ppt salinity (Bigford 1979). Bigford (1979) reported finding rock crabs with egg masses in Narragansett Bay from April to June, and Williams (1965) stated that ovigerous females occur during March in Florida and August in Massachusetts. Larvae would thus hatch from late spring through summer (Bigford 1979).

Fecundity of both Dungeness and rock crabs is relatively low. In both crabs the number of eggs carried is related to body size; large animals ordinarily have more eggs than small ones (MacKay 1942; Bigford 1979). Bigford (1979) provided a regression equation utilizing carapace width to estimate egg count for rock crabs:

$$\log_{10} \text{eggs} = -0.33459 + 3.01016 \log_{10} \text{carapace width in mm}$$

For the Dungeness crab, as many as 1.5 million eggs have been counted on one animal (MacKay 1942), while it is estimated that rock crab fecundity is only 4430 to 330,400 (Bigford 1979). Additionally, there are no records of repeated spawnings within 1 year for either Dungeness or rock crabs. However, rock crabs have been induced to reproduce, even to egg laying and hatching, by laboratory conditioning, offering some promise for culturing (Dr. Steve Rebach, University of Maryland—Eastern Shore, Princess Anne, MD, personal communication).

Much effort has been expended investigating culture techniques and procedures for Dungeness crabs (Poole 1966; Reed 1969; Hartman 1977; Fisher and Nelson 1978). Poole (1966) cultured Dungeness crab larvae in order to describe the various stages. For this he used 189-liter (50-gal) aquaria, fitted with air pumps and sub-bottom filters. Reed (1969), in studying the effects of temperature, salinity, and diet on Dungeness crab larvae, used 250-ml Erlenmeyer flasks held in constant-temperature water baths. Neither researcher was concerned with raising large numbers of larvae. Likewise, Fisher and Nelson (1978) in their investigations into the applications of antibiotics did not raise mass quantities of larvae. Hartman (1977), however, described a mass larval-rearing system for Dungeness crab larvae. After citing numerous examples of attempts to culture brachyuran crab larvae on a laboratory scale, Hartman (1977) proceeded to describe a system adaptable to either a flow-through or recirculating system. His design involved slow upwelling water columns, which tended to offset passive larval sinking and provided a constant flowing, well-oxygenated environment. Hartman (1977) noted that this system has unlimited scale-up potential, while emphasizing its low maintenance requirements.

The same facilities used by Hartman (1977) for culture of Dungeness crab larvae would also be applicable to culture of rock and jonah crabs. Larvae of both species have been raised successfully on a variety of diets including *Artemia* nauplii, *Balanus glandula* nauplii, *Skeletonema costatum,* and various combinations of these items (Poole 1966; Reed 1969; Sastry 1971, 1977; Shleser 1976; Hartman 1977). Similarly, juvenile crabs have been found to feed readily on assorted fish and fish scraps, chopped *Mercenaria mercenaria,* blue mussel *(Mytilus edulis),* and adult brine shrimp (Glude 1977; Bigford 1979).

A major expense in any culture operation is feed. The use of scrap or waste products for the culture of adult Dungeness crabs was investigated over a 3-year period by Welsh (1974). The overall objectives of this project were to evaluate utilization of fish and crustacean processing wastes (heads, racks, shells, etc.) to accelerate the meat-recovery rates of Dungeness crabs after molting and also to investigate holding the crabs over extended periods of time using these diets. Adult male crabs were caught during the late molting season (August and September in California) and used as test animals. A 23% meat yield was the criterion used to determine when a crab (both cultured and wild-caught) was suitable for market. During the project many problems were encountered and modifications made in the experimental design to counteract them. Rather than detailing all these changes, only the final results and suggestions will be presented. Foods that were evaluated included salmon heads (*Onchorhynchus* spp.), rockfish round (*Sebastes* spp.), fillet wastes of Dover sole *(Microstomus pacificus)* and sable fish *(Anoplopoma fimbria),* and hand-picked shrimp wastes (carapace, head, etc.). Welsh (1974) concluded that bottom-contact pens in water a minimum of 60 cm (24 in.) deep were superior to floating pens. In these pens, the best substrate was found to be equal parts of broken oyster shell and sand, 10 cm thick (4 in.). To obtain accelerated meat recovery, initiation of culture should occur between August 15 and October 1. Crabs should be segregated by size and stocked at a density of 6.0–10.7 crabs/m^2 (0.6–1.0/ft^2). Any crabs missing more than two appendages should not be held. Within the pens, competing crabs (primarily *Cancer antennaris* and *Cancer productus*) should be eliminated. The best strategy is to feed a diet composed of 90% Dover sole waste and 10% shrimp offal, at a rate of 10% of the crabs' body weight, twice weekly. This food should be presented in chunks no larger than 5 cm^3 (2 $in.^3$) and introduced simultaneously to all parts of the holding facility. Finally, crabs should be disturbed as little as possible between feedings. Using this strategy, Welsh (1974) believes that acceleration of postmolt meat recovery and long range holding of crabs can be accomplished. However, at this time no economic studies have been conducted to determine what benefits, if any, would result from large-scale culture efforts.

Taking a different approach to the food question, Rebach (1981) developed a pelletized food for rock crabs; for a description of the composition and preparation of the pelletized food, the reader is referred to the original article. Rebach (1981) found the pelletized diet to be quite satisfactory and in some cases superior to other commercially available foodstuffs. It also had several added benefits: uniformity of size, permitting accurate measurement of portions; dispensable by mechanized delivery systems; and no requirement for refrigeration or mold inhibitors (Rebach 1981).

Despite all the information available, complete culture from egg to market of either the Dungeness crab or the rock and jonah crabs appears to have a low potential. At this time in the United States, there are no known companies that actively culture or are considering the culture of any *Cancer* crabs. There are several reasons for this. In all species, cannibalism during larval stages and later crab stages is a problem (Reed 1969; Welsh 1974; Glude 1977; Bigford 1979; Dr. James Welsh, Humboldt State University, Arcata, CA, personal communication). Slow growth rates, requiring 2 to 3 years for crabs to reach market size, also mitigate against commercial-scale culture (MacKay 1942; Poole 1966; Glude 1977; S. Rebach, personal communication). In the case of the Dungeness crab, the need to rely on wild-caught brood stock has been mentioned as a deterrent to commercial-scale culture (J. Welsh, personal communication). This is probably true for rock and jonah crabs as well. The relatively low fecundity of rock crabs also would restrict their culture (Bigford 1979). On a strictly economic note, sharp price fluctuations and a relatively low market value make *Cancer* crabs unsuitable candidates for culture (Glude 1977).

Even though it is conceded that complete life-cycle culture of rock crabs is impractical at this time, there is potential for the production of soft-shelled rock crabs, similar to soft-shelled blue crabs (Haefner and Van Engel 1972, 1975; Haefner *et al.* 1973; S. Rebach, personal communication). In the case of rock crabs, however, production would take place during the winter months when there is no soft-shelled blue crab production (Haefner and Van Engel 1975). In fact, several years ago, soft-shelled rock crabs were commercially produced on Virginia's Eastern Shore. However, unstable and unpredictable supplies of premolt rock crabs led the operators to stop production. More recently (1981–1982) there was renewed interest in soft-shelled rock crabs. The use of closed, recirculating systems is currently being considered for use in shedding rock crabs.

Basically the same procedures used for shedding blue crabs would apply to rock crabs (see preceding section). Haefner and Van Engel (1972), Haefner *et al.* (1973), and Oesterling (1984) have provided instructions on all aspects of rock crab shedding, including harvest and identification of premolt crabs, timing, and facility construction, as well as details of the differences from blue crab shedding operations.

SPOT PRAWN

The spot prawn (*Pandalus platyceros* Brandt) supports a fishery along the U.S. Pacific coast. Butler (1970) lists its distribution as temperate regions of the eastern Pacific Ocean from San Diego, California, to Unalaska, Alaska, as well as Asiatic waters. Although there is currently no commercial culture operation for spot prawn (E. Prentice,

NMFS, Manchester, WA, personal communication), it is still considered to have potential for culture. The spot prawn is more suitable for a polyculture system than a monoculture operation (Kelly *et al.* 1977; Rensel and Prentice 1979, 1980; E. Prentice, personal communication.)

Spot prawn exhibit many traits conducive to culturing: relatively large size; gregariousness, not normally cannibalistic; easy maintenance of larvae and postlarvae with high survival rates; tolerance of a wide range of temperature and salinity; disease resistance; acceptance of a wide variety of foods; reproduction in captivity; and current market demand that cannot be met by harvest from the wild (Wickins 1972; Shleser 1976; Kelly *et al.* 1977; Rensel and Prentice 1979, 1980). It is considered a potential companion crop to pen-reared Pacific salmon (*Oncorhynchus* spp.) due to its adaptability to both vertical and horizontal growing surfaces, its scavenging on dead salmon, and the function it performs in net cleaning (Rensel and Prentice 1979, 1980).

Spot prawns are protandric hermaphrodites (Butler 1964, 1970; Wickins 1972; Kelly *et al.* 1977; Rensel and Prentice 1977), maturing first as males at age 1.5 years, passing through a transition or intersexual phase at 2.5 years, finally becoming a functional female at age 3.5 (Rensel and Prentice 1977). Both male and female spot prawns are capable of multiple spawnings (Rensel and Prentice 1977). Unfortunately for potential aquaculturists, there is a rather low fecundity of between 2000 and 5000 eggs per female at first spawning (Kelly *et al.* 1977; Rensel and Prentice 1977; Wickins and Beard 1978). Fecundity decreases to 10–1000 eggs per female at subsequent spawnings (Rensel and Prentice 1977).

The larval history of the spot prawn is well understood, and several authors have been successul in culturing the prawn from egg through juvenile stages (Price and Chew 1972; Wickins 1972; Kelly *et al.* 1977; Rensel and Prentice 1977). Price and Chew (1972) described the various larval stages. They took wild-caught ovigerous females into the laboratory for culturing in a recirculating seawater system. Larvae were hatched and held for 83 days on a diet of *Artemia* nauplii, during which time they passed through five larval and four postlarval stages. Price and Chew (1972) reported the larval period to be approximately 35 days at a temperature of 11°C (52°F). This was corroborated by Kelly *et al.* (1977) who reported a larval period of 26–35 days at temperatures ranging from 9°–12°C (48°–54°F). In their experiments, they found that circular, flat-bottomed fiberglass tanks supplied with a continous flow of filtered seawater, but no supplemental aeration, were effective for mass larval rearing. An average survival of 46% to postlarvae was observed by Kelly *et al.* (1977) when larvae were stocked at a density of approximately 0.11/cm^2 (0.71/in.^{2}) of floor space. Wickins (1972) suggested that the larval period may be reduced by culturing at "elevated" temperatures; he observed a larval period of 15–20 days at 14°C (57°F).

Larvae of spot prawn are positively phototactic and appear to molt mostly at night (Price and Chew 1972). As in other crustaceans with pelagic larvae, the earliest stages are most vulnerable to predation and death by other causes. Most critical is the first 2 weeks when larvae are molting from Stage I to Stage II (Wickins 1972; Kelly *et al.* 1977). Kelly *et al.* (1977) observed a synchronous molt from Stage I to Stage II larvae after 7 days. However, no other synchrony was observed, and subsequent molts occurred at varied intervals.

Both Price and Chew (1972) and Wickins (1972) fed larvae on *Artemia* nauplii. Price and Chew (1972) reported that larvae could live up to 11–13 days on stored yolk alone, but would feed immediately upon hatching if food was provided. Kelly *et al.* (1977) suggested that a combination *Artemia*–algae diet (they suggested *Phaeodactylum tricornutum*) may enhance growth rates.

Postlarvae readily accept a wide variety and combination of foods. Wickins (1972) fed postlarvae small pieces of *Mytilus* sp., frozen *Crangon*, live worms *(Lumbricillus)*, and a "compounded diet" consisting largely of Norwegian shrimp meal. In their food study, Kelly *et al.* (1977) tested 12 different combinations or types of food. Their findings showed that a combination diet of sea mussel *(Mytilus californianus)* and sea urchin *(Strongylocentrotus franciscanus)* gave not only the highest growth rate but also the best food conversion ratio (5.5:1). Either in combination or alone, sea mussel and sea urchin—as well as squid *(Loligo opalescens)*—resulted in survival rates of at least 80% (Kelly *et al.* 1977).

According to Butler (1970), the postlarvae of *Pandulus platyceros* remain in shallow water, sublittoral to 55 m (180 ft) of depth, until the end of their first year, when they move to deeper waters. This suggests that juveniles possess some degree of tolerance to a range of salinities and temperatures. Wickins (1972), however, found that the effect of salinity on growth was more pronounced than that of temperature (Table 5.1). He concluded that the optimum salinity was above 30 ppt and the optimum temperature was about 18°C (64°F). Kelly *et al.* (1977) found that mortality was less than 20% in salinities equal to or greater than 22 ppt; at 16 ppt, mortality was 100% after only 4 hours. They further reported that at salinities between 24 and 30 ppt, mortality did not exceed 6%, leading them to conclude that a minimum culture salinity should be 24 ppt. In their culture of larval spot prawn, Price and Chew (1972) maintained a salinity of 29–30 ppt.

TABLE 5.1. EFFECT OF SALINITY AND TEMPERATURE ON THE GROWTH OF JUVENILE *PANDALUS PLATYCEROS*

Salinity (ppt)	Growth[a]		
	15°C	18°C	20°C
22	63.5% (4)	100.2% (4)	65.9% (2)
26	201.1 (5)	272.1 (5)	90.1 (5)
30	205.5 (4)	360.0 (5)	151.9 (5)

Source: Wickins 1972. Reproduced courtesy of the Ministry of Agriculture, Fisheries and Food, Great Britain.

[a] Expressed as mean percentage increase in weight. Figures in parentheses are the number surviving in each group of five prawns.

Rensel and Prentice (1979, 1980) have conducted extensive research into the growth and survival of juvenile and adult spot prawns in various culture situations, both in the laboratory and field. Their 1980 study assessed the effects of environmental factors on growth and survival of prawns held in floating net pens at two sites in Puget Sound. They found that juvenile and yearling prawns avoid brightly illuminated areas and cease feeding when suddenly exposed to light. In comparing the growth rates they obtained for net-reared prawns to the data provided by Butler (1964) for wild prawns, they found similar increases in weight up to October. However, following October, cultured prawns continued to increase in weight, whereas wild prawns showed a decreased growth rate. They reasoned that water temperature and reduced food availability contributed to the decline in growth rate observed in wild populations. In these studies, the water temperatures from July through September never exceeded the optimum 18°C (64°F) proposed by Wickins (1972). However, during the course of the experiments, deaths of prawns were observed to coincide with massive phytoplankton blooms (dinoflagellates, *Ceratium* sp., and *Peridinium* sp.). Associated with these blooms were rapid rises in water temperature; weekly mean temperatures rose about 5°C (9°F).

Several feeding experiments were conducted by Rensel and Prentice (1979, 1980). Their 1979 work compared different diets for juveniles in the laboratory and in net pens. Four diets were evaluated on prawns held in flowing seawater tanks in the laboratory: mussel *(Mytilus edulis)* meat; chopped salmon; feces and pseudofeces from the Pacific oyster *(Crassostrea gigas);* and a no food control group. Mussel-fed prawns exhibited the best survival and growth rates of all laboratory-raised prawns (Table 5.2). Those fed on salmon were smaller than the mussel-fed group, but both equaled or surpassed the growth reported for natural populations by Butler (1964). Prawns fed on oyster wastes or receiving no food grew slower than the groups fed other diets.

TABLE 5.2. GROWTH AND SURVIVAL OF JUVENILE *PANDALUS PLATYCEROS* ON DIFFERING DIETS

				Start of Experiment		60 Days After Start of Experiment		End of Experiment		
Experiment	Location	Diet	No. of Replicates	No. of Prawns in Each Replicate	Mean Weight (g)	Mean Survival (%)	Mean Weight (g)	No. of Days	Mean Survival (%)	Mean Weight (g)
Prawns Alone	Laboratory Tanks	Mussel	4	25	0.72	82	2.52	90	74	3.14
		Salmon	4	25	0.62	83	2.27	90	71	2.61
		Oyster Wastes[a]	4	25	0.69	64	1.06	—	—	—
		No Food[a]	4	25	0.64	26	1.07	—	—	—
Prawns Alone	Net Pens	Mussel	2	200	0.64	98	3.14	365	69	10.30
		Salmon[b]	2	200	0.63	98	2.79	365	64	10.61
Prawns and Salmon	Net Pens	Variety of Foods[c]	1	100	0.64	93	3.42	206	93	8.60

Source: Rensel and Prentice 1979.

[a] Terminated at 60 days.

[b] Includes net-fouling organisms.

[c] Includes dead salmon, uneaten fish food (Oregon moist pellets), salmon feces, and net-fouling organisms.

In their net pen feeding experiments, Rensel and Prentice (1979) used two different strategies. In one experiment, only prawns were stocked in the pens and fed either mussel meat or salmon. In addition, the pens were vertically subdivided into three compartments, only two of which contained prawns, the third being left empty. This was done to study the net-cleaning ability of prawns and the possible food value of fouling organisms. The second net study involved stocking juvenile prawns with age "O" (approximately 20-g or 0.71-oz) coho salmon *Oncorhynchus kisutch)*. While salmon were fed "Oregon moist pellets," the prawns were left to scavenge on dead salmon or uneaten food, as well as fouling organisms. Prawns in the first net experiment grew faster than those in the lab fed similar diets; no difference was observed between the mussel or salmon-fed prawns (Table 5.2). Rensel and Prentice (1979) hypothesized that net-fouling organisms were responsible for this improved growth, as prawns were observed actively foraging on fouling organisms. The second net experiment provided even more interesting results. After 6.5 months, the growth of the prawns held with salmon exceeded that of prawns held alone (Table 5.2). Survival of the former group was high, 93%, but not significantly different from the prawn monoculture group. No adverse interactions between prawns and salmon were observed. In fact, adding prawns to salmon cultures offers several potential benefits to salmon farmers: the net cleaning of the prawns helps maintain good water circulation, assuring adequate dissolved oxygen levels, and helps in flushing wastes; scavenging on dead salmon reduces labor costs; and net maintenance costs in general could be reduced.

Although no detailed economic studies have been conducted, it is the feeling of Kelly *et al.* (1977) and E. F. Prentice (personal communication) that monoculture of spot prawn would be economically questionable. However, Prentice went on to state that a reevaluation of the economies of spot prawn culture may be warranted. Kelly *et al.* (1977), Rensel and Prentice (1979), and E. F. Prentice (personal communication) all agree that polyculture offers the best possibilities.

The culture of spot prawns is not without constraints or the need for additional information. E. F. Prentice (personal communication) suggests that studies are needed to accurately determine the number of spawns per animal; to assess the use of heated water to speed larval development; and to define the requirements for economically feasible operations. Rensel and Prentice (1979) suggest a need for evaluation of combination diets for net pen culture, as well as studies of stocking densities for prawn–salmon culture in order to assess maximum growth and survival rates. A relatively low fecundity, which necessitates holding large numbers of brood stocks, and a rather slow growth rate are the major biological constraints on prawn culture (Shleser 1976; Kelly *et al.* 1977; E. F. Prentice, personal communication).

SPINY LOBSTERS

Spiny lobsters are decapod crustaceans belonging to the family Palinuridae, which are principally tropical and subtropical in distribution (Williams 1965). They support major fisheries worldwide. In 1980, total U.S. landings of spiny lobster were 3.1 million kg (6.8 million lb), valued at $14.8 million (U.S. Department of Commerce 1981). Primarily two species of spiny lobsters are fished commercially: in Florida, *Panulirus argus* (Figure 5.8), and in California, *P. interruptus*. The Florida landings account for 95% of the total poundage and 93 % of the value (U.S. Department of Commerce 1981).

For both *P. argus* and *P. interruptus* a great deal of information is available. Kanciruk and Herrnkind (1976) included more than 1100 references concerning spiny lobsters in an indexed bibliography. Since their effort, many additional reports have been published regarding all aspects of spiny lobster biology, fisheries, economics, and aquaculture potential (Warner *et al.* 1977; Davis 1980; Labisky *et al.* 1980; Menzies and Kerrigan 1980; Prochaska and Cato 1980; Simmons 1980; Tamm 1980). It is generally concluded that the potential for culture of spiny lobsters is very low (Iversen 1968; Bardach *et al.* 1972; Ting 1973; Serfling and Ford 1975b; Tamm 1980; S. A. Serfling, San Diego,

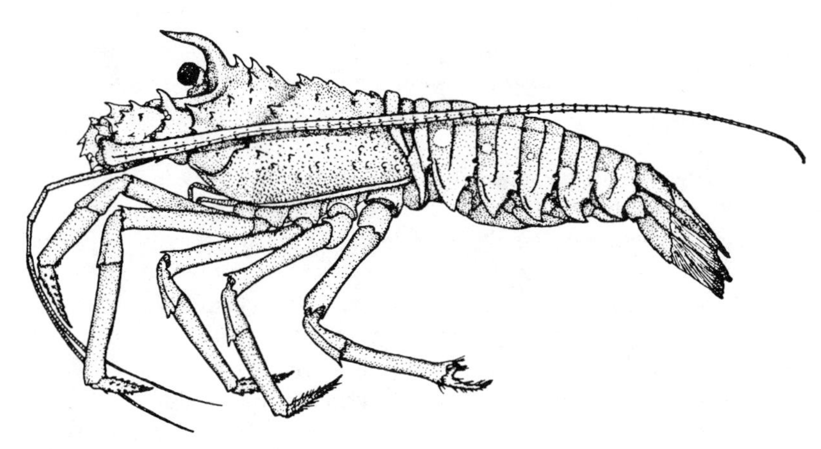

FIG. 5.8 Female *Panulirus argus* in lateral view. (After Williams 1965.)

CA, personal communication; D. E. Sweat, Marine Advisory Program, Largo, FL, personal communication). However, due to increasing market demand and declining harvest from natural populations, the feasibility of spiny lobster culture has been investigated.

The life histories of both *P. argus* and *P. interruptus* are well known. A general description will serve to illustrate the life cycles, which are similar for both species. A male passes sperm in a viscous fluid that becomes attached to the female's abdomen. This fluid hardens to form a sperm sac. As the female extrudes eggs she breaks the sperm sac with her legs. As the eggs pass by the sperm sac they are fertilized and then attached to the swimmerets on the abdomen. The number of eggs produced varies with carapace size: a female with a carapace length of 7.6 cm (3 in.) produces about 0.5 million eggs; one of 12.7 cm (5 in.) carapace length about 1.5 million eggs (Simmons 1980). The eggs hatch after about 1 month. The larvae of spiny lobster are flattened, leaf-shaped, planktonic organisms known as phyllosomes (singular is phyllosoma). After 3–6 months and 6–11 stages the phyllosoma metamorphoses into a puerulus stage or first postlarva. Pueruli are shaped like miniature adults, but are colorless and have a soft exoskeleton. They move to shallow waters to begin their benthic existence (Menzies and Kerrigan 1980). After up to 3 years and possibly 11 postlarval stages in this nursery area, juveniles mature and move to offshore reefs. *Panulirus interruptus* takes 7–9 years to reach minimum legal size, while *P. argus* may attain legal size in 3 years (Bardach *et al.* 1972).

It is generally conceded that culture from egg through larval stages to adult stages is impractical. The primary stumbling block to this is the inability to culture phyllosomes through the entire larval phase (Ingle and Witham 1969; Ting 1973; Serfling and Ford 1975a, b; Tamm 1980; D. E. Sweat, personal communication). Problems with maintaining large numbers of phyllosomes are due to their shape and appendage size, as well as to a lack of information on larval diet (Ting 1973; Tamm 1980).

Although the culture of spiny lobster larvae (phyllosomes) on a commercial scale is considered impractical, the rearing of pueruli and juveniles is a relatively straightforward affair. The feasibility of raising *P. argus* juveniles was demonstrated by Witham *et al.* (1968) and Sweat (1968), who reared *P. argus* pueruli to a carapace length of 50 mm (2 in.) in 28 months. Under his conditions, animals would attain the minimum legal size (76.3 mm or 3 in. carpace length) in 4–5 years. Serfling and Ford (1975b), using *P. interruptus,* maintained and reared pueruli to juveniles for a period of 14 months. They also investigated the use of elevated temperatures to accelerate the growth rate. At ambient temperatures of 12°–20°C (54–68°F), they estimated a grow-out period of 7 years to reach the legal size (82 mm or 3.25 in. carapace length). However, at 22°C (72°F), they estimated a period of 3 years to attain legal size; and at 28°C (82°F) only 2 years to legal size. They concluded that the control of temperature would be an important factor in any commercial attempt to culture *P. interruptus,* but pointed out that increased water temperatures required that more food be fed the lobsters.

The major problem in culturing pueruli or older spiny lobsters is acquiring sufficient numbers of individuals for stocking purposes. Several investigators have evaluated the use of "habitat" traps to capture pueruli (Witham *et al.* 1968; Ingle and Witham 1969; Serfling 1972; Serfling and Ford 1975a, b). Witham *et al.* (1968) first proposed the use of Witham habitat traps, named after their originator, for the capture of pueruli of *P. argus.* The Witham trap is constructed of a polyurethane float with "leaves" of a nonwoven nylon, monofilament mat, similar to an air conditioner filter (Witham *et al.* 1968). The rationale for using such traps was an attempt to duplicate the natural algal environment that pueruli seek out for settling and protection. Witham *et al.* (1968) tested both floating and benthic habitat traps, concluding that surface placement was superior to bottom placement. They also observed that the length of time the trap was in the water was not critical; that pueruli were most active at night; and, for *P. argus,* the greatest number of pueruli were caught during the new moon, with none caught during full moon. Pueruli were collected monthly with peak occurrences from February through May (Witham *et al.* 1968).

In evaluating various designs of habitat traps for *P. interruptus,* Serfling (1972) and Serfling and Ford (1975a, b) reached differing conclusions. When comparing Witham habitat traps to similar traps containing natural surfgrass *(Phyllospadix torreyi),* they found that although pueruli could be collected in all the designs tested, a simple nylon-mesh bag filled with *Phyllospadix torreyi* was superior. The nylon bags also were more durable, easier to use, and cost less than other traps. In further experiments on pueruli collection, Serfling and Ford (1975a) found that, contrary to Witham *et al.* (1968), there was no correlation between phases of the moon and pueruli settlement and that illumination (either natural or artificial) may serve as a settling stimulus. Near San Diego, California, pueruli of *P. interruptus* occurred from late May through mid-September, with the greatest abundance during the first week of August, corresponding to the seasonal period of highest water temperature (Serfling and Ford 1975a).

Ingle and Witham (1969) proposed a scheme for artificial cultivation of *P. argus*. It would first be necessary to locate protected embayments where pueruli could be reliably obtained. Then, numerous habitat traps would be used to attract and collect pueruli, as well as later stages. Traps should be checked daily and pueruli transported to holding ponds abundantly supplied with hiding places, such as tiles and bricks. These protective niches are important so that lobsters can hide during daylight hours, as well as during ecdysis. Food should be added regularly and wastes flushed, perhaps by natural tidal action. After the lobsters pass through the most vulnerable juvenile stages, three courses of action would be available: one might continue feeding in the ponds until lobsters are harvestable; one might move the lobsters to large enclosures in shallow, nearshore waters where they could forage for natural food, with some supplemental feeding; or, in the case of governmental agencies, the young lobsters could be used for restocking purposes. No attempts at implementing any of these schemes have been made, nor have any economic evaluations been conducted.

D. E. Sweat (personal communication) envisioned a similar strategy; however, he suggested that it may not be necessary to culture to legal harvest sizes. There exists the possibility, if certain legal constraints could be overcome, of marketing sublegal lobsters ("shorts") as high-priced gourmet items, similar to imported lobster tails.

Currently, the culture of spiny lobsters does not offer a potential alternative to wild harvest. Despite volumes of research, major hurdles must still be overcome. As already pointed out, a protracted larval life and uncertainties about nutritional requirements limit culture of spiny lobsters. Although culture starting with pueruli or later stages offers some possibility, the collection of sufficient numbers of pueruli still remains a problem.

THE SMALLER SHRIMPS

Aside from the penaeids (Chapter 3) and the caridean genus *Macrobrachium* (Chapter 2), there are several smaller shrimps for which there is some aquaculture potential. Although none is being produced on a commercial scale yet, the biological and technical information already available could make their large-scale culture feasible. Because of their small size, these species would not be grown for human food, but for nonfood purposes, such as bioassay, or ornamental use in aquaria.

Many species of glass or grass shrimp of the genus *Palaemonetes* are found in the United States. Many, perhaps all, are important food items in the diets of local fishes and most feed on detritus and/or microscopic plants and animals. Some are found strictly in fresh water, others in estuaries. Some are known for their broad tolerance to changes in environmental conditions such as salinity. *P. pugio,* for example, may be found in environments ranging from essentially freshwater to coastal salinities, and its larvae may develop in a salinity as low as 5 ppt (Anninos 1982). Most species have a conventional larval development somewhat similar to *Macrobrachium rosenbergii,* with a number of larval stages preceding metamorphosis. At least two species, *P. paludosus* and *P. kadiakensis,* both with wide distributions, differ significantly from the others in possessing an abbreviated larval development, which permits their young to reach metamorphosis without feeding (Dobkin 1963). Thus, the young can be produced without elaborate hatchery techniques much in the same way as crawfish are cultured.

Mass culture of *P. kadiakensis* has been demonstrated by Chew (1981) in Georgia. Into an earthen pond of approximately 0.04 ha (0.10 A), which was fertilized at intervals before and after stocking to promote plankton blooms, he introduced approximately 200 brood animals in mid-March. Within 2 months, young were seen, and feeding with a high-protein commercial catfish chow was begun. In mid-November, a yield of 26 kg (58 lb) or about 283,000 individuals was obtained. The calculated harvest for a hectare pond would weigh approximately 648 kg (580 lb/Ac), a respectable yield for a culture season of 6–7 months. Polyculture of *Palaemonctes* with finfish to provide a natural food source in gamefish ponds has been proposed, but would require substantial cover for the shrimp to survive. There are no diseases or parasites of any consequence that would hinder *Palaemonetes* culture. The wide range of *P. kadiakensis* west of the Allegheny Mountains (southern Ontario to the Gulf of Mexico and as far west as northern Mexico) and of its close relative *P. paludosus* east of the Allegheny Mountains (New England to Florida) suggests that these species may be cultured in ponds anywhere in the United States, though it is likely that annual production will be higher in the southern regions.

At present, there is no substantial large-scale culture of any *Palaemonetes* species in the United States. Despite relatively high prices for aquarium and bioassay specimens, it is unlikely that commercial culture will develop until a larger market is identified.

The other small shrimp of interest are tropical, marine representatives of the stenopidean and caridean groups. All are colorful and bring relatively high retail prices in the aquarium trade. Representatives of the benthic stenopedians include the red-and-white-banded coral shrimp, *Stenopus hispidus,* the golden-banded coral shrimp, *Stenopus scutellatus,* and the scarlet flame shrimp, *Microprosthema semilaevis*. The very large caridean group contains many potential aquarium species, including some currently on the market. The painted prawn *(Hymenocera elegans)*, the lady shrimp *(Lysmata grabhami)*, and the candy-striped dancing shrimp *(Hippolysmata wurdemanni)* are among the most frequently encountered. Some of these species practice varying degrees of parasite picking and hence are compatible with many marine finfish.

All these species have "normal" larval development with a series of free-swimming stages between hatching and metamorphosis, but none is so biologically unique that general larval-rearing techniques cannot be modified to suit them. Some species have rather small larvae requiring appropriate live food such as rotifers, while others can feed immediately after hatching on *Artemia* nauplii. *Hippolysmata wurdemanni* is relatively easy to rear and has been cultured in at least six laboratories known to the authors. It can reach adulthood in a few months from metamorphosis, feeding on a wide variety of prepared diets including commercial tropical fish food. Even at wholesale rates, this species is worth hundreds of dollars per pound. Yet, it is one of the less expensive marine ornamental shrimp. Again, limited demand and marketing problems will probably inhibit commercial-scale culture of this and similar species. Possibly the availability of large numbers through artificial propagation would cause the market value to drop which, in turn, could increase demand. All the tropical marine ornamental shrimp require conditions that essentially preclude their being cultured in natural environments. However, while they do require tropical temperatures, necessary conditions can be provided in properly equipped hatcheries anywhere, and the availability of artificial sea salts could permit production without regard to geography. At present, the authors know of no commercial production of these small, ornamental shrimps.

LITERATURE CITED

ANNINOS, P. 1982. A laboratory study of monogenetic embryonic adaptation to salinity and its subsequent effects upon larval development of the grass shrimp *Palaemonetes pugio* Holthuis. M.S. Thesis, Department of Oceanography, Old Dominion University, Norfolk, Virginia.

BARDACH, J. E., RYTHER, J. H., and MCLARNEY, W. O. 1972. Aquaculture: The Farming and Husbandry of Freshwater and Marine Organisms. John Wiley & Sons, New York.

BEARDEN, C. M., CUPKA, D. M., FARMER III, C. H., WHITAKER, J. D., and HOPKINS, S. 1979. Information on establishing a soft shell crab operation in South

Carolina. Rep. to the Fishermen. South Carolina Wildlife and Marine Resources Dept, Charleston, SC

BIGFORD, T. E. 1979. Synopsis of biological data on the rock crab, *Cancer irroratus* Say. NOAA Tech. Rep. NMFS Circ. 426. National Oceanic and Atmospheric Administration, Washington, DC.

BUTLER, T. H. 1964. Growth, reproduction, and distribution of pandalid shrimps in British Columbia. J. Fish. Res. Board Can. **21,** 1403–1452.

BUTLER, T. H. 1970. Synopsis of biological data on the prawn, *Pandalus platyceros* Brandt, 1851. FAO UN Fish. Rep., 4(57), 1289–1315.

CHEW, L. E. 1981. Culture of the grass shrimp. Prog. Fish Cult. **43**(1), 25–26.

CHURCHILL, E. P. 1919. Life history of the blue crab. Bull. U.S. Bur. of Fish. 36, 95–128.

COSTLOW, J. D. 1967. The effect of salinity and temperature on survival and metamorphosis of megalops of the blue crab *Callinectes sapidus*. Helgol. Wiss. Meeresunter. **15,** 84–97.

COSTLOW, J. D., and BOOKHOUT, C. G. 1959. The larval development of *Callinectes sapidus* Rathbun reared in the laboratory. Biol. Bull. **116,** 373–396.

CUPKA, D. M., and VAN ENGEL, W. A. (editors). 1982. Proc. Workshop on Soft Shell Blue Crabs, September 1979, Charleston, South Carolina. Virginia Inst. Mar. Sci. Contrib. No. 1003.

DARNELL, R. M. 1959. Studies of the life history of the blue crab (*Callinectes sapidus* Rathbun) in Louisiana waters. Trans. Am. Fish. Soc. **88**(4), 294–304.

DAUGHERTY, F. M. 1952. The blue crab investigation, 1949–50. Texas J. Sci. **1,** 77–84.

DAVIS, G. E. 1980. Juvenile spiny lobster management. Fisheries **5**(4), 57–59.

DOBKIN, S. 1963. The larval development of *Palaemonetes paludosus* (Gibbes, 1850) reared in the laboratory. Crustaceana **6**(1), 41–61.

DUDLEY, D. L. and JUDY, M. H. 1971. Occurrence of larval, juvenile, and mature crabs in the vicinity of Beaufort Inlet, S.C. NOAA Tech. Rep. NMFS-SSRF-637. National Oceanic and Atmospheric Association, Washington, DC.

FISCHLER, F. J. 1965. The use of catch-effort, catch-sampling, and tagging data to estimate a population of blue crabs. Trans. Am. Fish. Soc. **94**(4), 287–310.

FISHER, W. S., and NELSON, R. T. 1978. Application of antibiotics in the cultivation of Dungeness crab, *Cancer magister*. J. Fish. Res. Board Can. **35**(10), 1343–1349.

FUTCH, C. R. 1965. The blue crab in Florida. Salt Water Fish. Leafl. No. 1. Florida Board of Conservation Marine Lab. St. Petersburg, FL.

GLUDE, J. B. (editor). 1977. NOAA Aquaculture Plan, Stock No. 796–732. National Oceanic and Atmospheric Administration, U.S.Department of Commerce, U.S. Government Printing Office, Washington, D.C.

GUNTER, G. 1950. Seasonal population changes and distribution as related to salinity, of certain invertebrates of the Texas coast, including the commercial shrimp. Publ. Inst. Mar. Sci. Univ. Tex. **1**(2), 7–51.

HAEFNER, P. A., JR. and GARTEN, D. 1974. Methods of handling and shedding blue crabs, *Callinectes sapidus*. Mar. Resources Advisory Ser. No. 8. Virginia Institute of Marine Science, Gloucester Point, VA.

HAEFNER P. A., JR., and VAN ENGEL, W. A. 1972. Methods for shedding rock crabs in winter. Mar. Resource Info. Bull. **4**(15), 1–3. Virginia Institute of Marine Science, Sea Grant Marine Advisory Services, Gloucester Point, VA.

HAEFNER, P. A., JR., and VAN ENGEL, W. A. 1975. Aspects of molting, growth and survival of male rock crabs, *Cancer irroratus,* in Chesapeake Bay. Chesapeake Sci. **16**(4), 253–265.

HAEFNER, P. A., JR., VAN ENGEL, W. A., and GARTEN, D. 1973. Rock crab: a potential new resource. Mar. Resources Advisory Ser. No. 7. Virginia Institute of Marine Science, Gloucester Point, VA.

HARTMAN, M. C. 1977. A mass rearing system for the culture of brachyuran crab larvae. Proc. 8th Ann. Workshop World Maric. Soc. San Jose, Costa Rica.

INGLE, R. M., and WITHAM, R. 1969. Biological considerations in spiny lobster culture. Gulf Carib. Fish. Inst. Univ. Miami. Proc. **21,** 158–162.

IVERSEN, E. S. 1968. Farming the Edge of the Sea. Garden City Press, Letchworth, Hert. Great Britain.

JACHOWSKI, R. L. 1969. Observations on blue crabs in shedding tanks during 1968. Ref. No. 69–24. Seafood Processing Laboratory, University of Maryland, Crisfield, Maryland.

JAWORSKI, E. 1972. The blue crab fishery-Barataria Estuary. Rep. LSU-SG-72-01. Center for Wetlands Resources, Louisiana State University, Baton Rouge, Louisiana.

KANCIRUK, P., and HERRNKIND, W. F. (editors). 1976. An indexed bibilography of the spiny lobsters, family Palinuridae. Sea Grant Program, Rep. No. 8. Univ of Florida, Gainesville, FL.

KELLY, R. O., HASELTINE, A. W., and EBERT, E. E. 1977. Mariculture potential of the spot prawn, *Pandalus platyceros* Brandt. Aquaculture **10,** 1–16.

LABISKY, R. F., GREGORY, JR., D. R., AND CONTI, J. A. 1980. Florida's spiny lobster fishery: an historical perspective. Fisheries **5**(4), 28–37.

MACKAY, D. C. G. 1942. The Pacific edible crab, *Cancer magister*. Fish. Res. Board Can. Bull. No. 62.

MENZIES, R. A. and KERRIGAN, J. M.. 1980. The larval recruitment problem of the spiny lobster. Fisheries **5**(4), 42–46.

OESTERLING, M. J. 1982. Mortalities in the soft crab industry: sources and solutions. Mar. Resource Rep. No. 82–6. Virginia Institute of Marine Science, Gloucester Point, VA.

OESTERLING, M. J. 1984. Manual for handling and shedding blue crabs *(Callinectes sapidus)*. Spec. Rep. in Appl. Mar. Sci. Ocean Engineering No. 271. Virginia Institute of Marine Science, Gloucester Point, VA.

OGLE, J. T., PERRY, H. M. and NICHOLSON, L. 1982. Closed recirculating seawater systems for holding intermolt blue crabs: literature review, systems design and construction. Gulf Coast Res. Lab. Ocean Springs, MS Tech. Rep. Ser. No. 3.

OTWELL, W. S. 1980. Harvest and identification of peeler crabs. Sea Grant Pub. No. MAFS-26, Univ. of Florida, Gainesville, FL.

OTWELL, W. S., CATO, J. C. and HALUSKY, J. G. 1980. Development of a soft crab fishery in Florida. Sea Grant College Rep. No. 31. Florida

PAPARELLA, M. (editor). 1979. Information tips. Mar. Products Lab. Rep. 79–3. University of Maryland, Crisfield, Maryland.

PAPARELLA, M. 1982. Information tips. Mar. Products Lab. Rep. 82–2. University of Maryland, Crisfield, Maryland.

PERRY, H. M., OGLE, J. T., and NICHOLSON, L. C. 1982. The fishery for soft crabs with emphasis on the development of a closed recirculating seawater system for shedding crabs. Proc. Blue Crab Colloq. October 1979, Biloxi, Mississippi, pp. 137–152.

POOLE, R. L. 1966 A description of laboratory-reared zoeae of *Cancer magister* Dana, and megalopae taken under natural conditions, (Decapoda Brachyura). Crustaceana **11**(1), 83–97.

PRICE, V. A., and CHEW, K. K. 1972. Laboratory rearing of spot shrimp larvae *(Pandalus platyceros)* and descriptions of stages. J. Fish. Res. Board Can. **29,** 413–422.

PROCHASKA, F. J., and CATO, J. C. 1980. Economic considerations in the management of the Florida spiny lobster fishery. Fisheries **5**(4), 53–56.

REBACH, S. 1981. Pelletized diet for rock crabs. Prog. Fish Cult. **43**(3), 148–150.

REED, P. H. 1969. Culture methods and effects of temperature and salinity on survival and growth of Dungeness crab *(Cancer magister)* larvae in the laboratory. J. Fish. Res. Board Can. **26**(2), 389–397.

RENSEL, J. E., and PRENTICE, E. F. 1977. First record of a second spawning of the spot prawn, *Pandalus platyceros,* in captivity. U.S. Fish Wildl. Serv. Fish. Bull. **75**(3), 648–649.

RENSEL, J. E., and PRENTICE, E. F. 1979. Growth of juvenile spot prawn, *Pandalus platyceros,* in the laboratory and in net pens using different diets. U.S. Fish Wildl. Serv. Fish. Bull. **76**(4), 886–890.

RENSEL, J. E., and PRENTICE, E. F. 1980. Factors controlling growth and survival of cultured spot prawn, *Pandalus platyceros,* in Puget Sound, Washington. U.S. Fish Wildl. Serv. Fish. Bull. **78**(3), 781–788.

SANDOZ, M., and ROGERS, R. 1944. The effect of environmental factors on hatching, moulting, and survival of zoea larvae of the blue crab, *Callinectes sapidus* Rathbun. Ecology **25**(2), 216–228.

SASTRY, A. N. 1971. Culture of brachyuran larvae under controlled conditions. Proc. Joint Oceanogr. Assembly, Tokyo, Japan, September 1970, pp. 475–477.

SASTRY, A. N. 1977. The larval development of the rock crab, *Cancer irroratus* Say, 1817, under laboratory conditions (Decapoda, Brachyura). Crustaceana **32**(2), 155–168.

SERFLING, S. A. 1972. Recruitment, habitat preference, abundance and growth of the puerulus and early juvenile stages of the California spiny lobster *Panulirus interruptus* (Randall). M.S. Thesis, California State University, San Diego, California

SERFLING, S. A., and FORD, R. F. 1975a. Ecological studies of the puerulus larval stage of the California spiny lobster, *Panulirus interruptus.* U.S. Fish. Wildl. Serv. Fish. Bull. **73**(2), 360–377.

SERFLING, S. A., and FORD, R. F. 1975b. Laboratory culture of juvenile stages of the California spiny lobster *Panulirus interruptus* (Randall) at elevated temperature. Aquaculture **6,** 377–387.

SHLESER, R. A. 1976. Development of crustacean aquaculture. 10th Eur. Symp. Mar. Biol. Ostend, Belgium, September 1975, **1,** 455–471.

SIMMONS, D. C. 1980. Review of the Florida spiny lobster resource. Fisheries **5**(4), 37–43.

SPOTTE, S. 1979. Fish and Invertebrate Culture, Water Management in Closed Systems. John Wiley & Sons, New York.

SULKIN, S. D. 1974. Factors influencing blue crab population size: nutrition of larvae and migration of juveniles. Chesapeake Biol. Lab. Annu. Rep. Ref. No. 74–125. Center for Environmental and Estuarine Studies, Solomons, MD.

SWEAT, D. E. 1968. Growth and tagging studies on *Panulirus argus* (Latreille) in the Florida Keys. Tech. Ser. **57,** pp. 1–30. Florida Board of Conservation Marine Research Laboratory, St. Petersburg, FL.

TAGATZ, M. E. 1965. The fishery for blue crabs in the St. Johns River, Florida, with special reference to fluctuation in yield between 1961 and 1962. U.S. Fish Wildl. Ser. Spec. Sci. Rep. Fish. No. 501.

TAGATZ, M. E. 1968a. Biology of the blue crab, *Callinectes sapidus* Rathbun, in the St. Johns River, Florida. U.S. Fish. Wildl. Serv. Fish Bull. **67**(1), 17–33.

TAGATZ, M. E. 1968b. Growth of juvenile blue crabs, *Callinectes sapidus,* in the St. Johns River, Florida. U.S. Fish Wildl. Serv. Fish. Bull. **67**(2), 281–288.

TAGATZ, M. E., and HALL, A. B. 1971. Annotated bibliography on the fishing industry and biology of the blue crab, *Callinectes sapidus*. NOAA Tech. Rep. NMFS SSRF-640. National Oceanic and Atmospheric Admin., U.S. Dep. Commerce, Washington, DC.

TAMM, G. R. 1980. Spiny lobster culture: an alternative to natural stock assessment. Fisheries **5**(4), 59–62.

TING, R. Y. 1973. Culture potential of spiny lobster. Proc. 4th Annu. Workshop World Maric. Soc. pp. 165–170.

U.S. DEPARTMENT OF COMMERCE. 1981. Fisheries of the United States, 1980. Current Fishery Stat. No. 8100. Washington, DC.

U.S. DEPARTMENT OF COMMERCE. 1984. Fisheries of the United States, 1983. Current Fishery Stat. No. 8320, Washington, DC.

VAN ENGEL, W. A. 1958. The blue crab and its fishery in Chesapeake Bay. Part 1. Reproduction, early development, growth, and migration. Commer. Fish. Rev. **20**(6), 6–17.

VAN ENGEL, W. A. 1962. The blue crab and its fishery in Chesapeake Bay. Part 2. Types of gear for hard crab fishing. Commer. Fish. Rev. **24**(9), 7–10.

WALDRON, K. W. 1958. The fishery and biology of the Dungeness crab (*Cancer magister* Dana) in Oregon waters. Contrib. No. 24. Fish Commission of Oregon, Portland, Oregon.

WARNER, R. E., COMBS, C. L., and GREGORY, JR., D. R. 1977. Biological studies of the spiny lobster, *Panulirus argus* (Decapoda; Palinuridae) in south Florida. Gulf Caribb. Fish. Inst. Univ. Miami Proc. **29,** 166–183.

WARNER, W. W. 1976. Beautiful Swimmers, Watermen, Crabs and the Chesapeake Bay. Little, Brown & Company, Boston, Massachusetts.

WELSH, J. P. 1974. Mariculture of the crab *Cancer magister* (Dana) utilizing fish and crustacean wastes as food. Sea Grant Project HSU-SG-4. Humboldt State University, Arcata, CA.

WICKINS, J. F. 1972. Experiments on the culture of the spot prawn *Pandalus platyceros* Brandt and the giant freshwater prawn *Macrobrachium rosenbergii* (de Man). Fish. Invest. Minist. Agric. Fish. Food (GB) Ser. II. **27**(5), 1–23.

WICKINS, J. F., and BEARD, T. W. 1978. Prawn culture research. Lab. Leafl. 42. Ministry of Agriculture, Fisheries and Food, Directorate of Fisheries Research, Lowestoft, Great Britain.

WILLIAMS, A. B. 1965. Marine decapod crustaceans of the Carolinas. U.S. Fish Wildl. Serv. Fish. Bull. **65**(1), 1–298.

WILLIAMS, A. B. 1971. A ten-year study of meroplankton in North Carolina estuaries: annual occurrence of some brachyuran development stages. Chesapeake Sci. **12**(2), 53–61.

WILLIAMS, A. B. 1974. The swimming crabs of the genus *Callinectes* (Decapoda: Portunidae). U.S. Fish Wildl. Serv. Fish. Bull. **72**(3), 685–798.

WITHAM, R., INGLE, R. M., and JOYCE, JR., E. A. 1968. Physiological and ecological studies of *Panulirus argus* from the St. Lucie estuary. Techn. Ser. **53,** 11–31. Florida Board of Conservation Marine Research Laboratory, St. Petersburg, FL.

6

Oyster Culture

Victor G. Burrell, Jr.

Introduction
Natural History
  Reproduction and Development
  Feeding
  Habitat
The American Oyster *(Crassostrea virginica)*
  Culture
  Diseases, Parasites, and Predators
  Economic Overview and Outlook
The Pacific Oyster *(Crassostrea gigas)*
  Culture Methods
  Diseases, Predators, and Parasites
  Climatic Restrictions and Other Problems
  Economic Overview and Outlook
The European Oyster *(Ostrea edulis)*
  Culture
  Diseases, Predators, and Parasites
  Climatic Restrictions and Other Problems
  Economic Overview and Outlook
The Olympia Oyster *(Ostrea lurida)*
  Biology
  Culture Techniques
  Diseases, Parasites, and Predators
  Climatic Restrictions and Other Problems
  Economic Outlook and Overview
Oyster Hatcheries
  Hatchery Methods
  Larval Culture
Problems Facing the Oyster Culture Industry
  Outlook
Summary
Literature Cited

INTRODUCTION

Oyster culture may well be one of the oldest forms of aquaculture, dating back at least to the Roman Empire (Korringa 1976a). In the United States very rudimentary culture activities—such as moving oysters to take advantage of particular growing conditions or for convenience—were practiced in the eighteenth century or earlier (Brooks 1879). Many minimal culture activities—such as planting oyster shell to improve oyster grow out bottoms or to catch spat—still constitute some of today's oyster culture operations. Until the mid-1950s, this and harvesting natural beds had been adequate to meet demand.

Aquaculture accounts for 40% of the oysters marketed each year in the United States, amounting to 10.8 million kg (23.7 million lb), valued at $37 million, in 1978. This total includes oysters cultured with all degrees of husbandry from the very rudimentary to the most sophisticated (Thompson 1982).

Four species of oysters are presently under cultivation in the continental United States (Fig. 6.1). *Crassostrea virginica,* the American oyster, is the principal species on the East Coast and Gulf of Mexico and accounts for most U.S. oyster landings, while *Crassostrea gigas,* the Pacific oyster, is the prime culture species on the West Coast and second in volume produced. Two species of the genus *Ostrea* are also cultivated to a lesser extent. *Ostrea edulis,* the European oyster, is grown in Maine, and *Ostrea lurida,* the native West Coast Olympia oyster, is the basis for a small industry in Washington state and Oregon. Oyster landings have been declining since early in the twentieth century. (Fig. 6.2).

NATURAL HISTORY

Reproduction and Development

The cupped oyster *(Crassostrea)* and the flat oyster *(Ostrea)* differ in their reproduction strategy: male and female *Crassostrea* release sex products into the water where fertilization and early development take place, whereas male flat oysters *(Ostrea)* expel sperm via exhalant currents, which are then drawn in by the inhalant current of the female. Eggs in *Ostrea* are retained in the pallial cavity near the gills of females where fertilization and early development take place. The larvae are brooded in the pallial cavity for 8 to 10 days or longer and then expelled as free-swimming veligers. Both the cupped and flat oyster are protandrous and can undergo sex changes. They first develop as males, and then may change to females. It is possible for

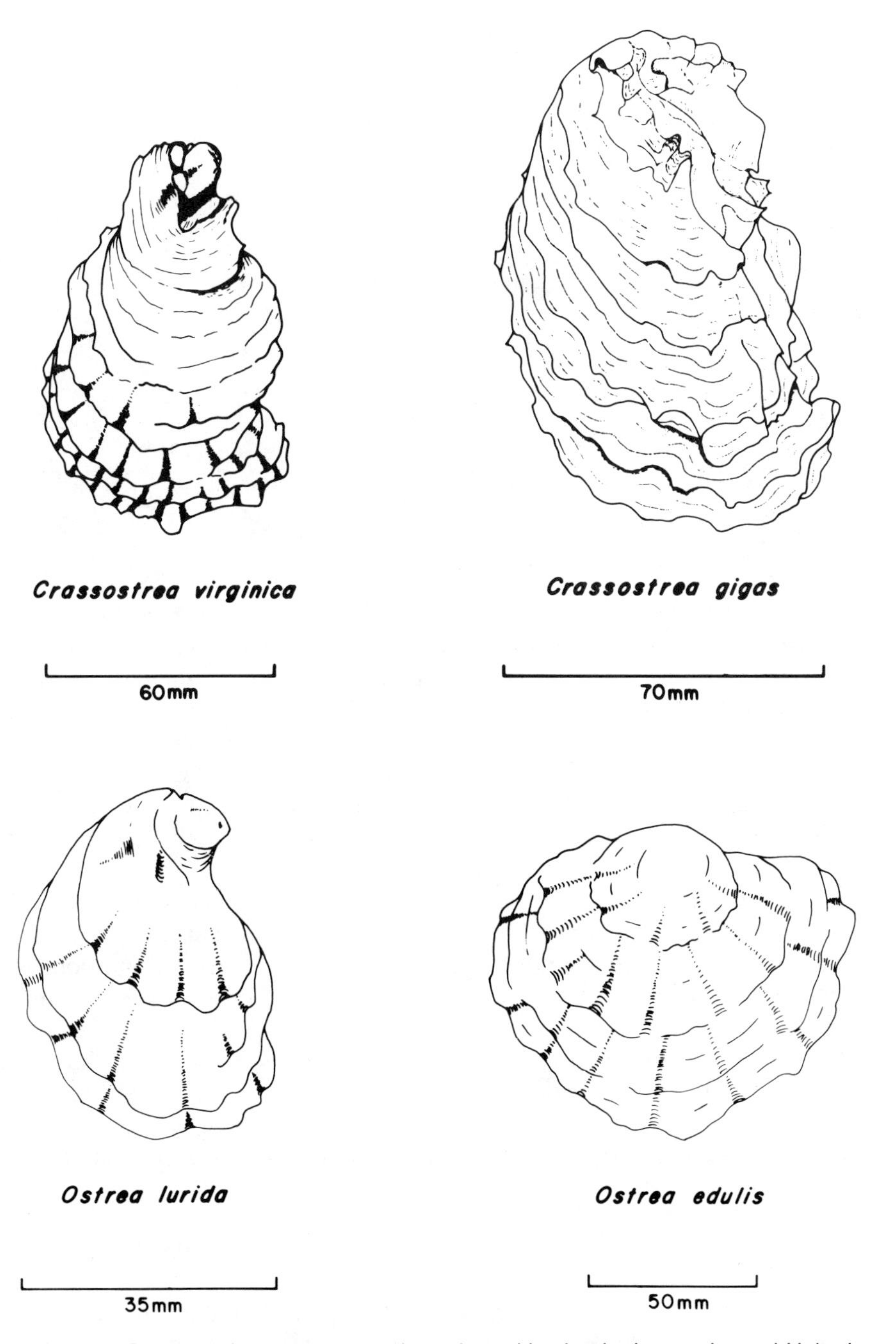

FIG. 6.1 Species of oysters presently under cultivation in the continental United States.

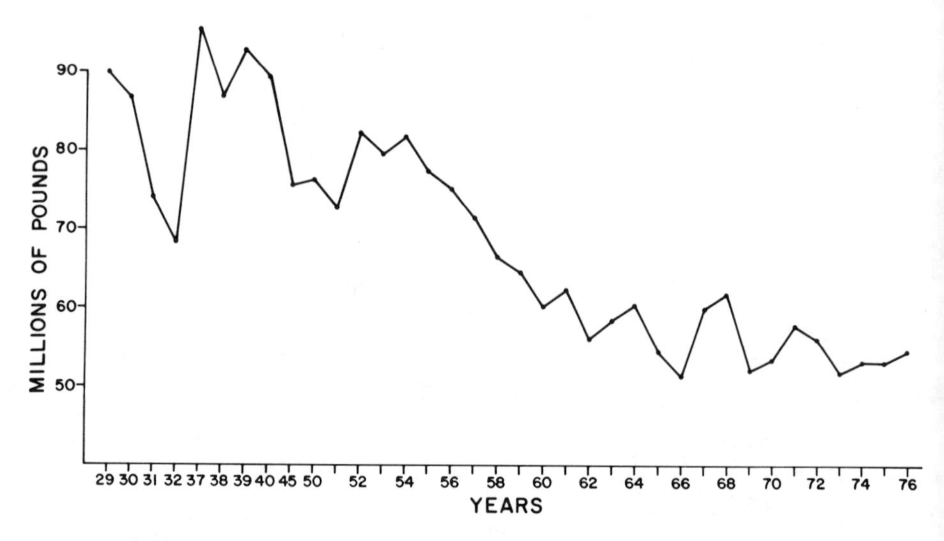

FIG. 6.2 Oyster landings for selected years, 1929–1976.

Ostrea to change sex more than once in a season, whereas *Crassostrea* usually remain the same sex throughout a season, but may change before the next.

Larval development is similar in both genera and may last up to 3 weeks. The first true larval stage, the trochophore, is motile, being propelled by a ring of beating cilia. The next stage, the veliger, is highly complex and possesses a strong swimming organ, the velum. With continued development the shell becomes more prominent, and a foot appears, which enables the larvae to crawl about the substrate in search of an attachment site. Once a suitable hard surface is found, the young oyster permanently cements itself to it and loses many of its larval organs such as the velum and foot. The American oyster may attach to almost any hard surface (glass, plastic, or wood), but the European oyster appears to prefer material containing calcium carbonate. Oyster or other mollusk shells are most often utilized by settling oysters, which at this stage are called spat; the process of attachment is called set or spatfall. Once attached, oysters remain in the same spot and mature.

Feeding

Oysters are filter feeders, drawing phytoplankton and other organic particles into the mantle cavity via an inhalant current. Suitable food and other material is passed into the digestive system through the mouth, and other unsuitable material is rejected by the mouth palps and expelled as pseudofeces (Galtsoff 1964; Korringa 1976a).

Habitat

As a general rule, *Crassostrea* requires temperatures greater than 19.5°C (67°F) for spawning, larvae need temperatures above 20°C (68°F) for proper development, and mature oysters grow well at 10°–30°C (50°–86°F) or higher (Castagna and Chanley 1973; Galtsoff 1964). *Ostrea* spawns at 15°–18°C (59°–64°F), larval development occurs best from 13° to 18°C (55° to 64°F), and mature oysters grow well at temperatures between 10° and 24°C, or 50° and 75°F, (Clime and Hammil 1979; Hidu and Richmond 1974; Korringa 1976b). Although *Crassostrea* prefers warmer water than does *Ostrea,* it can withstand continued below freezing temperatures, whereas *Ostrea* cannot.

Crassostrea virginica and *C. gigas* are estuarine species and tolerate a wide range of salinities. Larval development takes place best above 15–16‰ (Bardach *et al.* 1972; Haven *et al.* 1978a). *Ostrea edulis* and *O. lurida* thrive best at salinities above 25‰ (Bardach *et al.* 1972) Korringa 1976b).

Cultivation of oysters in the United States is usually restricted to firm, stable subtidal or intertidal bottoms. Cupped oysters tolerate more turbid water than flat oysters. *Crassostrea virginica* grows from the intertidal zone to depths greater than 30 m (98 ft), while *C. gigas* is most often found intertidally in nature but may be cultivated subtidally. *Ostrea edulis* is seldom present intertidally because it does not tolerate high temperatures or frost (Korringa 1976b; Bardach *et al.* 1972). *Ostrea lurida* is always intertidal in Puget Sound; however, some are found subtidally (K. Chew, University of Washington, Seattle, personal communication).

In nature oysters occur where hydrological conditions permit and suitable hard substrate for settlement of spat is present. In a great number of instances, best spat fall does not occur in the best growout area. Much effort has been expended to enhance production of seed in nature and to produce seed in hatcheries for transfer to suitable growing areas.

THE AMERICAN OYSTER *(Crassostrea virginica)*

The American oyster probably has been studied more than any other invertebrate species. Several very extensive bibliographies give evidence to the wide range of subjects that have been addressed (Baughman 1948; Galtsoff 1972; Joyce 1972; Breisch and Kennedy 1980). This account is taken largely from the works of Galtsoff (1964), Haven *et al.* (1978), Kennedy and Breisch (1981), Korringa (1976a), McKenzie (1981), and Yonge (1960).

Crassostrea virginica grows subtidally thoughout its range from Maine to the Gulf of Mexico. Large intertidal beds are also present on the eastern shore of Virginia and south of Cape Fear, North Carolina, to northeast Florida (Haven and Burrell, 1982).

The American oyster grows well in mesohaline (salinities 5–18‰) and polyhaline (salinities 18–30‰) zones in estuaries. While a temperate-zone species, this oyster withstands freezing temperatures as well as direct exposure to the summer sun. It spawns at temperatures above 19°C (66°F), and spawning may occur from May through October in the southern end of its range. Intertidal oysters tend to be intermittent spawners; that is, a single oyster may discharge eggs or sperm intermittently over a 3- or 4-month period. Subtidal oyster populations, however, generally exhibit one or two short periods of peak activity (Haven *et al.* 1978a). Larval development takes place externally, and free-swimming stages last from 2 to 3 weeks. Larvae develop best in salinities between 17.5 and 22.5‰ (Castagna and Chanley 1973).

Culture

Efforts to enhance production of the American oyster range from merely providing a clean surface for spat settlement to high-technology hatchery production of seed and grow out in a recirculating closed mariculture system (Epifanio and Mootz 1976). In some instances, the only culture technique employed is towing a crab dredge with an open bag over an oyster bed to turn up shell to provide a clean substrate for spat settlement.

The most common practice in farming American oysters consists of harvesting seed oysters in an area where oyster setting occurs and transplanting them to an area where growth and sometimes survival is made better (Fig. 6.3). Oyster or other molluscan shell, called cultch, is often planted in a setting area to provide a suitable substrate for attachment by larval oysters when they reach the plantigrade or setting stage. *Crassostrea* requires a cleaner surface for attachment than does *Ostrea,* so the planting of cultch must be timed to coincide with peak of spawning activity. Cultch density may range from 125,000 to 250,000 kg/0.4 ha (5000 to 10,000 bushels/Ac) according to bottom condition. Sometimes chicken wire bags containing shell (shell bags) are set out to catch seed (Haven and Garten 1972). Oyster seed are

FIG. 6.3 Seed oysters tonged from the James River, Virginia, are loaded aboard a "buy" boat. The James River seed beds are world famous.
Courtesy of Virginia Institute of Marine Science.

removed from beds by tongs or dredges, depending on local regulations, and then spread on growing grounds at densities of 12,500–25,000 kg/0.4 ha (500–1000 bushels/Ac), depending on the estimated number of seed per bushel. At Fisher Island, New York, seed are produced by hanging shell strings from rafts in ponds. Growers expect to harvest a bushel of market oysters for a bushel of seed planted. Good seed areas must have the necessary hydrography to deliver a suitable quantity of larvae capable of developing to the setting stage, a firm bottom with a minimum of resuspended sediment, be relatively free of disease, and have salinities low enough to protect spat from predation yet high enough for survival.

Since 1950, seed supplies have not been consistent or adequate to supply the needs of oyster growers in many areas. Although seed have been transported from regions of high seed production to areas of need, this has not been very successful and is not considered a solution to the problem. This has led to increased interest in hatchery produced oyster seed.

Grow-out areas generally must have the same attributes as seed beds, as well as ample currents and enough food for rapid growth. They must also be free of pollutants. American oysters reach market size (75 mm or 3 in.) in less than 2 years in the South but may take as long as 5 years in New England.

Where American oysters are grown intertidally, culture may consist only of distributing shell on grounds that have been harvested the past winter. This is done in summer and seed caught on these grounds may be harvested again in a year to a year and a half. Sometimes these beds are self-renewing and require little or no attention between harvests. In the past, intertidal harvest was done entirely by hand, but recently experimental mechanical harvesters were introduced in North Carolina and South Carolina (Fig. 6.4). Many of these oysters are steam-processed and canned as a cooked product or sold in the shell for roasting instead of being shucked as a fresh product.

The greatest American oyster production comes from subtidal beds. These beds may be well within estuaries where salinities are low enough to reduce losses from diseases and predation. Bottoms may be improved with oyster shell to support seed oysters, which are planted on them in spring and fall (Fig. 6.5). Subtidal oysters are harvested by a variety of methods. Hand tongs, box dredges, mechanical dredges, and patent tongs are used on leased grounds, but on public beds, oysters harvest methods usually are restricted. Mayland restricts harvest of public beds to use of hand tongs and box dredges operated from sailing vessels (Figs. 6.6 and 6.7). The majority of these oysters are hand-shucked and sold as raw oysters in one form or another.

Except in some areas in Maine, off-bottom culture of the American oyster in East Coast estuaries has not proven economically practical (Aprill and Maurer 1976; Walker and Gates 1981).

FIG. 6.4 The pick-up head of a mechanical oyster harvester designed and built at Clemson University in South Carolina to harvest intertidal oysters.

FIG. 6.5 Oyster shell is planted on oyster beds to stiffen up the bottom and to provide cultch for spat.

FIG. 6.6 A box dredge used to harvest market oysters on lease grounds in Virginia.
Courtesy of Virginia Institute of Marine Science.

FIG. 6.7 Maryland skipjacks are permitted to harvest oysters on public beds under sail power 3 days per week and with motor power on 2 days.

Diseases, Parasites, and Predators

Andrews (1979), Leibowitz (1978a), Farley (1978), and Sindermann (1977) have recently reviewed oyster diseases. Several diseases of the American oyster are caused by protozoans. These pathogens have greatly modified oyster culture practices in Maryland and Virginia. One *(Minchinia nelsoni)*, responsible for the Delaware Bay disease (MSX), is active in salinities above 15‰, with infections occurring May through October. Mortality rates exceeding 60% per year occurred in oyster beds in Chesapeake and Delaware bays in the late 1950s and early 1960s (Andrews and Wood 1967; Haskin *et al.* 1966). Efforts to breed disease-resistant stocks have met with limited success, and oysters spawned in areas where the disease is prevalent have shown some acquired resistance to MSX. At the present, however, the most effective means of coping with this disease where it can be epizootic is to plant in salinities below 15‰.

A similar disease caused by *Minchinia costalis,* called Seaside disease (SSO), is present in salinities greater than 20‰ from Long Island Sound to Chesapeake Bay. Deaths occur from mid-May to early July (Andrews *et al.* 1962).

A third source of heavy mortality in the American oyster is *Perkinsus marinus,* or "Dermo," (Andrews and Hewatt 1957; Quick and Mackin 1971; Ray 1954). This disease is responsible for mass deaths of oysters from Chesapeake Bay south to the Gulf of Mexico in salinities above 15‰. Seed oysters up to a year of age are not killed by "Dermo," but older oysters must be moved to lower salinites to avoid possible catastrophic losses in regions where this disease is epizootic.

Virus diseases of oysters have also been reported. Sindermann (1977) describes methods of diagnosing these maladies. Several diseases that occur in oyster larvae are known from hatchery experience and include bacterial and fungal infections (Liebowitz 1978b; Sindermann 1977). When present, almost total mortality occurs (Elston 1979a; Sindermann 1977). Diagnostic methods are described in Elston (1979b), Elston *et al.* (1981), Elston *et al.* (1982), and Sindermann (1977). Less than optimum culture conditions such as high organic load in culture waters may also contribute to poor larval vigor and encourage blooms of ciliates to invade shells of young *C. Virginica* (Elston 1979a). Elston (1979a) suggests that problems of larval disease may be minimized by careful selection of stocks, which should be disease free for at least a year and display normal growth and development. A short quarantine should be required during which seed animals should be tested for bacteriological pathogens as well as examined for indications of suboptimal culture conditions (ciliates and protozoans are abundant in cultures). In addition, strict maintenance of optimal culture conditions (for example, by monitoring incoming water quality, sterilizing algae food stocks, and controlling metabolite levels) helps reduce diseases in oyster larvae.

Predators of American oysters are numerous and include starfish, fish, gastropods, flatworms, and crabs.

Starfish. Specially constructed mops may be dragged over oyster beds to entangle starfish, which may then be hoisted from the water and killed by a hot-water dip. Quick lime applied to the oyster bed is also effective in controlling this pest (MacKenzie 1977).

Fish. Cownose rays *(Rhinoptera bonasus)* and black drum *(Pogonias cromis)* have inflicted heavy damage on some oyster beds. They are able to crush oysters with strong teeth. Cownose rays often smother oysters as they excavate hollows to lie in (Galstoff 1964; Merriner and Smith 1979).

Gastropods. Oyster drills *(Urosalpinx cinerea), Eupleura caudata,* and conchs *(Thais haemostoma)* are the greatest enemy of oysters. These gastropods consume their prey after boring a hole through the bivalve's shell by combined chemical and mechanical means. They cause serious losses in oyster grounds where salinities remain above 15‰. Various methods such as trapping, dredging, and rotation of growing areas have provided limited control of these pests, but an effective method to cope with predatory gastropods still remains to be found (Carriker 1955; Galtsoff 1964).

Flatworms. The oyster leech *(Stylochus ellipticus)* may cause serious losses on occasion, especially in young oysters (Galtsoff 1964).

Crabs. The blue crabs *(Callinectes sapidus)* and others, such as mud crabs, are able to crack the shell of young oysters with their claws to reach the meat inside. Heavy losses as a result of this predation may occur (Lunz 1947).

Economic Overview and Outlook

The majority of American oysters are harvested from public grounds and thus are not considered to be cultivated. However, several states (Maryland, Virginia, and Louisiana) have extensive shell- and seed-planting programs to maintain and rehabilitate state-owned beds (Table 6.1). Maryland, the largest producer of *C. virginica* (6.76 million kg or 14.9 million lb in 1976), at present has only few acres under lease and a moratorium on further leasing; the cultivated crop accounts for only about 5% of the oysters produced there in 1976. Virginia, which encourages leasing of grounds that are not natural producers of oysters, annually produced more than 6.49 million kg (14.3 million lb) in private culture during the mid-1950s. In 1976, however, Virginia private leases accounted for only 1.50 million kg (3.3 million lb). Haven *et al.* (1978a) attributed this decline to loss of grounds due to disease and to the unwillingness of many growers to plant marginal acreage because of the high cost of seed, labor, and services, especially when they are competing for the same market with state-subsidized Maryland shell stock. They felt that if some of the public grounds that are not now important producers, but have good historical potential, were made available for leasing, more oysters would be grown again. Oyster farmers in Virginia were able to outproduce the public beds seven to one over the period 1931–1960. This leads to the conclusion that in some areas production is seriously limited by lease restrictions.

TABLE 6.1. AMERICAN OYSTER LANDINGS FROM PUBLIC AND PRIVATE GROUNDS—1976

	Public Grounds ($kg \times 10^6$)	Private Grounds ($kg \times 10^6$)
Maryland	6.44	.30
Virginia	1.22	1.50
New England	.03	.06
Mid Atlantic	.01	1.59
South Atlantic	.13	.64
Gulf	5.62	4.13
Total	13.45	8.22

Source: Fisheries Statistics of the United States 1976. U.S. Department of Commerce Stat. Dig. 70.

At the present time, the American oyster industry, because it is made up of many small firms, is unable to address other factors that restrict expansion (Office of Fisheries Development NMFS 1977). These include such things as (1) controlling labor costs through mechanization of planting, harvesting, and processing; (2) developing new products that would have increased shelf life, appeal to the convenience-minded consumer, and garner a better share of the institutional business; and (3) mounting a strong marketing campaign to increase demand for oysters.

THE PACIFIC OYSTER *(Crassostrea gigas)*

This discussion of Pacific oyster culture is taken chiefly from Breese and Malouf (1975), Chew (1979), Korringa (1976b), Quayle (1969), and Quayle and Smith (1976).

This species was successfully introduced as seed from Japan before 1920 (Chew 1979). Two strains, Miyagi and Kumamoto, originating from prefectures of the same names in Japan are now being cultured on the West Coast of the United States (Chew 1979; Woelke 1955).

Culture Methods

No wild-stock fishing of *C. gigas* exists on the West Coast of the United States. Culturists either import seed from Japan, use hatchery-produced seed, or catch natural seed on suspended cultch. The majority of *C. gigas* seed was imported from Japan until the early 1970s. Since that time, naturally occurring spat fall has supplied an increasingly greater proportion of grower needs. Hatchery-produced seed are being used by many culturists, particularly those with a vertically integrated operation.

Spat fall of commercial intensity occurs sporadically in Willapa Bay, and more regularly in northern Hood Canal (Dabob and Quilcene Bay), Washington. Gonadal products develop in warm water in spring and spawning starts in June. Natural seed are caught on strings consisting of Pacific oyster shell strung on galvanized wire suspended from racks or rafts (Fig. 6.8). Biologists with the Washington State Department of Fisheries monitor larvae present in the water during summer and provide this information to culturists so that they may put out clean cultch when the setting stage is reached (Korringa 1976a; Westley 1968).

Spatted cultch on shell strings is moved from racks or rafts the following spring to grow-out areas. The most common method of culture is on intertidal beds and is called bottom culture. If this method

FIG. 6.8 Putting out shell strings for catching natural seed in Dabob Bay, Washington.
Courtesy of K. Chew.

is employed, cultch with attached seed are removed from strings and planted on beds fairly high in the intertidal zone and left for about 22 months (Fig. 6.9). The oysters are then taken up, the clusters broken up and replanted in the lower intertidal zone for fattening. Some growers do not break up clusters or replant. One summer season is sufficient for fattening, and harvests start in fall. Oysters are taken from the bed in each of these operations by hand or dredge and shoveled by hand or washed overboard by a high-pressure hose on the fattening beds (Korringa 1976a). Hatchery seed are treated in the same manner.

In earlier years, when seed were regularly imported from Japan, they usually arrived in March and were held in the original shipping cases in the upper intertidal zone until the water warmed. The construction of the shipping boxes is such that water flows freely through them, bathing the young oysters on each tide. Late in April these oysters are moved to growing grounds. If bottom culture is practiced, 25 to 100 cases, each containing 10,000–14,000 spat, are planted per 0.4 ha (1 Ac). Culture methods thereafter are the same as that for naturally caught spat.

Three forms of off-bottom culture are used by a small number of West Coast oyster growers. These are rack, stake, and raft culture. The advantages of off-bottom methods are that areas with soft bot-

FIG. 6.9 A bed of *C. gigas* growing intertidally in Washington state.
Courtesy of K. Chew.

toms may be used, some predators such as drills cannot reach the young oysters, and growth is much better than with bottom culture (Brett *et al.* 1972).

Racks, constructed of creosoted material, are located in the intertidal and upper subtidal zones of estuaries. They consist of posts driven into the bottom and joined at the top by stringers fastened both lengthwise and crosswise. Spatted cultch strung on wire with a length of plastic tubing separating each shell are hung from the stringers so that the lowermost shell is suspended above the bottom (Fig. 6.10).

Stake culture involves suspending a spatted shell from a single stake in the intertidal zone. A 25 mm (1 in.)-diameter stake, 45 cm (18 in.) long, is driven about two-thirds its length into the bottom and a spatted shell hung from a 5-cm (2-in.) nail driven part way into the stake near the top (Fig. 6.11).

Rafts may be of any size and constructed from a variety of materials; the chief requirements are that they be sturdy enough to withstand sea conditions in the area in which they are moored and have enough buoyancy to maintain strings of suspended oysters above the bottom as they increase in weight with growth. Strings are similar to those used in rack culture. Sites for mooring rafts must provide protection from storm winds and waves, be sufficiently deep to assure that reasonably long strings will not rest on the bottom at low tide, and have waters that are conducive to rapid growth. One 25-mm seed grown on rafts will reach market size in a 9-month (March to Novem-

FIG. 6.10 Rack culture in Humboldt Bay, California.
Courtesy of K. Chew.

ber) growing season. Quayle (cited in Brett *et al.* 1972) found that oyster production per unit area with raft culture greatly exceeded that with bottom culture with only moderate increases in cost.

Diseases, Predators, and Parasites

High disease-related mortality of Pacific oysters occurs in summer in some areas during some years. The causative organism is not known,

FIG. 6.11 Stake culture in Coos Bay, Oregon.
Courtesy of K. Chew.

but it may be a bacterium such as *Vibro* (Chew 1979). These oysters may be more susceptible to disease during stress associated with the gametongenic cycle, or environmental stress alone may cause deaths (Perdue *et al.* 1981). Elston (1979a,b, 1980) and Leibowitz *et al.* (1978) have described several diseases of larval *C. gigas*. These include viral, fungal, and bacterial infections. Control of these pathogens has been discussed by Leibowitz (1978b), and it is a concern of hatchery operators.

The Japanese oyster drill *(Ocenebra japonica)*, several starfish, crabs of the genus *Cancer*, the flatworm *Pseudostylochus ostreophagus*, and the bat ray fish *(Holorhinus californicus)* are important predators of the Pacific oyster. The same control methods suggested for the American oyster can be used to mitigate losses of *C. gigas* to predators, although some, such as using mops to rid beds of starfish, are not utilized on the West Coast.

Fouling organisms, such as barnacles, compete with oysters for settling space and food, but are more of a concern in off-bottom culture than in bottom culture because they add weight to the oyster that must be supported by the suspension structure and be lifted at harvest time.

Parasites of *C. gigas* include the cyclopoid copepod *Mytilicola orientalis)* and ciliates such as *Ancistrocama* and *Trichodina*.

Climatic Restrictions and Other Problems

Water temperature may not rise sufficiently in some areas during some years to trigger spawning of Pacific oysters. When spawning does not occur, developed gametes are reabsorbed, and during this period, oyster quality is poor. Spawned oysters feed rapidly and restore glycogen levels much quicker than do those with reabsorbed gametes. In the northern Hood Canal area, stratification occurs in summer with a layer of warmer water overlying much colder deeper water. Intensity of spat set has been directly correlated with water temperature and thickness of this warmer layer. When this layer is disrupted by winds, intensity of set is drastically curtailed, and in some years this results in a spat failure (Westley 1968).

Economic Overview and Outlook

Wild stocks of *Crassostrea gigas* are not harvested commercially on the West Coast, so culture of some form accounts for all production. In 1976, Washington state produced 2.48 million kg (5.46 million lb), followed by California with 0.32 million kg (0.70 million lb) and Oregon with 0.12 million kg (0.26 million lb) of meat. A small Pacific oyster industry exists in Hawaii and Alaska, but commercial production is small at this time. There are four commercial hatcheries in Washington, three in California, and one in Oregon. Each state has research laboratories produing seed: two in Washington, one in California, and one in Oregon (Chew 1979).

Annual Pacific oyster production in the United States peaked in 1954–1956 when it exceeded 4.99 million kg (10.98 million lb) of meat; since that time production has decreased to an annual level of approximately 2.72 million kg (5.98 million lb) since about 1970. This drop in the production of West Coast oysters can be traced to several factors. First, low-cost imports from Japan and Korea initially captured the entire market for canned oysters. Second, short shelf life and poor access to inland markets have restricted distribution of Pacific oysters. Third, competition with American oysters for the fresh trade has curtailed markets. Finally, almost no effort has been put forth to promote sales of Pacific oysters because the industry is made up of small companies with little money for this type of promotion. High production costs and limited markets have forced most growers to utilize only their best grounds, where grow out is more economical, in order to reduce production costs (Chew 1979; Glude 1974).

Specialty products such as soups, stews, and breaded oysters prepared for institutional service would take advantage of the increase in the number of people eating out. Yearling oysters, marketed for half-shell consumption, would require a shorter grow-out time than is now usual and could bring a high market price. Innovative culture methods, such as off-bottom grow out and use of the eyed-larvae technique instead of cultch-set seed from hatcheries, offer potentially more favorable production economics. Finally, an aggressive marketing program to regain old consumers and attract new consumers would increase demand for all oyster products.

THE EUROPEAN OYSTER *(Ostrea edulis)*

Culture of the European oyster in the United Staes takes place almost exclusively in Maine. The account of this species is taken chiefly from Clime and Hamill (1979), Hidu and Richmond (1974), Korringa (1976b), Packie *et al.* (1976), and Walne (1974).

Culture

European oysters, when ripe, spawn as water temperature reaches 15°–18°C (59°–64°F). The seed is not collected in Maine from natural beds because there are no large natural beds and temperatures in high-salinity areas do not rise sufficiently to permit consistent spawns. Therefore the entrepreneur depends entirely on hatchery stock.

Growers may purchase either small seed (less than 10 mm) or larger seed (10–25 mm). Those buying smaller seed or hatcheries selling larger seed place 2- to 3-mm seed in surface trays for 3 to 6 weeks beginning in early or mid-April. As they grow, the oysters are sorted and moved to larger mesh trays (usually two sorts in the nursery phase), and the density is decreased in trays at this time (Fig. 6.12). Trays must be cleaned to control fouling (Fig. 6.13). This may be done once a week or even daily.

All grow out in Maine is off bottom in trays or nets that are suspended from rafts or long lines (Figs. 6.14 and 6.15). This phase of

FIG. 6.12 Tending nursery trays of hatchery-reared shellfish in Maine.
Courtesy of H. Hidu.

FIG. 6.13 Removing fouling in European oysters with a hose at Maritec, Inc., South Bristol, Maine.
Courtesy of H. Hidu.

FIG. 6.14 Tray modules are prepared for deployment on long lines at Maritec, Inc., South Bristol, Maine.
Courtesy of T. Archambault.

FIG. 6.15 Air drying of lantern nets to control fouling at Culture Fisheries, Inc., Damariscotta River, Maine.
Courtesy of R. Gillmor.

culture begins in late spring when spat are 10 mm or more in length. Oysters are graded several times during summer, and slow growers are separated from faster growers, which can reach 30–40 mm (1–1.5 in.) in length. Trays carrying all first-year oysters and slower-growing second-year oysters are lowered to the bottom or suspended from a sheltered raft for overwintering. In spring, these oysters will again be suspended in the water column to continue grow out. Care must be taken to locate the young oysters where they will not be subject to temperatures below 0°C (32°F) for periods as long as 6 weeks.

Market-sized oysters should be held where they can be retrieved in the coldest months.

Diseases, Predators, and Parasites

Maine's European oyster populations have no known diseases. Drills, crabs, and starfish prey on European oysters. Off-bottom culture prevents adult predators from reaching oysters, but motile starfish and crab larvae may settle in grow-out structures and must be removed or killed.

Climatic Restrictions and Other Problems

Problems with winter ice and temperatures have been mentioned. Low summer temperatures are a boon in that spawning does not take place until perhaps late summer and brooded larvae are not present to prevent early and mid-summer consumption of this species grown in Maine. Fouling is a serious problem for all off-bottom culturists and requires much labor to control.

A "red tide" organism *Gonayulax tamarensis* is present on occasion in Maine estuarine waters. This dinoflagellate, when eaten by oysters, imparts a toxin that causes paralytic shellfish poisoning in humans, other mammals, and birds. The Maine Department of Marine Resources closely monitors levels of this organism, and no outbreaks of this malady have occurred in humans.

Economic Overview and Outlook

Two recent surveys of Maine oyster culturists, one in 1979 by the University of Maine at Orono and the second in 1982 by Phillip Avrill of the Maine Department of Marine Resources, indicate that the industry is small but stable. Avrill found that 27 growers were carrying about 5,500,000 European oysters from yearling nursery to adults. Dr. Herb Hidu, the prime mover in the effort to get this new industry started, feels that only time will tell whether oyster culture in Maine will expand or decline (H. Hidu, personal communication).

Off-bottom culture, as practiced in Maine, is expensive because of the labor involved to handle the product, the cost of culture gear, and the dependence on hatchery-reared seed. Recently, however, European or "Belon" oysters cultivated in Maine (Fig. 6.16) commanded the highest price, $1.25 each, of any of 11 oyster groups on the menu of New York's Oyster Bar and Restaurant at Grand Central Station (Laitin 1982). With a reasonable production volume and market share, the oyster cottage industry should thrive in Maine.

THE OLYMPIA OYSTER *(Ostrea lurida)*

This account of the Olympia oyster is taken primarily from reports by Breese and Wick (1972), Hopkins (1937), Korringa (1976b), and Yonge (1960). T. Nosho with the University of Washington Sea Grant Program provided current information on conditions in this fishery. *Ostrea lurida* is a small species seldom larger than 5 cm in shell length. It is grown in Puget Sound, Washington, and Netarts and Yaquina

FIG. 6.16 A tray of Maine "Belon" oysters *(O. edulis)* ready for market. *Courtesy of T. Archambault.*

bays in Oregon (Breese and Wick 1972). It grows best at fairly high salinities (>25‰). Spawning occurs from 13° to 16°C (55° to 61°F) and may last from 2.5 months in British Columbia to 7 months in southern California. Larvae are brooded 8–10 days, and the free-swimming stage may be approximately 4 weeks. Larvae grow well at 17°–18°C (63°–64°F); adults grow well at 16°–25°C (61°–77°F).

Culture Techniques

Seed grounds and grow-out grounds for Olympia oysters are impounded areas within the intertidal zone (Fig. 6.17). Dikes, constructed from creosoted timbers or concrete, prevent the water from ebbing out and the oysters from being subject to extremes of temperature. Grounds higher in the intertidal zone are best for catching seed, while lower impoundments are better for fattening for market. Soft bottoms within the impoundments are firmed up using gravel and shell.

Growers catch *O. lurida* seed in two areas. Some seed are caught in channels, but most are caught on shell planted within diked areas. Some growers use seed flats, which are raftlike structures with chicken

FIG. 6.17 Impoundments are used to grow Olympia oysters *(O. lurida)* in Skookum Inlet in southern Puget Sound, Washington. These are part of the old Tom Nelson Oyster Farm now operated by a grandson Glen Rau.
Courtesy of T. Nosho.

wire bottoms, to suspend cultch off the bottom. Cultch, usually oyster shell or manilla clam shells *(Venerupis japonica)*, is put out in June when spawning is well underway. Cemented wood lathe or egg crate fillers have been used in the past to capture seed. Seed caught on shell suspended from flats is moved after 2 months to impoundments high in the intertidal zone. Seed caught at channel sites are moved the following spring. Oysters are kept in high-level impoundments for 2 years and then moved to low intertidal diked areas for a period of 1 to 2 years.

When Olympia oysters are ready for harvest, they are loaded onto a scow or sink float with hand forks. A sink float is a raftlike structure with a wire-lined well in the center that permits the oysters to be held in water until they are readied for shipment. The oysters are then transported to the culling house where market-size oysters are separated from small oysters and fouling organisms such as "cups" *(Crepidula fornicata)*, oyster drills, and debris. Market oysters are packed in 50-kg or 2-bushel bags and shipped to wholesalers who shuck and pack them in jars and tins primarily for the restaurant trade.

Diseases, Parasites, and Predators

An infection, especially among older Olympia oysters, sometimes results in heavy mortality in spring. The causative agent is unknown, but *Hexamita* may be involved. Another disease of mysterious origin may strike on occasion. This is thought to be of microbial nature and results in many deaths (Korringa 1976b). The parasitic copepod *Mytilicola orientalis* is responsible for sporadic deaths of Olympia oysters (Sindermann 1977).

Predators include ducks, whelks *Thais lamellosa),* starfish, and moon snails *(Polynices lewesii).* The oyster drill *Urosalpinx cinerea* and the cup or slipper shell *(Crepidula fornicata)* were introduced with the American oyster, and the drill *Ocenebra japonica* and the flatworm *Pseudostylochus ostreophagus* were imported from Japan with the Pacific oyster. The latter, *P. ostreophagus,* has been known to kill 90% of setting spat (T. Nosho, personal communication). Mud shrimp *(Upogebia pugettensis)* and ghost shrimp *(Callianasa* sp.) may cause leaks in dikes through their burrowing efforts.

Climatic Restrictions and Other Problems

Olympia oysters grow where temperatures are below 30°C (86°F) in summer and above freezing in winter, thus they must be in beds continually covered by water.

Economic Outlook and Overview

Landings of *O. lurida* were probably less than 450 kg (1000 lb) in 1981 (T. Nosho, personal communication). Poor spat fall has caused growers to shift their operations almost entirely to *C. gigas.*

The Olympia oyster, while commanding a high price, does not appear to have very much potential as a mariculture species other than on a cottage industry basis. The cost of building and maintaining dikes is prohibitive; the species relatively slow growth and small size, which increases shucking costs, further add to production expenses. Habitat changes such as those brought on by timbering activities and waterfront development also have reduced the area suitable for culture of Olympia oysters.

OYSTER HATCHERIES

Interest in controlled production of seed oysters dates back to the late 1800s. More recently, shortages of naturally produced seed have led to a concentrated effort in many areas of the country to develop reliable and economical means to supply the needs of growers through hatcheries (Fig. 6.18). Modern hatchery techniques have been described by Breese and Malouf (1975) for the Pacific oyster, by Dupuy

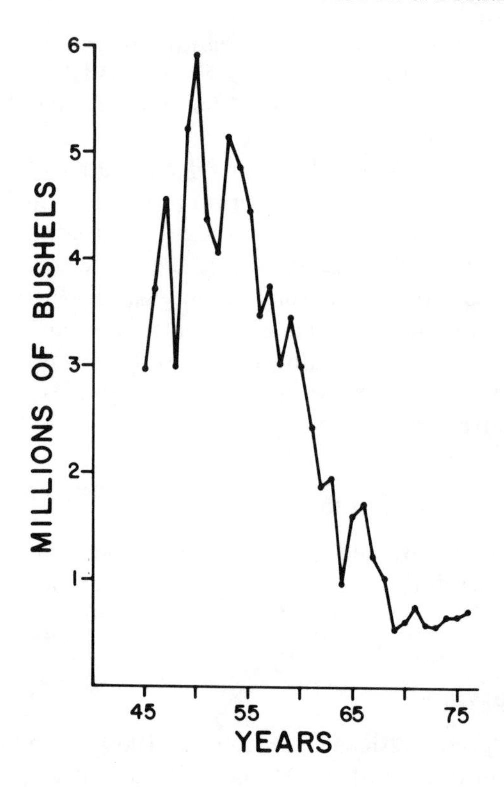

FIG. 6.18 Production of oyster seed declined drastically after 1950.
Source: Fishery Statistics of the United States, U.S. Department of Commerce.

et al. (1977) for the American oyster, by Hidu and Richmond (1974) and Hidu *et al.* (1981) for American and European oysters, and by Walne (1974) for the European oyster. In addition, Castagna and Kraeuter (1981) discuss hatchery techniques in general and Loosanoff and Davis (1963) describe general rearing methods for several molluscan species. Many other references deal with specific techinques and treat such subjects as genetic manipulation, disease, nutrition, and economics (Stanley *et al.* 1981; Newkirk and Haley 1982; Elston 1979a,b; Ukeles 1969; Lipshultz and Kranz 1978, 1980; Lipovsky 1980).

Hatchery Methods

Basic culture technology, which is very similar in all bivalve hatcheries, is described in this section. The basic methods may be modified to accommodate the species involved, the area, and the type of hatchery.

Selection of a hatchery site is important. An ideal site would have the following attributes:

1. Water should have the proper salinity and fertility for the species to be cultured.
2. Sediment and organic detritus should be minimal to avoid an excessive filtering problem.
3. The area should be relatively pollution free. Some farming and industrial chemicals in very low concentrations are highly toxic to oyster larvae. An area that has a thriving oyster population generally would make a good hatchery site.
4. Ideally, ambient temperatures should reach those required for the species to spawn; however, the prime reason for some hatcheries in the first place is that spawning temperatures do not occur on a regular basis in an area.

Brood stock should be selected from healthy, fast-growing animals in peak condition. Young oysters (1½–2 years old) are preponderantly males, so at least 30% of the brood stock should be in this category; older stock (over 2½ years) provides females.

Brood stock is sometimes conditioned to prepare them for spawning (Fig. 6.19). This involves placing the oysters in a flume where they are fed algal food, and the water temperature is raised to the level at which sex products are produced. The temperature and time required for oysters to reach spawning condition vary among species. Spawning is initiated generally in *Crassostrea* by rapidly raising and lowering water temperature. Sometimes to bring about spawning it is necessary to introduce eggs or sperm into the water holding the oysters. In *Ostrea,* spawning usually occurs spontaneously once the animals attain the proper condition.

As soon as *Crassostrea* oysters begin to spawn, they are isolated in separate containers where spawning is allowed to continue. Eggs and sperm are removed from the containers holding adult oysters once

FIG. 6.19 Hatchery brood stock of *C. virginica* being conditioned for spawning in running water flumes. *Courtesy of Virginia Institute of Marine Science.*

spawning ceases. The eggs are then passed through a sieve to remove unwanted material, and a suspension of sperm is added to them. Unless a specific cross is desired, it is best to pool eggs from several females and sperm from several males. Further care must be taken to add the proper amount of sperm to the eggs. Too few will not achieve sufficient fertilization, and too many may result in abnormal embryos caused by several sperm penetrating the same egg membrane.

European and Olympia oysters do not shed their eggs once fertilization takes place, but spawning is indicated by piles of eggs around the shell margins. Early larval development in *Ostrea* takes place in the pallial cavity of the female, and the larvae are released in swarms several days after spawning. The progress of larval development in *Ostrea* before swarming may be determined microscopically by examining larvae that at times dribble out or are obtained by rapidly squeezing the valves of the female together to cause a small number of larvae to be released (Hidu and Richmond 1974).

Larval Culture

Once straight hinged size is reached in Crassostrea and brooded larvae released in the case of *Ostrea,* larval oysters are placed in culture tanks where further growth takes place (Fig. 6.20). Larvae are fed either cultured or natural algae present in the water pumped into the hatchery. There are actually three methods of supplying algal food for the larvae. One is the Wells-Glancey method (Wells 1969) in which ambient water is "clarified" in a milk separator to remove zooplankton and large algal cells, and then incubated 12–24 hr to produce a bloom of small algal species, which is then fed to the larvae. Hidu *et al.* (1969) also used bay water, but strained it through a fine pore filter and fed it directly to the larvae. The Wells-Glancey method

FIG. 6.20 Larval rearing tanks at the Coast Oyster Company hatchery in Quilcene, Washington.

only requires a change in the culture water every 2 or 3 days, whereas the sieved technique mandates a change every day (Dupuy *et al.* 1977). A third method, in which pure cultures of algae are fed singularly or in combination to larvae, is presently used by most hatcheries. Pure cultures of such species as *Isochrysis galbana, Phaeodactylum* sp., *Platymonas* sp., *Monochrysis lutheri, Dunaliella terctiolecta,* and new isolates reported by Dupuy *et al.* (1977)—*Pyramonas virginica, Pseudo-isochrysis paradoxa,* and *Chlorella* sp.—have been found to promote superior growth and survival of larval oysters. Recently, Kranz *et al.* (1982) has used a tropical species "Tahitian *Isochrysis,*" which can be grown at laboratory temperatures year-round, concentrated, and stored for later use.

The majority of hatcheries use a batch-culture technique to produce algae to feed larvae (Figs. 6.21 and 6.22). This consists of producing a large volume of algal-rich water by a series of steps. A pure strain of algae is added to a small vessel that contains sterilized seawater and a nutrient solution. This culture is allowed to grow and provide the inoculant for larger similarly charged vessels, which are then used to start even larger cultures. The entire batch process generally entails four steps and takes approximately 4 weeks to achieve the desired algae density in the final volumes to be fed to the oyster larvae. Algal culture vessels are illuminated with banks of fluorescent tubes, and larger vessels are aerated to keep the algae in suspension. Carbon dioxide or nitrogen gas may be used to control the culture pH or reduce the concentration of dissolved oxygen, which inhibits growth when in too high concentrations.

Accurate counts of both larvae and algae assure that the proper amount of food is provided for best larval growth and survival; however, many experienced operators estimate needs without counts.

After a period as free-swimming organisms, the larvae develop a pigmented spot (eye spot) and are ready to set (attach and metamorphose into the sessile stage). The planktonic stage is of varying lengths

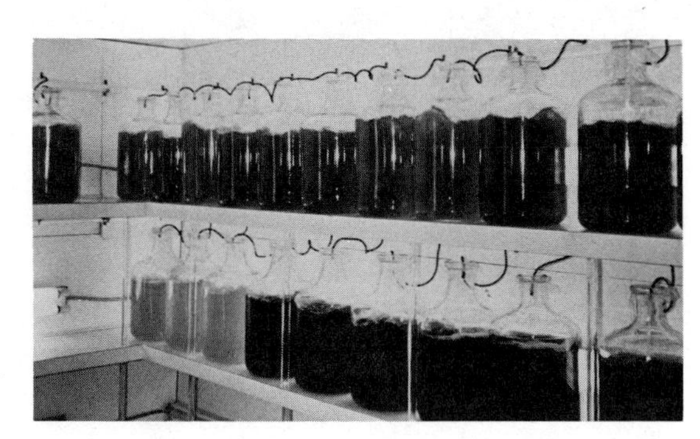

FIG. 6.21 Pure cultures of algae being reared to inoculate large algal culture tanks.
Courtesy of Virginia Institute of Marine Science.

FIG. 6.22 Algal culture tanks at Bristol Shellfish Farm, Round Pond, Maine. *Courtesy of J. Manzi.*

according to species and temperature. Setting may take place over a period of several days. Hidu *et al.* (1978) and Hidu and Richmond (1974) have found that the addition of water in which adult oysters have pumped may induce a majority of *O edulis* and *C. virginica* to set in a matter of an hour or two. Raising the temperature may also achieve the same results (M. Castagna, personal communication).

Setting substrate varies according to whether or not "cultchless" or cultched seed is to be produced. Generally, West Coast oysters are set on oyster shell and East Coast operations produce cultchless seed on a variety of materials. Dupuy *et al.* (1977) allowed larvae to set on mylar sheets and, after a period of growth, removed the seed by flexing the sheet. Hidu *et al.* (1981) used calcium carbonate particles as a setting surface for cultchless seed. Kranz *et al.* (1982) used vinyl-coated wire trays dipped in concrete and sprinkled with oyster shell chips to collect oyster set (Fig. 6.23). Most commercial East Coast hatcheries use crushed shell chips or "mini" cultch (M. Castagna, personal communication).

Cultchless spat must be held in flumes, trays, or some other protective device until they reach 25 mm (1 in.) in diameter. Kranz (1981) found that concrete-coated trays permitted ease of cleaning and mov-

FIG. 6.23 Cement-coated plastic wire trays used to receive hatchery-grown spat and for subsequent grow out to planting size. *Courtesy of G. Kranz.*

ing and better utilization of space than did other systems used to grow seed to planting size.

Hatcheries on the West Coast place containers (wire or plastic bags or plastic baskets) holding cleaned oyster shell in large setting tanks to which the eyed larvae are introduced (Fig. 6.24). The seed are held for a period of time (longer in winter) and then moved to growing

FIG. 6.24 Cultch-filled bags in setting tanks at the Coast Oyster Company, Quilcene, Washington. Eyed larvae are added to the tanks and later moved to growing grounds.

grounds (Breese and Malouf 1975). In a method known as the "eyed larvae technique" or "remote setting," which was developed on the West Coast, the oyster grower purchases eyed *C. gigas* larvae from a hatchery and sets it at his own headquarters. The eyed larvae are concentrated on a screen and then put on a damp cloth and wrapped in wet paper towels. This package is then placed in a plastic cooler (with the ice package in the lid) and transported to the setting location where the larvae are placed in setting tanks as is done at hatcheries. Larvae kept cool at 1°–4°C (34°–40°F) can be held for 7 days without appreciable loss of set. This method reduces the cost of transporting seed significantly (W. P Breese, personal communication).

It is essential to maintain absolute cleanliness in hatcheries to avoid bacterial contamination of cultures, which may lead to complete loss of a batch. Detailed records must be kept for each step of the operation so that the history of each batch is known and problems may be pinpointed should they arise.

Dupuy *et al.* (1977) estimated that an oyster hatchery in the Chesapeake Bay region built to their specifications would require $500,000 to establish and operate for the first 18 months. They also presented detailed cash flow information. Im and Langmo (1977) estimated costs of hatchery production of Pacific oyster seed in the Pacific Northwest. They estimated that the cost to build and equip a hatchery would be $184,572 to $301,470, excluding land purchase. Operating costs ranged from $108,500 to $183,500 per year according to production method and number of seed produced.

PROBLEMS FACING THE OYSTER CULTURE INDUSTRY

Oyster culture has been and continues to be among the few prospering aquaculture industries in the United States. However, pressing problems face oyster growers. High production costs are restricting expansion by established culturists and are making it more difficult than previously for newcomers to enter the business. Land for shoreside facilities is in demand by housing, recreational, and manufacturing companies, which can and do pay premium prices to obtain choice sites (Burrell 1982). Many competing uses for areas suitable for oyster culture are not compatible with oyster grow out. These include industrial and agricultural uses and those associated with marinas and housing projects. New forestry practices and other coastal activities also have had major impacts on drainage systems, which have taken their toll on oyster grow-out areas (Maggioni and Burrell 1982).

The oyster industry is labor intensive, and not only are wages going up, but the labor supply is not always sufficient. Many young people are selecting other trades because of the rigorous nature, seasonality, and relatively low pay associated with many aspects of oyster production (Haven 1972). Demand for oysters has not kept up with the growth in population; per capita consumption in 1981 was only about 0.14 kg (0.31 lb.) whereas it was almost 0.77 kg (1.69 lb.) in 1910 (Anderson and Power 1952; Thompson 1981). This decrease in consumption may in part be due to high oyster prices, but certainly an almost total lack of marketing effort has contributed to the decline (Glude 1974). Mechanization has not received as much attention in oyster culture as it has in agricultural operations. This to some degree is a reflection of the lower value of the oyster crop compared with that of some land crops (Jordan and Webb 1976; Haven 1982).

Regulations restricting water and bottom tenure have decreased financing options for aquaculture ventures, thus hindering development of larger operations (Haven *et al.* 1978a; Jensen 1981; Matthiessen 1971). The oyster culture industry comprises many relatively small firms (Office of Fisheries Development 1977). This may be a result of some of the problems listed above, such as leasing restrictions, high labor costs, and lack of mechanization. This, too, must also contribute to the industry's problems since small firms are unable to mount extensive marketing programs, provide funds for engineering research, or develop new products as larger firms can do.

Outlook

Hatcheries appear to be the answer to meeting seed requirements, where this is a problem. They also make possible the development of disease-resistant as well as fast-growing, hardy, good-flavored, well-shaped oysters (Newkirk 1980; Perdue *et al.* 1981). It may be possible through hatchery manipulation or special feeding regimes to grow oysters that are in top market condition year-round (Ingle *et al.* 1981; Hidu *et al.* 1981). Hatcheries need to develop more efficient means of getting newly attached spat to planting size and of producing large quantities of algal food.

Off-bottom grow out, while not now economically practical may be the way of the future in some areas (Aprill and Maurer 1976; Walker and Gates 1981; Shaw 1969; Bardach *et al.* 1972). Quayle (cited in Brett *et al.* 1972) has shown that off-bottom culture of *C. gigas* can give many times the yield that bottom culture does in the same area. Use of heated effluent from power plants offers the potential for year-round growth in cooler climates. Advances in depuration techniques may make it possible to use marginally polluted areas for oyster cultivation (Haven *et al.* 1978b).

Mechanization is slowly coming to the oyster culture industry. Mechanical harvesters have been in use on the West Coast for some time and are now being introduced to the East Coast (Haven 1981; Jordan and Webb 1975). Shucking methods remain a problem for a fresh-shucked product; however, heat-assisted methods are making hand-shucking more efficient (Prier 1981).

Glude (1974) notes that a real potential exists for developing markets for specialty products such as yearling oysters cultivated for the half-shell trade; however, a concerted effort to reestablish the oyster in consumers' diet is not in the offing. Other shellfish products have been created for the institutional trade and as convenience items for home use. The oyster industry needs to exploit this field (Paparella 1981).

Oyster aquaculture operated as a cottage industry appears practical for the upper New England region, and methods developed for use there may well translate to other regions. Large food industries may never enter oyster aquaculture *per se;* however, small firms should look at joint ventures with larger ones. This would provide greater market opportunities and more efficient transportation, marketing, and warehousing.

SUMMARY

Four species of oysters are cultured in the United States: *Crassostrea virginica* on the East and Gulf Coasts, *Ostrea edulis* on the Northeast Coast, *Crassostrea gigas* and *Ostrea lurida* on the West Coast. Most seed is produced on natural beds; however, hatchery-reared seed is being used by more and more growers, particularly on the West Coast where seed have been mostly Japanese imports and in Maine where natural sets are not sufficient. Most oysters are grown on the bottom, but off-bottom culture has been introduced successfully on the West Coast and in Maine.

Landings of oysters and seed have shown a downward trend in recent years. The reasons for this decline are complex but include seed shortages, high labor costs, lack of mechanization, loss of growing grounds to pollution, foreign imports, and loss of markets. A reversal in these trends is possible but depends upon development of labor-saving devices in all phases of the industry, development of new products, initiation of a dynamic marketing program, and further development of culture techniques (for example, better control of predators and disease and hatchery methods that increase seed production).

ACKNOWLEDGMENTS

I am indebted to my colleagues Wilbur Breese, Mike Castagna, Ken Chew, Bill Dupaul, Dexter Haven, Herb Hidu, George Kranz, John Manzi, and Terry Nosho who generously provided information, photographs, and commentary on the various species and techniques treated in this chapter. I thank Louise Hodges and Cassie Linx for typing several drafts and the final manuscript and Karen Swanson for drafting figures.

LITERATURE CITED

ANDERSON, A. W. and E. A. POWER. 1952. Fishery Statistics of the United States 1952. U. S. Fish Wildl. Serv. Stat. Dig. 34. U. S. Department of Interior, Washington, D.C.

ANDREWS, J. D. 1979. Oyster diseases in Chesapeake Bay. Mar. Fish. Rev. **41,** 45–53.

ANDREWS, J. D., and HEWATT, W. G. 1957. Oyster Mortality Studies in Virginia II. The fungus disease caused by *Dermocystidium marinum* in oysters of Chesapeake Bay. Ecol. Monogr. **27,** 1–26.

ANDREWS, J. D., and WOOD, J. L. 1967. Oyster Mortality Studies in Virginia VI. History and distribution in *'Minchinia nelsoni'* a pathogen of oysters in Virginia. Chesapeake Sci. **8**(1), 1–13.

ANDREWS, J. D., WOOD, J. L., and HOESE, H. D. 1962. Oyster Mortality Studies in Virginia III. Epizootiology of a disease caused by *Haplosporidium costale* Wood and Andrews. J. Invertebr. Pathol. **4,** 327–343.

APRILL, G., and MAURER, D. 1976. The feasibility of oyster raft culture in east coast estuaries. Aquaculture **7,** 147–160.

BARDACH, J. E., RYTHER, J. H., and MCLARNEY, W. O. 1972. Aquaculture. The Farming and Husbandry of Freshwater and Marine Organisms. John Wiley & Sons, New York.

BAUGHMAN, J. L. 1948. An annotated bibliography of oysters with pertinent material on mussels and other shellfish and an appendix on pollution. Texas A&M Research Foundation, College Station, Texas.

BREESE, W. P., and MALOUF, R. E. 1975. Hatchery manual for the Pacific oyster. Sea Grant Program Publ. No. ORESU-H-75-002. Oregon State University, Corvallis, Oregon.

BREESE, W. P., and WICK, W. Q. 1972. Oyster farming; culturing, harvesting, and processing a product of the Pacific Coast area. Mar. Sci. Edu. Publ. No. SG 13. Extension Marine Advisory Program. Oregon State University, Corvallis, Oregon.

BREISCH, L. L., and KENNEDY, V. S. 1980. A selected bibliography of world-wide oyster literature. Sea Grant Publ. No. UM-SG-TS-80-11. University of Maryland, College Park, Maryland.

BRETT, J. R., CALAPRICE, J. R., GHELARDI, R. J., KENNEDY, W. A., QUAYLE, D. B., and SHOOP, C. T. 1972. A brief on mariculture. Fish. Res. Board Can. Tech. Rep. 301.

BROOKS, W. K. 1879. Abstract of observations upon artificial fertilization of oyster eggs and embryology of American oyster. Am. J. Sci. **18,** 425–427.

BURRELL, V. G., JR. 1982. Status of the South Atlantic Oyster Industry. Proc. North Am. Oyster Workshop, World Maric. Soc. Publ. No. 1, pp. 125–127.

CARRIKER, M. R. 1955. Critical review of biology and control of oyster drills *Urosalpinx* and *Eupleura*. U. S. Fish Wildl. Ser. Spec. Sci. Rep. Fish. 148.

CASTAGNA, M., and CHANLEY, P. 1973. Salinity tolerance of some marine bivalves from inshore and estuarine environments in Virginia waters in the western mid-Atlantic coasts. Malacologia **12,** 47–96.

CASTAGNA, M., and KRAEUTER, J. N. 1981. Manual for growing the hard clam. Spec. Rep. Appl. Mar. Sci. Ocean Eng. No. 249. Virginia Institute of Marine Science, Gloucester, Virginia.

CHEW, K. K. 1979. The Pacific Oyster *(Crassostrea gigas)* in the West Coast of the United States. *In* Exotic Species in Mariculture, R. Mann (editor), Massachusetts Institute of Technology Press, Cambridge, Massachusetts.

CLIME, R., and HAMILL, D. 1979. Growing Oysters and Mussels in Maine. Coastal Enterprises, Bath, Maine.

DUPUY, J. L., WINDSOR, N. T., and SUTTON, C. F. 1977. Manual for design and operation of an oyster hatchery. Spec. Rep. Appl. Mar. Sci. Ocean Eng. No. 142. Virginia Institute of Marine Science, Gloucester, Virginia.

ELSTON, R. 1979a. Economically important bivalve diseases and their control. Riv. Ital. Piscicult. Ihispatol. A **14**(2), 47–54.

ELSTON, R. 1979b. Virus like particles associated in the lesions in larval Pacific oysters *(Crassostrea gigas)*. J. Invertebr. Pathol. **33,** 71–74.

ELSTON, R. 1980. Untrastructure of a serious disease of hatchery reared larval oysters, *Crassostrea gigas,* Thunberg. J. Fish Diseases **3,** 1–10.

ELSTON, R., ELLIOTT, E. L., and COLWELL, R. R. 1982. Shell fragility, growth depression and mortality of juvenile American and European oysters *Crassostrea virginica* and *Ostrea edulis)* and hard clams *(Mercenaria mercenaria)* associated with surface coating *Vibrio* spp. bacteria. J. Fish Diseases **5**(4), 265–284.

ELSTON, R., LEIBOVITZ, L., RELYEA, D., and ZATILA, J. 1981. Diagnosis of vibiosis in a commercial oyster hatchery epizootic: diagnostic tools and management features. Aquaculture **24,** 53–62.

EPIFANIO, C. E., and MOOTZ, C. A. 1976. Growth of oysters in a recirculating maricultural system. Proc. Nat. Shellfish. Assoc. **65,** 32–37.

FARLEY, C. A. 1978. Virus and virus-like lesions in marine mollusks. Mar. Fish. Rev. **40**(10), 18–20.

GALTSOFF, P. S. 1964. The American oyster *Crassostrea virginica* Gmelin. U.S.. Fish Wildl. Serv. Fish. Bull. **64,** 1–480.

GALTSOFF, P. S. 1972. Bibliography of Oysters and Other Marine Organisms Associated with Oyster Bottom and Estuarine Ecology. G. K. Hall & Co., Boston, Massachusetts.

GLUDE, J. B. 1974. Recent developments in shellfish culture on the U. S. Pacific coast. Proc. 1st U. S.-Japan Meeting on Aquaculture, Tokyo, Japan, pp. 89–95. NOOA Tech. Rep. UMFS Circ. **338,** 1–113.

HASKIN, H. H., STAUBER, L. A., and MACKIN, J. A. 1966. *'Minchinia nelsoni'* N.Sp. (Haplosporida, Haplosporidiidae): Crusative agent of the Delaware Bay oyster epizootic. Science **153**(3742), 1414–1416.

HAVEN, D. S. 1972. Oysters. *In* A Study of the Commercial and Recreational Fisheries of the Eastern Shore of Virginia, Accormack and North Hampton Counties. V. G. Burrell, M. Castagna, and R. K. Dias (editors), Spec. Sci. Rep. Appl. Mar.

Sci. Oceangr. Eng. No. 20. Virginia Institute of Marine Science, Gloucester, Virginia.

HAVEN, D. S. 1981. Modernizing the Oyster Industry. *IN* Oyster Culture in Maryland 1980 D. Webster (editor). Sea Grant Rep. No. UM-SG-MAP-81-01. University of Maryland, College Park, Maryland.

HAVEN, D. S., and BURRELL, JR., V. G. 1982. The oyster—a shellfish delicacy. Leaflet 11, Atlantic States Mar. Fish. Commiss. 6 p.

HAVEN, D. S., and GARTEN, D. 1972. Shell bags for catching oyster spat. Mar. Resources Advisory Ser. No. 6. Virginia Institute of Marine Science, Gloucester, Virginia.

HAVEN, D. S., HARGIS, JR., W. J., and KENDALL, P. D. 1978a. The oyster industry of Virginia: Its status problems and promise. Spec. Pap. Mar. Sci. No. 4. Virginia Institute of Marine Science, Gloucester, Virginia.

HAVEN, D. S., PERKINS, F. O., MORALES-ALAMO, R., and RHODES, M. W. 1978b. Bacterial depuration of the American oyster *(Crassostrea virginica)* under controlled conditions. VI. Biological and technical studies. Spec. Sci. Rep. No. 88. Virginia Institute of Marine Science, Gloucester, Virginia.

HIDU, H., CHAPMAN, S. R., and DEAN, D. 1981. Oyster mariculture in subboreal (Maine, United States of America) waters: Cultchless setting and nursery culture of European and American oysters. J. Shellfish Res. **1**(1), 57–67.

HIDU, H., DROBECK, K. G., DUNNINGTON, JR., E. A., ROOSENBERG, W. H., and BECKETT, R. L. 1969. Oyster hatcheries for the Chesapeake Bay Region. NRI Spec. Rep. No. 2. Natural Resources Institute, University of Maryland, College Park, Maryland.

HIDU, H., and RICHMOND, M. S. 1974. Commercial oyster aquaculture in Maine. Mar. Sea Grant Bull. **2,** 1–59.

HIDU, H., VALLEAU, W. G., and VEITCH, F. P. 1978. Gregarious setting in European and American oysters response to surface chemistry vs waterborne pheromones. Proc. Nat. Shellfish. Assoc. **68,** 11–16.

HOPKINS, A. E. 1937. Experimental observation in spawning, larval development, and setting in the Olympia oyster, *Ostrea lurida.* U. S. Dep. Commerce Bur. Fish. Bull., **48,** 439–503.

IM, K. H., and LANGMO, D. 1977. Economic analysis of producing Pacific oyster seed in hatcheries. Proc. Nat. Shellfish. Assoc. **67,** 17–28.

INGLE, R. M., MEYER, D. G., and LANDRUM, M. P. 1981. Preliminary notes on a pilot plan for the feeding of adult American oysters. Adelanto Corporation, Apalachicola, Florida.

JENSEN, W. P. 1981. Leased bottom and the Maryland oyster fishery. *In* Oyster Culture in Maryland 1980, (D. Webster (editor), Sea Grand Publ. No. UM-SG MAP 81-01. University of Maryland, College Park, Maryland.

JORDAN, A. G., and WEBB, B. W. 1975. Development of equipment for the mechanical harvesting oysters in S. C. Prog. Rep. Annual meeting of the American Society of Agriculture Engineers, Pap. No: 75-5014.

JOYCE, E. A. 1972. A partial bibliography of oysters, with annotations. Spec. Sci. Rep. No. 34. Florida Department of Natural Resources, St. Petersburg, Florida.

KENNEDY, V. S., and BREISCH, L. L. 1981. Maryland's oysters: Research and Management. Sea Grant Publ. No. UM-SGTS-81-04. University of Maryland, College Park, Maryland.

KORRINGA, P. 1976a. Farming the Cupped Oysters of the Genus *Crassostrea.* Elsevier Scientific Publishing Company, New York.

KORRINGA, P. 1976b. Farming the Flat Oysters of the Genus *Ostrea*. Elsevier Scientific Publishing Company, New York.

KRANZ, G. 1981. Production of seed oysters. *In* Oyster Culture in Maryland 1980, D. Webster (editor), Sea Grant Publ. No. UM-SG MAP 81-01. University of Maryland, College Park, Maryland.

KRANTZ, G. E., BAPTIST, G. J., and MERITT, D. W. 1982. Three innovative techniques that made Maryland oyster hatcheries cost-effective. Annu. Meet. Nat. Shellfish. Assoc., Baltimore, Maryland, June 1982, (Abstr.).

LAITIN, J. 1982. New York oyster bar. Seafood Bus. **80**(1), 77–79.

LEIBOVITZ, L. 1978a. Shellfish diseases, MFR Paper 1300. Mar. Fish. Rev. **40**(3), 61–64.

LEIBOVITZ, L. 1978b. Bacteriological studies of Hog Island shellfish hatcheries: An abstract. Mar. Fish. Rev. **40**(10), 8.

LEIBOVITZ, L., ELSTON, R., VIPOVSKY, V. P., and DONALDSON, J. 1978. A new disease of larval Pacific oysters *(Crassostrea gigas)*. Proc. World Maric. Soc. **9,** 603–615.

LIPOVSKY, V. P. 1980. An industry review of the operations and economics of molluscan shellfish hatcheries in Washington State. Proc. World Maric. Soc. **11,** 577–579.

LIPSCHULTZ, F., and KRANTZ, G. 1978. An analysis of oyster hatchery production of cultched and cultchless oysters utilizing linear programming optimization techniques. Proc. Nat. Shellfish. Assoc. **68,** 5–10.

LIPSCHULTZ, F., and KRANTZ, G. E. 1980. Production optimization and economic analysis of an oyster *(Crassostrea virginica)* hatchery on the Chesapeake Bay, Maryland, USA. Proc. World Maric. Soc. **11,** 580-591.

LOOSANOFF, V. L., and DAVIS, H. C. 1963. Rearing of bivalve mollusks. Advan. Mar. Biol. **1,** 1–136.

LUNZ, G. R. 1947. *Callinectes* versus *Ostrea*. J. Elisha Mitchell Sci. Soc. **57**(2), 273–283.

MACKENZIE, C. F. 1977. Use of quicklime to increase oyster seed production. Aquaculture **10**(1), 45–51.

MACKENZIE, C. F. 1981. Biotic potential and environmental resistance in the American oyster *(Crassostrea virginica)* in Long Island Sound. Aquaculture **22,** 229–268.

MAGGIONI, G. J., and BURRELL, JR., V. G. 1982. South Carolina oyster industry. Proc. North Am. Oyster Workshop, Seattle, Washington, March 1981, World Maric. Soc. Spec. Publ. No. 1, pp. 132–137.

MATTHIESSEN, G. C. 1971. A review of oyster culture and the oyster industry in North America. Contrib. No. 2528. Woods Hole Oceanographic Institution, Woods Hole, Massachusetts.

MERRINER, J. V., and SMITH, J. W. 1979. A report to the oyster industry of Virginia on the biology and management of the Cownose ray (*Rhinoptera bonasus* Mitchill) in lower Chesapeake Bay. Spec. Sc. Rep. Appl. Mar. Sci. Oceangr. Eng. No. 216. Virginia Institute of Marine Science, Gloucester, Virginia.

NEWKIRK, G. F. 1980. Review of the genetics and the potential for selective breeding of commerically important bivalves. Aquaculture **19,** 209–228.

NEWKIRK, G. F., and HALEY, L. E. 1982. Phenotypic analysis of the European oyster (*Ostrea edulis* L: Relationship between length of larval period and post setting growth rate. J. Exp. Mar. Biol. Ecol. **59,** 177–184.

OFFICE OF FISHERIES DEVELOPMENT, NATIONAL MARINE FISHERIES SERVICE. 1977. A comprehensive review of the commercial oyster industries in the United States, 1977–240-848/166. U. S. Government Printing Office, Washington, D.C.

PACKIE, R., H. HIDU, and M. S. RICHMOND. 1976. The suitability of Maine waters for culturing American and European oysters, *Crassostrea virginica* Gmelin and *Ostrea edulis* L. Maine Sea Grant, Orono, Publication TR 10-76.

PAPARELLA, M. 1981. Looking ahead. In Oyster Culture in Maryland 1980. D. Webster (editor), Sea Grant Publ. No. UM-SG MAP 81-01. University of Maryland, College Park, Maryland.

PERDUE, J. A., BEATTIE, J. H., and CHEW, K. K. 1981. Some relationships between gametogenic cycle and summer mortality phenomenon in the Pacific oyster *(Crassostrea gigas)* in Washington State. J. Shellfish Res. **1**(1), 9–16.

PRIER, R. 1981. Shucking by machine. *In* Oyster Culture in Maryland 1980. D. Webster (editor), Sea Grant Publ. No. UM-SG MAP 81-01. University of Maryland, College Park, Maryland.

QUAYLE, D. B. 1969. Pacific oyster culture in British Columbia. Fish. Res. Board Can. Bull. **169,** 1–193.

QUAYLE, D. B., and SMITH, D. W. 1976. A guide to oyster farming. Publication Marine Resources Branch, Department of Recreation and Travel Industry, Victoria, British Columbia, Canada.

QUICK, J. A., JR. and J. G. MACKIN. 1971. Oyster parasitism by *Labyrinthomyxa marina* in Florida. Professional Pap. Ser. No. 13. Florida Department of Natural Resources, St. Petersburg, Florida.

RAY, S. M. 1954. Biological studies of *Dermocystidium marinum,* a fungus parasite of oysters. Rice Institute Pamphlet, Monogr. Biology, Special Issue.

SHAW, W. N. 1969. The past and present status of off bottom oyster culture in North America. Trans. Am. Fish. Soc. **98**(4), 755–761.

STANLEY, J. G., ALLEN, S. K., and HIDU, H. 1981. Polyploidy induced in the early embryos of the American oyster with cytochalasin B. J. Shellfish Res. Abstr. 1980 Annu. Meet. June 1980.

SINDERMANN, C. J. 1977. Disease Diagnosis and Control in North American Marine Mariculture. Developments in aquaculture and Fisheries Science vol. 6. Elsevier, New York.

THOMPSON, B. G. 1982. Fisheries of the United States 1981. Current Fish. Stat. No. 8200. National Oceanographic and Atmospheric Administration, National Marine Fisheries Service, Washington, DC.

UKELES, R. 1969. Nutritional requirements in shellfish culture. Proc. Conf. Artificial Propagation Commer. Valuable Shellfish—Oysters, University of Delaware, Newark.

WALKER, N. P., and GATES, J. M. 1981. Financial feasibility of high density oyster culture in saltmarsh ponds with artificially prolonged tidal flows. Aquaculture **22,** 11–20.

WALNE, P. R. 1974. Culture of bivalve mollusks. Fishing New Books, Surrey, England.

WELLS, W. F. 1969. Early oyster culture investigations of the New York State Conservation Commission (1920–1926). New York Conservation Department Division Marine and Coastal Resources, Albany, NY. (Reprint).

WESTLEY, R. E. 1968. Relation of hydrography and *Crassostrea gigas* settling in Daboh Bay, Washington. Proc. Nat. Shellfish. Assoc. **58,** 42–45.

WOELKE, C. E. 1955. Introducing the Kumamato oyster *Ostrea (Crassostrea) gigas* to the Pacific coast fishery. Res. Pap. 1(3):1–9. Washington Department of Fisheries, Seattle, WA.

YONGE, C. M. 1969. Oysters. Willmer Bars & Haram, London.

7

Clam Aquaculture

John J. Manzi

Introduction
Clam Fisheries and Aquaculture Production
Basic Biology
Culture Techniques
  Gametes to Set: The Hatchery
  Set to Seed: The Nursery
  Seed to Market: Growout Systems
Parasites and Diseases
Constraints
Status and Economic Overview
Summary
Literature Cited

INTRODUCTION

Although the specific origins of molluscan aquaculture are lost in antiquity, references to shellfish culture date back to 2000 B.C. in Eastern civilizations (Iverson 1968) and to around 400 and 100 B.C. in Greek and Roman civilizations, respectively (Milne 1972). Bardach *et al.* (1972) suggested that the culture of clams predates oyster culture in Japan, and Bourne (1981) stated that references to clam culture are made in Japanese literature as early as the eighth century. Regardless of its origins, the culture of clams has been and continues to be of great interest to aquafarmers throughout the world.

Commercial clam fisheries in the United States have relied almost entirely on wild-stock harvests. Of the total 54.5 million kg (120 million lb) of clam meats harvested in 1981, less than 10% were derived, directly or indirectly, from aquaculture activities. Despite present low yields through aquaculture, clams have high potential for commercial aquafarming in this country. This potential readily becomes apparent when one considers criteria for aquaculture candidate species. Webber and Riordan (1976) analyzed these criteria and summarized them into three categories: marketing, biology, and site–technology interaction. Based on these criteria, several clam species are suitable for aquaculture and one, the hard clam *(Mercenaria mercenaria),* seems particularly appropriate (Fig. 7.1).

Hard clams are in strong demand (natural harvests have declined from 2 million bu in 1950 to 1.2 million bu in 1980) and are highly valued. Even when deflated by the Consumer Price Index ($n_o = 1967$), the price of hard clams tripled between 1958 and 1981. Recent data (USDOC 1982) show that the unadjusted exvessel price for this species increased fourfold from 1967 to 1981. Consumer acceptance and demand are high and extensive transportation and marketing networks exist. The biology of *M. mercenaria* is well known since it has been a constant subject of research for the last half century. Its environmental requirements and ecological interactions have been studied, disease and parasite diagnosis and treatment protocols established, and some genetic research performed. Finally, culture procedures have been instituted to the extent that several commercial facilities exist to produce and supply seed, and commercial-scale growout systems are in operation, or being tested, in several states.

The prognosis for successful clam mariculture in the United States is thus encouraging. The organisms, in general, are amenable to high-density cultivation and suitable to the myriad of sites available along U.S. coasts. This chapter explores the subject of intensive clam aquaculture, reviews previous and ongoing operations, and develops criteria for evaluating the potential of clam aquaculture in the United States.

CLAM FISHERIES AND AQUACULTURE PRODUCTION

Clam populations support significant commercial fisheries in the United States. Fourteen species of clams are harvested in 18 states (Dressel and FitzGibbon 1978). Commercially valuable clam genera and the locations of major fisheries are presented in Table 7.1. The majority of the fisheries are small, localized, or intermittent, and accounted for only a fraction of total annual U.S. clam landings over the last two decades. Ritchie (1977) indicated that between 1966 and

TABLE 7.1. COMMERCIAL CLAM FISHERIES OF THE UNITED STATES

Clam Genera	Fisheries
Veneridae	
Hard Clam, *Mercenaria*	Maine to Florida
Native Littleneck, *Protothaca*	California, Oregon, Washington
Manila Clam, *Tapes (Venerupis)*	Washington
Butter Clam, *Saxidomus*	Alaska, California, Oregon, Washington
Sunray Venus Clam, *Macrocallista*	Florida
Solenidae	
Razor Clam, *Ensis*	Massachusetts to Delaware
Razor Clam, *Siliqua*	Alaska, California, Oregon, Washington
Solecurtidae	
Jacknife Clam, *Tagelus*	California
Arctidae	
Ocean Quahog, *Arctica*	Rhode Island to Delaware
Mactridae	
Surf Clam, *Spisula*	New York to Virginia
Marsh Clam, *Rangia*	North Carolina
Horse Clam, *Tresus*	California, Oregon, Washington
Saxicauidae	
Geoduck, *Panope*	Washington
Myidae	
Soft-shell Clam, *Mya*	Maine to Maryland; Alaska, California Oregon, Washington

1975 clam fisheries in the United States were dominated by four species: the hard clam *(M. mercenaria)*, the surf clam *(Spisula solidissima)*, the soft-shell clam *(Mya arenaria)*, and the ocean quahog *(Arctica islandica)*. These four fisheries accounted for 99% of total landings and 98% of the exvessel value for the entire U.S. commercial clam industry in 1975. This trend has continued, with the latest fishery data (USDOC 1982) indicating the dominance of these four fisheries in 1981 (Table 7.2).

Less than 2% of commercial clam harvests are attributed to Pacific

TABLE 7.2. COMMERCIAL CLAM LANDINGS—1981

Species	Landings (1000 lb)	(%)	Value (1000 $)	%
Hard Clam	18,118	15.02	51,169	47.47
Ocean Quahog	36,107	29.93	10,184	9.50
Soft-shell Clam	8,072	6.69	13,906	12.98
Surf Clam	46,100	38.21	23,466	21.90
Other	12,234	10.15	8,420	7.86
Total	120,631	100.00	107,145	100.00

Coast landings. Glude (1974) suggested that clam production along the Pacific Coast is limited primarily by lack of suitable habitat, particularly in Oregon and California. Ritchie (1977) added pollution, dredging, and recreational demands as other limiting factors. Washington accounts for approximately 95% of Pacific Coast landings, which Ritchie (1977) attributed to large areas of suitable habitat, good water quality, commercial management of productive grounds, and a balance between recreational and commercial utilization. The geoducks, manila clams, native littlenecks, butter clams, and razor clams constitute the majority of Washington's commercial landings. Although the commercial clam fishery of Alaska is small, a very large resource exists. The principle obstacle to exploiting the resource is the prevalence of paralytic shellfish poisoning (Nosho 1972).

The commercial production of clams through aquaculture activities has not developed rapidly in the United States despite this country's pioneering technological development in clam hatcheries and nursery systems. In 1975, FAO estimated world production of clams through aquaculture at approximately 37 million kg (81 million lb) (JSA 1980). Estimates for U.S. production of clams through aquaculture show an increase from 1.05 million kg (2.3 million lb) in 1972 (JSA 1980) to 4.05 and 4.14 million kg (8.9 and 9.1 million lb) for 1979 and 1980, respectively (USDOC 1981). This production however, is less than 9% of the national wild-stock harvests for the same periods. Thus, while clam aquaculture technology has developed significantly, the transfer of this technology to commercial activity has been retarded. Menzel (1971) commented over 10 years ago that most techniques had been established for the commercialization of clam mariculture and awaited only appropriate application. Today, to large extent, that statement is still true. Several native species and one nonendemic species *(Tapes japonica)* appear economically suited for commercial mariculture activities in the United States (Fig. 7.1). Hatcheries, already in operation, either presently produce seed of these species or have the capabilities to do so. Nursery technology and grow-out protocols have been developed, and several prototype and small commercial operations have been initiated.

The organisms illustrated in Fig. 7.1 do not exhaust the list of candidate clam species that have reasonable aquacultural potential. Other endemic marine (e.g., *Ensis, Siliqua, Chione, Tagelus*), brackish-water (e.g. *Rangia, Macoma*), and freshwater (e.g., *Anodonta, Lampsilis*) genera have representatives that exhibit culture potential. In addition, a large number of exotic clam species can be considered for culture in the United States, as the accidental introduction and subsequent success of *Tapes* on the Pacific Coast indicates. Genetic

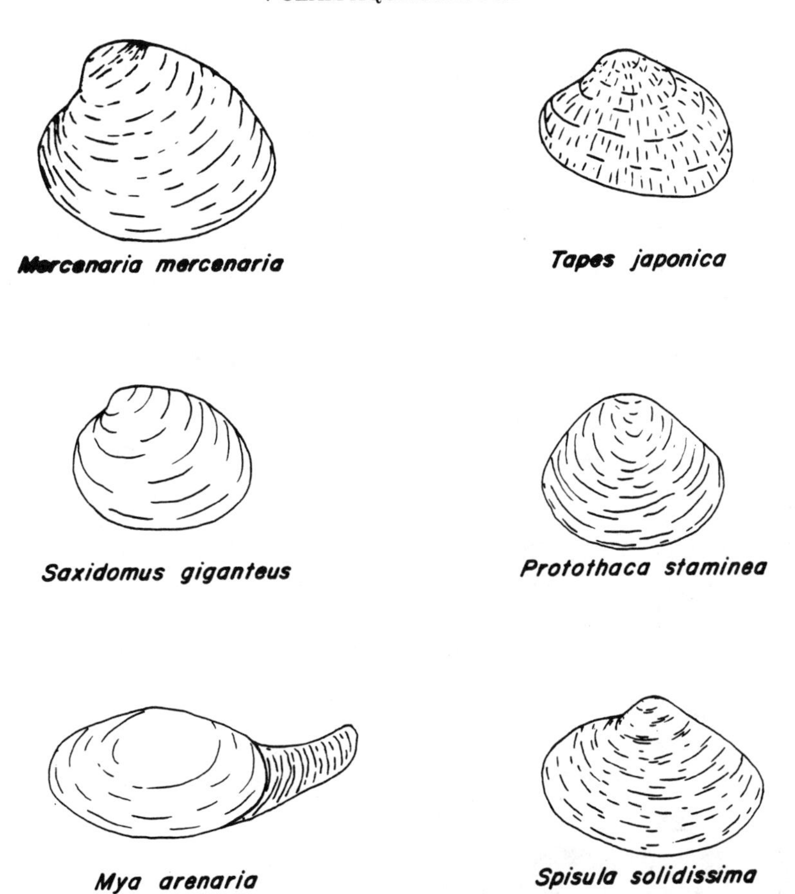

FIG. 7.1 Six candidate species for clam aquaculture in the United States: the hard clam, or quahog, *(M. mercenaria);* manila clam *(T. japonica);* butter clam *(S. giganteus);* littleneck clam *(P. staminea);* soft-shell clam *(M. arenaria);* and surf clam *(S. solidissima).*
Courtesy of K. Swanson.

manipulations, particularly through selection and hybridization (Menzel 1964; Davis 1968; Shultz 1970; Longwell and Stiles 1973; Longwell 1974; Menzel 1977), offer still additional candidates. Because of this diversity in potential culture species and the limited space available here, this chapter will concentrate only on clam species of high commercial importance *(M. arenaria, S. solidissima, A. islandica, M. mercenaria,* and *T. japonica),* with primary attention on the latter two species as representatives of clam culture on the Atlantic and Pacific coasts, respectively.

BASIC BIOLOGY

The term "clam" refers to a group of bivalve mollusks most of which are classified under the order Heterodonta (a notable exception are the Unionidae). This group is generally characterized by eumellibranch gills; similar adductor muscles; heterodont or schizondont dentition; equal valves; and smooth or concentrically ribbed, equilateral shells with an elongate, subrectangular, or ovoid shape. Clams are ciliary feeders with mechanisms for sorting and sieving food particles on ctenidia. The majority of species are marine and burrow in sand or mud with a compressed muscular foot. Clams are dioecious, although hermaphrodism and protandry occur, and fertilization is external. The resulting zygote develops into free-swimming trochophore and veliger larval stages (Fig. 7.2), which metamorphose into the adult form. For more information on systematics, anatomy, and embryology refer to Wilbur and Yonge (1964, 1966), Hyman (1967), and Purchon (1968).

The five major clam types of commercial interest range over the Atlantic, Gulf, and Pacific coasts of the continental United States. The hard clam, or quahog *(M. mercenaria)* is found from the Gulf of

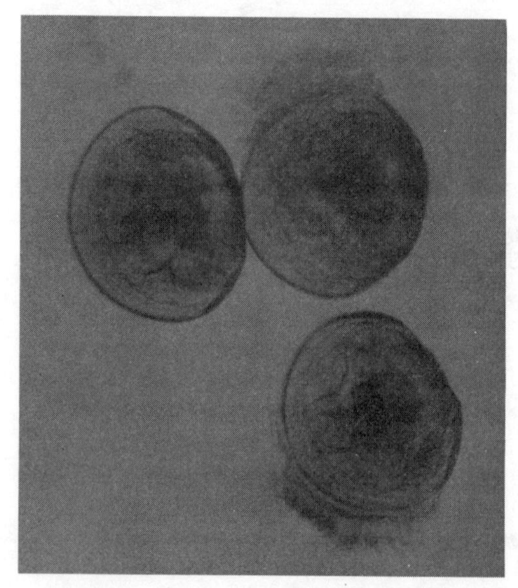

FIG. 7.2 Three-day-old veliger larvae (110–130 μm) of the hard clam, *Mercenaria mercenaria.* *Courtesy of National Marine Fisheries Service, Milford, CT.*

St. Lawrence to the Gulf of Mexico and has been introduced on the Pacific coast of California. A subspecies (or possible hybrid), *M. mercenaria texana,* is prominent on the Atlantic coast of Florida and in the Gulf of Mexico. *M. campechiensis,* which is confused with and more importantly sold as *M. mercenaria,* occurs as far north as New Jersey but is found in commercial densities only in the southern part of its range, i.e., the Atlantic coast of Florida and the Gulf of Mexico (Menzel 1971). The soft-shell clam *(M. arenaria)* occurs naturally from Labrador to South Carolina but has been successfully introduced to the Pacific Coast and is now present in commercial quantities in Washington (Goodwin and Jones 1976) with smaller populations in Oregon and California. The Atlantic surf clam *(S. solidissima)* has a range from Nova Scotia to South Carolina (Serchuk *et al.* 1978; Ropes 1979) and presently supports the country's largest clam fishery from extensive nearshore and offshore beds off New Jersey and the Delmarva peninsula. The ocean quahog, or mahogany clam *(A. islandica),* is intermittently distributed in subtidal waters (normally deeper than 15 m) from Newfoundland to North Carolina (Murawski and Serchuk 1979; Merrill and Ropes 1969). The manila clam *(T. japonica)* was apparently accidentally introduced on the Pacific Coast in the early part of this century (Quayle 1941) and is now found from British Columbia to California (Glude 1974; Ritchie 1977; Magoon and Vining 1981).

Throughout their range, these clams all require relatively high salinities. *Mercenaria mercenaria* is rarely found where salinities fall below 20 ‰, and *S. solidissima* and *A. islandica* are truly marine, or euhaline, in their requirements. *Mercenaria mercenaria* appears to have the greatest tolerance to temperature as reflected by its geographic range and habitat. The other four species are limited to cooler waters and, in the case of *A. islandica* and *S. solidissima,* to subtidal or offshore habitats. All species burrow into the sediment and are infaunal benthonts at least in their adult stage.

Clams can reach sexual maturity at a very early age. Motile sperm and mature ova have been observed in *M. mercenaria* less than a year old and under 20 mm (0.8 in.) in anterior-posterior length (Menzel 1971; Eversole *et al.* 1980). *Mya arenaria* (Zuraw *et al.* 1967) and *Tapes* (Mann 1979) also appear to reach sexual maturity at relatively small sizes, while *S. solidissima* (Jones 1980) and *A. islandica* (Thomson *et al.* 1980) are considerably larger (and older) at sexual maturity. When prompted by appropriate environmental stimuli (notably temperature), sexually mature clams spawn releasing gametes to the surrounding waters. The resulting zygotes develop into veliger larvae (70–110 μm in size) in 24–48 hours. The veliger, a herbivorous zooplankter, is translocated primarily through tidal and wind-generated currents. It apparently has the ability, however, to influence its distribution by vertical swimming and sinking in response to tidal influences (Wood and Hargis 1971).

The length of the larval cycle is species dependent and is influenced within species by environmental parameters. The larvae of most clam species, however, reach the pediveliger stage and begin metamorphosis (set) within 7–14 days (Loosanoff and Davis 1963; Loosanoff *et al.* 1966). Recent set remain epifaunal for short periods before adopting infaunal niches more typical of juvenile and adult clams. Growth in clams is regulated primarily by environmental parameters (e.g., temperature, food, water quality) and physiological disposition (e.g., spawning activity), and in many species growth is indeterminant. Most clams grow fairly rapidly in the first 1 to 4 years of life and then slow to a fairly slow but constant growth rate. Many clam species are very longevous with reports of *Mercenaria* and *Spisula* over 25 years old and *Arctica* over 100 years old (Thompson *et al.* 1980).

CULTURE TECHNIQUES

Clams have been a subject of research and culture for many years. The result of this attention has been the successful rearing of many clam species through their life cycles and the establishment of routine larviculture protocols for several commercially important clam species. Loosanoff and Davis (1963) described larval rearing techniques for 19 bivalve species including the hard clams *(M. mercenaria* and *M. campechiensis),* the razor clam *(Ensis directus),* the surf clam *(S. solidissima),* the soft clam *(M. arenaria)* and the manila clam *(T. japonica).* Landers (1976) described the laboratory culture of *A. islandica,* the mahogany clam or ocean quahog. Aside from some idiosyncrasies of spawning and setting and differences in larval morphometrics and length of larval period, the culture of most clam species can be accomplished by following techniques established for general bivalve hatchery operations (Loosanoff and Davis 1963; Walne 1964; Davis 1969; Lucas 1976; Walne 1979; Dupuy *et al.* 1977; Castagna and Kraeuter 1981).

Gametes to Set: The Hatchery

Hatchery production of seed clams is a well-established and relatively routine operation. Figure 7.3 presents a schematic representation of two traditional bivalve hatchery methodologies in common use

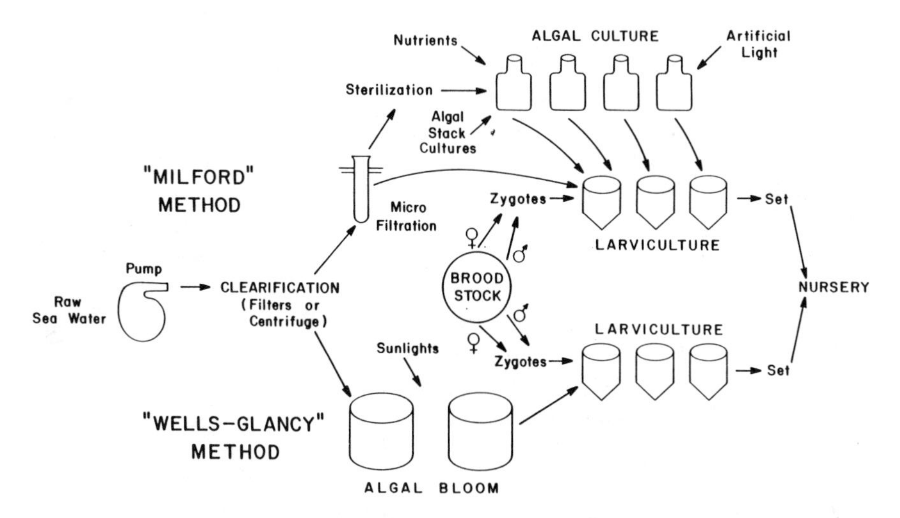

FIG. 7.3 Schematic representation of the Milford and Wells-Glancy methods of bivalve larval culture.
Courtesy of K. Swanson

today: the Milford and the Wells-Glancy methods. The primary difference between these methods is in the techniques employed for algal production. The Wells-Glancy method (Wells 1927) relies on blooms of natural phytoplankton populations initially segregated by filtration or centrifugation. The Milford method (Loosanoff and Davis 1950; Loosanoff 1954) introduced brood-stock conditioning and the controlled culture of specific phytoplankters as rations in bivalve larviculture. Large commercial hatcheries in the United States generally use a combination of these methods: the Milford method for larval rearing and the Wells-Glancy method for postset populations. Several hatcheries bypass the use of raw seawater completely by employing saltwater wells. This alleviates the need for filtration and sterilization and provides a uniformly stable culture medium for both larvae and algae. It does require, however, a complete dependence on cultured algae for hatchery operations.

In general, hatcheries divide their operations into four distinct functions: (1) brood-stock maintenance and conditioning, (2) algal culture, (3) larval culture, and (4) postset maintenance. Assuming that the hatchery has been properly sited with good access to high-quality water, the maintenance of brood stock is reduced to providing sufficient quantities of ambient water to the population. However, maintaining brood stock that are capable of spawning at any time of the year requires the ability to provide low-temperature water in the summer and high-temperature water in the winter. *Mercenaria mer-*

cenaria can be prevented from spawning during the summer months if they are removed from ambient conditions before they have fully developed gonads and stored in flow-through or recirculating seawater at approximately 12°C (53.6°F). Brood stock stored in this manner or removed from ambient winter conditions can be conditioned to spawn by gradually increasing seawater temperatures to 18°–20°C (64°–68°F) and maintaining this temperature for 2–4 weeks (supplemental feeding is sometimes necessary). Using combinations of cool storage and conditioning techniques can thus provide spawnable brood stock throughout the year (Loosanoff and Davis 1950, 1951).

The success of larviculture and subsequently the success of a hatchery is to a great extent dependent on the quantity and quality of algal production. Despite their importance, the nutritional requirements of adult and larval clams are only partially understood. Ukeles (1971) provided an excellent review of nutritional aspects of shellfish culture and reiterated the rather pragmatic nature of algal culture support systems for lamellibranch larviculture. Early culture research indicated that certain phytoplankters were better than others as rations for larvae. Cole (1937) concluded that only minute naked flagellates (Chlamydomonaceae, Cryptomonadaceae, and Chrysomonadaceae) were utilized by swimming bivalve larvae. Davis and Guillard (1958) and Walne (1970) indicated that clam larvae do well on such naked flagellates as *Isochrysis* and *Monochrysis* but do best with mixtures of these two and *Dunaliella* and/or *Platymonas*. Today, naked Chrysophytes are the most commonly used algae in bivalve culture. *Isochrysis galbana* and *Pavlova lutheri* are used widely but certain local isolates are often mixed in and other recently isolated species are gaining popularity (e.g., Tahitian *Isochrysis*).

Hatcheries using blooms of natural phytoplankton to produce larval culture media try to limit phytoplankton size by filtering incoming water. Water thus filtered through a 5- or 10-μm bag or subjected to centrifugation and then forced to bloom will produce relatively dense concentrations of mixed nannoplankters (~3–5 μm in diameter). These natural "batch" cultures are then used to provide feed to bivalve larvae. Although cost effective and relatively easy to perform this Wells-Glancy technique is not used by many large commercial facilities. Selective monoculture of algae (the Milford method) is employed by most large-scale commercial hatcheries in the United States. Axenic algal stock cultures are maintained in test tubes or small Erlenmyer flasks and are used as inocula for larger cultures. Each culture stage serves as inocula for progressively larger cultures until final culture volumes are attained (Fig. 7.4). These cultures are intermittently cropped to provide food for larvae. The final production cultures can

FIG. 7.4 Typical carboy (A) and large-volume container (B) for culture of algae in a commercial clam hatchery using the Milford method.

often be kept in a log growth phase for weeks by appropriate cropping and thus provide many times their original volume in food for bivalve larvae. These high-density algal cultures are dependent on nutrient supplements (i.e., sources of N, P and other elements) added to sterile seawater cultures. These can be simple commercial fertilizers (Loosanoff and Davis 1963) or complex formulations for specific algal groups (Guillard 1958; Ukeles 1971). In comparison with the Wells-Glancy method, the Milford method requires much more energy, labor, and material per unit volume of algae produced, but its common use in commercial facilities certainly testifies to its value and to some degree its cost effectiveness (DePauw 1981; Epifanio 1979).

Larviculture begins with the spawning of sexually mature male and female clams. Properly conditioned adults can be stimulated to spawn in many ways, but normally temperature stimulation is sufficient to promote the release of gametes. Hatcheries normally use a spawning table, which is either a shallow sink or a trough receiving temperature-regulated seawater. The same effect is also obtained by surrounding seawater held in glass baking dishes with temperature controlled tap water. Spawners are placed in the seawater, allowed to acclimate for a few minutes, and then subjected to relatively rapid increases in water temperature. The temperature variation alone is often sufficient to stimulate spawning, but additional stimulation in the form of a sperm suspension or dense phytoplankton infusion is sometimes necessary. Released eggs and sperm are accumulated and mixed, and the resulting zygotes distributed into larval rearing tanks at densities of approximately 30/ml. At culture temperatures between 20° and 25°C (68° and 77°F) the zygotes develop into straight-hinge veliger larvae in 24–48 hr (Table 7.3).

Larvae are reared in a variety of containers from 1-liter beakers to 30,000-liter tanks. Commercial hatcheries generally use 400- or 1600-liter fiberglass tanks (about 100 or 400 gal) with circular tops and sides and conical bottoms (Fig. 7.5). Larvae are stocked at densities of 1–15/ml and fed algal suspensions to produce culture concentrations of 10,000–100,000 cells/ml/day. Depending on technique, larval cultures are changed as often as once per day or as little as twice per week. Culture water is drained through various fine screens (nylon or stainless steel) to retain and segregate larvae by size. Culture containers are cleaned, filled with water and food, and the larvae resuspended. Various water-treatment procedures are often included in the culture-changing process to reduce the incidence of pathogens. These

TABLE 7.3. SIZE OF OVA, VELIGER AND PEDIVELIGER LARVAE OF SOME COMMON CLAM SPECIES

Species	Ova[a] (μm)	Veliger[a] (μm)	Pediveliger[b] (μm)
M. mercenaria	70	105	190–230
M. campechiensis	70	105	175–215
T. japonica	70	95	200–235
E. directus	70	90	210–260
S. solidissima	80	90	230–270
A. islandica	70	110	185–200

[a] Approximate anterior-posterior dimension.
[b] Most commonly reported range of anterior-posterior dimension in metamorphosing larvae.

FIG. 7.5 Banks of 400-liter larval rearing conicals in a commercial clam hatchery.

include the use of ultraviolet radiation, ozonation, sodium hypochlorite and acid rinses, and even routine use of broad spectrum antibiotics. The length of the larval cycle depends on species, temperature, and quality and quantity of algae. At 25°C (67°F) and sufficient quantities of appropriate algae as food, *M. mercenaria* larvae begin to metamorphose in 6–8 days. Most clam species can be reared through the larval cycle within 2 weeks under proper conditions, but larval periods as long as 15–20 days are not uncommon at lower culture temperatures and/or limited algal rations.

Most clam species begin to set or metamorphose at about 200 to 240 μm (Table 7.3). At this point of development they are called pediveligers, indicating the presence of both a functional foot and velum. Commercial hatcheries normally maintain recent set in larval rearing containers or in shallow static-water tanks. The cultures are usually fed higher concentrations of algae, and containers are changed daily. In lieu of feeding cultured algae, some facilities provide a low flow of prefiltered (25 μm) raw seawater to the culture containers or shallow water tables (Castagna and Kraeuter 1981). These techniques, however, are limited by ambient water conditions and require considerable maintenance. In a good growth milieu, postset can grow at a rate of 20–30 μm/day, providing 600-μm clams in 2–3 weeks. At this time hatcheries usually transfer stocks to nursery facilities.

Set to Seed: The Nursery

Nurseries are constructed to provide semicontrolled conditions for intermediate growth between hatcheries and field grow out. Many investigators believe that present field grow-out techniques require a minimum initial seed size of 8–10 mm (Castagna and Kraeuter 1977; Manzi *et al.* 1980, 1981; Claus 1981). Seed producers, however, consider 4- to 6-mm seed to be the maximum profitable size and actually prefer to sell seed at much smaller sizes (H. Eastman, Bristol Shellfish Farms, Roundpond, ME and S. Czyzyk, Bluepoints Co., West Sayville, NY, personal communication). Consequently, commercial and noncommercial grow-out facilities are usually forced to incorporate some nursery protocols into grow-out operations. Aside from providing intermediate growth, nurseries are also used for acclimation and quarantine of imported seed stocks and for wet storage of animals between various stages of field grow out.

Nurseries are classified into two major types: field systems, in which young seed are placed in protected areas in the natural environment, and onshore systems, in which seawater is pumped to land-based seed-support units.

Field Nurseries. Several techniques are used in field system nurseries to provide a protected suitable environment for young postset; these include bottom culture, tray or rack culture, and suspended culture.

Castagna and Kraeuter (1981) described bottom nurseries for *M. mercenaria* in shallow subtidal or low intertidal areas utilizing aggregate cover and either baffles and pens or plastic mesh tenting. The aggregate cover and plastic mesh provide protection from predators, and the tenting or baffles allow some control over siltation. Anderson *et al.* (1982) suggested protective netting for intertidal plants of small *T. japonica* on the basis of experiments that showed covered plots exhibited much higher recoveries than unprotected areas.

Tray or rack nursery culture utilizes either intertidal or subtidal grounds. Trays constructed of frames and mesh of various materials are either placed directly on the bottom or stacked in tiers. These systems provide good protection and growth for dense populations of bivalve postset (Spencer and Gough 1978) but tend to require appreciable maintenance because of fouling and siltation.

Suspended nursery culture utilizes a wide variety of devices includ-

FIG. 7.6 A chain of rafts used for hard clam culture in a South Carolina estuary.

ing rafts, suspended trays, lantern and pearl nets, and upflow or vertical flow systems. Rafts have been used by commercial facilities for many years. Although many different designs have been employed, they all attempt to hold high densities of postset within the water column in either a single layer or in a stacked or tiered configuration (Lucas and Gerard 1981; Shaw 1981). Rafts are usually very simple in design, consisting of flotation devices and suspended culture units. They are normally anchored, either singly or in chains or pods, in areas providing optimum water characteristics (Figs. 7.6 and 7.7). The culture units are suspended below the raft and thus maintained in the warmer more productive surface waters and out of contact with benthic predators.

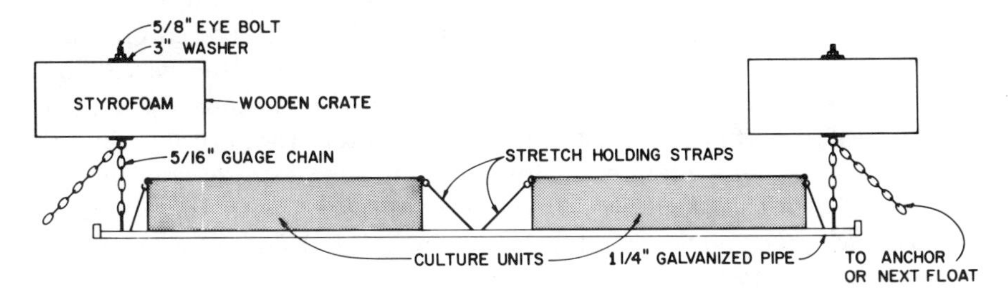

FIG. 7.7 Cross-sectional construction drawing of a typical bivalve culture raft.
Courtesy of K. Swanson.

In recent years, three-dimensional culture units consisting of prefabricated plastic trays, lantern nets, or pearl nets have replaced the more traditional single-layer wood trays and mesh baskets (Shaw 1981). One recent technique that is receiving considerable attention is upflow culture. This system is used primarily in land-based nurseries but has been modified for suspended field nurseries (Lintell 1981). Water is pulled up through a packed column of seed by an airlift. The seed rests on a mesh bottom, which provides water access, and airlift flows exit at the top of the column, thus creating a vertical water flow through the seed (see the following discussion about onshore nurseries for more detail on upflow culture). Field nurseries provide simple, low-energy input systems for preliminary growth of postset. They do, however, have several distinct disadvantages including susceptibility to storm damage, limited access, minimum control over stocked populations, incomplete predator protection, and high maintenance requirements.

Onshore Nurseries. Land-based nurseries are a relatively long-standing tradition in bivalve mariculture operations in the United States. Most bivalve hatcheries support postset holding and grow-out facilities as a natural extension of their operations. These nurseries utilize raceway (shallow trays or tanks) and/or upflow systems to provide semicontrolled environments for early seed growth. In areas with high natural phytoplankton populations, seawater is pumped directly to raceway or upflow culture systems.

Raceways are any of an assortment of shallow trays or tanks (5–100 cm in depth, or 2–40 in.) through which seawater is directed in a horizontal flow. They can be almost any shape (although most are long and narrow) and are constructed of many different materials including fiberglass, epoxy-coated wood, and cement. In shallow raceways, clam seed are spread over the bottom and water is introduced at one end. The water passes over the clams along the length of the raceway and is drained from a standpipe at the opposite end (Fig. 7.8). In deeper trays and tanks, clam seed are often layered on individual mesh sheets forming a tiered rack within the raceway. This three-dimensional design allows for a much greater capacity over a given tank bottom area. Castagna and Kraeuter (1981) recommend a modified raceway system or water table for early postset. Recent research attests to the efficacy of raceway nurseries with such bivalve seed as *M. mercenaria* (Hadley and Manzi, 1984) and *S. solidissima* (Rhodes *et al.* 1981).

Upflow systems for land-based nurseries have become increasingly popular in recent years (Bayes 1979, 1981; Lucas and Gerard 1981;

FIG. 7.8 Covered shallow raceways (1.2×5 m) for nursery grow out of seed clams (note forced upflow cylinders at right).

Manzi and Whetstone 1981; Rodhouse *et al.* 1981). These systems utilize a vertical water flow that passes up through the bivalve seed rather than across the seed as in raceways. Two upflow systems are in common use: forced flow and passive flow (Fig. 7.9). Forced-flow systems introduce water under pressure into a closed bottom cylinder with an intermediately positioned screen of an appropriate mesh size to hold small postset. Water is forced vertically through the screen and eventually exits at the top of the cylinder. Passive-flow systems use open-ended cylinders positioned in a reservoir of appropriate size and configuration. A screen of an appropriate mesh forms the bottom of each cylinder and supports the contained seed mass, which is several centimeters thick. Seawater is pumped into the reservoir and is drawn up through the seed populations of each upflow cylinder by drains located in the upper cylinder. The water accelerates rapidly as it travels through the interstitial spaces of the packed clam seed and decelerates immediately upon reaching the overlying pool between the surface of the seed pack and the drain (Fig. 7.10). Wastes and silt are thus swept through the seed pack and settle as a loose layer at the surface of the seed.

Both active- and passive-upflow systems are supplied with enough

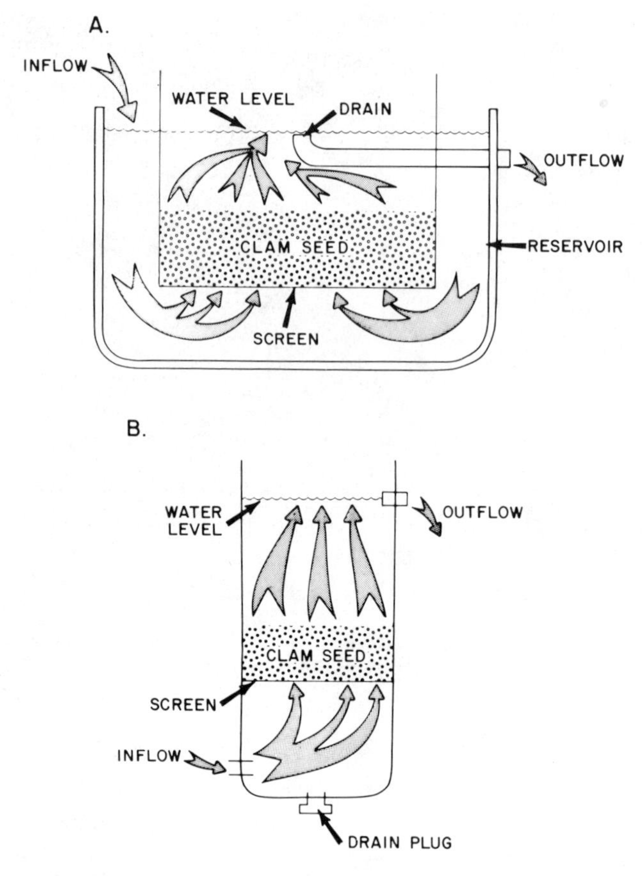

FIG. 7.9 Schematic representation comparing (A) passive and (B) forced flow upwelling or upflow systems.
Courtesy of K. Swanson.

water flow to partially fluidize their contained populations. The conditions created by upflow systems provide rapid growth at extremely dense concentrations of seed. Manzi *et al.* (in press) reported rapid growth of clam seed at densities as high as 15.0 kg/m^2 (3 lb/ft^2) and Bayes (1981) indicated that as much as 1000 kg (2200 lb) of seed can be maintained in a 1.5-m^3 (~50 ft^3) upflow container. Upflow systems appear attractive for commercial nursery application because they have the potential to significantly reduce costs. They are inexpensive to fabricate and require less space and water per unit biomass to operate. In addition, maintenance requirements for these systems are much less than for traditional raceways and thus they may provide a way to reduce nursery labor costs, which are significant at present.

FIG. 7.10 Batteries of passive upflow culture systems in a hard clam nursery.

Seed to Market: Grow-out Systems

Grow-out systems in the United States range from the high technology of closed recirculating systems (Epifanio *et al.* 1976; Epifanio 1979) to the simple broadcasting of seed onto natural shellfish beds. In theory, all these systems work in that they successfully produce market-sized organisms. In practical terms, however, the relevant question is do they produce sufficient quantities of organisms at a reasonable unit cost. Very high technology systems often yield high survival and rapid growth of seed but do so at costs that normally exceed wild-stock prices. Low-technology systems function at low unit costs but normally suffer high seed mortality and/or slow seed growth. Clam grow-out systems that appear to be enjoying reasonable success are those that are intermediate in this range, i.e., those applying sufficient technology to reduce mortality but still relying heavily on natural systems.

Castagna and Kraeuter (1977) and Kraeuter and Castagna (1980) described a simple aggregate cover technique in shallow subtidal waters for the culture of *M. mercenaria*. Baffles, pens, or net tents are used in combination with crushed stone to protect and support seed at densities as high as 4300/m^2 (400/ft^2). This technique has been used in Virginia for the past several years and appears to provide cost-effective predator control for high-density clam culture if proper siting is available. Anderson *et al.* (1982) and Glock and Chew (1978) reported a similar technique for intertidal planting of the Manila clam (*T. japonica*) in Washington. Seed, approximately 10 mm in size, are planted at a density of 1000/m^2 (93/ft^2) on natural gravel substrates and covered with 1.25-cm (0.5-in.) mesh plastic netting for seed retention and predator control. Seed are thinned to approximately one-half of their original density as they progress through the grow-out period. Grow-out takes 2–3 years depending primarily on site suitability (ambient productivity, beach slope, substrate, etc.).

A grow-out system for the surf clam (*S. solidissima*), which required energy input, was analyzed by Goldberg (1980). Final grow out of seed from 15 to 55 mm was performed in raceway tanks, $10 \times 1.3 \times 1$ m ($33 \times 4.3 \times 3.3$ ft), each receiving 50 liters/min (13.2 gal/min) of raw seawater. Seed planted at densities as high as 500/m^2 (46/ft^2) could attain a mean size of 55 mm within 5 months (May–November) in Milford, Connecticut. This and other raceway grow-out systems (Scura *et al.* 1979) show marginal economic potential particularly if associated with inexpensive water modification systems to modulate ambient conditions for year-round growth. Such systems include the use of heated effluents (Peterson and Seo 1977) and waste recycling (Mann and Ryther 1977).

Trays have been used in subtidal, intertidal, and raft culture of hard clams (*M. mercenaria*) for many years and in many sections of the country (Chestnut 1952; Haven and Andrews 1957; Menzel 1964; Godwin 1968; Eldridge *et al.* 1979). Results, for the most part, have indicated both high survival and rapid growth in tray culture when site selection was appropriate. A commercial-scale hard clam mariculture project using intertidal tray culture has recently been established in South Carolina (Manzi *et al.* 1981; Manzi *et al.*, in press). Nursery-reared seed (8–10 mm) are planted at initial densities of 4400/m^2 (409/ft^2) in 6-m^2 (64-ft^2) intertidal units composed of steel reinforcing rod frames supporting vinyl-coated wire baskets, which are in turn lined with plastic netting (Figs. 7.11 and 7.12). Seed densities are reduced to 1100/m^2 (102/ft^2) as mean population size reaches about 25 mm for final grow-out to market size (45–50 mm).

The commercial culture of clams appears to have realistic mariculture potential in the United States. Hatcheries exist to supply seed, and nursery technology has progressed to the point where economic intermediate growth of seed has been attained. Systems for the field grow-out of seed to market size have been, and continue to be, tested

FIG. 7.11 Deploying and planting intertidal hard clam grow-out units.

and improved. Finally, private commercial ventures are in existence, several of which are approaching economic success.

PARASITES AND DISEASES

Natural populations of bivalve mollusks express disease and parasitic syndromes, which affect abundance. This has been well documented in species of commercial interest whose abundance and recruitment are routinely monitored. Aquaculture, by definition, often creates abnormally high concentrations of organisms; these in turn stimulate parasite and disease expression. The study of diseases and parasites in bivalve mollusks has thus been, for the most part, con-

FIG. 7.12 A "pod" of intertidal culture units in a South Carolina tidal creek.

fined to commercially important species, particularly those that are aquaculture candidate species. Literature on clam parasites and diseases is relatively sparse, particularly in comparison to that available for oysters. However, many diseases and parasites described for clams are similar to those reported for oysters (Sprague 1970; Sindermann 1974). The reader is encouraged to review Chapter 6 on oysters to supplement information presented here.

A number of diseases and parasites have been reported in juvenile and adult clams. Sindermann (1970) in his review of diseases of marine fish and shellfish listed the principal diseases of clams, categorized by causative agent (microbial, helminths, and parasitic crustacea), as well as tumors and other abnormalities. Microbial diseases include the haplosporidian *Perkinsus* (=*Dermocystidium*) *marinus*, the coccidia *Hyaloklossia* and *Pseudoklossia*, a gregarine (*Nematopsis*), and several ciliates (e.g., the peritrich *Trichodina* and the thigmotrich *Ancistrocoma*). Diseases caused by helminths result from different trematode life history stages including sporocysts, cercaria, metacercaria, and adults. Larval cestodes have been reported in several clam species from the West Coast. Parasitic crustaceans that use clams as host organisms include at least three species of parasitic copepod *Mytilicola* and a number of genera of pinnotherid crabs. Other abnormalties reported in adult clams include papillary tumors (reported endemic in Chesapeake Bay *Mya*), hyperplastic polypoid and papillary lesions, and polypoid tumors. Nonpathogenic abnormalities of adult clams reviewed by Sindermann (1970) included shell malformations and siphon abnormalities, e.g., functional and nonfunctional supernumerary siphons.

A recent survey for endemic diseases of cultured shellfish in Long Island Sound (Meyers 1981) showed a relatively low incidence of parasitism and histological abnormalities in juvenile and adult *M. mercenaria*. Intracytoplasmic inclusions in the gills and diverticula were the only abnormalities occurring in more than 5% of the examined organisms. Other recent studies reported epizootic neoplasms in *Macoma balthica* (Christensen *et al.* 1974); neoplasia in *Mya arenaria* (Brown *et al.* 1977); and Chlamydiae, mycoplasmas, and rickettsiae in clams from Chesapeake Bay (Harshbarger *et al.* 1977). In general, juvenile and adult clams appear to show lower diversity and incidence of disease and histological abnormalities than oysters, at least to the extent discernible from available comparative literature.

The development of bivalve larviculture techniques and the subsequent emergence of commercial shellfish hatcheries have produced a need for information on larval and postset disease control procedures. Although there appear to be a number of diseases and parasites that infect bivalve larval cultures, two—vibriosis and larval mycosis—are particularly important. Both are ubiquitous and both are effective disease agents capable of causing rapid mass mortality in larval cultures.

Tubiash *et al.* (1965) described a bacillary necrosis in bivalve larvae that was probably a *Vibrio*. Vibriosis has since been described in several areas including Connecticut (Brown and Losee 1978) and New York (Leibovitz 1979; Elston and Leibovitz 1980) and has been implicated as a pathogen in commercial hatcheries from Maine to California. It is thought that vibriosis occurs when adverse environmental conditions stimulate critical infection levels of *Vibrio* spp. Lysis and necrosis of tissues occur in 4–5 hr after exposure and complete mortality of culture populations usually occurs within 18 hr (Sindermann 1974). Diagnosis is by direct microscope examination and confirmation by culture and challenge. Specific diagnosis of *Vibrio* spp. as an etiologic agent can be made with more rapid methods (trypan-blue dye exclusion and antibody tests) described by Elston *et al.* (1981). Several antibiotics, including combistrep, chloramphenicol, polymyxin B, erythromycin, and neomycin, have been used effectively as therapeutics. Most commercial facilities destroy infected cultures and thoroughly sanitize equipment rather than treat infected cultures with antibiotics.

Larval mycosis is a systemic fungal invasion of bivalve larvae (Davis *et al.* 1954). Vishniac (1955) described the pathogen as *Sirolpidium zoophthorum*. No specific treatment is known, and hatcheries normally discard infected cultures and institute sterilization procedures.

The most common symbionts in bivalve larval cultures are ciliates. When ciliate concentrations cannot be controlled by water changes and thinning of larval populations, infected cultures are normally discarded and their containers sterilized.

CONSTRAINTS

Aside from the limitation of venture capital when the general economy is sluggish, the primary constraints to hard clam mariculture can be categorized as (1) natural or environmental restrictions, (2) imposed or regulatory restrictions, and (3) limitations resulting from lack of knowledge.

Natural constraints include the climatic restrictions associated with each particular species selected for mariculture. Most clams require relatively high salinities (>20‰) and a relatively narrow temperature range for rapid growth and development. Throughout its geographical range, a particular species may exhibit different growth rates and/or react differently when exposed to extreme extensions of its preferred environmental conditions (Pratt and Campbell 1956; Ansell 1968). In addition to temperature and salinity, other aspects of water quality must be considered in site selection. Among the more important of these are silt loads and siltation rates, types and relative amounts of natural phytoplankton species and their rates and cycles of productivity, presence or potential presence of pollutants, quantity and rate of water exchange, current and tidal range, substrate characteristics, and the incidence of endemic parasites and disease organisms.

The presence and density of natural predators, although not a restriction of aquaculture activities per se, will limit certain types of grow-out procedures or dictate specific predator-exclusion protocols. A number of species have been documented as major predators of eastern U.S. clam populations (Belding 1912; Haven and Andrews 1957; Menzel *et al.* 1976; Mackenzie 1977; Castagna and Kraeuter 1977, 1981; Manzi and Whetstone 1981; Whetstone and Eversole 1978). These include the rays *(Rhinoptera, Dasyatus,* and *Myliobatis),* conchs or whelks (*Busycon* spp.), moon snails (*Polinices* sp.), drills *(Urosalpinx, Eupleura, Thais),* starfish *(Asterias),* crabs *(Callinectes* spp., *Minippe, Neopanope, Panopeus, Cancer),* and a number of fish, as well as certain wading birds, ducks, and such shore scavengers as gulls and fish crows. Anderson *et al.* (1982) cites several predators of *T. japonica* cultured in Washington, including the moon snail *(Polinices lewisi),* crabs *(Cancer, Hemigrapsus* spp., *Pugettia),* fish *(Lepidopsetta, Parophyrys, Platichthys, Rhachochilis),* starfish (*Pycnopodia, Evasterias, Piaster* spp.), and at least three common duck species. Only two mammals pose serious threats as clam predators, the raccoon and man. Of the two, the latter poses greater potential problems through acts of vandalism and poaching.

The importance of natural or environmental restrictions actually reflects the ramifications of site selection for a given species. The importance of carefully evaluating potential sites, as a preliminary activity for the potential aquaculturist, cannot be overstressed. It is probably safe to say that more aquaculture ventures fail because of inappropriate siting than for any other cause.

The second category of principal constraints to aquaculture in the United States is that of imposed or regulatory restrictions. In a recent report to the U.S. Fish and Wildlife Service (Aspen Research and Information Center 1981), it was observed that over 120 statutory programs of the federal government significantly relate to aquaculture development. This includes over 50 federal statutes (and implementing regulations) that have direct compliance requirements in such aquaculture-related activities as the importation of nonindigenous species, water construction permits, water availability and quality, health and sanitation, occupational safety, etc. Added to these regulations are the protective and restrictive land and water programs that restrict siting and water use. The majority of laws and regulations that specifically authorize, permit, or control aquaculture are found at the state level. These include the following general areas:

1. Species Management—exotic species importation, leasing, harvest management, fisheries research, endangered species, licensing, fishery conservation, etc.
2. Water Management—dams and reservoirs, dredging and filling, navigation and harbor management, facility construction, boating management, estuaries management, groundwater and watershed protection, boundary waters, etc.
3. Land Management—coastal-zone management, submerged land and wetland management, industrial siting, zoning, regional planning, dredge and fill, forest and game management, eminent domain, agricultural land use, recreation, soil conservation, public lands, etc.
4. Health and Safety—facility design and construction, import restrictions, quarantine, food and drug regulations, disease control, sanitation, processing, inspection, occupational health and safety.
5. Pollution Control.
6. Commerce and Labor—aquaculture/agriculture assistance and loan programs, crop insurance, marketing, licensing, certification, taxation, investment protection, economic development, employment regulations, etc.

Aquaculture activities in most states are often restricted because existing regulations, appropriately enacted for the management of wild-stock fisheries, inadvertently limit operational protocols. Examples of such regulations include minimum size limitations of organisms in possession; restrictions on mechanical harvesting, importation of exotics, and holding (wet storage), transportation, and culture of game species; lease maintenance requirements; and the opening and closing of legal fishing seasons.

Thus the often lengthy and convoluted procedures for obtaining required permits, certificates, and licenses present an appreciable impasse and often discourage potential investment in aquaculture projects. In addition, the unavailability of suitable habitat for commercial lease and the all too frequent confrontations with traditional fishery interests frequently confound the initiation of aquaculture in certain states.

The last category of constraints to clam aquaculture in the United States is that associated with gaps in scientific and technical knowledge. Although much research has been performed on clams and clam culture (McHugh *et al.* 1982), the literature is scattered, with few unifying, comprehensive reviews. The culturist is thus faced with the substantial task of locating information pertinent to his specific requirements and location. Certain areas of research lack definitive work, particularly in some areas of direct interest to aquaculturists (e.g., genetics, nutrition, and disease and parasite control). Additional applied research is needed in such areas as hatchery and nursery engineering, construction, and operation; harvesting and processing; and pilot or demonstration operations.

A discussion of constraints associated with clam culture in the United States would not be complete without some mention of financial limitations. Many aquaculturists, from both the private and public sectors, believe that the largest present need in aquaculture is financial (JSA 1980). In the economic climate of the early 1980s, characterized by high interest rates and scarce capital, the high risk associated with aquaculture does not easily draw venture capital. Low interest, guaranteed loans, tax incentives, subsidized crop insurance, etc., would do much to encourage investment in aquaculture. Fortunately many states have developed, or are in the process of developing, aquaculture plans as a preliminary to legislative action. Almost without exception, these plans call for financial incentives for aquaculture. The federal government in the National Aquaculture Act of 1980 (P.L. 96–362) declared aquaculture to be in the national interest and advocates a national policy to encourage its development. Although it does not provide for the direct incentives that the aquaculture business and research communities need, the Act does mandate the development and implementation of a National Aquaculture Development Plan—the first step for more specific national legislative action.

STATUS AND ECONOMIC OVERVIEW

Total U.S. aquaculture production lags considerably behind that of such nations as Japan, China, and Russia. Latest estimates indicate that the United States produces about 3% of its domestic consumption of fish and shellfish in aquaculture operations, whereas China produces 50 times and Japan 12 times this quantity (Aspen Research and Information Center 1981). As stated earlier, U.S. aquaculture production of clams is only about 9% of the total national harvest, or less than 4.55 million kg (10 million lbs) annually. Thus, while clam aquaculture technology has developed significantly in the United States, there has been little reflection of this development in aquaculture product yield.

The hard clam *(M. mercenaria)* is the most commonly cultured clam in the United States. Private and/or public hard clam mariculture activities are ongoing in at least 14 Atlantic and Gulf Coast states and two Pacific Coast states. There are 17 commercial hatcheries producing hard clam seed either on a regular basis or coincidentally with the culture of other bivalve seed (Table 7.4). These hatcheries range

TABLE 7.4. PARTIAL LIST OF COMMERCIAL HATCHERIES SUPPLYING CLAM SEED IN THE UNITED STATES

Aquacultural Research Corporation
P.O. Box 597
Dennis, MA 02638
(617) 385-3933

Bluepoints Company, Inc.
West Sayville
Long Island, NY 11796
(516) 589-0123

Bristol Shellfish Farms
Moxie Cove Road
Round Pond, ME 04564
(207) 529-5634

Coast Oyster Company
Hatchery Division
P.O. Box 635
Ocean Park, WA 98640
(206) 665-4075

Cozy Harbor Seafarms
Pratts Island Road
West Southport, ME 04569

Frank M. Flower & Sons
P.O. Box 92
Bayville, NY 11709
(516) 628-2077

Huskey's Clam Hatchery
P.O. Box 186
Atlantic, NC 28511

International Shellfish Enterprises, Inc.
P.O. Box 200
Moss Landing, CA 95039
(408) 633-3063

Intertide Corporation
North Harpswell, ME 04079
(207) 729-4245

Long Island Oyster Farms
Eatons Neck Road
Northport, NY 11768
(516) 757-1600

Marine Bioservices Co.
High Island
South Bristol, ME 04568
(207) 644-8537

North Carolina Shellfish Enterprises
P.O. Box 277
Harkers Island, NC 28513
(919) 728-2160

Pacific Mariculture Inc.
P.O. Box 336
Moss Landing, CA 95039
(408) 633-3548

Pigeon Point Shellfish Hatchery
921 Pigeon Point Road
Pescadero, CA 94060
(415) 879-0391

Shellfish Incorporated
West Sayville
Long Island, NY 11796
(516) 589-5770

Shinnecock Indian Tribal Project
P.O. Box 670
Southampton, NY 11968
(516) 283-3776

Westport Fisheries
Rt. 88
Westport, MA 02790
(617) 636-5468 from relatively small pragmatic facilities, producing less than 20 million seed per year, to large, highly organized facilities producing over a billion bivalve postset per year. Most of these facilities operate on a seasonal basis, closing during the winter months to perform routine maintenance or renovations. Some facilities maintain limited production during the winter months and only a very few remain in full production throughout the year. Total seed production figures for the United States are not available, but seed supplies (at least of the smaller sizes) are high and are not, in general, a limiting factor for the industry. Many hatcheries, particularly larger facilities, are vertically integrated into larger seafood businesses. Thus, the vast majority of their seed production is used internally and is not available for sale to outside interests.

Available data indicate that at least 20 commercial clam aquaculture ventures are in operation on the East Coast and six on the West Coast of the United States. Total production of these facilities is probably much higher than the reported total clam aquaculture production. This occurs because several companies do not differentiate between wild-stock and cultured products and because certain operations carry controlled culture through only a part of the life cycle, relying on unprotected broadcast planting for final grow-out to market size. The total value of current U.S. investment in clam aquaculture is not available but must be relatively modest in relationship to the value of the fishery.

Private and public research and development in clam aquaculture have enjoyed a recent resurgence, although the total effort is still lacking in several critical areas (e.g., genetics and nutrition). New York and New England have the highest concentration of commercial hatcheries, and a number of research projects are supported in this area by federal, state, and local governments, universities, and private industry. This research includes studies on raft culture and predator control, hatchery and nursery design and development, basic biology, and disease identification and control. In the mid-Atlantic region, New Jersey, Maryland, and Virginia have a strong research program concentrating on hatchery methodology, basic biology (including some nutritional aspects), and field grow-out systems. Among the South Atlantic states, South Carolina and Georgia have programs analyzing nursery and field grow-out systems and clam genetics. On the West Coast, Washington has programs in field grow-out systems, ecology, and predator control and California in hatchery development and nursery methodology. This partial review of research topics not only reflects the range of current studies but also the broad interest in clam aquaculture throughout the United States.

The economics of clam aquaculture are not well documented. Industry figures are generally unavailable or are reported without indications of the basis for calculations. Commercial hatchery managers report costs of production ranging from $1.00 to $8.00 per thousand 1-mm seed and $3.50 to $20.00 per thousand 8-mm seed. A recent survey of clam seed prices (Manzi and Whetstone 1981) indicated price ranges of $1.50–4.00 per thousand for seed below 1 mm in size and $12.50–30.00 per thousand for seed between 8 and 10 mm. A recent analysis (JSA 1980) reported the following approximate costs for a pilot facility capable of producing 2 million clams per year: $140,000, establishment costs amortized over the first 3 years; $12,000–14,000, production costs per million clams; $50,000–70,000, annual labor costs; $7,000–$15,000, harvesting costs per million clams; and $5000, marketing costs per million clams. This amounts to between $145,000 and $185,000 per year for the first 3 years of operation (assuming establishment costs are amortized over only this period). This figure does not include costs of real estate, taxes, interest, or loss of potential earnings. These costs are to a great degree theoretical, and their validity in relation to actual costs is unknown until more data are derived from the operating experience of aquaculturists.

Models have recently been used to extrapolate growth and mortality rates in bivalves in order to assess the economics of a culture system (Askew 1978). Such analysis provides projections of potential gross revenues against time, which can be used to compare operational alternatives (planting schedules, species options, scale options, etc.). Growth and survival data have also been used to optimize nursery utilization by adjustments of importation schedules of acquired seed (Brown *et al.,* 1983). These and other methods of economic analysis are helping to provide the data necessary for valid evaluations of aquaculture methodologies. They are, however, no substitutes for cost analyses from commercial aquaculture operations. More data, derived from actual operating experiences of aquaculturists, is required for valid economic assessment (Smith and Roberts 1976).

SUMMARY

The prognosis for clam aquaculture in the United States is encouraging. Clams in general and certain species in particular are prime aquaculture candidates with strong, established markets and diminishing wild-stock populations. Techniques for the culture of several species have been developed and optimized in recent years, and commercial quantities of hatchery-reared seed are available from a number of private hatcheries. The basic biology of major species is well known, and research programs on clams are in progress in almost every coastal state of the nation. Pilot projects indicate the feasibility of nursery and field grow-out protocols developed over recent years, and new materials make these methods more cost efficient. Diseases and parasites, although not studied extensively, are under control in hatchery and nursery culture and have not been a common problem in field grow-out systems.

Constraints to the development of clam aquaculture in the United States are, for the most part, the same as those hindering other types of aquaculture. These include natural or environmental restrictions, imposed or regulatory restrictions, and limitations resulting from lack of knowledge. These constraints, coupled with the financial limitations imposed by a weak economy and the lack of publicly supported incentives, do little to encourage aquaculture development.

Although considerable, the constraints to aquaculture have not precluded investment in commercial aquaculture ventures. The financial picture is so favorable with certain species, including clams, that entrepreneurs have continued to support or have recently initiated aquaculture operations in many regions of the country. The commercialization of clam aquaculture has been aided by a resurgence of support by federal, state, and local governments for research activities in aquaculture. This development, together with the attributes of clams discussed in this chapter, make clams prime candidates for imminent commercial aquaculture success.

ACKNOWLEDGMENTS

I wish to thank those colleagues who generously shared their time and knowledge to help in the preparation of this chapter. I am particularly indebted to M. Castagna, E. W. Rhodes, K. Chew, P. Eldridge, H. Q. M. Clawson, and V. G. Burrell, Jr., for their reviews and comments. N. Hadley, M. Maddox, and F. Stevens provided help with literature searches and reviews; K. Swanson drafted the figures; and C. Linx prepared the typescript.

LITERATURE CITED

ANDERSON, G. J., MILLER, M. B., and CHEW, K. K. 1982. A guide to manila clam aquaculture in Puget Sound. Sea Grant Rep. WSG82-4, University of Washington, Seattle, Washington.

ANSELL, A. D. 1968. The rate of growth of the hard clam *Mercenaria mercenaria* (L.) throughout its geographical range. J. Cons. Int. Explor. Mer **31**(3), 364–409.

ASKEW, C. G. 1978. A generalized growth and mortality model for assessing the economics of bivalve culture. Aquaculture **14,** 91–104.

ASPEN RESEARCH AND INFORMATION CENTER. 1981. Aquaculture in the United States: regulatory constraints. Final Rep. Contr. No. 14-16-009-79-095. U.S. Fish and Wildlife Service, Washington, DC.

BARDACH, J. E., RYTHER, J. H., and MCLARNEY, W. O. 1972. Aquaculture, the Farming and Husbandry of Freshwater and Marine Organisms. Wiley-Interscience, New York.

BAYES, J. C. 1979. How to rear oysters. Proc. 10th Annu. Shell. Conf. Shellfish Assoc. Great Britain, pp. 7–17.

BAYES, J. C. 1981. Forced upwelling nurseries for oysters and clams using impounded water systems. *In* Nursery Culturing of Bivalve Molluscs, C. Claus, N. DePauw, and E. Jaspers (editors), Spec. Publ. No. 7, pp. 73–83. European Mariculture Society, Bredene, Belgium.

BELDING, D. L. 1912. The quahog fishery of Massachusetts. Mass. Dept. Nat. Resour. Div. Mar. Fish. **2,** 1–41.

BOURNE, N. 1981. Clam farming—fact or fancy. B.C. Shellfish Maricult. Newsl., **1**(3), 1–5.

BROWN, C., and LOSEE, E. 1978. Observations on natural and induced epizootic of vibriosis in *Crassostrea virginica* larvae. J. Invertebr. Pathol. **31,** 41–47.

BROWN, J. W., MANZI, J. J., CLAWSON, H. Q. M., and STEVENS, F. S. 1983. Moving out the learning curve: an analysis of hard clam, *Mercenaria mercenaria,* nursery operations in South Carolina. Mar. Fish. Rev. **45**(4–6), 10–15.

BROWN, R. S., WOLKE, R. E., SAILA, S. B., and BROWN, C. W. 1977. Prevalence of Neoplasia in 10 New England populations of the soft-shelled clam, *Mya arenaria.* Ann. New York Acad. Sci. **298,** 522–534.

CASTAGNA, M., and KRAEUTER, J. N. 1977. *Mercenaria* culture using stone aggregate for predator protection. Proc. Nat. Shellfish. Assoc. **67,** 1–6.

CASTAGNA, M., and KRAEUTER, J. N. 1981. Manual for growing the hard clam *Mercenaria.* Spec. Rep. No. 249, Virginia Institute of Marine Science, Gloucester Point, Virginia.

CHESTNUT, A. F. 1952. Growth rates and movements of hard clams, *Venus mercenaria.* Gulf Caribb. Fish. Inst. Univ. Miami Proc. **4,** 49–59.

CHRISTENSEN, D. J., FARLEY, C. A., and KERN, F. G. 1974. Epizootic neoplasms in the clam *Macoma balthica* (L.) from Chesapeake Bay. J. Nat. Cancer Inst. **52,** 1739–1749.

CLAUS, C. 1981. Trends in nursery rearing of bivalve mulluscs. *In* Nursery Culturing of Bivalve Molluscs. C. Claus, N. DePauw, and E. Jaspers (editors), Spec. Publ. No. 7, pp. 1–33. European Mariculture Society, Bredene, Belgium.

COLE, H. A. 1937. Experiments in the breeding of oysters in tanks with special reference to the food of the larvae and spat. Fish. Invest., Minist. Agric. Fish. GB, Ser. II **15,** 1–24.

DAVIS, H. C. 1968. Shellfish hatcheries present and future. Am. Fish. Soc. **98,** 18.

DAVIS, H. C. 1969. Shellfish hatcheries: present and future. Trans. Am. Fish. Soc. **98**(4), 743–750.

DAVIS, H. C., and GUILLARD, R. R. 1958. Relative value of ten genera of microorganisms as food for oyster and clam larvae. U.S. Fish Wildl. Serv. Fish. Bull. **136**(58), 293–304.

DAVIS, H. C., LOOSANOFF, V. L., WESTON, W. H., and MARTIN, C. 1954. A fungus disease in clam and oyster larvae. Science **120,** 36–38.

DEPAUW, N. 1981. Use and production of microalgae as food for nursery bivalves. *In* Nursery Culturing of Bivalve Molluscs, C. Claus, N. DePauw, and E. Jaspers (editors), Spec. Publ. No. 7, pp. 35–69. European Mariculture Society, Bredene, Belgium.

DRESSEL, D. M., and FRITZGIBBON, D. S. 1978. The United States molluscs shellfish industry. *In* Drugs and Food from the Sea. P. N. Kaul and C. J. Sindermann (editors) pp. 251–261. Univ. of Oklahoma, Norman, Oklahoma.

DUPUY, J. L., WINDSOR, N. T., and SUTTON, C. E. 1977. Handbook for design and operation of an oyster seed hatchery. Spec. Rep. No. 142. Virginia Institute of Marine Science, Gloucester Point, Virginia.

ELDRIDGE, P. J., EVERSOLE, A. G., and WHETSTONE, J. M. 1979. Comparative survival and growth rates of hard clams, *Mercenaria mercenaria,* planted in trays subtidally and intertidally at varying densities in a South Carolina estuary. Proc. Nat. Shellfish. Assoc. **69,** 30–39.

ELSTON, R., and LEIBOVITZ, L. 1980. Pathogenesis of experimental vibriosis in larval American oysters, Crassostrea virginica. Can. J. Fish. Aquatic Sci. **37,** 969–978.

ELSTON, R., LEIBOVITZ, L., RELYEA, D., and ZATILA, J. 1981. Diagnosis of vibriosis in a commercial oyster hatchery epizootic: diagnostic tools and management features. Aquaculture **24**(1,2), 53–62.

EPIFANIO, C. E. 1979. Growth in bivalve molluscs: nutritional effects of two or more species of algae in diets fed to the American oyster *Crassostrea virginica* and the hard clam *Mercenaria mercenaria.* Aquaculture **18,** 1–12.

EPIFANIO, C. E., MOUTZ LOGAN, C., and TURK, C. L. 1976. Culture of six species of bivalves in a recirculating seawater system. Proc. 10th Eur. Symp. Mar. Biol. **1,** 97–108. Universa Press, Wetteren, Belgium.

EVERSOLE, A. G., MICHENER, W. K., and ELDRIDGE, P. J. 1980. Reproductive cycle of *Mercenaria mercenaria* in a South Carolina estuary. Proc. Nat. Shellfish. Assoc. **70,** 22–30.

GLOCK, J. W., and CHEW, K. K. 1978. Growth, recovery, and movement of Manila clam, *Venerupis japonica,* planted under protective devices and on open beaches at Squaxin Island, Washington. Final Rep. College of Fisheries, University of Washington, Seattle, Washington.

GLUDE, J. B. 1974. Recent developments in shellfish culture, NOAA Nat. Oceanic Atmos. Admin. Tech. Rep. NMFS Circ. No. 338.

GODWIN, W. 1968. The growth and survival of planted clams, *Mercenaria mercenaria,* on the Georgia coast. Contrib. Ser. No. 9. Georgia Fish and Game Commission, Athens, GA.

GOLDBERG, R. 1980. Biological and technological studies on the aquaculture of yearling surf clams. Part I: Aquacultural production. Proc. Nat. Shellfish. Assoc. **70,** 55–60.

GOODWIN, C. L., and JONES, C. 1976. Standing crop estimates of soft-shell clams in Skagit and Port Susan Bays. Summary Rep. Washington Department of Fisheries, Seattle, Washington.

GUILLARD, R. R. 1958. Some factors in the use of nannoplankton cultures as food for larval and juvenile bivalves. Proc. Nat. Shellfish. Assoc. **48,** 134–142.

HADLEY, N. H., and MANZI, J. J. 1984. Growth of seed clams *(Mercenaria mercenaria)* at various densities in a commercial scale nursery system. Aquaculture **36,** 369–378.

HARSHBARGER, J. C., HANGE, S. C., and OTTO, S. V. 1977. Chlamydiae, mycoplasmas, and rickettsia in Chesapeake Bay bivalves. Science **196,** 666–668.

HAVEN, D., and ANDREWS, J. D. 1957. Survival and growth of *Venus mercenaria, Venus campechiensis,* and their hybrids in suspended trays and natural bottoms. Proc. Nat. Shellfish. Assoc. **47,** 43–49.

HYMAN, L. H. 1967. The Invertebrates, Vol. 6, Mollusca I. McGraw-Hill Book Co., New York.

IVERSON, E. S. 1968. Farming the Edge of the Sea. Garden City Press Ltd., Letchworth, United Kingdom.

JSA. 1980. National aquaculture plan. U.S. Dept. of Commerce Joint Subcommittee on Aquaculture, National Oceanic and Atmospheric Administration. Draft Document.

JONES, D. S. 1980. Annual cycle of reproduction and shell growth in bivalves, *Spisula solidissima* and *Arctica islandica.* Ph.D. Thesis. Princeton University, Princeton, New Jersey.

KRAEUTER, J. N. and M. CASTAGNA. 1980. Effects of large predators on the field culture of the hard clam, *Mercenaria mercenaria.* U.S. Fish Wildl. Serv. Fish. Bull. **78**(2), 538–541.

LANDERS, W. S. 1976. Reproduction and early development of the ocean quahog, *Arctica islandia,* in the laboratory. Nautilus **90,** 88–92.

LEIBOVITZ, L. 1979. A study of vibriosis at a Long Island shellfish hatchery. New York Sea Grant Reprint Ser. NYSG-RR-79-02. New York Sea Grant Institute, Albany, New York.

LINTELL, J. L. 1981. Shellfish culture. A view from a nursery. Proc. 12th Annu. Shellfish Conf. Shellfish Assoc. Great Britain, pp. 3–10.

LONGWELL, A. C. 1974. Some impressions regarding genetics and the fisheries of Japan. Proc. 1st U.S.-Japan Meet. Aquaculture, Tokyo, Nat. Mar. Fish. Serv. Circ. 338, U.S. Department of Commerce, Washington, DC.

LONGWELL, A. C., and STILES, S. S. 1973. Oyster genetics and the probable future role of genetics in aquaculture. Malacol. Rev. **6,** 151–177.

LOOSANOFF, V. L. 1954. New advances in the study of bivalve larvae. Am. Sci. **42,** 607–624.

LOOSANOFF, V. L., and DAVIS, H. C. 1950. Conditioning *V. mercenaria* for spawning in winter and breeding its larvae in the laboratory. Biol. Bull. **98,** 60–65.

LOOSANOFF, V. L., and DAVIS, H. C. 1951. Delaying spawning of lamellibranchs by low temperature. J. Mar. Res. **10,** 197–202.

LOOSANOFF, V. L., and DAVIS, H. C. 1963. Rearing of bivalve molluscs. Adv. Mar. Biol. **1,** 1–136.

LOOSANOFF, V. L., DAVIS, H. C., and CHANLEY, P. E. 1966. Dimensions and shapes of larvae of some marine bivalve Mollusks. Malacologia **4**(2), 351–435.

LUCAS, A. 1976. Aspects of the rearing and cultivation of *Venerupis.* Proc. 7th Annu. Shellfish Conf. Shellfish Assoc. Great Britain, pp. 32–36.

LUCAS, A., and GERARD, A. 1981. Space requirement and energy cost in some type of bivalve nurseries. *In* Nursery Culturing of Bivalve Molluscs. C. Claus, N. DePauw, and E. Jaspers (editors), Spec. Publ. No. 7, pp. 151–170. European Mariculture Society, Bredene, Belgium.

MCHUGH, J. L., SUMMER, M. W., FLAGG, P. J., LIPTON, D. W., and BEHRENS, W. J. 1982. Annotated bibliography of the hard clam *(Mercenaria mercenaria).* NOAA Tech. Rep. NMFS SSRF-756, U.S. Department of Commerce, Washington, DC.

MACKENZIE, C. L., JR. 1977. Predation on hard clam *(Mercenaria mercenaria)* populations. Trans. Am. Fish. Soc. **106**(6), 530–537.

MAGOON, C., and VINING, R. 1981. Introduction to Shellfish Aquaculture. Washington Dept. Natural Resources, Seattle, WA.

MANN, R. 1979. The effect of temperature on growth, physiology, and gametogenesis in the manila clam *Tapes philippinarum*. J. Exp. Mar. Biol. Ecol. **38,** 122–133.

MANN, R., and RYTHER, H. 1977. Growth of six species of bivalve molluscs in a waste recycling-aquaculture system. Aquaculture **11,** 231–245.

MANZI, J. J., BURRELL, JR., V. G., and CARSON, W. Z. 1980. A mariculture demonstration project for an alternative hard clam *(Mercenaria mercenaria)* fishery in South Carolina. Proc. 11th Annu. Meet. World Maric. Soc. **11,** 79–89.

MANZI, J. J., BURRELL, JR., V. G., and CLAWSON, H. Q. M. 1981. Commercialization of hard clam *(Mercenaria mercenaria)* in South Carolina: Preliminary report. Proc. 12th Annu. Meet. World Maric. Soc. **12,** 181–189.

MANZI, J. J., BURRELL, JR., V. G., CLAWSON, H. Q. M., STEVENS, F. S., and MADDOX, M. B. Commercialization of hard clam *(Mercenaria mercenaria)* mariculture in South Carolina: One year later. J. Shellfish Res. (in press).

MANZI, J. J., and WHETSTONE, J. M. 1981. Intensive hard clam mariculture: A primer for South Carolina watermen. South Carolina Sea Grant Consort. Marine Advisory Publ. No. 81-01.

MENZEL, R. W. 1964. Seasonal growth of northern and southern quahogs, *Mercenaria mercenaria* and *M. campechiensis,* and their hybrids in Florida. Proc. Nat. Shellfish. Assoc. **53,** 111–119.

MENZEL, R. W. 1971. Quahog clams and their possible mariculture. Proc. World Maric. Soc. **2,** 23–36.

MENZEL, R. W. 1977. Selection and hybridization in quahog clams (*Mercenaria* spp.). Proc. 8th Annu. Meet. World Maric. Soc. **8,** 507–521.

MENZEL, R. W., COKE, E. W., HAINES, M. L., MARTIN, R. E., and OLSEN, L. A. 1976. Clam mariculture in northwest Florida: Field study on predation. Proc. Nat. Shellfish. Assoc. **65,** 59–62.

MERRILL, A. S., and ROPES, J. W. 1969. The general distribution of the surf clam and ocean quahog. Proc. Nat. Shellfish. Assoc. **59,** 40–45.

MEYERS, T. R. 1981. Endemic diseases of cultured shellfish of Long Island, New York: Adult and juvenile american oysters *(Crassostrea virginica)* and hard clams *(Mercenaria mercenaria)*. Aquaculture **22**(4), 305–330.

MILNE, P. H. 1972. Fish and shellfish farming in coastal waters. Whitefriars Press Ltd., London.

MURAWSKI, S. A., and SERCHUK, F. M. 1979. Shell length-meat weight relationships of ocean quahogs, *Arctica islandica,* from the middle Atlantic Shelf. Proc. Nat. Shellfish. Assoc. **69,** 40–46.

NOSHO, T. Y. 1972. The clam fishery of the Gulf of Alaska. *In* A Review of the Oceanography and Renewable Resources of the Northern Gulf of Alaska. D. H. Rosenberg (editor), Institute of Marine Science Rep. R72-23, pp. 351–360. University of Alaska, Fairbanks, Alaska.

PETERSON, R. E. and K. K. SEO. 1977. Thermal aquaculture. Proc. World Maric. Soc. **8,** 491–503.

PRATT, D. M., and CAMPBELL, D. A. 1956. Environmental factors affecting growth in *Venus mercenaria*. Limnol. Oceanogr. **1,** 2–17.

PURCHON, R. D. 1968. The Biology of the Mollusca. Pergamon Press, Oxford.

QUAYLE, D. B. 1941. The Japanese "little neck" clam accidentally introduced into

British Columbia waters. Progr. Rep. No. 48, pp. 17–18. Fish. Res. Board, Pacific Coast Station, Vancouver, British Columbia.

RHODES, E. W., GOLDBERG, R., and WIDMAN, J. C. 1981. The role of raceways in mariculture systems for the bay scallop, *Argopecten irradians irradians,* and the surf clam, *Spisula solidissima. In* Nursery Culture of Bivalve Molluscs. C. Claus, N. DePauw, and E. Jaspers (editors), Spec. Publ. No. 7, pp. 227–251. European Mariculture Society, Bredene, Belgium.

RITCHIE, T. P. 1977. A comprehensive review of the commercial clam industries in the United States. Sea Grant Program Publ. No. DEL-SG-26-76. Nat. Marine Fish. Serv. Publ. No. NOAA-S/T77-2752. University of Delaware, Newark, Delaware.

RODHOUSE, P. G., OTTWAY, B., and BURNELL, G. M. 1981. Bivalve production and food chain efficiency in an experimental nursery system. J. Mar. Biol. Assoc. U. K. **61,** 243–256.

ROPES, J. R. 1979. Biology and distribution of surf clams *(Spisula solidissima)* and ocean quahogs *(Arctica islandica)* off the northeast coast of the U.S. Proc. Northeast Clam Ind. Management for the Future. Univ. of Mass. Coop. Ext. Serv., University of Massachusetts, Amherst, Mass.

SCURA, E. D., KULJIS, A. M., YORK, R. H., and LEGOFF, R. S. 1979. The commercial production of oysters in an intensive raceway system. Proc. World Maric. Soc. **10,** 624–630.

SERCHUK, F. M., MURAWSKI, S. A., HENDERSON, E. M., and BROWN, B. E. 1978. The population dynamics basis for management of offshore surf clam populations in the Middle Atlantic. Proc. Northeast Clam Ind. Management for the Future. Univ. of Mass., Coop. Ext. Serv., University of Massachusetts, Amherst, Massachusetts.

SHAW, W. N. 1981. The role nurseries play in the culture of molluscs in Japan. *In* Nursery Culturing of Bivalve Molluscs. C. Claus, N. De Pauw, and E. Jaspers (editors), Spec. Publ. No. 7, pp. 269–281. European Mariculture Society, Bredene, Belgium.

SHULTZ, F. T. 1970. Genetic potentials in aquaculture. *In* Food and Drugs from the Sea, H. W. Youngken (editor), Proc. 1969. Mar. Technol. Soc., Washington, D.C., pp. 118–134.

SINDERMANN, C. J. 1970. Principal Diseases of Marine Fish and Shellfish. Academic Press, New York.

SINDERMANN, C. J. 1974. Diagnosis and control of mariculture diseases in the United States. Tech. Rep. No. 2. Middle Atlantic Coastal Fisheries Center, National Marine Fisheries Service, Highlands, New Jersey.

SMITH, F. J., and ROBERTS, K. J. 1976. Aquaculture economics research needs. Sea Grant Tech. Rep. No. 5. South Carolina Sea Grant Program, Charleston, South Carolina.

SPENCER, B. E. and C. J. GOUGH. 1978. The growth and survival of experimental batches of hatchery-reared spat of *Ostrea edulis* L. and *Crassostrea gigas* Thimberg, using different methods of tray cultivation. Aquaculture **13,** 293–312.

SPRAGUE, V. 1970. Some protozoan parasites and hyperparasites in marine bivalve molluscs. Symp. Diseases Fishes Shellfishes, pp. 511–526. American Fisheries Society Spec. Publ. No. 5.

THOMPSON, I., JONES, D. S., and DREIBELBIS, P. 1980. Annual internal growth banding and life history of the ocean quahog *Arctica islandica* (Mollusca: Bivalvia). Mar. Biol. **57,** 25–34.

TUBIASH, H. S., CHANLEY, P., and LEIFSON, E. 1965. Bacillary necrosis, a disease of larval and juvenile bivalue Mollusks. J. Bacteriol. **90**(4), 1036–1044.

UKELES, R. 1971. Nutritional requirements in shellfish culture. Proc. Conf. Artificial Propagation Commer. Valuable Shellfish, University of Delaware, pp. 43–64.

U.S. DEPT. OF COMMERCE. 1981. Fisheries of the United States, 1980. Current Fisheries Stat. No. 8100. NMFS, Washington, DC.

U.S. DEPT. OF COMMERCE. 1982. Fisheries of the United States, 1981. Current Fisheries Stat. No. 8200. Nat. Marine Fish. Serv., Washington, DC.

VISHNIAC, H. S. 1955. The morphology and nutrition of a new species of *Siropidium*. Mycologia **47,** 633–645.

WALNE, P. R. 1964. The culture of marine bivalve larvae. *In* Physiology of the Mollusca, K. Wilbur and C. M. Yonge (editors), Vol. I, pp. 197–210. Academic Press, New York.

WALNE, P. R. 1970. Studies on the food value of nineteen genera of algae to juvenile bivalves of the genera *Ostrea, Crassostrea, Mercenaria,* and *Mytilus*. Fish. Invest. Minist. Agric. Fish Food GB Ser. II, **26**(5).

WALNE, P. R. 1979. Culture of bivalve Molluscs—50 years' experience at Conwy, 2nd Ed. Whitefriars Press Ltd., London.

WEBBER, H. H., and RIORDAN, P. F. 1976. Criteria for candidate species for aquaculture. Aquaculture **7,** 107–123.

WELLS, W. F. 1927. Report of the experimental shellfish station. Report of the Rep. No. 16, pp. 1–22. New York State Conservation Department.

WHETSTONE, J. M., and EVERSOLE, A. G. 1978. Predation on hard clams, *Mercenaria mercenaria,* by mud crabs, *Panopeus herbstii*. Proc. Nat. Shellfish. Assoc. **68,** 42–48.

WILBUR, K. M., and YONGE, C. M. 1964–1966. Physiology of Mollusca, Vols. I and II. Academic Press, New York.

WOOD, L., and HARGIS, JR., W. J. 1971. Transport of bivalve larvae in a tidal estuary. 4th Eur. Mar. Biol. Symp., pp. 29–44. Cambridge Univ. Press, Oxford.

ZURAW, E. A., LEONE, D. E., and GRISCOM, C. A. 1967. Ecology of molluscs and the culture of *Mya arenaria*. Rep. No. U413-67-206. General Dynamics/Electric Boat Division, New London, Connecticut.

8

Mussel Aquaculture in the United States

Richard A. Lutz

Introduction
European Mussel Culture Technology
  Bottom Culture
  Bouchot Culture
  Raft Culture
  Long-Line Culture
Feasibility of European Culture Techniques in the United States
Current Mussel Aquaculture Production
Recent Research Efforts
Life Cycle of the Blue Mussel
Experimental Culture—East Coast
  Growth, Survival, and Recruitment Studies
  Harvest and Transplantation of Mussel Seed
Experimental Culture—West Coast
Presence of Pearls
Predation, Parasitism, and Accumulation of Algal Biotoxins
Mussel Aquaculture in Heated Effluents
Aquaculture Carrying-Capacity Model
Economics
Summary
Literature Cited

INTRODUCTION

Mussel *(Mytilus edulis)* aquaculture in the United States is presently in its infancy. The first commercial operation (Abandoned Farm, Inc.) involved with culturing this species in U.S. waters was established in the spring of 1973, largely in response to a growing concern over increased commercial utilization of natural mussel resources. Before I discuss the status of U.S. mussel aquaculture and associated areas of research, a very brief background that outlines some of the events leading to the expansion of mussel culture activities in U.S. waters is presented.

From the early 1970s to early 1980s annual mussel landings (from natural populations) within the United States more than tripled (Fig. 8.1). Such increased harvesting over the decade led to considerable concern regarding the maximum sustainable yield (MSY) of the natural fishery. Estimation of MSY has proved extremely difficult due to the lack of quantitative estimates of available stocks, as well as the paucity of data accurately quantifying the growth dynamics of natural populations of this species (Lutz 1976, 1980a,b; Lutz *et al.* 1977; Chalfant *et al.* 1980). It is possible, however, to indirectly assess the effect of increased production on fishery yields through examination and interpretation of fishery statistics during World War II when considerable numbers of mussels (a "marginal" food in the United States upon which no ration points were placed) were harvested and canned (Herrington and Scattergood 1942; Scattergood and Taylor 1949a,b,c.) Total mussel landings from 1942 onwards are summarized in Fig. 8.1. While the increased production during the war years does not require an explanation, the sudden drop in 1947 does. Dow and Wallace (1954) attributed this sharp decline (at least in Maine, whose mussel landings accounted for over two-thirds of the total U.S. harvest) to two principal factors: (1) the availability of other protein foods considered to be more desirable; and (2) the lack of available good-quality mussels. Concerning the second factor, they stated that "canning operations during the preceding war years had cropped off almost completely the readily available supply."

In 1980 approximately 1161 MT (2.5 million lb) of mussel meats were landed in the United States (Fig. 8.1), not including mussels harvested from aquaculture operations. Such a production level approximates that realized during World War II. If this high production level is sustained for several years, the need for conservation of known resources and for additional sources of supply will be obvious. It has been, in fact, the difficulty in maintaining supply, in the face of growing demand, that has stimulated much of the academic and private industrial research on the practicality, feasibility, and possible advantages of the culture of mussels in U.S. waters (Lutz 1974, 1978, 1979, 1980b; Hurlburt and Hurlburt 1975, 1980; Lutz *et al.* 1977).

Since 1974, various workers have shown that when aquacultural techniques are employed, a marketable mussel of 50 mm (2.0 in.) or

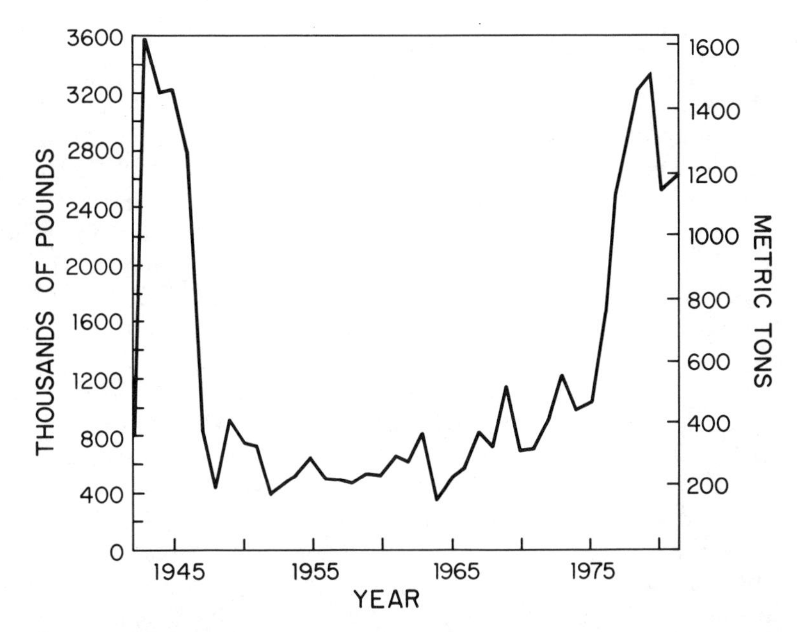

FIG. 8.1 Quantity (by wet meat weight) of annual U.S. mussel *(Mytilus edulis)* landings since World War II. Data for 1979 and 1980 are preliminary estimates and do not include landings for the West Coast. For conversion purposes the U.S. government considers meat weight to be equal to one-forth the organisms live weight in the shell. One bushel, which contains approximately 27.2 kg (60 lb) live weight, thus is considered to contain 6.8 kg (15 lb) of wet meat weight.
After Clifton (1980).

more in length can be obtained in numerous environments throughout the United States in 12–15 months (Lutz 1974, 1980; Chaves 1975; Chaves and Chew 1975; Lutz *et al.* 1977; Incze *et al.* 1978). Reported estimates of the average growth rate of *M. edulis* in natural populations, whether submerged or intertidal, in the North Atlantic region vary from 1.9 to 16.0 mm (0.07 to 0.63 in.) per year (Field 1922; Mossop 1921, 1922; Andrews 1972; Chalfant *et al.* 1980). The growth rate (shell length increment per time unit) of cultured mussels may be 3 to 25 times that of mussels in natural benthic populations.

The initiation in U.S. coastal waters of mussel aquaculture, with its associated high organism growth rates, should greatly increase the productivity of the total mussel industry. The potential for such increased productivity is dramatically illustrated by comparison of the 1947 production of mussels in the United States with that of an average Spanish mussel raft in the Galician Bays of Spain. In 1947 the annual U.S. production was approximately 371 MT (818,000 lb) of wet meat, a figure that, according to Dow and Wallace (1954), may largely reflect overutilization of the species during the preceding years. By comparison, one typical Galician raft produces some 22.7 MT (50,000 lb) of meat per year (Edwards 1968), which is equal to over 6% of the total 1947 U.S. production. It would thus take only 17 Galician rafts to exceed the entire U.S. production in 1947. When Spanish raft culture production is compared with the 1980 U.S. production of approximately 1161 MT (2.5 million lb) (Fig. 8.1), one may calculate that it is necessary to harvest only 51 Galician rafts to exceed the 1980 U.S. production, and there are presently over 3000 such rafts in Spain (Bardach *et al.* 1972).

In *Aquaculture in the United States: Constraints and Opportunities* (National Research Council 1978), which attempted to define the status and potential of U.S. aquaculture, the status of mussel culture was summarized in four brief sentences: "The small demand for mussels *(Mytilus edulis)* in the United States has provided little incentive for mussel farming. Slowly increasing demand may accelerate development of a culture system. Mussels are a major aquaculture industry in some countries, such as Spain and The Netherlands. However, the labor-intensive methods used in those countries are not easily adaptable to the United States, and modifications will have to be developed." While there is much truth in this statement, it certainly does not adequately portray either the status of mussel cultivation in the United States in the mid-1980s or the extensive research progress made in this country since the early 1970s. In this chapter I will attempt to more adequately summarize the current status of mussel culture and associated research efforts in the United States. Certain figures and portions of the text of this chapter have been reproduced, with permission, from Lutz (1980a); this book may be consulted for further details.

EUROPEAN MUSSEL CULTURE TECHNOLOGY

Mussel aquaculture by various methods is practiced extensively in Spain, France, The Netherlands, and Sweden, and has been described in detail for these and other localities by numerous workers (Havinga 1956a,b, 1964; Andreu 1958, 1968a,b,c; Ryther 1968; Mason 1971, 1972, 1976; Bardach *et al.* 1972; Lutz 1974, 1980a,b; Hurlburt and Hurlburt 1975, 1980; Korringa 1976; Haamer 1977; Myers 1980). At present, there exist four principal methods, all of which have met with varying degrees of success in the different countries. These are (1) bottom culture in The Netherlands and Germany; (2) "bouchot" culture (culture on posts or stockades) in France; (3) raft culture in Spain; and (4) long-line culture in Sweden. Before the feasibility of adapting these techniques to U.S. waters is considered, details of each method, as well as the peculiarities of the various regions where they are practiced, are briefly summarized (for further details, see Hurlburt and Hurlburt 1975, 1980).

Bottom Culture

Off the coasts of Germany and The Netherlands, the culture of mussels takes place directly on the bottom of shallow, partially diked (or enclosed) seas. Individual growing areas are leased directly from the government. Mussel seed is dredged from natural populations of juvenile mussels with shell lengths between approximately 8 and 13 mm. This seed is then transplanted to growing areas (culture plots) located in waters with depths of 3–6 m (10–20 ft). Most Dutch mussels are grown in the Waddenzee. This sea, on the northwest coast of The Netherlands, has a predominantly muddy bottom and is protected from the North Sea by encircling islands. Dutch mussel culture, from seed to market, is highly mechanized and, for this reason, the mussels produced are highly competitive with those produced using different culture methods in other countries. However, this high degree of mechanization has certain drawbacks. An important one involves the amount of rough handling of the live mussel. Such harsh treatment renders the mussels less "durable," particularly during live transport (Hurlburt and Hurlburt 1980).

Bouchot Culture

The long, gradually sloping beaches that extend far out to sea along much of the Atlantic coast of France provide excellent regions for the oldest (dating back to the thirteenth century) method of mussel culture, the "bouchot" system. These beaches and the adjacent ocean floor are unprotected from storms and other vagaries of nature. The water temperature ranges annually from 4° to 21°C (39° to 70°F), and the salinity varies from 29 to 34‰ depending on the season. The tremendous tides are perhaps the most unusual feature of much of this coastline. In some northern areas, the tidal range sometimes exceeds 15 m (49 ft), which has both advantages and disadvantages for mussel farmers. At low tide, the sand or mud of the ocean floor is exposed for many kilometers.

In the northern regions, where tidal ranges are the greatest, oak poles, approximately 3 m (10 ft) long and 20 cm (8 in.) in diameter, are driven into the ocean floor, leaving 1.5 to 2 m (5 to 6 ft) exposed above the sand. The bottom 30 cm (12 in.) of the exposed portion of the pole is wrapped with smooth plastic to discourage starfish, crabs, and other predators. The poles, which are arranged in rows, are spaced at intervals of approximately 1 m (39 in.); the distance between rows is generally about 3 m (10 ft), although this may vary from area to area. Along these coasts, culture activities are carried out at low tide using horses, tractors, oxcarts, or even bicycles to reach the bouchots, which may be several kilometers from shore. In the southern regions, where tidal fluctuations are not as extreme, most of the culture activities are conducted at low tide by boat. The culture of mussels takes place on the top 1.5 m (5 ft) of the poles; the bottom of the poles are seldom exposed to air. The mussel-growing areas are leased to the farmers by the government (Hurlbur and Hurlbur 1975, 1980).

Raft Culture

Raft mussel culture operations have existed in rias (bays) along the northwest Atlantic coast of Spain for about the last 30 years. The rias extend far inland and the slope of the land to the water is reasonably steep. The largest rias are 24 km (15 mi), long, 3–10 km (2–6 mi) wide, and up to 60 m (200 ft) deep, with an average depth of about 30 m (100 ft). They are protected from the full force of the ocean by islands at their mouths. Annual surface water temperatures range from 9° to 21°C (48° to 70°F), and the salinity varies little from 35‰. The average tidal range in the rias is 3–4 m (10–13 ft).

The rafts used for mussel culture are rather simple structures (for illustrations, see Hurlbur and Hurlbur 1980). The first ones were constructed using the hulls of old fishing vessels. More recent structures generally have from four to six concrete or steel floats or pontoons. A few new ones are constructed of styrofoam and fiberglass. On top of these floats there is a wooden (eucalyptus) lattice framework constructed of 5-cm (2-in.) square timbers, spaced about 45 cm (18 in.) apart, to which the ropes are fastened. The overall size of the rafts varies, but an average one might be 23 m^2 (220 ft^2) supporting about 700 ropes, each approximately 9 m (30 ft) long. The wooden lattice frames are supported by stays running down from masts. The rafts are anchored along the sides of the rias with large concrete moorings in about 11 m (36 ft) of water at low tide; the ropes never touch the bottom, eliminating the majority of problems associated with starfish and other benthic predators.

Until 1974, Spain was the world's leading producer of mussels (The Netherlands is currently the leading producer); the annual Spanish yield was in excess of 130,000 MT (286 million lb). Although Spanish production dropped substantially during the rest of the 1970s, production from raft culture activities in the early 1980s was in excess of 50,000 MT (110 million lb) per year, averaging approximately 22.7 MT (50,000 lb) of meat per raft per year. According to Mason (1971), this productivity is approximately 200 times greater than that of any other type of husbandry or culture in which animals are grown naturally with no supplemental or artificial feeding (Hurlburt and Hurlburt 1975, 1980).

Long-Line Culture

The most recently developed method of mussel culture and that which has considerable potential for large-scale utilization in U.S. waters has been termed "long-line" culture. The largest pilot commercial operations utilizing long-line systems are located in shallow fjord basins along the northwest coast of Sweden near Strömstad. Water depths where culture activities are concentrated are generally less than 10 m (33 ft). Like Spanish raft culture systems, long-line operations are based upon suspended techniques in which various types of substrates are hung vertically in the water column for both spat collecting and subsequent growth of the mussels. Unlike the Spanish operations, no large rafts are required, and the relatively simple long-line system is adaptable to mechanization (various mechanical harvesting devices for these long-line systems are presently being designed and evaluated in both Sweden and the United States).

Like so many well-designed, innovative techniques, the long-line system is relatively simple in principle. It consists of a series of buoyed horizontal lines from which many vertical lines are hung (Figs. 8.2 and 8.3). The horizontal lines are approximately 60 m (200 ft) long. Large, 200-liter (50-gal), plastic air-filled drums are attached for flotation at intervals of approximately 6 m (20 ft). The distance between drums may vary, since additional flotation drums are added as necessary to compensate for additional weight resulting from growth of the mussels and increased fouling on the lines. Each horizontal line is securely anchored at each end with large concrete blocks or other suitable anchoring systems (Fig. 8.2). The distance between horizontal lines is generally about 3 m (10 ft), although the lines are sometimes arranged in pairs—i.e., alternating distances of 1 and 3 m (3 and 10 ft) between lines. The vertical substrate lines are spaced at intervals of approximately 50 cm (20 in.) along the horizontal lines. In Swedish systems, the length of the verticals has been standardized to facilitate the eventual mechanization of the system; the standard

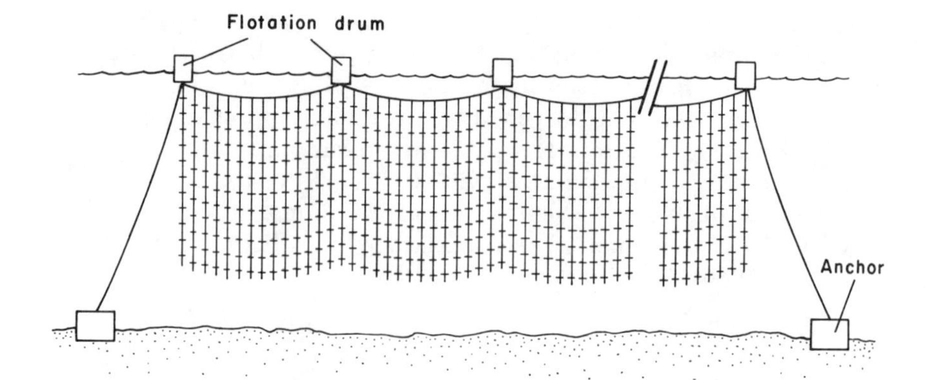

FIG. 8.2 Schematic diagram of one long-line unit with associated anchoring and flotation systems. See text for typical dimensions of lines and spacings.
After Haamer (1977); reprinted from Hurlburt and Hurlburt (1980) with permission from Elsevier Scientific Publishing Company, © 1980.

FIG. 8.3 Underwater view of cultured mussels in a long-line system.
Courtesy of J. Haamer; reprinted with permission from Elsevier Scientific Publishing Company, © 1980.

length is 6.5 m (21 ft) because of the shallow (<10 m) depths of the growing waters. Thirteen or fourteen vertical substrates are generally hung between each of the flotation drums, although the number may vary depending on the specific design of the long-line system.

One advantage of the long-line system is its ability to withstand the rigors of cold, harsh winters. In those Swedish waters where mussel culture operations are currently located, the tidal range is only about 1 m (39 in.). As a result of minimal tidal currents, ice presents little or no problem, despite the fact that many culture sites freeze solid during the winter (Hurlburt and Hurlburt 1980).

FEASIBILITY OF EUROPEAN CULTURE TECHNIQUES IN THE UNITED STATES

With the previous section as a background, the feasibility of adapting the various existing systems of mussel culture to U.S. waters will be briefly considered. Of the various culture systems, those utilizing suspended (i.e., raft and long-line systems) and bottom culture techniques appear to be the most adaptable to U.S. waters (see also Hurlburt and Hurlburt 1980).

In general, the United States does not possess the extensive tidal flats or large tidal range of northern France, where the "bouchot" system is employed. There are, however, certain U.S. coastal waters where bouchot culture might be practiced, but little is known about the effects of predation and interspecific competition in U.S. waters when mussels are cultured on posts.

Much of the U.S. coastline (particularly in the Northeast and the state of Washington) affords numerous deep protected bays, estuaries, and inlets characterized by high summer productivity and large tidal exchange. The five large rias of the northwest coast of Spain are deep, drowned river valleys with steep granite sides and a maximum 4-m (13-ft) tidal range producing a maximum tidal current of 50 cm/sec (20 in./sec) (Andreu 1968a). They are longer, wider, but also less protected, than inlets characteristic of much of the U.S. coast, such as those in Maine and Washington. Although the surface areas of these rias are very large—the largest, Ria de Arosa, is 33 km (20.6 mi) long and covers 320 km^2 (115 mi^2)—the rafts are located only along the coastline in waters with maximum depths of slightly over 10 m (40 ft) at low tide (MacLeod 1975). Primary productivity measured by Fraga and Vives (cited in Andreu 1968a) within the rias was 1.1–16.6 mg carbon (C)/liter/hr, with an average production of 10.5 mg C/liter/hr. The estuaries of the Damariscotta and the Sheepscot rivers along the central coast of Maine are, respectively, 15 km (9 mi) and 19 km (12 mi) long. Mean depths of these estuaries at mean low water are 8.7 and 9.4 m (28.5 and 30.8 ft), respectively (B. McAlice, unpublished data); tidal ranges approach 4 m (13 ft) and tidal velocities exceed 50 cm/sec (20 in./sec). Data obtained in the surface waters of a deep harbor located between the mouths of these two rivers have indicated primary productivity values as high as 39.8 mg C/liter/hr during the warm summer (August) months. Lowest productivity values are encountered in January and February and are approximately one-tenth of the July and August values (C. Yentsch, unpublished data). Hydrographic and biologic conditions thus suggest a good prospectus for mussel culture in these waters, and both experimental and commercial mussel culture operations are presently underway in this area. Results from these operations and those elsewhere in the United States have provided further encouraging data on the excellent potential of certain U.S. waters for mussel culture when suspended techniques are employed (Lutz 1974, 1980a,b; Hurlburt and Hurlburt 1975, 1980; Chaves 1975; Chaves and Chew 1975; Myers 1976, 1980; Incze and Lutz 1980; Miller 1980).

While both types of suspended mussel culture systems (raft and long-line) may be adaptable to U.S. waters, carbon copies of a typical Spanish raft have a number of disadvantages for use in the United States. Such rafts are large, bulky, awkward, and aesthetically unappealing to large numbers of coastal residents and visitors. They could not be used where there is moving ice or where there are heavy storms. Furthermore, the rafts do not lend themselves to mechanization, so culture activities are labor intensive. Variations of the Swedish long-line system would appear to offer the greatest potential for use in North American waters. As mentioned, the Swedish system of culture is readily adaptable to mechanization, an extremely important consideration in light of the high U.S. labor costs. Several individuals and industries in various areas along the east coast of the United States (primarily in Maine and Rhode Island) are currently utilizing such systems in pilot commercial operations (Hurlburt and Hurlburt 1980).

CURRENT MUSSEL AQUACULTURE PRODUCTION

In the United States in 1982 there were four private industrial firms actively engaged in the aquaculture of blue mussels *(Mytilus edulis):* (1) Abandoned Farm, Inc.—2 ha (5 Ac) in the Damariscotta River, Lincoln County, Maine; (2) Penn Cove Mussels—6 ha (15 Ac) in Puget

Sound, Coupeville, Washington; (3) Great Eastern Mussel Farms*—2 ha (5 Ac) in the Sheepscot River, Edgecomb, Maine and 26 ha (65 Ac) in Stonington, Maine; (4) Blue Gold Sea Farms, Inc.—24 ha (60 Ac) in Narragansett Bay, Middletown, Rhode Island and 42 ha (60 Ac) in Penobscot Bay, Maine. The 1982 annual combined production from these four commercial operations was approximately 2600 MT (~105,000 bu) based on whole shell weight or 771 MT (~1.7 million lb) of wet meat with an estimated annual value to the firms of approximately $1.9 million. While a few smaller firms have initiated mussel culture activities, they have not been considered here because their current production is insignificant relative to that of the four major operations.

All of the U.S. firms currently involved in mussel aquaculture have used or are currently using suspended culture systems and Great Eastern Mussel Farms is presently engaged in bottom culture of mussels. The designs and constructions of the utilized suspended culture systems vary greatly from one firm to another. Penn Cove Mussels has used predominantly rafts (Figs. 8.4–8.6), although a long-line unit has recently been deployed (Fig. 8.7). Abandoned Farm, Inc., initiated operations in 1973 with a raft constructed of four 12.8-m (42-ft) power poles set 0.9 m (36 in.) on centers (Fig. 8.8), but more recently has converted to an unique system that involves culturing mussels on ropes suspended from scrap tires filled with urethane foam (Figs. 8.9–8.12). These "tire-units" are attached to one another, forming long arrays which have relatively low profiles above the water (Fig. 8.10). At the time of harvest, individual tire units are hoisted mechanically with the aid of an engine mounted directly on the company's workboat (Fig. 8.12). Blue Gold Sea Farms, Inc., utilizes a slightly modified Swedish long-line system (Figs. 8.13–8.15). Until 1981, Great Eastern Mussel Farms utilized a long-line culture system located in the Sheepscot River, Maine. More recently this company seeded approximately 26 ha (65 Ac) of subtidal bottom off Stonington, Maine (Fig. 8.16) with juvenile mussels and during 1981–1982 harvested approximately 750 MT (30,000 bu) from this leasehold (Fig. 8.17). An additional 11 ha (27 Ac) of bottom in the vicinity of Stonington has recently been leased by the company and the success of these pilot-scale "bottom culture" operations will be evaluated by the firm during the next few years.

In addition to selling mussels that are aquacultured in the strict

*Since this paper was written, Great Eastern Mussel Farms has greatly expanded. The company, as of 1984, leases approximately 200 Ac in various environments along the Maine coast and is actively engaged in bottom culture activities. The annual aquaculture production of the firm is approximately 1500 MT (60,000 bu) and is based primarily on bottom culture techniques.

FIG. 8.4 Mussel culture rafts utilized by Penn Cove Mussels in Puget Sound, Coupeville, Washington.
Courtesy of P. Jefferds.

FIG. 8.5 Mussel culture raft utilized by Penn Cove Mussels in Puget Sound, Coupeville, Washington.
Courtesy of P. Jefferds.

FIG. 8.6 Close-up view of the Penn Cove Mussels culture raft depicted in Fig. 8.5.
Courtesy of P. Jefferds.

FIG. 8.7 Long-line mussel culture system recently deployed by Penn Cove Mussels in Puget Sound, Coupeville, Washington. *Courtesy of P. Jefferds.*

FIG. 8.8 Initial mussel culture raft placed in the water in 1973 by Abandoned Farm, Inc. The raft was constructed of four 12.8-m (42-ft) power poles set 0.9 m (3 ft) on centers. Flotation was furnished by two 2.7-m (9-ft) styrofoam logs boxed in at each end.

FIG. 8.9 Scrap tires filled with urethane foam serve as flotation devices for mussel culture raft units at Abandoned Farm, Inc. Each tire has six culture ropes spaced evenly around the circumference. *Reprinted from Myers (1980) with permission from Elsevier Scientific Publishing Company, © 1980.*

FIG. 8.10 One "long-line" of tire units at Abandoned Farm, Inc. Tires and plastic containers are additional flotation devices that have been added above the original tires to support the growing columns of mussels.
Courtesy of B. Porter.

FIG. 8.12 Hoisted tire unit at Abandoned Farm, Inc. containing six 6-m (20-ft) columns of growing mussels of marketable size.
Reprinted from Myers (1980) with permission from Elsevier Scientific Publishing Company, © 1980.

FIG. 8.11 Underwater view of the upper portion of one tire unit at the time of its initial emplacement by Abandoned Farm, Inc. Note the six polypropylene ropes spaced evenly around the circumference of the flotation tire. Birch dowels through the lay of the rope will prevent growing masses of mussels from sliding down the units.
Reprinted from Myers (1980) with permission from Elsevier Scientific Publishing Company, © 1980.

FIG. 8.13 Aerial view of long-line arrays of Blue Gold Sea Farms, Inc., located in Narragansett Bay, Rhode Island.
Courtesy of Blue Gold Sea Farms, Inc.

FIG. 8.14 An array of long-lines used by Blue Gold Sea Farms, Inc., in Narragansett Bay, Rhode Island.
Courtesy of Blue Gold Sea Farms, Inc.

FIG. 8.15 Close-up view of a typical long-line unit currently used by Blue Gold Sea Farms, Inc., in Narragansett Bay, Rhode Island.
Reprinted from Hurlburt and Hurlburt (1980) with permission from Elsevier Scientific Publishing Company, © 1980.

FIG. 8.16 Area in Stonington, Maine, currently leased by Great Eastern Mussel Farms for pilot-scale "bottom culture" operations. Approximately 65 acres of subtidal bottom have been seeded with juvenile mussels.
Courtesy of Great Eastern Mussel Farms.

sense (i.e., those grown from young spat to market size using aquacultural techniques), some firms are currently selling as a "cultured" product mussels that have been harvested from natural populations and subsequently placed for 24 to 48 hr in suspended culture systems. The quality of such mussels varies greatly, and many often contain large quantities of annoying pearls (see discussion later in this chapter for a consideration of problems associated with the pearl incidence in mussels). It is interesting to note, however, that despite the vari-

FIG. 8.17 Harvesting marketable mussels from pilot-scale bottom culture operation off Stonington, Maine.
Courtesy of Great Eastern Mussel Farms.

able quality of these briefly suspended "cultured" mussels, they are currently being sold at approximately the same price as true aquacultured mussels. The production figures cited at the beginning of this section are only for mussels aquacultured in the strict sense. Similarly, in the remainder of this chapter, large mussels suspended for only brief periods of time prior to sale are not considered an aquacultured product.

RECENT RESEARCH EFFORTS

Research on various aspects of mussel culture in U.S. waters has been extensive over the past decade. Efforts on the West Coast have been centered at the University of Washington where attempts have been made to assess the biological and economic feasibility of culturing mussels in Puget Sound (for details, see Waterstrat *et al.* 1980). Results obtained over the past few years indicate that conditions in these waters are conducive to mussel culture (Chaves 1975; Chaves and Chew 1975; Waterstrat 1979); mussels have been found to grow to market size (shell length $\geq$ 50 mm or 2 in.) in a period of 1 year when aquaculture techniques are employed. Research has also been conducted on the feasibility of obtaining mussel seed from hatcheries; growth of this seed is currently being compared with growth of mussel seed from natural populations. Finally, market studies have been conducted to ascertain how best to present a new and different product (the cultured mussel) to consumers.

Mussel culture studies on the East Coast have been centered primarily in New England. Research at the University of Maine has contributed to an increased understanding of factors responsible for pearl formation (Lutz 1974, 1975, 1978, 1980c; Lutz and Hidu 1978); growth, survival, and recruitment (Lutz 1974; Lutz and Porter 1977; Incze *et al.* 1978; Incze and Lutz 1980; Incze *et al.* 1980); and condition (Incze and Lutz 1980; Lutz *et al.* 1980) in cultured populations of mussels in various areas. Research programs operated jointly by the University of Maine and the University of New Hampshire have explored various economic and market development aspects of mussel culture (Haley *et al.* 1977; Miller 1977; Clifton 1980). Recent studies at Harvard (Hurlburt 1977) and the University of New Hampshire (Alonzo 1977) have resulted in new mussel recipes and handling techniques designed for large-scale restaurant preparation and home consumption of mussels. Problems associated with storage and processing of cultured mussels are being resolved (Slabyj and Hinkle 1976; Slabyj and Carpenter 1977; Slabyj *et al.* 1978; Slabyj 1980). The use of heated effluent waters from nuclear power plants for mussel culture has been assessed by a number of workers (Hess *et al.* 1976; Lutz and Porter 1977; Lutz and Hess 1979; Lutz *et al.* 1980). In these studies the growth and survival of mussels experimentally cultured at varying distances from the heated discharge waters of a 855-megawatt nuclear power reactor have been extensively monitored. In addition, the uptake and retention of gamma-ray-emitting radionuclides by these mussels has been quantified. In the following sections, some of the more significant findings of these various areas of mussel aquaculture research are presented. First, however, a very brief account of the life cycle of *Mytilus edulis* is presented for the benefit of those readers who may not be familiar with the early life history of this species.

LIFE CYCLE OF THE BLUE MUSSEL

The life cycle and settling behavior of *M. edulis* have been described by numerous workers (Stafford 1912; Field 1922; Werner 1939; Sullivan 1948; Bayne 1964, 1965; Chanley and Andrews 1971; de Schweinitz and Lutz 1976; Lutz and Hidu 1979).

Fertilization is external; the diameter of the fertilized egg is approximately 68 μm (Fig. 8.18). At 20°C (68°F) the first cleavage occurs approximately 45 min after fertilization; subsequent embryological development proceeds by way of unequal, holoblastic spiral cleavage, with a ciliated trochophore stage being reached approximately 24 hr after fertilization (Fig. 8.18). At about 44 hr, the first signs of larval shell formation are apparent. A fully formed shell of the "veliger" stage is present by 48 hr, with length, height and hingeline dimensions of approximately 95, 70, and 70 μm, respectively.

The "planktonic" veliger stage generally lasts approximately 2–4 weeks, during which time the larvae actively feed in the water column (Fig. 8.19). When larvae reach 140–150 μm in total length, a rounded umbo appears, marking the transition from the "straight-hinge" to the "umbo" stage of development (for shell morphology terminology, see Chanley and Andrews 1971). As the larvae grow past 210 to 230 μm in length, the umbo gradually extends from the hinge and shoulders as a low knob (see Chanley and Andrews 1971). de Schweinitz and Lutz (1976) have recently depicted the relationship between larval shell length and height for *M. edulis* veligers at all

FIG. 8.18 Photomicrographs depicting the embryological development of *Mytilus edulis* from the fertilized egg (top left) to the first shelled veliger stage (bottom right). All micrographs were taken at the same magnification.

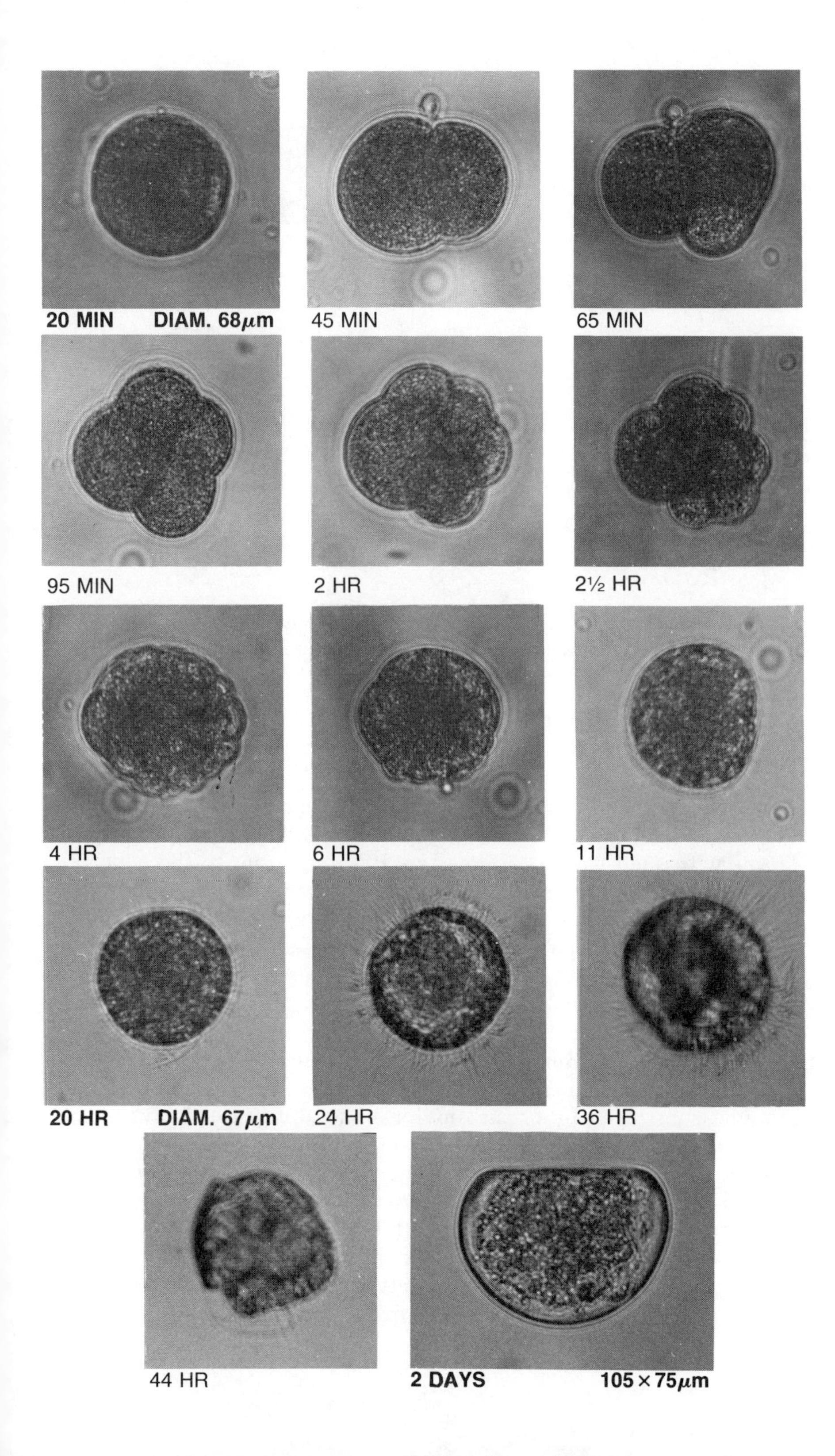

20 MIN
DIAM. 68μm
45 MIN
65 MIN
95 MIN
2 HR
2½ HR
4 HR
6 HR
11 HR
20 HR
DIAM. 67μm
24 HR
36 HR
44 HR
2 DAYS
105×75μm

FIG. 8.19 Photomicrographs of veliger larvae of *Mytilus edulis* from the early straight-hinge stage to metamorphosis. Shell length and height dimensions are given in micrometers below the individual specimens. Larvae are oriented with the anterior end to the left.
Reprinted with permission from de Schweinitz and Lutz (1976).

stages of development; the equation of least squares calculated by these authors is: height = 1.08 length − 37.22 ($r^2 = 0.99$). Eyespots develop in larvae with shell lengths as small as 205 μm, although the majority of larvae develop such structures when shell length reaches 220–230 μm (Fig. 8.20). Larvae as large as 245 μm are occasionally observed to lack an eyespot. A well-defined foot is present by the time larvae achieve lengths of 195–210 μm, and active extension of this organ begins at the organisms reach lengths between 215 and 240 μm (Fig. 8.21).

Upon reaching approximately 260 μm in length the pediveliger larvae are capable of metamorphosing. However, in the absence of a suitable substrate, the organisms have the ability to "delay" metamorphosis for up to 40 days at 10°C (50°F) (Bayne 1965, 1976). Upon contact with a suitable substrate, such as mussel culture ropes, the pediveligers undergo a process known as "settlement," characterized by a crawling behavior that ends with the secretion of byssal threads that attach the larva to the substrate and signals the beginning of a sessile existence. It is this act of settlement upon which those engaged in mussel culture must capitalize.

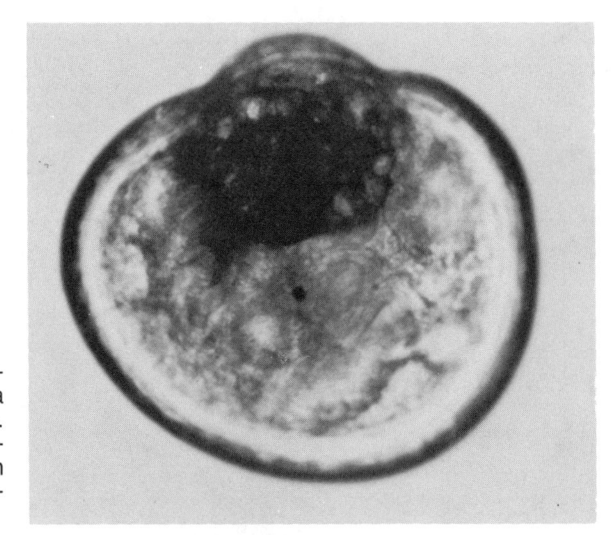

FIG. 8.20 Mature *Mytilus edulis* larva with a shell length of 265 μm. Note well-defined eyespot (small dark dot) in approximately the center of the larva.

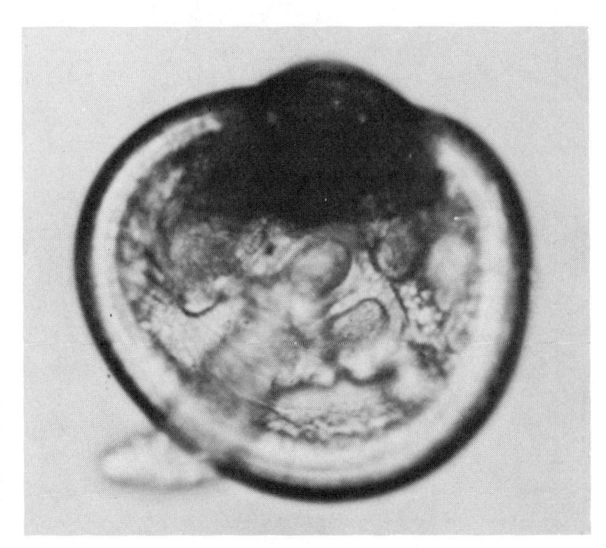

FIG. 8.21 Mature *Mytilus edulis* larva (pediveliger) with a shell length of 270 μm. Note extension of well-defined foot.

EXPERIMENTAL CULTURE—EAST COAST

Growth, Survival, and Recruitment

Experimental mussel culture activities along the east coast of the United States began in 1972. Since that time data on growth, survival, and recruitment of mussels using experimental suspended culture systems have been collected from Machiasport, Maine, south to

Narragansett Bay, Rhode Island, with the majority of work being concentrated in waters along the coast of Maine. Most of these studies have used small experimental rafts (for a photographic illustration of a typical experimental raft, see Lutz *et al.* 1980, p. 172). The locations of these rafts along the East Coast are depicted in Fig. 8.22.

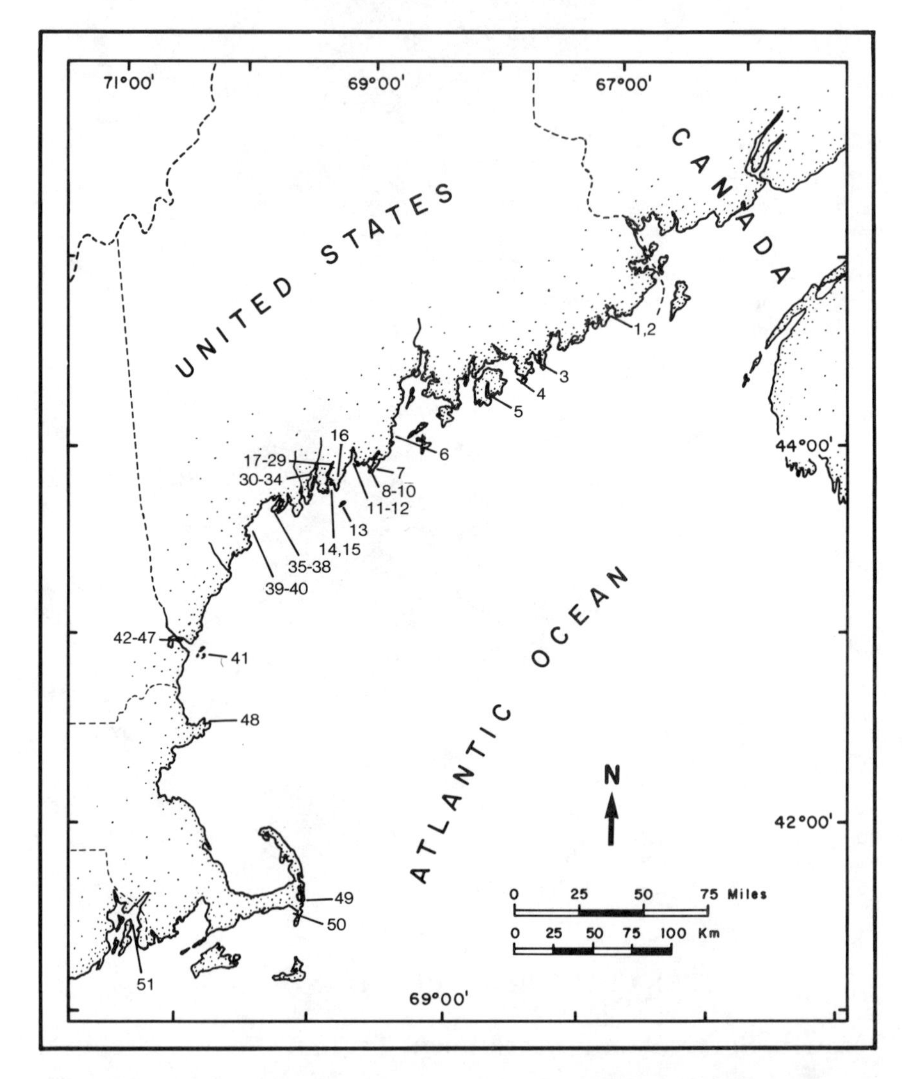

FIG. 8.22 Locations of experimental mussel rafts along coast of the northeastern United States. Specific locations and growth data are provided in Table 8.1. *After Incze and Lutz (1980); reprinted with permission from Elsevier Scientific Publishing Company, © 1980.*

Growth of mussels in these experimental systems has been monitored using seed mussels (with standardized initial shell lengths) taken from commercial spat collectors and transplanted to screened, flow-through containers attached to the underside of small experimental rafts; this procedure permits frequent measurements of the same organisms (Lutz and Porter 1977; Incze *et al.* 1978, 1980; Incze and Lutz 1980). Although this method creates growth conditions that differ somewhat from those associated with commercial culture substrates (where there is probably greater competition for food and space), it provides data that can be used for comparing growth potentials in various environments. The mean times required for mussels to reach market size in experimental raft studies conducted from 1973 to 1978 are summarized in Table 8.1. These results, obtained from numerous coastal and estuarine sites, indicate that mussels of market size, with a shell length of 50 mm (2 in.) or more, can be produced using suspended culture systems in most of the environments studied in 12–15 months after the culture substrates are seeded with a natural spatfall (Lutz 1974; Incze *et al.* 1978; Incze and Lutz 1980). With the exception of a few locations, survival of mussels in the experimental rafts was generally greater than 90%. Mussel seed recruitment on experimental Manila rope collectors attached to experimental rafts was found to be relatively consistent in those environments studied for several years. Incze and Lutz (1980) recently concluded from these studies that seed mussels can be reliably procured by natural recruitment on collector substrates in at least some northeastern environments. For further details on recruitment in the various locations summarized in Fig. 8.22 and Table 8.1, see Incze and Lutz (1980).

Growth curves generated from data obtained at two typical experimental sites over a period of several years are depicted in Fig. 8.23. The similarity of the annual curves at each site demonstrates that the variability in growth from one year to the next is not great enough to cause extreme fluctuations in culture production, a factor of critical economic concern to the mussel grower. The different locations of these two sites—one in a coastal environment (Johns Bay, Maine) near the open sea and the other in the middle portion of an estuary (the Damariscotta River, Lincoln County, Maine)—suggest that extensive areas of the coastline could prove environmentally suitable to mussel growth and production in suspended culture systems. This conclusion is supported by the rapid growth of mussels also observed during 1-year studies conducted at the other East Coast sites listed in Table 8.1. The large number of apparently suitable growing environments greatly increases the probability of successfully developing an extensive and

TABLE 8.1. LOCATIONS OF EXPERIMENTAL SITES SHOWN IN FIG. 8.22, SHOWING ENVIRONMENTAL CLASSIFICATION AND GROWTH OF EXPERIMENTALLY CULTURED MUSSELS[a]

Site No.	Location	Environment	Mean Length/ No. of Months After Settlement[a]
1	Machiasport, Maine	coastal	45 mm/14–15 mo.
2	Starboard, Maine	coastal	45 mm/14 mo.
3	Steuben, Maine	coastal	NA
4	Sullivan, Maine	coastal	NA
5	Somes Sound, Maine	fjord-like basin (8 km from the sea)	50 mm/13 mo.
6	Rockport Harbor, Maine	coastal	50 mm/12–13 mo.
7	Tenant's Harbor, Maine	coastal	50 mm/13 mo.
8–10	St. George River, Maine	estuarine (4–7 km from the sea)	50 mm/12–13 mo.
11	Upper Muscongus Bay, Maine	coastal	50 mm/13 mo.
12	Upper Muscongus Bay, Maine (different site than site 14)	coastal	50 mm/13 mo.
13	Monhegan Island, Maine	offshore island	43 mm/12 mo.
14	Johns Bay, Maine (1975–1977)	coastal	50 mm/12–13 mo.
15	Johns Bay, Maine (1978) (different site than site 17)	coastal	50 mm/12–13 mo.
16	Pemaquid River, Maine	estuarine (2 km from the sea)	50 mm/13 mo.
17–20	Lower Damariscotta River, Maine	estuarine (1–3 km from the sea)	50 mm/12 mo.
21–24	Middle Damariscotta River, Maine	estuarine (4–16 km from the sea)	50 mm/12–13 mo.
25–29	Upper Damariscotta River, Maine	estuarine (18–25 km from the sea)	M
30	Sheepscot River, Maine	estuarine (15 km from the sea)	45 mm/13–14 mo.
31–33	Sheepscot River (Montsweag Bay), Maine	estuarine (14 km from the sea)	Variable[b]
34	Cross River, Maine	estuarine (14 km from the sea)	48 mm/13 mo.
35–37	Harpswell Neck, Maine	coastal (peninsula)	50 mm/13–15)
38	Cundys Harbor, Maine	coastal	50 mm/13 mo.
39	Cousins Island, Maine	coastal	50 mm/12–13 mo.
40	Cumberland Foreside, Maine	coastal	50 mm/12–13 mo.
41	Isles of Shoals, New Hampshire	offshore island	50 mm/13–14 mo.
42–47	Piscataqua River, New Hampshire	estuarine (2–18 km from the sea)	42–45 mm/11 mo.
48	Rockport, Massachusetts	coastal	45 mm/11–12 mo.
49	Pleasant Bay, Massachusetts	shallow coastal bay	M
50	Sesuit Harbor, Massachusetts	shallow coastal bay	50 mm/12–13 mo.
51	Portsmouth, Rhode Island	coastal	50 mm/12–13 mo.

Source: After Incze and Lutz 1980.
[a] M = high mortality; NA = data not available.
[b] Affected by thermal discharges of a nuclear power plant (see Lutz and Porter, 1977).

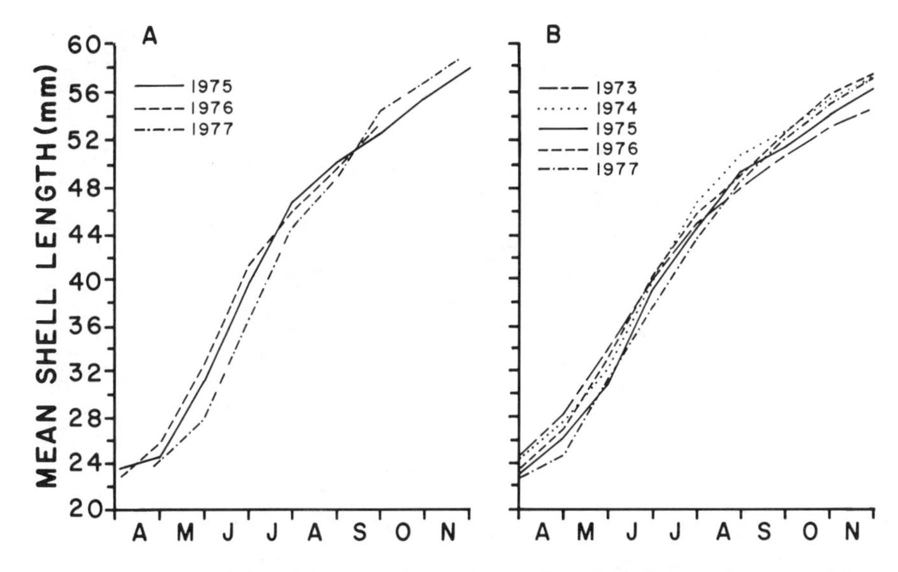

FIG. 8.23 Annual growth curves of experimentally cultured mussels at two sites: (A) a coastal embayment in Johns Bay, Maine (Fig. 8.22, site 14); and (B) the mid portion of an estuary in the Damariscotta River, Maine (Fig. 8.22, site 23). *Reprinted from Incze and Lutz (1980) with permission from Elsevier Scientific Publishing Company, © 1980.*

viable mussel culture industry along the East Coast of the United States.

Harvest and Transplantation of Mussel Seed

In Spanish mussel aquaculture operations, seed mussels are generally harvested from natural substrates and subsequently transplanted to the suspended culture systems where the organisms are reared to market size; approximately half of the seed currently used in Galicia is collected by hand near the mouths of the rias (Korringa 1976). There is some evidence, however, that similar harvesting and transplanting practices may not be effective in U.S. waters.

In one experiment, wild seed mussels, which had survived storage for several days in a crate moored to a dock in a tidal estuary, were transferred to an experimental raft in the Damariscotta River in Maine, where their growth and survival could be monitored (Incze and Lutz 1980). Mortality subsequent to the transfer increased rapidly, and the mussels grew much less rapidly than did control seed taken from local spat collectors and maintained on an adjacent raft (Fig. 8.24). In a second experiment, seed mussels were taken from the ledges of a

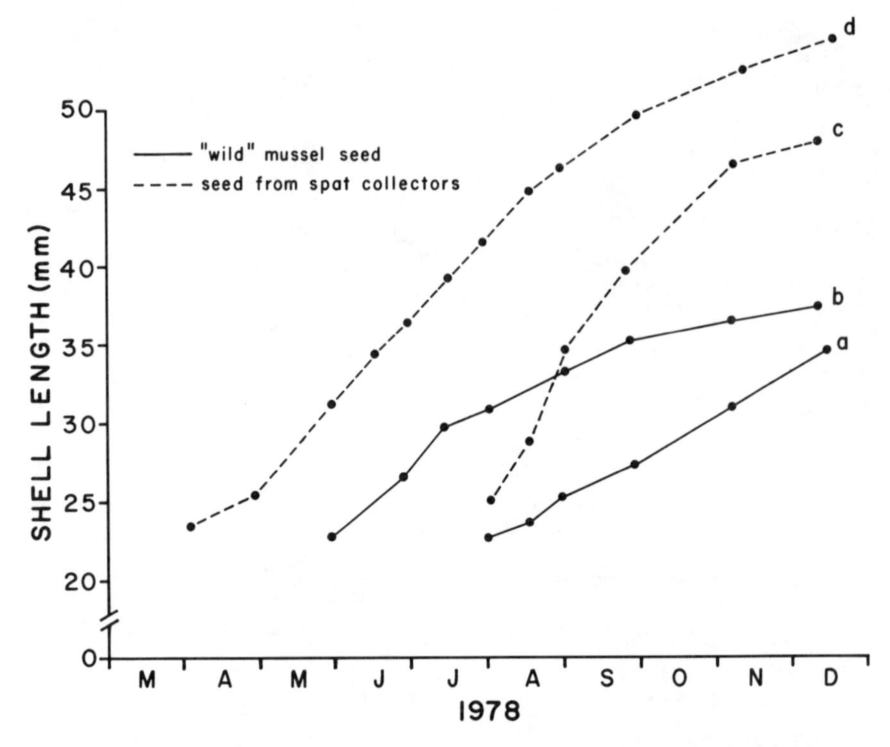

FIG. 8.24 Growth of "wild" mussel seed (*a* and *b*) and mussel seed taken from spat collectors (*c* and *d*). Seed *a* was removed from an intertidal ledge and kept densely packed in crates for several days before being transplanted to an experimental raft. Seed *b* was removed from an exposed rocky headland and transplanted immediately to an experimental raft. Seed *c* was removed from a spat collector in mid-summer following predation of larger seed mussels by eider ducks in late winter and early spring. Seed *d* was taken from a spat collector in late March. Seed mussels in groups *c* and *d* had settled on artificial collectors in August of the previous year (1977). Seed *a* appeared to have settled in June of 1978.

Reprinted from Incze and Lutz (1980) with permission from Elsevier Scientific Publishing Company, © 1980.

rocky headland near the mouth of the Damariscotta River and transferred immediately to an experimental raft in the middle portion of the estuary (temperature and salinity conditions in the two environments did not differ substantially). These mussels also exhibited markedly decreased growth compared with control seed (Fig. 8.24). Although it was not possible to assign specific ages to the specimens in the experiments, it was clear from their markedly corroded shells that they were not recently settled mussels; their small size may have been a result of a physiologically stressful history, genetic factors, or some other factor(s), which apparently continued to limit their potential for growth even under the comparatively favorable conditions afforded by the suspended culture systems.

Thus, not all mussels of appropriate size constitute "desirable" seed with a potential for rapid growth in mussel aquaculture systems. It appears that only newly settled populations should be used for seed-harvesting activities, and that these seed should be transplanted to the culture systems as soon as possible. If proposed aquaculture operations are to rely on wild seed for transplantation purposes, extensive and accessible areas with dense, regular spatfalls on natural substrates should first be identified (Incze and Lutz 1980).

EXPERIMENTAL CULTURE—WEST COAST

The status of experimental mussel culture research on the west coast of the United States has been summarized by Waterstrat *et al.* (1980). Pilot culture research activities, which have taken place largely without the benefit of an existing industrial structure, have been conducted mostly in Puget Sound in the state of Washington. Academic research investigations have included studies on mussel cultivation in natural and artificial upwelling systems, the toxicity of mussels resulting from ingestion of red tide organisms, shellfish hatchery techniques, and seed recruitment.

Studies at the University of Alaska have focused on improving mussel growth rates through the use of artificial upwelling systems. The high primary productivity in such systems can greatly accelerate mussel growth even in cold water. Additional studies at the University of Alaska have been involved with the development of reliable chemical techniques for determination of the level of toxins in mussels resulting from ingestion by the mussels of the extremely toxic dinoflagellate *Gonyaulax catenella* (the causative agent responsible for "red tides," or "toxic dinoflagellate blooms," along extensive areas of the West Coast). The most promising technique tested to date is a colorimetric assay involving oxidative degradation of saxitoxin and the qualification of the resulting quanidine. The method, however, may successfully be used to assay mussels only after extensive chromatographic purification (Waterstrat *et al.* 1980).

Research on the use of shellfish hatcheries has been conducted at both Oregon State University's Marine Science Center and the University of Washington. Some of the earliest studies (Breese *et al.* 1963) provided methods to stimulate spawning of mussels under hatchery conditions, and later hatchery studies (Courtright *et al.* 1971) resulted in the formulation of a synthetic seawater for use with mussel bioassays. In more recent years, an experimental hatchery has been established by the University of Washington to assess the potential of producing commercial quantities of mussel seed. To date at this hatchery, 14 batches of larvae have been reared to settlement. A total of 914,000 pediveliger larvae (length $\geq$ 270 μm) was produced from 206,400,000 fertilized eggs, for an overall survival rate of 0.4%. Survival rates were estimated to be 10.1% from egg to the straight-hinge stage and 3.3% from the straight-hinge stage through settlement. Survival of pediveliger larvae to the plantigrade stage varied significantly, ranging from 2.9 to 31.8% in individual rearing tanks. The overall survival rate from fertilized egg to the plantigrade stage was 0.2% (Waterstrat *et al.* 1980).

Intensive studies designed to monitor the recruitment of mussel seed in various areas of Puget Sound were undertaken by the University of Washington and the Washington State Department of Fisheries (WDF) in 1978. Significant differences have been found in larval abundance and recruitment of mussels at different locations (12 separate sites). Peak larval abundances in the various locations ranged from a high of 152 larvae per liter to a low of 0.13 larvae per liter. Differences in larval settlement were equally striking, ranging from 0 to 64,000 plantigrades per collector, each of which consisted of $20.0 \times 10.0 \times 2.5$ cm ($10 \times 4 \times 1$ in.) sections of rubberized curled hair, sampled at weekly intervals from March through October. Further details concerning these seed recruitment studies are presented by Waterstrat *et al.* (1980).

PRESENCE OF PEARLS

The presence of pearls embedded directly within the mantle epithelium of mussels in various natural populations along the east coast of the United States has been a serious deterrent to commercial utilization of large quantities of mussels. In the early 1940s, when large numbers of mussels were harvested, the pearls present in commercially important mussel populations along the northeastern coast from Eastport, Maine, to Cape Cod, Massachusetts, were quantitatively compared by Scattergood and Taylor (1949b). As a result of their extensive survey, they recommended in 1942 the closing of six areas in Maine (areas that accounted for approximately one-fifth of the total estimated Maine mussel production at that time) due to the presence of pearls. Canning operations were advised in January 1943 to avoid mussels taken from these areas.

The presence of pearls in mussels is by no means restricted to populations along the northeastern coast of the United States. According to Giard (1907), the first documented account of pearls in mussels was given by Olaus Worm in 1655 based upon examination of mussels taken near Copenhagen. Other European researchers subsequently have reported pearl-infested mussels in a number of localities along the coasts of Denmark, France, and England (Garner 1857; d'Hamonville 1894; Dubois 1901, 1909; Jameson 1902; Herdman 1904; Giard 1907). Stafford (1912) isolated pearls from specimens of *M. edulis* sampled from certain areas along the Gaspé coast of Canada. In all of these studies, considerable variation in both the number and size of pearls with geographical location has been reported. Such variable incidence has resulted in the designation of distinct pearl-producing areas and is of considerable biological interest. It is interesting to note that the presence of pearls is reportedly not a problem to the commercial utilization of mussels found along the west coast of the United States. Factors responsible for the observed differences between various areas have been the research subject of several workers (Garner 1872; Dubois 1901, 1909; Jameson 1902; Scattergood and Taylor 1949b) and still are not well understood.

The independent studies of Garner (1872) and Dubois (1901) showed that pearl formation in mussels resulted from infection by a parasitic trematode. Jameson (1902) described the organism; according to Odhner (1905), it is probably *Gymnophallus bursicola*. Stunkard and Uzmann (1958) described morphologically similar metacercariae (probably *G. bursicola*) isolated from the mantle tissues of mussels from New York, Connecticut, and Massachusetts waters. Work in Maine (Lutz 1974, 1975, 1978, 1980c; Lutz and Hidu 1978) suggests that the presence of pearls in these waters also may be the result of infection by this gymnophallid. Lutz (1980c, p. 196) presents a photographic illustration of the distome parasite that may be responsible for the presence of pearls in mussels from Maine waters.

After the original description of the trematode by Jameson (1902), considerable work was done to ascertain the life cycle of the parasite (Odhner 1905; Nicoll 1906; Giard 1907; Dubois 1907a,b, 1909). The suggested life cycle involves either the common eider duck, *Somateria mollissima,* or the American scoter, *Oidemia nigra,* or both, as the definitive host(s) and the blue mussel, *M. edulis,* as an intermediate host (Dubois 1907b; Jameson and Nicoll 1913; Stunkard and Uzmann 1958). Whether a second intermediate host is required remains uncertain (Nicoll 1906; Dubois 1909; Stunkard and Uzmann 1958). In general, therefore, the life cycle of the trematode is not well established and attempts to further unravel the cycle have been frustrated largely because of taxonomic chaos. This is emphasized by the statement of Stunkard and Uzmann (1958, p. 298) that "specific identification (of gymnophallids) is so uncertain that we prefer to list the worms by host and location rather than propose names that might further confuse the taxonomic situation."

During the past century, various workers have speculated on factors responsible for the presence or absence of pearls in various populations of mussels (and, hence, factors controlling the distribution patterns of the trematode that initiates pearl formation). Of these, salinity has been alluded to by two workers. Jameson (1902, p. 143) stated that the "most favorable places [for pearl formation] seem to be estuaries or landlocked channels," while Dubois (1909) observed that peral-producing populations were frequently located at the mouths of rivers. d'Hammonville (1894), studying a population of "moulières perlières" in Billiers (a borough in Brittany, France, near the mouth of the Vilaine), could detect no differences between various environmental parameters (sediment, current velocity, seaweed distribution, and plankton) of the pearl-infested area and those of an adjacent area containing "moulières sans perles." Lack of contact of the mussel with the sediment has been suggested as contributing to reduced pearl incidence by both Jameson (1902) and Nicoll (1906), based on studies of the creeping motion of the supposed cercaria of the parasite responsible for the initiation of pearl formation. Furthermore, Jameson (1902) observed reduced pearl incidence in mussels taken from stakes or floating objects, although Dubois (1909) found pearls within *Mytilus galloprovincialis* on ropes in the experimental culture parks operated by the University of Lyon. A final factor that may influence the incidence of pearls, and one that has received only limited attention, is age. d'Hammonville (1894) commented that pearls were found only in the largest and less regularly formed mussels ("celles qui sont le moins régulièrement formées") at Billiers. These stunted and larger individuals well may have been the older organisms in the population. Similarly, Jameson (1902) observed that pearls were seldom found in mussels less than 40 mm (1.6 in.) in length and attributed the relatively small size of pearls in the Billiers population to the active mussel fishery in the area, which prevented individual mussels from reaching "a great age." The effect of age is summarized by Jameson's (1902, p. 162) statement that the "general experience of everybody acquainted with pearl-fisheries is that the largest pearls were found in the oldest and thickest shells, which proves how intimately growth of pearl and shell are associated."

In an extensive series of studies, Lutz and Hidu (1978; see also Lutz 1978, 1980c) have quantified the relationship between pearl incidence and age for a number of cultured and natural populations of *M. edulis* along the northeastern coast of the United States. In these studies, marked differences in pearl incidence were observed between raft and shore populations of mussels of similar sizes, as well as between mussels of similar lengths obtained from rafts left in the water for various periods of time. The quantitative relationship between number of pearls and age for two populations (raft and shore) located in Maine waters is shown in Fig. 8.25 (Lutz and Hidu 1978). These workers concluded that no significant age-independent differences in pearl incidence exist between rafted and shore-based mussels from adjacent populations. They attributed the marked differences between the two populations with regard to the presence of pearls in organisms of similar lengths to differences in the growth rates and age structure within the two population types. They further reasoned that, if the rate of pearl formation and pearl growth remains relatively constant

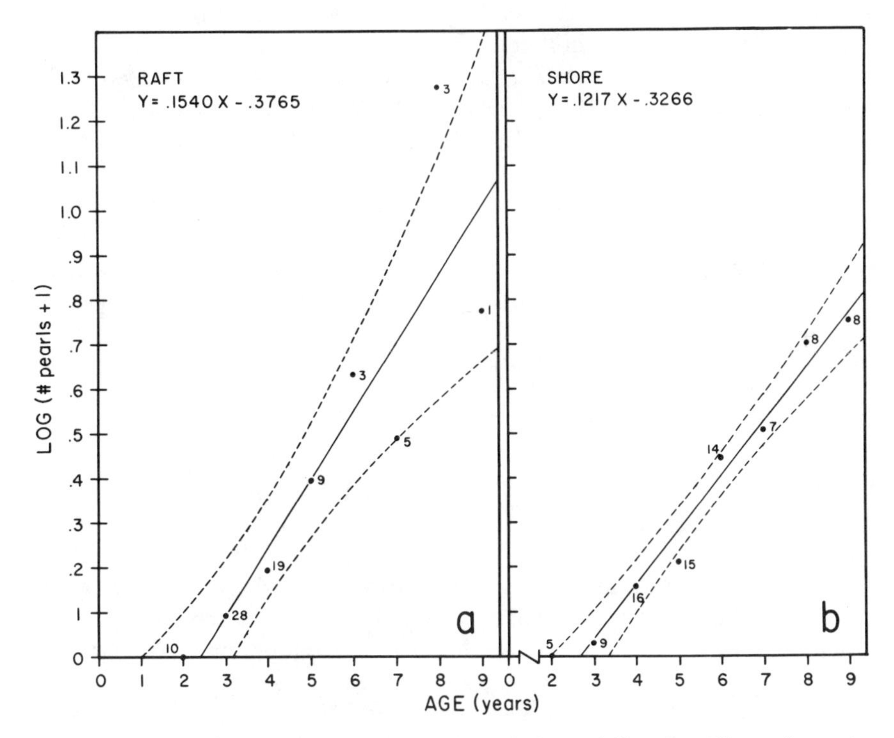

FIG. 8.25 Regressions of logarithmically transformed [log (x+1)] pearl counts on age for raft- and shore-based mussels, *Mytilus edulis*. (A) raft population in South Bristol, Maine; (B) shore population located at the mean low-water level in South Bristol, Maine. Dashed lines represent 95% confidence belts.
Reprinted from Lutz (1980c) with permission from Elsevier Scientific Publishing Company, © 1980.

throughout the life of a mussel, one might expect little or no difference between old and young individuals with regard to the number of extremely small pearls. They found, however, that this was generally not the case. There were greater numbers of small pearls in shore-based individuals and in mussels sampled from the rafts left in the water for the longest periods, suggesting that there is a general increase in the number of pearls being formed and/or a decrease in the growth rate of pearls as the individual ages.

The implications of the results of the various stuidies conducted by Lutz and Hidu (1978) for mussel aquaculture are profound. As a result of the 1942 survey conducted by Scattergood and Taylor (1949b), it was recommended that areas be closed where samples of mussels had either nine pearls with a diameter of 1 mm or greater or more than three pearls with a diameter of 1.5 mm larger in a 84-g (3-oz) sample of freshly drained meats. In the studies conducted by Lutz and Hidu, no pearls with diameters greater than 1 mm were found in specimens less than 5 years old. As mentioned earlier in this chapter, a marketable mussel (shell length ≥ 50 mm) can be obtained in a period of 12 to 15 months in U.S. waters using suspended culture techniques (Lutz 1974, 1979, 1980a,b; Chaves 1975; Chaves and Chew 1975; Lutz and Porter 1977; Incze *et al.* 1978). Whatever the factors responsible for pearl formation, the harvesting of mussels from suspended culture systems that have been in the water for less than 5 years should provide a quality product even in areas where the presence of large and numerous pearls in natural stocks of mussels has proved to be a deterrent to the commercial utilization of this species. A more detailed discussion concerning the incidence of pearls in cultured, as well as natural, populations of mussels has been presented by Lutz (1980c).

PREDATION, PARASITISM, AND ACCUMULATION OF ALGAL BIOTOXINS

Several species of sea ducks have caused considerable damage to mussel stocks in commercial and experimental suspended culture systems in U.S. waters. On the East Coast, the eider duck *(Somateria mollissima)* and the common scoter *(Oidemia nigra)* are responsible for the majority of mussel losses (Incze and Lutz 1980). In West Coast waters, *O. nigra* also creates serious problems for the mussel grower, as does the white-winged scoter *(Melanitta deglandi)*, the surf scoter *(Melanitta perspicillata)*, and Burrow's golden eye *(Bucephala islandica)* (Waterstrat *et al.* 1980). These migratory birds prey heavily on small mussels (see Graham 1975) and are capable of completely decimating stocks in a single growing season, thus seriously reducing the potential harvest from commercial culture operations. Underwater sirens, surface alarms, protective nets and other attempts to control duck predation have proven relatively ineffective and often incompatible with environmental regulations. Much further research is necessary to develop an adequate means of controlling losses from these predators.

Predation by starfish is a potential problem in mussel culture operations, even when suspended systems are utilized. Various species of starfish (*Asterias forbesi, Asterias vulgaris,* and *Asterias rubens* along the East Coast and *Pisaster ochraceus* and *Evasterias troschelii* along the West Coast) have been observed settling on mussel spat collectors shortly after the mussels have settled. The starfish grow rapidly and decimate the mussel populations on the collectors within 3 to 4 months in certain localities. In some commercial operations, if starfish are a problem, it may be necessary to remove them manually from collectors (MacLeod 1975).

Several potential predators may occur in bottom culture systems. In addition to the various species of starfish mentioned in the preceding paragraph, several species of whelks (e.g., *Thais lapillus*) are known predators of natural benthic populations of *M. edulis* (see MacLeod 1975), and may create serious problems in bottom culture operations. Other predatory organisms that could significantly affect bottom culture efforts include several species of crabs (e.g., *Cancer irroratus, Cancer borealis, Carcinus maenas, Callinectes sapidus*) and a number of carnivorous fish (e.g., *Rhacochilus vacca, Embiotoca lateralis*). Starfish, crabs, ducks, etc., are common predators in many European bottom culture operations, and predatory losses must be expected as part of the culture process (Incze and Lutz 1980).

The parasitic copepod *Mytilicola intestinalis,* which affects the quality of mussels in many areas of Europe (Mann 1956; Korringa 1959; Andreu 1963) is not known to occur along the east coast of the United States (Sindermann and Rosenfield 1967), although it has been reported along the West Coast (Waterstrat *et al.* 1980). In European studies it was found that the incidence of this parasite in mussels was considerably reduced when off-bottom culture techniques in fast-moving waters were employed (Hepper 1955; Andreu 1963). To date, this parasite has not posed a serious problem to U.S. mussel culture operations.

Several workers have speculated that certain species of haplosporidians may be responsible for mortalities of mussels occasionally observed in various North American waters (Sprague 1965; Taylor 1966;

Sindermann and Rosenfield 1967; Quayle 1978; Li and Clyburne 1979; Incze and Lutz 1980; Waterstrat *et al.* 1980). It is not clear, however, how common such infestations might be, and causal relationships between observed mortalities and haplosporidian infestations have not been demonstrated. At present, significant infestations by these protozoan parasites appear to be relatively isolated.

The accumulation of algal biotoxins in mussels, which can result in paralytic shellfish poisoning in humans (see Lutz and Incze 1979; Yentsch and Incze 1980) is a continuing problem, which periodically affects mussel culture activities on both coasts of the United States. The species of dinoflagellates reportedly responsible for the sporadic "red tides" which cause intoxications in mussels are *Gonyaulax catenella* and *Gonyaulax polyedra* on the West Coast and *Gonyaulax excavata* on the East Coast. Considerable research is being conducted on factors responsible for such toxic dinoflagellate blooms and extensive monitoring programs exist. For a detailed discussion of the accumulation of algal biotoxins in both natural and cultured stocks of mussels, see Yentsch and Incze (1980).

MUSSEL CULTURE IN HEATED EFFLUENTS

Since the late 1960s numerous studies have been conducted attemping to define the potential usefulness of heated effluent waters for the culture of various commercially important species (Nash 1968; Huguenin and Ryther 1974; Brackiel 1979; Blank 1979; Botsford *et al.* 1979; Carlberg *et al.* 1979; Eble *et al.* 1979; Ingram 1979; Kuroda 1979; Lutz and Hess 1979; Shaw 1979; Thain 1979; Lutz *et al.* 1980). Bivalve molluscs are among those species that have been studied in some detail, and preliminary results have been most encouraging from an aquacultural perspective. The elevated temperatures associated with power-plant effluent systems are generally compatible with biological requirements and lead to acceleration of gametogenesis, increased growth rates, and extension of growing seasons of these sessile invertebrates (Price 1975; Hess *et al.* 1976; Price *et al.* 1976; Lutz and Hess 1979). Relevant studies, as well as the advantages and disadvantages associated with the use of power-plant effluents for bivalve culture, have been enumerated by Huguenin and Ryther (1974).

Research on the culture of mussels in heated effluents in the United States has been limited to detailed analyses of two separate effluent systems of one nuclear power reactor located in a relatively cold-water natural environment near Wiscasset, Maine. The results of the studies have been outlined by Lutz *et al.* (1980) and are briefly summarized as follows. During periods when the generating facility employed a surface outflow effluent system, growth rates and survival of mussels were generally positively correlated with linear distance from the effluent, reaching maximum values at two control localities within adjacent estuaries. Mussel mortality ranged from 89 to 100% in the immediate vicinity of the heated outfall as surface water temperatures rose above 25°C (77°F). After installation of a multi-port diffuser effluent system, growth and survival of mussels at all sites located within the sphere of influence of the power-plant effluent were significantly greater than during previous years when the surface outfall system was employed, although both growth and survival values remained low relative to control sites. When either effluent system was employed, trace amounts of ^{58}Co, ^{60}Co, ^{134}Cs, ^{137}Cs, ^{54}Mn, ^{95}Zr, ^{95}Nb, and ^{40}K were detected in both the shells and soft tissue of the mussels cultured in these waters. Recruitment of mussel spat in the heated waters of both effluent systems are several orders of magnitude less than that found in control environments. Lutz *et al.* (1980) concluded that the potential of artificially heated waters for the commercial culture of *M. edulis* in U.S. waters appears to be minimal and suggested that for optimal growth of mussels commercial culture operations should be restricted to environments in which the upper limit of annual temperature ranges does not exceed 20°C (68°F).

AQUACULTURE CARRYING-CAPACITY MODEL

A dense population of mussels is an extremely heterotrophic community, requiring an environment that can provide abundant supplies of food and oxygen and assimilate or otherwise remove and process metabolic wastes (Nixon *et al.* 1971). The accelerated growth and superior condition of mussels in suspended culture systems is due to their maintenance under the optimal conditions available in the environment. Simply stated, this involves maximizing exposure to food, while minimizing physical trauma and predation. The rapid growth of this sessile organism is subsidized by physical energy flows in nature; mussels require considerable volumes of water to meet their metabolic demands, and these volumes are provided primarily by tidal circulation. Although these free energy flows help to make mussel aquaculture an extremely efficient means of protein production, limits to the productivity of natural waters necessarily limit the amount of shellfish biomass that can be supported before competition for available nutrients limits overall growth. The success of suspended culture techniques is directly dependent upon the production of food and oxygen in the growing waters and upon adequate circulation (Incze and Lutz 1980; Lutz 1980; Incze *et al.* 1982).

A consideration of possible biological constraints to mussel aquaculture in an estuary or bay involves an extension of the ecological concept of "carrying capacity" to the culture or husbandry of mussels. By evaluating the ability of a body of water to support dense aggregations of shellfish, the optimal production density can be determined. This evaluation requires an assessment of primary productivity, levels of suspended particulate matter, tidal velocities and exchange, and the effects of temperature, salinity, and particle size on rates of filtration. The importance of these findings to the efficiency of mussel aquaculture is obvious (Lutz *et al.* 1977; Incze and Lutz 1980). Aspects of this problem have been investigated in Europe (Jørgensen 1949, 1952, 1960, 1966; Fraga and Vives 1960; Davids 1964; Dral 1968; Widdows and Bayne 1971; Bayne *et al.* 1976; Widdows 1978a,b; Widdows *et al.* 1979), and some studies of ecological energy flows in natural populations of mussels and other bivalves have been conducted in the United States (Odum and Smalley 1959; Kuenzler 1961; Ryther 1969; Dame 1972; Incze and Lutz 1980; Incze *et al.* 1982). Incze *et al.* (1982; see also Incze and Lutz 1980) presented a preliminary model for determining the carrying capacity of a given body of water for the culture of mussels when a long-line aquaculture system is employed. The major components of this model are summarized in the remainder of this section.

The aquacultural gear in the hypothetical, large-scale long-line system, upon which the carrying capacity model is based, is arranged in such a way that the space between individual long-lines is 10 m (33 ft). This distance provides sufficient room for maneuvering of workboats and gear and prevents entanglement of mooring, long-line, and culture lines. Individual culture units on each long-line are kept at least 2 m (6.5 ft) apart to prevent entanglement. The lengths (depths) of the culture lines are 8 m (26 ft) and the system is set up in such a manner that tidal currents flow approximately parallel to the long-lines. A typical system would contain approximately 550 culture units. This hypothetical system is depicted in Fig. 8.26.

The model assumes that all particles are filtered from the water with equal efficiency by the mussels, regardless of whether or not those particles are nutritionally useful (Bayne *et al.* 1976). Therefore, the concentration of total seston is considered to be a critical element of the model, and a concentration of one-half of ambient is considered to be the approximate threshold for maximum ingestion rates in estuarine waters, at least along the northeastern coast of the United States [see Incze and Lutz (1980) for a detailed discussion concerning

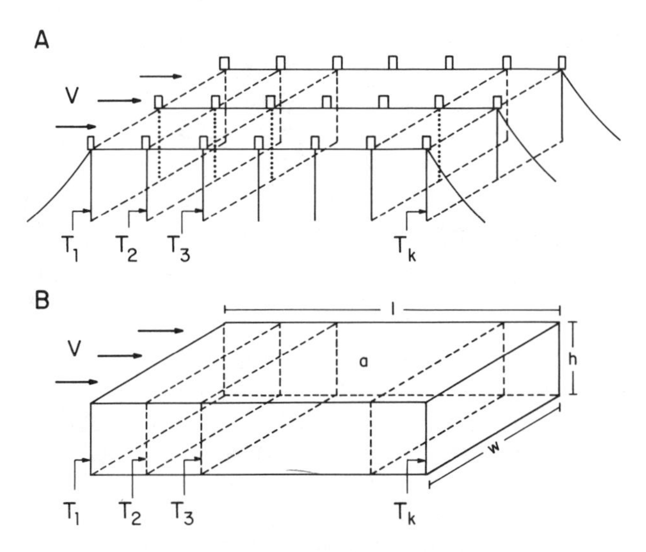

FIG. 8.26 Model for estimating impact of intensive mussel cultivation on seston concentration and thus for estimating optimal culture density. (A) small model of long-line system showing tiers of mussels (T_1, T_2 . . . T_k) perpendicular to the flow of current (V); (B) box diagram showing dimensions of system used in developing the mathematical carrying-capacity model. See text for equations and solutions.
Reprinted from Incze and Lutz (1980) with permission from Elsevier Scientific Publishing Company, © 1980.

utilization of this value in the model]. The value used for seston concentration can be altered easily in the future as more is learned about this parameter in nature.

The hypothetical culture system (Fig. 8.26) is three dimensional, having a depth h, a width w, and a length l, with a water surface area a and a face area $(A = hw)$ "facing" the current. For any system, the width and depth are constants for any one solution of the model, while the length is determined by the outcome of the model. Since mussels are cultured on discrete culture units, the solution to the model is via a series of finite sums involving one "tier" of culture units at a time. The "tier" consists of a "line" of culture units extending the full width w of the culture site (face A). Thus, tier 1 consists of the first culture unit from each of the long-lines. Although the long-lines are parallel to the current, the tiers (as they are presented in this model) are perpendicular. The model attempts to determine the concentration of seston as water enters each tier as a function of water flow rates, original seston concentration (before entering the first tier of mussels), and the filtration of particles by all mussels in the up-current tiers. The model is dependent upon the following assumptions:

(1) the concentration of particles per liter is homogeneous as it enters each tier; (2) flow is normal to the face of each tier; (3) flow through the system is laminar; and (4) each mussel filters 2.4 liter/hr.

Terms utilized in the model may be defined as follows: A (m^2) is the area of the culture system normal to the flow $(A = wh)$; V (m/hr) represents the flow rate of the water mass entering normal to face A; N (liter/hr) is the volume of water entering through face A per unit time ($N = Va \times 10^3$); n_k (mg/liter) is the concentration of particles per liter flowing into tier T_k, where $k = 1,2...$; and M is the number of mussels suspended in each of the tiers.

In each tier, M mussels will filter 2.4 M liter/hr as the water passes through. Since there are N liter/hr passing through each tier, then $2.4M/N$ is the fraction of particles filtered out of the water as it passes through each tier of mussels. In tier T_1, there are $n_1 N$ mg of particles flowing in per hr, so that $(2.4M/N)\, n_1 N = 2.4Mn_1$ mg will be filtered out. This leaves $n_1N - 2.4Mn_1 = n_1 (N - 2.4M)$ mg to flow into the next tier, T_2. Thus,

$$n_2 = \frac{n_1(N - 2.4M)}{N} \tag{1}$$

Likewise, following the same argument for the number of particles flowing into T_3, we get

$$n_3 = \frac{n_2(N - 2.4M)}{N} \tag{2}$$

Substituting Eq. (1) for n_2 into Eq. (2) yields

$$n_3 = n_1\left[\frac{N - 2.4M}{N}\right]^2 \tag{3}$$

By induction, the concentration of particles flowing into tier T_k is

$$n_k = n_1\left[\frac{N - 2.4M}{N}\right]^{k-1}, \qquad k = 2, 3, 4, 5, \ldots \tag{4}$$

Question: For what value of K (i.e., in which tier) will the concentration of seston be one-half of the original concentration, n_1, flowing into T_1?

Solution: Set $n_k = 1/2n_1$ and solve for k:

$$n_1\left[\frac{N-2.4M}{N}\right]^{k-1} = \tfrac{1}{2}n_1;$$

$$(k-1)\ln\left[\frac{N-2.4M}{N}\right] = \ln \tfrac{1}{2} = -\ln 2.$$

$$k = 1 - \ln 2 \Big/ \ln\left[\frac{N-2.4M}{N}\right] \tag{5}$$

(round this number down to the nearest whole number).

Thus with the requirement that seston concentrations not be reduced by more than one-half, the number of tiers that can be included in an ideal system (one which conforms to the initial set of assumptions) can be determined. Furthermore, under a given set of environmental conditions, the geometry and density of the culture system can be designed to yield optimum growth conditions.

By modifying the form of the model, one can also approach the problem from a different perspective. If the size of the culture operation that is desired is known in advance, and the mean seston concentrations for coastal waters are known, then the current velocity required to support such a system can be determined and an appropriate site sought. By rewriting Eq (4) we get:

$$(n_k/n_1)^{1/(k-1)} = \frac{N-2.4M}{N};$$

$$N(n_k/n_1)^{1/k-1} = N - 2.4M;$$

$$N[1-(n_k/n_1)^{1/k-1}] = 2.4M. \tag{6}$$

$$VA \times 10^3 = N = \frac{2.4M}{1-(n_k-n_1)^{1/k-1}} \tag{7}$$

and

$$V = \frac{2.4M}{A \times 10^3\,[1-(n_k-n_1)^{1/k-1}]} \tag{8}$$

Since, for this treatment of the model, k is known, and n_k can be decided, the required current velocity V (m/hr) can be determined.

While this model provides a useful framework for examining potentials for large-scale, high-density culture of mussels in U. S. waters, further work is needed before it can be applied with confidence to real systems in natural environments. Specifically, (1) the constants used in the model require further investigation, particularly the critical seston concentrations and size-specific filtration rates and energy needs of mussels; (2) seasonal variations in the nature and abundance of particles in natural waters must be compared to seasonal variations in the metabolic requirements of mussels in various U. S. waters to make certain that proper constants are used in the model to define the carrying capacity under critical limiting conditions; and (3) the usefulness of various particles in nature for mussel growth and metabolism must be investigated further so that a better ecological understanding of factors affecting growth potentials for aquaculture production may result. For further details concerning this model and its applications, see Incze *et al.* (1982).

ECONOMICS

Clifton (1980) summarized in detail the costs for an actual small-scale mussel culture operation in the northeastern United States (Table 8.2). His economic analysis is for a firm utilizing modified long-line and Spanish raft culture systems and capable of yielding 99,200–124,000 kg (4000–5000 bu) of mussels in a single growing season of

TABLE 8.2. COST STRUCTURE IN 1977 OF A SMALL-SCALE U.S. MUSSEL CULTURE OPERATION THAT USES SUSPENDED CULTURE TECHNIQUES (MODIFIED LONG-LINE AND SPANISH RAFT CULTURE SYSTEMS)

	Cost per Bushel ($)	% Cost per Bushel	Cost per 100 kg ($)
Farming Costs			
Fixed Overhead			
Land lease	1.71	6.7	6.84
License and permits	0.05	0.2	0.20
Electricity	0.28	1.1	1.12
Insurance	0.95	3.7	3.80
Depreciation (plus two arrays and vessel (+.90)	0.55(1.45)	5.7	5.80
Interest	0.24	0.9	0.96
Supplies	0.40	1.6	1.60
Miscellaneous	0.38	1.5	1.52
Variable Overhead			
Boat expense	0.48	1.9	1.92
Fuel	0.12	0.5	0.48
Truck	0.48	1.9	1.92
Packaging	1.20	4.7	4.80
Wages and Salaries			
Plant labor; 3	4.76	18.7	19.04
Biologist/harvester, ½ time	1.31	5.1	5.24
Plant manager	1.86	7.3	7.64
Subtotal:	15.67	61.5	62.88

TABLE 8.2. *(Continued)*

	Cost per Bushel ($)	% Cost per Bushel	Cost per 100 kg ($)
Marketing Costs			
Salary	0.90	3.5	3.60
Commission, 10% of sales	2.25[a]	8.8	9.00
Subtotal:	3.15	12.3	12.60
(Combined) Subtotal:	18.82	73.8	75.48
Administrative Costs			
Fixed Overhead			
Office supplies	0.24	0.9	0.96
Telephone	0.29	1.1	1.16
Services	0.30	1.2	1.20
Professional fees	0.24	0.9	0.96
Travel	0.60	2.3	2.40
Wages and Salaries			
President[b]	1.24(3.81)	14.9	15.24
Secretary/Bookkeeper (½ time)	1.19	4.7	4.76
Subtotal:	6.67	26.0	26.68
Total Cost:	25.49	99.8	102.16
Selling Price:	22.50		

Production-Function Relations

Output to Labor ratio[c] – Q/L_1 = 1,333 bushels per man-year ($30,000/man-year)
– Q/L_2 = 667 bushels per man-year ($15,000/man-year)
Capital to Labor ratio[d] – K/L_1 = $10,000 per man
Capital to Output ratio – K/Q = $7.14 per bushel per year

Source: Reprinted from Clifton (1980) with permission from Elsevier Scientific Publishing Company, © 1980.
[a] Based on a wholesale price of $22.50/bushel.
[b] Clearly the first figure is substantially below a market rate and reflects the development state of the operation. The figure in parentheses is based on an annual salary of $16,000 (along with other adjustments).
[c] Based upon estimates of current capacity constraints. Floating arrays of mussel ropes @ 10 bushels/unit yield an annual output of 4,000 to 5,000 bushels. Current processing capacity – 100 bushels/week. L_1, production labor; L_2, production and supervisory labor.
[d] Estimated cost of existing arrays plus processing machinery.

15 months. Relevant biological parameters include an annual surface water temperature range from −1.5°C to 17°C (29° to 63°F) and an annual variation in salinity from 25 to 32‰. Clifton (1980) concluded that at an average selling price of $0.91/kg ($22.50 per bu), which was the firm's average price, mussel cultivation was not profitable. He further estimated that a 10% positive net return would require an average wholesale price of $1.14/kg ($28.30/bu), which was over three times the 1977 average wholesale price of natural stocks of mussels. For further details of the economics of this small-scale operation, as well as an assessment of possible factors responsible for the unhealthy economic picture, see Clifton (1980).

In an attempt to present a realistic appraisal of the economic feasibility of mussel cultivation on a larger scale, Clifton (1980) presented a detailed summary of the cost structure of a hypothetical mussel culture operation (utilizing modified long-line and Spanish raft culture systems) with an annual capacity of 1000 MT (40,000 bu). The costs of this proposed medium-scale operation are summarized in Table 8.3. The major differences between this hypothetical operation and the one summarized in Table 8.2 are (1) the inclusion of taxes as an element of cost; (2) the addition of certain marketing expenses not found in the cost structure of the firm summarized in Table 8.2; (3) salaries for two vice-presidents; and (4) costs due to substantial changes in the planned degree of mechanization. While the basic cost structure of the hypothetical 40,000-bu facility is below that of a small-scale, labor-intensive facility, Clifton cautions that it is doubtful, given current harvest levels and prices of natural stocks of mussels, that such a quantity of cultured mussels could be sold at the proposed price of $27.50/bu. In a final sobering statement directed at those individ-

TABLE 8.3. COST STRUCTURE OF A HYPOTHETICAL MEDIUM-SCALE U.S. MUSSEL CULTURE OPERATION USING SUSPENDED CULTURE TECHNIQUES (MODIFIED LONG-LINE AND SPANISH RAFT CULTURE SYSTEMS)[a]

	Cost per Bushel ($)	% Cost per Bushel		Cost per 100 kg ($)
		T − (P + A)[b]	T[c]	
Farming Costs				
Fixed Overhead				
Land lease (facility lease)	0.24	1.5	1.0	0.96
Taxes	0.05	0.3	0.2	0.20
Licenses and permits	0.01	—	—	0.03
Electricity	0.12	0.7	0.5	0.48
Insurance	0.08	0.5	0.3	0.30
Depreciation (total)	(1.70)	(10.5)	(7.0)	(6.90)
boats and ship @ 10 years straight-line depreciation ($3600 per year)	0.09	0.6	0.4	0.36
arrays	1.45	9.0	5.9	5.80
machinery	0.12	0.7	0.5	0.46
sheds	0.04	0.2	0.2	0.18
Interest	0.52	3.2	2.1	2.10
Supplies	0.38	2.3	1.6	1.50
Miscellaneous	0.30	1.8	1.2	1.20
Variable Overhead				
Boat and ship expense	0.22	1.4	0.9	0.90
Fuel	0.08	0.5	0.3	0.30
Truck: local, long distance	0.42	2.5	1.7	1.67
Packaging	1.60	9.5	6.5	6.40
Wages and Salaries				
Marine labor; 3	(0.78)	(4.8)	(3.2)	(3.12)
plus fringes and taxes	0.97	6.0	4.0	3.88

TABLE 8.3. *(Continued)*

	Cost per Bushel ($)	% Cost per Bushel T − (P + A)[b]	% Cost per Bushel T[c]	Cost per 100 kg ($)
Marine foreman	(0.34)	(2.1)	(1.4)	(1.35)
plus fringes and taxes	0.42	2.5	1.7	1.67
Production labor; 5	(1.04)	(6.4)	(4.2)	(4.16)
plus fringes and taxes	1.29	8.0	5.3	5.16
Shore foreman	(1.34)	(2.1)	(1.4)	(1.35)
plus fringes and taxes	0.42	2.6	1.7	1.67
Subtotal:	8.82	53.8	36.0	35.22
Marketing Costs				
Promotional, discounts, allowances	5.50	—	22.5	22.00
Commissions @ 15% on net discount	3.30	20.4	13.5	13.20
Media advertising	2.75	—	11.2	11.00
Subtotal:	11.55	20.4	47.2	46.20
(Combined) Subtotal:	20.37	74.2	83.2	81.42
Administrative Costs				
Fixed Overhead				
Office supplies	0.08	0.5	0.3	0.32
Telephone	0.10	0.6	0.4	0.40
Services	0.08	0.5	0.3	0.32
Professional fees	0.04	0.2	0.2	0.16
Travel	0.15	0.9	0.6	0.60
Wages and Salaries				
President	(0.88)	(5.4)	(3.6)	(3.50)
plus fringes and taxes	1.09	6.7	4.4	4.34
Vice President—Sales	(0.56)	(3.4)	(2.3)	(2.24)
plus fringes and taxes	0.69	4.2	2.8	2.78
Vice President—Production	(.56)	(3.4)	(2.3)	(2.24)
plus fringes and taxes	0.69	4.2	2.8	2.78
Clerical and Secretarial	(0.94)	(5.8)	(3.8)	(3.75)
plus fringes and taxes	1.16	7.2	4.7	4.65
Subtotal:	$4.08	25.0	16.5	16.35
Total:	$24.45	99.2	99.9	97.77
(Excluding marketing costs)	$12.90			
Selling Price:	$27.50			

Production-Function Relations

Output to Labor ratio − $Q/L_1 = 5{,}000$ bushels per man-year or $137,500 per man-year
($112,500 per man-year @ $22.50 per bushel price)
− $Q/L_2 = 4{,}000$ bushels per man-year
− $Q/L_3 = 2{,}857$ bushels per man-year[d]

Capital to Labor ratio − K/L_1 = $51,000 per man

Capital to Output ratio − K/Q = $10.20 per bushel per year

Source: Reprinted from Clifton (1980) with permission from Elsevier Scientific Publishing Company, © 1980.

[a] Numbers in parentheses duplicate other items in columns and are not added to arrive at column totals.

[b] Based on a total cost figure of $16.20; excluding promotional and advertising costs for purposes of comparison

[c] Includes all costs

[d] Includes production, supervisory, and administrative labor.

uals determined to pursue their fortunes in aquacultural ventures, Clifton emphasizes:

> Mussel aquaculture in the United States is a possibility just over the horizon. It is a business opportunity which, given time, should generate a fair rate of return and add a delectable item to the range of popular seafoods in the country. It is not an opportunity to become a millionaire overnight.

SUMMARY

An increasing consumer demand for *Mytilus edulis,* coupled with a limited standing crop of this species, is presently necessitating a shift from production based on harvest of natural stocks to production based on aquaculture. Certain methods of mussel aquaculture practiced extensively in Europe appear to be adaptable to U.S. waters. The most recently developed method and that which appears to offer considerable potential for use in the United States is long-line culture. This system is readily adaptable to mechanization, which is an extremely important consideration in light of the high cost of labor in the United States. All of the U.S. firms currently involved in mussel aquaculture have used or are currently using this method of culture. However, bottom culture offers considerable opportunities in areas with favorable environmental conditions.

There are presently four private industrial firms actively engaged in the aquaculture of blue mussels. In 1982 annual production from these four operations combined was approximately 2600 MT (~105,000 bu) (whole shell weight) or 771 MT (~1.7 million lb) of wet meat, with an estimated annual value (to the firms) of approximately $1.9 million. Most of the production from these operations is currently from long-line and bottom culture systems.

Mussel culture research activities over the past decade have been extensive. Long-range research in the United States has been focused on the problems and needs of culture operations; it is crucial for future growth of the mussel culture industry that such research efforts be continued. In the short run, however, it is necessary that research and government policy be focused on effective management of existing natural populations. As the infant mussel aquaculture industry develops, measures should also be taken by the federal government to ensure that the expanding U. S. mussel industry is not faced with unfair competition from abroad.

Experimental mussel culture activities along the east coast of the United States began in 1972. Since that time, data on growth, survival, and recruitment of mussels using experimental suspended culture systems have been collected from Machiasport, Maine, to Narragansett, Rhode Island, with the majority of work being concentrated along the coast of Maine. Numerous environments along the East Coast have been identified as suitable areas for mussel culture, suggesting the strong possibility of developing an extensive and viable mussel culture industry in the United States. Research conducted to date concerning the harvest and transplantation of mussel seed suggests that only newly settled populations should be used for seed-harvesting activities, and that these seed should be transplanted to culture systems as soon as possible. If proposed aquaculture operations are to rely on wild seed for transplantation purposes, extensive and accessible areas with dense, regular spatfalls on natural substrates should first be identified.

On the West Coast, pilot research activities have been centered primarily in Washington and, more specifically, in Puget Sound. Research efforts have included studies concerned with mussel cultivation in natural and artificial upwelling systems, the toxicity of mussels resulting from ingestion of red tide organisms, shellfish hatchery techniques, and seed recruitment.

The presence of pearls in mussels from various natural populations along the east coast of the United States has been a serious deterrent to commercial utilization of large quantities of mussels. In recent studies marked differences in pearl incidence have been demonstrated between cultivated and natural stocks of mussels, with far fewer and smaller pearls in the aquacultured product. Results obtained to date strongly suggest that the harvesting of mussels from suspended culture systems that have been in the water for less than 5 years should provide a quality product even in areas where the presence of large and numerous pearls in natural stocks of mussels has hindered commercial utilization of this species.

A number of predators have caused considerable damage to mussel stocks in commercial and experimental suspended culture systems in U.S. waters. Such predators include starfish, crabs, snails, and various species of sea ducks. Although many steps have been taken to reduce the impact of these predators, predatory losses presently must be considered as a normal part of the culture process.

The accumulation within mussels of toxins from dinoflagellate blooms (red tides) is a continuing problem facing U.S. mussel culture operations and represents a serious constraint to expansion of the industry in certain regions. Research is continuing into factors responsible for periodic red tide outbreaks but solutions to the problem are currently not within reach.

The potential for using artificially heated effluent systems (e.g., from power plant effluents) in the commercial culture of mussels in U.S. waters appears to be minimal. For optimal growth of mussels it is suggested that commercial culture operations be restricted to environments in which the temperature does not exceed 20°C.

Although mussel aquaculture is an extremely efficient means of protein production, the total biomass of mussels in a culture system is limited by the capacity of the growing waters to provide food and oxygen and to remove and process metabolic wastes. A model has been presented that provides a useful framework for examining potentials for large-scale, high-density culture of mussels in U.S. waters. Further work is needed, however, before it can be applied with confidence to functioning systems in natural environments.

Mussel culture using suspended culture techniques in U.S. waters, is very labor intensive and, at present, the economic feasibility of suspended culture operations (on various scales) is somewhat questionable. In many instances, high initial costs restrict individuals from entering into mussel culture ventures in the United States. Capital equipment for a large-scale operation is expensive, and loans are difficult to obtain for this high-risk business. In summary, considerable caution should be exercised by those individuals determined to pursue their fortunes in the rapidly expanding mussel aquaculture industry.

ACKNOWLEDGMENTS

This article is Publication No. NJSG-84-139 of the New Jersey Marine Sciences Consortium and Publication No. D-32401-3-84 of the New Jersey Agricultural Experiment Station, Cook College, Rutgers University. New Brunswick, New Jersey, supported by state funds and NOAA Sea Grants NA81AA-D-00065 and NA83AA-D-00034.

LITERATURE CITED

ALONZO, R. S. 1977. Mussels of many: a handbook for food service managers. Sea Grant Bull. No. UNH-SG-AB-103. University of New Hampshire, Durham, New Hampshire.

ANDREU, B. 1958. Sobre el cultivo del mejillón en Galicia. Biologia, crecimiento y producción. Invest. Pesq. 745–746:44–47.

ANDREU, B. 1963. Propagación del copépodo parasito *Mytilicola intestinalis* en el mejillón cultivado de las rias gallegas (NW de Espāna). Invest. Pesq. **24,** 3–20.

ANDREU, B. 1968a. The importance and possibilities of mussel culture. Paper presented at the Seminar on Possibilities and Problems of Fishery Development in Southeast Asia. September, 1968, Working Paper 5.

ANDREU, B. 1968b. Pesqueria y cultivo de mejillónes y ostras en Espāna. Publ. Tec. Junta Estud. Pesca, Madrid **7,** 303–320.

ANDREU, B. 1968c. Fishery and culture of mussels and oysters in Spain. Proc. Symp. Mollusca **3,** 835–846.

ANDREWS, J. T. 1972. Recent and fossil growth rates of marine bivalves, Canadian Arctic, and Late-Quaternary Arctic marine environments. Palaeogeogr., Palaeoclimatol., and Palaeoecol. **11,** 157–176.

BACKIEL, T. 1979. On fish culture and on the use of heated effluents in Poland. *In* Power Plant Waste Heat Utilization in Aquaculture, B. L. Godfriaux, A. F. Eble, A. Farmanfarmaian, C. R. Guerra and C. A. Stephen (editors), pp. 215–230. Allanheld, Osmun and Co., Montclair, New Jersey.

BARDACH, J. E., RYTHER, J. H., and MCLARNEY, W. O. 1972. Aquaculture: the Farming and Husbandry of Freshwater and Marine Organisms. John Wiley & Sons, New York.

BAYNE, B. L. 1964. Primary and secondary settlement in *Mytilus edulis* L. (Mollusca). J. Anim. Ecol. **33,** 513–523.

BAYNE, B. L. 1965. Growth and the delay of metamorphosis of the larvae of *Mytilus edulis* (L.). Ophelia **2,** 1–47.

BAYNE, B. L. 1976. The biology of mussel larvae. *In* Marine Mussels: Their Ecology and Physiology, B. L. Bayne (editor), pp. 81–120. Cambridge University Press, Cambridge, United Kingdom.

BAYNE, B. L., THOMPSON, R. J., and WIDDOWS, J. 1976. Physiology: I. *In* Marine Mussels: Their Ecology and Physiology, B. L. Bayne (editor), Cambridge University Press, Cambridge, United Kingdom.

BLANK, K. 1979. Direct use of cooling and blow-down water for intensive fish farming results obtained in a pilot facility in the Niederaussem Power Plant. *In* Power Plant Waste Heat Utilization in Aquaculture, B. L. Godfriauz, A. F. Eble, A. Farmanfarmaian, C. R. Guerra and C. A. Stephens (editors), pp. 235–242. Allanheld, Osmum and Co., Montclair, New Jersey.

BOTSFORD, L. W., VAN OLST, J. C., CARLBERT, J. M., and GOSSARD, T. W. 1979. Economic evaluation of the use of thermal effluent in lobster culture. *In* pp. 79–95. Power Plant Waste Heat Utilization in Aquaculture, B. L. Godfriaux, A. F. Eble, A. Farmanfarmaian, C. R. Guerra and C. A. Stepehens (editors), Allanheld, Osmun and Co., Montclair, New Jersey.

BREESE, W. P., MILLEMAN, R. E., and DIMICK, R. E. 1963. Stimulation of spawning in the mussels, *Mytilus edulis* Linnaeus and *Mytilus californianus* Conrad, by Kraft Mill effluent. Biol. Bull. Woods Hole, Mass. **125,** 197–205.

CARLBERG, J. M., VAN OLST, J. C., and FORD, R. F. 1979. Pilot-scale systems for the culture of lobsters in thermal effluent. *In* Power Plant Waste Heat Utilization in Aquaculture, B. L. Godfriaux, A. F. Eble, A. Farmanfarmaian, C. R. Guerra and C. A. Stephens (editors), pp. 69–79. Allanheld, Osmun and Co., Montclair, New Jersey.

CHALFANT, J. S., ARCHAMBAULT, T., WEST, A. E., RILEY, J. G., and SMITH, N. 1980. Natural stocks of mussels: Growth, recruitment and harvest potential. *In* Mussel Culture and Harvest: A North American Perspective, R. A. Lutz (editor), pp. 38–68. Elsevier, Amsterdam.

CHANLEY, P., and ANDREWS, J. D. 1971. Aids for identification of bivalve larvae of Virginia. Malacologia **11,** 45–119.

CHAVES, L. A. 1975. Experimental culture of mussels at Seabeck Bay. Sea Grant Publ. No. WSG-TA-75-14. University of Washington, Seattle, Washington.

CHAVES, L. A., and CHEW, K. K. 1975. Mussel culture studies in Puget Sound, Washington. Proc. World Maric. Soc. **6,** 185–191.

CLIFTON, J. A. 1980. Some economics of mussel culture and harvest. *In* Mussel Culture and Harvest: A North American Perspective, R. A. Lutz (editor), pp. 312–338. Elsevier, Amsterdam.

COURTRIGHT, R. C., BREESE, W. P., and KRUEGER, H. 1971. Formulation of a synthetic seawater for bioassays with *Mytilus edulis* embryos. Water Res. **5,** 877–888.

DAME, R. F. 1972. The ecological energies of growth, respiration and assimilation in the intertidal American oyster *Crassostrea virginica.* Mar. Biol. **17,** 243–250.

DAVIDS, C. 1964. The influence of suspensions of micro-organisms of different concentrations on the pumping and retention of food by the mussel (*Mytilus edulis* L.) Neth. J. Sea Res. **2,** 233–249.

DE SCHWEINITZ, E. H., and LUTZ, R. A. 1976. Larval development of the northern horse mussel *Modiolus modiolus* (L.), including a comparison with the larvae of *Mytilus edulis* L. as an aid in planktonic identification. Biol. Bull. Woods Hole, Mass. **150,** 348–360.

D'HAMONVILLE, B. 1894. Les moules perlières de Billiers. Bull. Soc. Zool. Fr. **19,** 140–142.

DOW, R. L., and WALLACE, D. E. 1954. Blue mussels *(Mytilus edulis)* in Maine. Dep. of Sea and Shore Fish. Bull. Maine Department of Marine Resources, Augusta, Maine.

DRAL, A. D. G. 1968. On the feeding of mussels (*Mytilus edulis* L.) in concentrated food suspension. Neth. J. Zool. **18,** 440–441.

DUBOIS, R. 1901. Sur la mécanisme de la formation des perles fines dans le *Mytilus edulis.* C. R. Hebdomad. Séances Acad. Sci. **133,** 603–605.

DUBOIS, R. 1907a. Sur les métamorphoses du Distome parasite des *Mytilus* perliers. C. R. Séances Soc. Biol. Paris **63,** 334–336.

DUBOIS, R. 1907b. Action de la chaleur sur le distome immature de *Gymnophallus margaritarum.* C. R. Séances Soc. Biol. Paris **63,** 502–504.

DUBOIS, R. 1909. Contribution à l'étude des perles fines de la nacre et des animaux que les produisent. Ann. Univ. Lyon **29,** 1–126.

EBLE, A. F., STOLPE, N. E., EVANS, M. C., DEBLOIS, N., and PASSANZA, T. 1979. Diseasonal waste heat aquaculture: A three-year review. *In* Power Plant Waste Heat Utilization in Aquaculture. B. L. Godfriaux, A. F. Eble, A. Farmanfarmaian, C. R. Guerra and C. A. Stephens (editors), pp. 139–158. Allanheld, Osmun and Co., Montclair, New Jersey.

EDWARDS, E. 1968. A review of mussel production by raft culture. Resource Rec. Pap. Fish. Devel. Div., Bord Iascaigh Mhara.

FIELD, I. A. 1922. Biology and economic value of the sea mussel, *Mytilus edulis.* U. S. Bur. Fish. Bull. 38-127-259.

FRAGA, R., and VIVES, F. 1960. Retención de particulas orgánicas por el mejillón on los riveras flotantes. Reun. Sobre Prod. Pesq. Barcelona **4,** 71–73.

GARNER, R. 1857. On the pearls of the Conway River, North Wales, with some observations on the natural productions of the neighboring coast. Br. Assoc. Adv. of Sci., Part 2, pp. 92–93.

GARNER, R. 1872. On the formation of British pearls and their possible improvement. J. Linn. Soc. London **11,** 426–428.

GIARD, A. 1907. Sur les Trématodes margaritigènes du Pas-de-Calais (*Gymnophallus somateriae* Levinsen et *G. bursicola* Odhner). C. R. Séances Soc. Biol. Paris **63,** 416–420.

GRAHAM, F., JR. 1975. Gulls, a Social History. Random House, New York.

HAAMER, J. 1977. Musselodling. Bokförlaget Forum AB, Stockholm, Sweden.

HALEY, R., REINHOLD, K., and SCHLOBOHM, S. 1977. Report of the test market introduction of Bright Seas cultured mussels in Augusta, Maine, Nashua, New Hampshire and Portsmouth, New Hampshire. University of New Hampshire, Durham, Marine Program (unpublished manuscript).

HAVINGA, B. 1956a. Oyster and mussel culture. Rapp. P. V. Réun. Cons. Int. Explor. Mer **140,** 5–6.

HAVINGA, B. 1956b. Mussel culture in the Dutch Waddensea. Rapp. P. V. Réun. Cons. Int. Explor. Mer **140,** 49–52.

HAVINGA, B. 1964. Mussel culture. Sea Frontiers **10,**155–161.

HEPPER, B. T. 1955. Environmental factors governing the infection of mussels *Mytilus edulis* by *Mytilocola intestinalis*. Fish. Invest. Minis. Agric. Fish. Food, GB Ser. II. **20,** 1–21.

HERDMAN, W. A. 1904. Recent investigations on pearls in shellfish. Proc. Trans. Liverpool Biol. Soc. **17,** 88–97.

HERRINGTON, W. C., and SCATTERGOOD, L. W. 1942. Sea mussels, a potential source of attractive low-cost seafood from the Atlantic coast. U. S. Fish Wildl. Serv., Fish. Leafl. No. 11.

HESS, C. T., SMITH, C. W., and PRICE, A. H. 1976. Using heated effluent from a 835 MWe nuclear power reactor for shellfish aquaculture. *In* Waste Heat Management and Utilization, pp. 41–72. Mechanical Engineering Department, University of Miami, Miami, Florida.

HUGUENIN, J. E., and RYTHER, J. A. 1974. The use of power plant waste heat in marine aquaculture. Proc. 10th Annu. Conf. Marine Tech. Soc., pp. 431–445. Contrib. No. 3381, Woods Hole Oceanographic Institution, Woods Hole, Massachusetts.

HURLBURT, S. 1977. The Mussel Cookbook. Harvard University Press, Cambridge, Massachusetts.

HURLBURT, C. G., and HURLBURT, S. W. 1975. Blue gold: Mariculture of the edible blue mussle *(Mytilus edulis)*. Mar. Fish. Rev. **37**(10), 10–18.

HURLBURT, C. G. and HURLBURT, S. W. 1980. European mussel culture technology and its adaptability to North American waters. *In* Mussel Culture and Harvest: A North American Perspective, R. A. Lutz (editor), pp. 69–98. Elsevier, Amsterdam.

INCZE, L. S., PORTER, B., and LUTZ, R. A. 1978. Experimental culture of *Mytilus edulis* in a northern estuarine gradient: Growth, survival and recruitment. Proc. World Maric. Soc. **3,** 523–541.

INCZE, L. S., and LUTZ, R. A. 1980. Mussel culture: An east coast perspective. In Mussel Culture and Harvest: A North American Perspective, R.A. Lutz (editor), pp. 99–140. Elsevier, Amsterdam.

INCZE, L. S., LUTZ, R. A., and WATLING, L. E. 1980. Relationships between potential food particles, environmental temperatures, and survival and growth of *Mytilus edulis* in a north temperate estuary. Mar. Biol. **57,** 147–156.

INCZE, L. S., LUTZ, R. A., and TRUE, E. 1982. Modeling carrying capacities for bivalve molluscs in open, suspended-culture systems. Proc. World Maric. Soc. **12** (in press).

INGRAM, M. V. 1979. Waste heat aquaculture in the United Kingdom—A general review and particulars of Marine Farm Ltd., Hinkley Point, Bridgewater, Somerset, England. *In* Power Plant Waste Heat Utilization in Aquaculture. B. L. Godfriaux, A. F. Eble, A. Farmanfarmaian, C. R. Guerra and C. A. Stephens (editors), pp. 231–234. Allanheld, Osmun and Co., Montclair, New Jersey.

JAMESON, H. L. 1902. On the origin of pearls. Proc. Zool. Soc., London **1,** 140–165.

JAMESON, H. L., and NICOLL, W. 1913. On some parasites of the scoter duck *(Oidemia nigra)* and their relation to the pearl-inducing trematode in the edible mussel *(Mytilus edulis)*. Proc. Zool. Soc. London, **11,** 53–63.

JØRGENSEN, C. B. 1949. The rate of feeding by *Mytilus* in different kinds of suspension. J. Mar. Biol. Assoc. UK **28,** 333–344.

JØRGENSEN, C. B. 1952. Efficiency of growth in *Mytilus edulis* and two gastropod veligers. Nature (London) **170,** 714.

JØRGENSEN, C. B. 1960. Efficiency of particle retention and rate of water transport in undisturbed lamellibranchs. J. Cons. Cons. Int. Explor. Mer **26,** 94–116.

JØRGENSEN, C. B. 1966. Biology of Suspension Feeding. Pergamon Press, Oxford.

KORRINGA, P. 1959. Checking *Mytilocola* advance in the Dutch Waddensea. Cons. Int. Explor. Mer, Rep. 87.

KORRINGA, P. 1976. Farming Marine Organisms Low in the Food Chain. Elsevier, Amsterdam.

KUENZLER, E. J. 1961. Structure and energy flow of a mussel population in a Georgia salt marsh. Limnol. Oceanogr. **6,** 191–204.

KURODA, T. 1979. Japanese aquaculture with thermal water from power plants—present condition and problems. *In* Power Plant Waste Heat Utilization in Aquaculture. B. L. Godfriaux, A. F. Eble, A. Farmanfarmaian, C. R. Guerra and C. A. Stephens (editors), pp. 247–259. Allanheld, Osmun and Co., Montclair, New Jersey.

LI, M. F., and CLYBURNE, S. 1979. Mortalities of blue mussels *(Mytilus edulis)* in Prince Edward Island. J. Invertebr. Pathol. **33,** 108–110.

LUTZ, R. A. 1974. Raft cultivation of mussels in Maine waters: Its practicability, feasibility and possible advantages. Mar. Sea Grant Bull. No. 4. University of Maine, Walpole, Maine.

LUTZ, R. A. 1975. *Mytilus edulis:* age determination, pearl incidence, and commercial raft cultivation implications. Ph.D. Thesis, University of Maine, Orono, Maine.

LUTZ, R. A. 1976. Annual growth patterns in the innter shell layer of *Mytilus edulis.* J. Mar. Biol. Assoc. UK **56,** 723–731.

LUTZ, R. A. 1978. Pearl incidence in *Mytilus edulis* and its commercial raft cultivation implications. Proc. World Maric. Soc. **9,** 509–522.

LUTZ, R. A. 1979. Bivalve molluscan mariculture: A *Mytilus* perspective. Proc. World Maric. Soc. **10,** 595–608.

LUTZ, R. A. (editor). 1980a. Mussel Culture and Harvest: A North American Perspective. Elsevier, Amsterdam.

LUTZ, R. A. 1980b. Introduction: Mussel culture and harvest in North America. *In* Culture and Harvest: A North American Perspective, R. A. Lutz (editor), pp. 1–17. Elsevier, Amsterdam.

LUTZ, R. A. 1980c. Pearl incidence: Mussel culture and harvest implications. *In* Mussel Culture and Harvest: A North American Perspective. R. A. Lutz (editor). pp. 193–222. Elsevier, Amsterdam.

LUTZ, R. A., and HESS, C. T. 1979. Biological and radiological analysis of the potential of nuclear power plant effluent waters for shellfish culture. *In* Power Plant Waste Heat Utilization in Aquaculture. B. L. Godfriaux, A. F. Eble, A. Farmanfarmaian, C. R. Guerra and C. A. Stephens (editors), pp. 109–138. Allanheld, Osmun and Co., Montclair, New Jersey.

LUTZ, R. A., and HIDU, H. 1978. Some observations on the occurrence of pearls in the blue mussel, *Mytilus edulis.* Proc. Nat. Shellfish. Assoc. **68,** 17–37.

LUTZ, R. A., and HIDU, H. 1979. Hinge morphogenesis in the shells of larval and early post-larval mussels (*Mytilus edulis* L. and *Modiolus modiolus* (L.). J. Mar. Biol. Assoc. UK **59,** 111–121.

LUTZ, R. A., INCZE, L. S., CHEW, K. K., CLIFTON, J. A., HALEY, R., MILLER, B. A., BROWNELL, W., CHAVES-MICHAEL, L., BLUMENSTOCK, M. W., HAYES, K. P., WELDON, A. DEARBORN, R. K., and LLOYD, D. 1977. A comprehensive review of the commercial mussel industries in the United States. Stock No. 003-020-00133-5. U. S. Government Printing Office, Washington, DC.

LUTZ, R. A., INCZE, L. S., PORTER, B., and STOTZ, J. K. 1980. Seasonal variation in the condition of raft-cultivated mussels (*Mytilus edulis* L.). Proc. World Maric. Soc. **11,** 262–268.

LUTZ, R. A., and PORTER, B. 1977. Experimental culture of blue mussels *Mytilus edulis* in heated effluent waters of a nuclear power plant. Proc. World Maric. Soc. **8,** 427–445.

MACLEOD, L. L. 1975. Experimental blue mussel *(Mytilus edulis)* culture in Nova Scotia waters. Interim Rep. I. Fish. Mar. Serv., Pictou, Nova Scotia.

MANN, H. 1956. The influence of *Mytilicola intestinalis* on the development of the gonads of *Mytilus edulis*. Rapport et procesverbaux des Reunions. Rapp. P. V. Reun. Cons. Int. Explor. Mer **140,** 57–58.

MASON, J. 1971. Mussel cultivation. Underwater J. **3,** 52–59.

MASON, J. 1972. The cultivation of the European mussel, *Mytilus edulis* Linnaeus. Oceanogr. Mar. Biol. Annu. Rev. **10,** 437–460.

MASON, J. 1976. Cultivation. In Marine Mussels: Their Ecology and Physiology, B. L. Bayne (editor), pp. 385–410. Cambridge University Press, Cambridge, United Kingdom.

MILLER, B. A. 1977. Development of a sustained edible blue mussel industry in the Gulf of Maine. *In* A Cooperative University Institutional Sea Grant Program Proposal, Vol. II, pp. 57–82. University of New Hampshire, Durham, New Hampshire and University of Maine, Orono, Maine.

MILLER, B. A. 1980. Historical review of U. S. mussel culture and harvest. *In* Mussel Culture and Harvest: A North American Perspective, R. A. Lutz (editor), pp. 18–37. Elsevier, Amsterdam.

MOSSOP, B. K. E. 1921. A study of the sea mussel (*Mytilus edulis* Linn.). Contrib. Can. Biol. Fish. **2,** 17–48.

MOSSOP, B. K. E. 1922. The rate of growth of the sea mussel (*Mytilus edulis* L.) at St. Andrews New Brunswick, Digby Nova Scotia, and Hudson Bay. Trans. R. Can. Inst. **14**(3), 3–22.

MYERS, E. A. 1976. Commercial culture of Maine mussels. *In* A Cooperative University Institutional Sea Grant Program Proposal, Volume IIA, Cooperative Projects. University of New Hampshire, Durham, New Hampshire and University of Maine, Orono, Maine.

MYERS, E. A. 1980. The evolution of a commercial mussel culture operation. *In* Mussel Culture and Harvest: A North American Perspective, R.A. Lutz (editor), pp. 266–311. Elsevier, Amsterdam.

NASH, C. E. 1968. Power stations as sea farms. New Sci. **40,** 367–369.

NATIONAL RESEARCH COUNCIL. 1978. Aquaculture in the United States: Constraints and Opportunities. U.S. Gov. Printing Office, Washington, D.C.

NICOLL, W. 1906. Notes on trematode parasites of the cockle *(Cardium edule)* and mussel *(Mytilus edulis)*. Ann. Magazine Nat. History **17,** 148–155.

NIXON, S. W., OVIATT, C. A., ROGERS, C., and TAYLOR, K. 1971. Mass and metabolism of a mussel bed. Oecologia **8,** 21–30.

ODHNER, T. 1905. Die Trematoden des arkischen Gebietes. Fauna Artica (Römer Schaudinn) **4**(2), 291–372.

ODUM, E. P. and SMALLEY, A. E. 1959. Comparison of population energy flow of a herbivorous and a deposit-feeding invertebrate in a salt marsh eco-system. Proc. Nat. Acad. Sci. USA **45,** 617–622.

PRICE, A. 1975. Obstacles to and needs for continued R and D for the utilization of thermal effluent in aquaculture. *In* Utilization of Thermal Effluent in Aquaculture: Identification of Research and Development Needs. University of Massachusetts Aquaculture Engineering Laboratory, Wareham, Massachusetts.

PRICE, A. H., HESS, C. T., and SMITH, C. W. 1976. A field study of *Crassostrea virginica* cultured in the heated effluent and discharged radionuclides of a nuclear power reactor. Proc. Nat. Shellfish. Assoc. **66,** 54–68.

QUAYLE, D. B. 1978. A preliminary report on the possibilities of mussel culture in British Columbia. Fish. Mar. Tech. Rep. No. 815. Resource Service Branch, Pacific Biological Station, Nanaimo, British Columbia, Canada.

RYTHER, J. H. 1968. The status and potential of aquaculture, particularly invertebrate and algae culture. Volume 1, Part II. Invertebrates and Algae Culture. American Institute of Biological Sciences.

RYTHER, J. H. 1969. The potential of the estuary for shellfish production. Proc. Nat. Shellfish. Assoc. **59,** 18–22.

SCATTERGOOD, L. W., and TAYLOR, C. C. 1949a. The mussel resources of the North Atlantic region. Part I. The survey to discover the locations and areas of the North Atlantic mussel-producing beds. Commer. Fish. Rev. **11**(9), 1–10.

SCATTERGOOD, L. W., and TAYLOR, C. C. 1949b. The mussel resources of the North Atlantic region. Part II. Observations on the biology and the methods of collecting and processing the mussel. Commer. Fish. Rev. **11**(10), 8–20.

SCATTERGOOD, L. W., and TAYLOR, C. C. 1949c. The mussel resources of the North Atlantic region. Part III. Development of the fishery and the possible need for conservation measures. Commer. Fish. Rev. **11**(11), 1–10.

SHAW, W. N. 1979. Overview of ongoing Sea Grant projects in the United States utilizing heated water effluents and its future potential. *In* Power Plant Waste Heat Utilization in Aquaculture. B. L. Godfriaux, A. F. Eble, A. Farmanfarmaian, C. R. Guerra and C. A. Stephens (editors), pp. 243–246. Allanheld, Osmun and Co., Montclair, New Jersey.

SINDERMANN, C. J., and ROSENFIELD, A. 1967. Principal diseases of commercially important marine bivalve Mollusca and Crustacea. U. S. Fish Wildl. Serv. Fish. Bull. **66,** 335–385.

SLABYJ, B. M. 1980. Storage and processing of mussels. *In* Mussel Culture and Harvest: A North American Perspective. R. A. Lutz (editor), pp. 247–265. Elsevier, Amsterdam.

SLABYJ, B. M., and CARPENTER, P. N. 1977. Processing effect on proximate composition and mineral content of meats of blue mussels *(Mytilus edulis)*. J. Food Sci. **32,** 1153–1155.

SLABYJ, B. M., CREAMER, D. L., and TRUE, R. H. 1978. Seasonal effect on yield, proximate composition and quality of blue mussel meats *(Mytilus edulis)* obtained from cultivated and natural stock. Mar. Fish. Rev. **40**(8), 18–23.

SLABYJ, B. M., and HINKLE, C. 1976. Handling and storage of blue mussels in shell. Res. Life Sci. **23**(4), 1–13.

SPRAGUE, V. 1965. Observations on *Chytridiopsis mytilovum* (Field), formerly *Haplosporidium mytilovum* Field *(Microsporida?)*. J. Protozool. **12,** 385–389.

STAFFORD, J. 1912. On the fauna of the Atlantic coast of Canada, third report—Gaspé, 1905–1906. Contrib. Can. Biol. Fish. 1906–1912, pp. 45–67.

STUNKARD, H. W., and UZMANN, J. R. 1958. Studies on digenetic trematodes of the genera *Gymnophallus* and *Parvatrema*. Biol. Bull. Woods Hole, Mass. **115,** 276–302.

SULLIVAN, C. M. 1948. Bivalve larvae of Malpeque Bay, P.E.I. Bull. Fish. Res. Board Can. **77,** 1–36.

TAYLOR, R. L. 1966. *Haplosporidium tumefacientis* sp. n., the etiologic agent of a disease of the California sea mussel, *Mytilus californianus* Conrad. J. Invertebr. Pathol. **8,** 109–121.

THAIN, B. P. 1979. The farming of Dover sole and turbot in warmed waters: the work of the British White Fish Authority in 1976, 1977 and 1978. *In* Power Plant Waste Heat Utilization in Aquaculture, B. L. Godfriaux, A. F. Eble, A. Farmanfarmaian, C. R. Guerra and C. A. Stephens (editors), pp. 97–107. Allanheld, Osmun and Co., Montclair, New Jersey.

WATERSTRAT, P. R. 1979. Prospects for the development of a mussel culture industry in Puget Sound. M.S. Thesis, University of Washington, Seattle, Washington.

WATERSTRAT, P., CHEW, K., JOHNSON, K., and BEATTIE, J. H. 1980. Mussel culture: A West Coast perspective. *In* Mussel Culture and Harvest: A North American Perspective, R. A. Lutz (editor), pp. 141–165. Elsevier, Amsterdam.

WERNER, B. 1939. Über die Entwicklung und Artunterscheidung von Muschellarven des Nordseeplanktons, unter besonderer Beruschsichtigung der Schalenentwicklung. Zool. Jahrb. Abt. Allg. Zool. Physiol. Tiere **66,** 1–54.

WIDDOWS, J. 1978a. Combined effects of body size, food concentration and season on the physiology of *Mytilus edulis*. J. Mar. Biol. Assoc. UK **58,** 109–124.

WIDDOWS, J. 1978b. Physiological indices of stress in *Mytilus edulis*. J. Mar. Biol. Assoc. UK. **58** 125–142.

WIDDOWS, J., and BAYNE, B. L. 1971. Temperature acclimation of *Mytilus edulis* with reference to its energy budget. J. Mar. Biol. Assoc. UK **51,** 827–843

WIDDOWS, J., FIETH, P., and WORRALL, C. M. 1979. Relationships between seston, available food and feeding activity in the common mussel *Mytilus edulis*. Mar. Biol. **50,** 195–207.

YENTSCH, C. M., and INCZE, L. S. 1980. Accumulation of algal biotoxins in mussels. *In* Mussel Culture and Harvest: A North American Perspective, R. A. Lutz (editor), pp. 223–246. Elsevier, Amsterdam.

9

Abalone: The Emerging Development of Commercial Cultivation in the United States

Neal Hooker
Daniel E. Morse

Introduction
  The Declining Fishery
  Developing Aquaculture Industries
Principal U.S. Species
  Species Distribution and Range
  Haliotis rufescens (Red Abalone)
  Haliotis corrugata (Pink Abalone)
  Haliotis cracherodii (Black Abalone)
  Haliotis fulgens (Green Abalone)
  Haliotis kamtschatkana (Pinto Abalone)
  Haliotis kamtschatkana assimilis (Threaded Abalone)
  Haliotis sorenseni (White Abalone)
  Haliotis walallensis (Flat Abalone)
  Interspecific Hybrids
Biology
  Anatomical Description and Function
  Reproduction
  Larval Development
  Settlement and Metamorphosis
  Nutrition and Growth
  Diseases and Parasites
  Predation
Conventional Cultivation Technology
  Control of Reproduction
  Larval Development
  Larval Settlement, Metamorphosis, and Early Postlarval
    Development and Growth

Growth of Juveniles
Barriers to Efficient Production Using Conventional Technology
Defective Juveniles and Seed with Low Early Survival
Coastal Pollution
Unreliable Supplies of Seed and Juveniles of Warm-Water Species
Slow Growth and Inefficient Food Conversion
Restricted Success of Ocean Grow Out
Biochemical and Genetic Engineering Techniques to Improve Production
Biochemical Induction of Settlement, Metamorphosis, and Juvenile Development
New Bioassay for Evaluation of Potential Hatchery Locations and Seawater Sources
Control of Reproduction and Seed Production in Thermophilic and Other Abalone Species
Increased Nutritional Efficiency and Acceleration of Growth
Ocean Ranching
Japan
United States
Summary and Prospects for Future Industrial Development
Literature Cited

INTRODUCTION

Abalones are large, herbivorus marine snails of the genus *Haliotis,* highly prized for the delicately flavored white meat of their large muscular foot (Fig. 9.1). Worldwide demand for this valuable resource has risen steadily, driving retail prices in the mid-1980s to $44–66/kg ($20–30/lb) for meat of the preferred species in California and Japan. Total U.S. consumption of abalone is valued at approximately $30 million (retail) annually; the world market for abalone, centered in Japan and other Asian countries, is estimated at an annual retail value of approximately $300–400 million (National Marine Fisheries Service 1982; Food and Agricultural Organization 1975). The product—purchased in fresh, frozen, canned, and dried forms—is sliced and cooked as steaks, cooked with other foods, or consumed raw (Food and Agricultural Organization 1975). The shells of several species, which are used for decorative purposes, jewelry, and as a traditional medicine in Asia, also have high market value.

Worldwide production of abalone, primarily from fisheries, totals approximately 20,000 MT (44 million lb) annually (Food and Agricultural Organization 1975). The principal producing countries include Mexico (ca. 34% of world production), Japan (29%), Australia (20%), South Africa (6%), the United States (5%), Korea (3%), and New Zea-

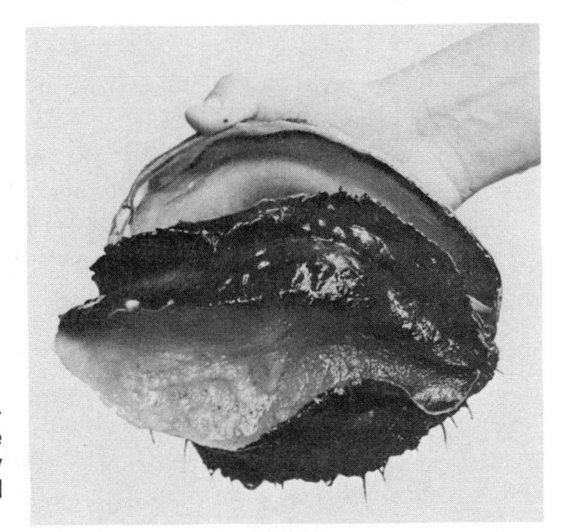

FIG. 9.1 A young adult red abalone, *Haliotis rufescens.* The large foot, visible here, typically yields more than 1 lb of dressed meat.

land (3%); smaller supplies are produced in Europe, northwestern Africa, and China. Approximately 15 regional species (of the 80 different abalone species worldwide) account for the bulk of the commercial product. An unrelated but similar snail of lower commercial value (*Concholepas concholepas;* common name, "loco") is produced in relatively large supplies in Chile and Peru; it is not included in the statistics above.

The Declining Fishery

The major source of supply of abalone is presently from fisheries. In the United States, abalone are found only along the Pacific Coast; they are fished commercially only in California, although fisheries and aquaculture studies also have been undertaken in Oregon, Washington, and Alaska, as well as in Canada.

The commercial abalone fishery in California was begun in the 1850s by Chinese immigrant workers, who recognized the value of the resource then abundant in the shallow intertidal zone. Their success is illustrated by the record of landings in 1879 of greater than 4 million lb (in the shell); by 1900, the depletion of the near-shore abalone required enactment of legislation prohibiting the sale of abalone taken from the littoral zone. In starting the harvesting of abalone from deeper waters, the Chinese were succeeded by Japanese-American hard-hat divers, who centered their fishing along the Monterey coast. These divers held a virtual monopoly on the commercial abalone fishery until the start of World War II, when they were incarcerated in detention centers inland. With the resumption of commercial harvesting after the war, the use of diving helmets was replaced by compressed-air regulators; air-compressors were located on the service vessels. In California, the use of scuba apparatus for commercial harvesting of abalone is prohibited by state law. (In Japan, where the use of any compressed-air device is prohibited for this purpose, abalone are harvested with long poles and hooks manipulated from boats, and to a lesser extent by free diving.)

With the increased efficiency of commercial harvesting, and the resulting intensification of commercial fishing activities, natural stocks of abalone were rapidly depleted from the central California coastal waters in which the shellfish previously had been abundant (Burge *et al.* 1975; Cicin-Sain *et al.* 1977; Oliphant 1979; Cox 1960). In recent years, this depletion has been greatly exacerbated by the unchecked and dramatic increase in the number of sea otters *(Ehydra lutris)*, a species now protected by federal and California law as both an endangered and marine mammal species (Smith 1982; Cicin-Sain *et al.* 1982). The otter is so effective a predator of the shellfish as to preclude any further commercial fishery for abalone within the otter's central California range (Burge *et al.* 1975; Woodhouse *et al.* 1977; Hardy *et al.* 1982; Silva 1982; Cicin-Sain *et al.* 1982). The waters of northern California, from Point Lobos north to the Oregon border (including a major portion of the sea otter's range), have been legally closed to commercial fishing for abalone since 1945.

As a result of the closure of northern California waters and the steady depletion of abalone from central California, the focus of the abalone industry in the United States moved progressively southward. The commercial fishery was based in Monterey through 1929, but it had moved south to Morro Bay by 1940. Between 1940 and 1960, the fishery spread further southward along the coast from Morro Bay to Los Angeles; by 1960, the center and principal focus of commercial activity was established at Santa Barbara, where it remains to the present time (Cox 1962; Cicin-Sain *et al.* 1977). Proximity to the highly productive commercial abalone fishery sites (and new ocean ranching sites) of the Santa Barbara Channel Islands is a principal reason for establishment of commercial abalone operations in Santa Barbara. The Channel Islands were opened to commercial harvesting of abalone in 1945; they remain designated and protected for that activity by recent federal legislation establishing several of these islands as national parks and monuments. These offshore islands in the Santa Barbara Channel are now the principal commercial source of abalone in the United States.

In the late 1960s, commercial landings of abalone in California began to decline rapidly. Before this time, the annual commercial catch had remained above 1.8 million kg (4 million lb) in the shell for decades. In 1966, the commercial harvest was just under 2.27 million kg (5 million lb); by 1979, the harvest had fallen to less than 450,000 kg (1 million lb) (National Marine Fisheries Service 1982). The principal causes of this major decline include overfishing by commercial and recreational divers along the mainland coast; pollution along the mainland coast from agricultural and large urban and industrial areas; predation by otters (Cicin-Sain *et al.* 1982); and competition for food and space from large populations of sea urchins (Burge *et al.* 1975; Cicin-Sain *et al.* 1977). As a result of this decline in the commercial abalone fishery, more than one-half of the abalone presently consumed in the United States is imported, coming principally from Mexico; the retail value of these imports in 1979 was more than $15 million (National Marine Fisheries Service 1982). The decline in the abalone fishery in the United States is not unique to this country, but is part of a worldwide trend resulting primarily from the overharvesting of available natural stocks (Conte 1981); Japan also now imports a significant proportion of its abalone consumption annually (National Marine Fisheries Service 1982; Food and Agricultural Organization 1975).

In summary, then, depletion and closure of the abalone fishery in the northern part of the range (Northern and Central California coast) has forced the U.S. industry southward; more recent depletion and pollution of mainland coastal areas has forced the industry further offshore, to the protected and productive islands of the Santa Barbara Channel. These factors also have combined in recent years to focus increasing industrial interest on abalone aquaculture, especially of warm water-tolerant (thermophilic) species adapted to the southern waters of the Santa Barbara Channel and the Channel Islands.

Developing Aquaculture Industries

The declining yields of the abalone fisheries, combined with the escalating price of the product, have stimulated interest in abalone aquaculture for both the enhancement of depleted stocks and the direct supply of product to meet the demand of the market. Over the past 15 years, a growing number of privately financed companies have been established in California, aimed at developing the technologies necessary to make abalone cultivation economically viable. These companies, located along the coast from Monterey to Port Hueneme, presently include Monterey Abalone Farms, Monterey; Pacific Ocean Farms, Carmel; Estero Bay Mariculture (formerly California Marine

Associates), Cayucos; California Sea Farms, Santa Barbara; and the Ab Lab, Port Hueneme. In addition, industry and public associations participating with these firms in abalone production and farming include the West Coast Aquaculture Foundation, Monterey; the California Abalone Association, Santa Barbara; and the Santa Barbara Mariculture Foundation, Santa Barbara. World Research Inc., of San Diego, also operates the nonprofit Ocean Studies Institute, which conducts research and development with abalone.

Supporting these industrial efforts, the California Department of Fish and Game has conducted research (Burge *et al.* 1975), assisted industrial entrants, and joined in a cooperative effort with researchers at the University of California in a preliminary study of the feasibility of replenishment of natural stocks by transplantation of hatchery-reared juvenile abalone to ocean-planting sites (Tegner *et al.* 1981). An unsung hero of these efforts has been Mr. Earl Ebert, Chief Biologist and Director of the Granite Canyon Shellfish Laboratory in Monterey, California, who has selflessly worked with new entrants in the commercial sector, developing and modernizing much of the basic technology in his own research (cf., Burge *et al.* 1975).

The University of California also has conducted research, in close cooperation with the industry, aimed at the development of new and cost-effective technologies for more efficient production (Morse *et al.* 1977a,b, 1978, 1979 a, b; Morse 1980, 1981, 1984 a–c); ocean seedling (Tegner *et al.* 1981) improved management of the industry (Cox 1962; Conte 1981; Bowden 1981; Cicin-Sain *et al.* 1977, 1982); implementation of new techniques, market development, and other assistance through extension services (Conte 1981). These programs have been supported by the University of California, the California State Resources Agency, California Department of Fish and Game, the U.S. Department of Commerce (National Oceanic and Atmospheric Administration)–University of California Sea Grant Program, and the U.S. Department of Commerce (NOAA) National Marine Fisheries Service. Private corporations, industrial associations, and public research groups also are being assisted in their research and development efforts in abalone cultivation by federal and state funds, as well as by direct support from the industrial sector.

It is a commonly held misconception that the United States lags far behind other countries—particularly Japan—in the development of abalone aquaculture. This is not the case, however. Industrialization of abalone cultivation is still only in the research and development phase in Japan, Mexico, France, Great Britain, Australia, South Africa, China, and other countries, as it is in the United States. The relatively primative status of technological development worldwide is reflected in the large governmental subsidies still required to sustain production and seeding operations, and continuing research and development in countries other than the United States. Indeed, significant innovations in cultivation technology recently have been developed—and continue to be developed—in the United States as well as in Japan, France, Australia, Mexico, and the other major producing countries, all with the assistance of cooperative governmental and industrial efforts. In the last few years, the efficient transfer of these new technological developments from publicly supported research laboratories to the private sector has sparked the establishment and growth of new abalone aquaculture industries in the United States, particularly in southern California. In this chapter, we review the basic biology and requirements for cultivation of various abalone species, remaining barriers to economically efficient production, recent development of new technologies, and prospects for future industrial development. Information pertaining to the United States is emphasized.

PRINCIPAL U.S. SPECIES

Species, Distribution, and Range

There are now approximately 80 living species of abalone inhabitating many of the temperate and semitropical coasts of the world; still more species of this successful group are found in the fossil record dating back to the Upper Cretaceous more than 65 million years ago. In North America, there are seven principal species (and several additional, recognized subspecies) of abalone inhabiting the western coast of the continent; all of these are found in U.S. waters, and are large enough to be of commercial interest.

The different species of abalone in U.S. waters can be recognized principally by their size, shell shape and color (Fig. 9.2), and the appearance and color of the epithelium (the epipodium) and tentacles surrounding the foot in a conspicuous, ruffled fringe (Cox 1960). These species also differ in depth and geographic range, in their temperature requirements for survival, reproduction, and optimal growth, and in the consistency, flavor, and color of their meat. As a result, the species differ in the market value of their meat and shell and in their suitability for cultivation in different areas. The range and depth distributions of the different abalone species are largely determined by the unique temperature requirements of each species (and by the distribution of suitable algal food species). These distributions thus generally exhibit the phenomenon of "warm-water submergence," with

FIG. 9.2 Shells of North American abalone species, including juveniles and adults, in dorsal view *(upper)* and ventral view (lower). In each illustration—top row: red abalones, showing banding of dorsal shell coloration from change in algal diet in the juvenile specimen shown on the left. Center row, from left to right: threaded abalone, white abalone, and green abalone. Bottom row, from left to right: pink abalone, black abalone (young adult), flat abalone, and pinto abalone.

individuals of a given species inhabiting progressively deeper, cooler waters in the more southern portions of their geographic range.

Of the North American species, the red abalone *(Haliotis rufescens)* is the largest abalone worldwide; it may attain a length of 30 cm (11.8 in.). As this largest abalone has been the traditional focus of the U.S. commercial fishery, and is the first species with which the developing abalone aquaculture industry in the United States has achieved some limited success, we first describe briefly the characteristics of this important abalone; the other major U.S. species and subspecies then are described, in alphabetical order by species name. A brief discussion of interspecific hybridization then follows. Additional information is available from Cox (1960, 1962), Mottet (1978), and Morris *et al.* (1980).

Haliotis rufescens (Red Abalone)

Ranging from Sunset Bay, Oregon, in the north to Turtle Bay, Baja California, Mexico, in the south, red abalone is found from the lower intertidal zone on rocky shorelines to depths of over 180 m (590 ft). It is most abundant along the central California coast, where it is found at depths of 6–17 m (20–56 ft). Red abalone will grow to a size exceeding 28 cm (11 in.), and commercial harvesting of animals greater than 18 cm (7.25 in.) is permitted (Cox 1960). Shell color varies from brick red to bands of aquamarine and white; these colors are a function of the algal diet (Leighton 1961; Olsen 1968a,b). The phycoerythrobilin pigment of red algae is responsible for the red coloration of the shell; brown and green algae produce the lighter shades. The red abalone is a highly prized commercial species both for its size and the quality of its meat; it once made up the largest part of the U.S. commercial abalone harvest. Natural stocks of this species, however, have been markedly reduced, and red abalone accounted for only 43% of the commercial catch in 1976; the absolute weight of the harvest also has been reduced to a small fraction of what had been landed in the past (Oliphant 1979). Most mariculture efforts with abalone in the United States have centered first on *H. rufescens,* as have experimental programs aimed at reseeding depleted natural stocks.

Haliotis corrugata (Pink Abalone)

Ranging from Point Conception, California, to Turtle Bay in Baja California, *H. corrugata* is named for its characteristically corrugated shell. The shell is circular in shape and highly arched, with scalloped edges. Pink abalone occurs along the rocky shores of protected bays and coves, as well as on the exposed coast, at depths of 6–60 m(20–

200 ft); it is most abundant to a depth of 30 m (100 ft). Adults may grow to a size of 25 cm (10 in.), but a length of 15–18 cm (6–7 in.) is more common. Adults are sedentary, forming scars on the rocks where they remain attached in wait for drift algae (Cox 1960). The pink abalone is a prized commercial species; in the past it has made up nearly 50% of othe commercial California catch, although its numbers now have been depleted. In 1976, it represented 25% (by weight) of the California commercial catch (Oliphant 1979).

Haliotis cracherodii (Black Abalone)

Black abalone is principally found intertidally from Coos Bay, Oregon, to Cabo San Lucas at the tip of the Baja California Peninsula, Mexico. The black abalone lives under rocks and in crevices on rocky shorelines; it subsists largely on diatoms and fragments of algae washed up in the surf. Populations of black abalone recently have been found in dense aggregations around shallow marine hot springs, where they have been observed to graze heavily on the thick mats of chemosynthetic sulfide bacteria supported by the mineral-rich hydrothermal effluent (Stein 1981). Adults of this species may reach a length of 20 cm (8 in.), but the more common adult size is 8–13 cm (3–5 in.). The shell of the black abalone is smooth, without raised respiratory pores; it is dark blue to black in color. This was not an important commercial species until relatively recently because of the darker color of the meat and its reputation for poor flavor in the California market (Cox 1960). However, with depletion of other abalone stocks, and the opening of a strong market in the Orient for canned black abalone, this situation reversed rapidly. In 1974, black abalone made up 44% of the California catch (McAllister 1976); continued heavy commercial harvesting lowered this proportion to 20% by 1976 (Oliphant 1979).

Haliotis fulgens (Green Abalone)

A warm-water species, green abalone is found from Point Conception to Magdalena Bay. It occurs from the lower intertidal zone to 18 m (60 ft), but is principally found in less than 10 m (40 ft) of water. Green abalone may grow to 25 cm (10 in.), although most adults are 13–20 cm (5–8 in.). The shell of this species may be olive green to reddish brown, with gently spiraling ribs on the exterior. The shell of the green abalone is most noted for its striking irridescent greenish interior; as a result, this shell has considerable market value. Because of its good meat quality, shallow-water distribution, and relative abundance in the warmer, southern California waters, the green abalone is a principal species harvested by the large sport-diving fishery. However, because the commercial harvesting of abalones is restricted to depths greater than 6.1 m (20 ft), this species accounted for only 7% of the commercial harvest in California in 1976 (Oliphant 1979).

Haliotis kamtschatkana (Pinto Abalone)

The range of pinto abalone, a northern, cold-water species, extends as far north as Sitka, Alaska, and as far south as Point Conception on the California coast. In Alaska the pinto abalone is found under rocks and in crevices from low-tide level to a depth of 7 m (23 ft); in the warmer waters of central California it is typically found at a depth of 11–15 m (36–50 ft) inhabiting the more exposed rock substrates. This is a small species; adults are commonly less than 10 cm (4 in.) in length, although some may grow to a length of 15 cm (6 in.). The shell is a distinctly elongated oval in shape; it is characteristically thin, with a mottled green-brown coloration bearing lighter colored spots. Because of its small size, and the low recovery of meat in proportion to its shell, this species is not sought commercially in the United States (Cox 1960; Mottet 1978).

Haliotis kamtschatkana assimillis (Threaded abalone)

A southern subspecies or variety of the northern pinto abalone, the threaded abalone ranges from Point Conception, California, southward to Turtle Bay on the Baja California Peninsula of Mexico. Found at depths of 21–30 m (70–100 ft), threaded abalone attain a maximum size of 15 cm (6 in.) in length, almost most adults are under 10 cm (4 in.). While occasionally taken commercially, this subspecies has never constituted more than a small fraction of the harvest (Cox 1960).

Haliotis sorenseni (White Abalone)

Ranging from Point Conception to Turtle Bay, Baja California, the deep-water white abalone is usually found at depths of 24–30 m (80–100 ft). With its deeply arched thin shell, large size, and excellent meat quality, the white abalone is the most highly valued species in California. However, as a result of its deep-water habitat, it has never constituted a high percentage of the commercial catch (Cox 1960). In 1976, white abalone accounted for less than 5% of the commercial U.S. harvest (Oliphant 1979).

Haliotis walallensis (Flat Abalone)

The flat abalone is found from British Columbia to La Jolla, California, from shallow subtidal depths to 21 m (70 ft). The shell is oval, narrow, and brick red in color with occasional mottling; the very slight arch of the shell gives this species its common name. Although individuals may grow to 18 cm (7.25 in.), most adults found are small, in the range of 8–13 cm (3.25–5.25 in.). Because of its sparse occurrence and small size, the flat abalone is not harvested commercially (Cox 1960; Mottet 1978).

Interspecific Hybrids

The abalone species described in this section do, with low frequency, produce interspecific hybrids both in their natural environment and after controlled breeding in laboratories and hatcheries (Owen *et al.* 1971; Vacquier *et al.* 1981; B. Owen and R. J. Meyer, manuscript in preparation; D. Leighton and J. McMullen, personal communication; Leighton and Lewis, 1982). These hybrids may be recognized both by external and biochemical characteristics (Owen *et al.* 1971; B. Owen and R. J. Meyer, manuscript in preparation; W. Swint, personal communication). Although the hybrids themselves generally are sterile or exhibit reduced fecundity, there has been some interest and success in the production of specific hybrids for commercial purposes. Such hybrids may exhibit desirable growth characteristics, and provide a convenient means for recognizing animals grown in ocean-ranching operations from hatchery-produced seed (J. McMullen, D. Leighton, personal communication).

BIOLOGY

Anatomical Description and Function

Abalones are large marine snails, generally classified in the genus *Haliotis*. They are hierarchically classed as archaeogastropods, as their anatomy and development are those of the typical snails. Detailed reviews of the biology of abalones have been published by Cox (1962) and Mottet (1978); see also Crofts (1929) and Ino (1952).

The most conspicuous external characteristic of this genus is its single, large, oval shell, covering the entire dorsum of the animal's body (see Fig. 9.2). This shell, in the shape of one large, flattened whorl, is perforated with a single row of prominent holes or respiratory pores. The hard shell, composed principally of calcium carbonate crystals embedded in a protein matrix, is secreted by the mantle (a thin fold of tissue covering nearly the entire inside surface of the shell).

The diffraction of light produced by the geometric regularity of these calcium carbonate crystals results in the pearly, iridescent hues of green, pink, and other colors characteristic of the inside of the valuable abalone shell. The outer or dorsal surface color is derived from algal pigments (Ino 1952; Leighton 1961; Olsen 1968a,b). Growth of the shell proceeds by secretion of new material from the mantle edge at the lip of the shell, and at the holes or pores; damage to the shell from breaks or invasive, boring organisms also can be repaired by localized secretion from the mantle. Similar secretion can produce colorful and irregular "abalone pearls" around foreign matter within the soft tissue. Lusterous pearls of considerable hardness, size, and beauty are most often found in the conical appendage (W. Swint, personal communication).

The abalone shell is perforated by a line of holes lying immediately above the gills; these function as respiratory pores and as excurrent channels for excretion and release of gametes. As the shell grows larger, new respiratory pores develop in the growing, leading edge of the shell above the head of the animal, while the older, most posterior pores, which no longer lie over the gills, become sealed with calcium carbonate. The characteristics of the shell, such as color, texture, shape, and the number and shape of the respiratory pores, often may be used to distinguish the different abalone species (Cox 1960; see Principal U.S. Species section). However, identification of these characteristics often may be complicated by the presence of encrusting organisms and by variability in shell characteristics within the different abalone species and hybrids (cf. Cox 1960; Owen *et al.* 1971).

The very large foot, or right shell muscle, appears to fill the ventral side of the shell (see Fig. 9.1). This muscle is the valued meat, which in many species is so large as to yield several sliced steaks per animal. This large, columnar muscle is attached dorsally to the inside of the shell; its removal during processing of the meat leaves a characteristic scar on the ventral shell surface. The "sole," or ventral surface of the foot, aided by powerful waves of muscular contraction and by copious secretion of mucus, is responsible for the strong suction and adherence of the animal to rock or other hard substrate, and for its gliding locomotion. When the anterior portion of the foot is raised from the substrate, in a typical feeding posture, it is used to secure algae and bring this food to the mouth. When abalone are taken unawares, they may be pulled from the substrate without great effort; once alarmed, however, they quickly contract the massive columnar foot muscle, pulling the shell close to the substrate and adhering with such powerful suction as to require the leverage of an "abalone iron" for their removal.

The small head, attached anteriorly to the foot, includes the mouth, two stalked eyes, asnd two long and retractable sensory cephalic tentacles. The epipodium, a ruffled flap of tactile sensory tissue surrounding the circumference of the foot (except at the head), also bears numerous short, sensory tentacles; this specialized tissue thus allows chemical and tactile sensing of the local environment around the animal. The distinctive color and shape of the epipodium and epipodial tentacles can be used reliably to distinguish the various species and hybrids of abalones in the United States (Cox 1960; Owen *et al.* 1971).

When the abalone is observed dorsally (from the top) with the shell removed, the two parallel gills (ctenidia) can be seen to the left of the large shell muscle. Water is taken in under the front margin of the shell, and is moved past the gills and out the respiratory pores by the action of strong cilia on the gills. The excretory and sex products also are released into the area between the mantle and the body around the gills (branchial chamber). These products are expelled through the respiratory pores with the aid of water currents forced by muscular contraction, or pumping, of the foot.

The important digestive gland (known as the hepatopancreas, or liver) and the gonad are contained concentrically within the prominent conical appendage, which may be seen upon reflection of the foot and epipodiuim from the ventral side of the shell. The digestive gland makes up the bulk of the posterior fundus of this organ as well as the inside of the cone; the gonad (ovary or testis) surrounds the digestive gland superfically and extends to the tip of the cone anteriorly. The conical appendage is capable of changing greatly in size depending upon the reproductive state of the animal, swelling with its content of eggs or sperm. The color of this conical appendage also allows easy recognition of the sex of the abalone when the animal is gravid (ripe with eggs or sperm); the green color of the eggs or cream-white color of the sperm can be seen in the distended gonad.

The digestive system of the abalone is long and complex, as it is in many herbivores. Food is brought in through the mouth, where it is ground by the long, filelike, toothed radula. The macerated food then is slowly broken down further and absorbed during its passage through the esophagus, crop, stomach, spiral caecum, and intestine. Fecal and other excretory material is released into the branchial chamber and expelled through the respiratory pores by contraction of the shell muscle.

Reproduction

Abalone are primarily dieocious, with individuals of separate sexes; occasionally, however, hermaphroditic animals are found (Murayama 1935; Morse *et al.* 1977b). The breeding season of abalones in North

America varies from species to species, generally being positively correlated with water temperatures. Thus, red abalone may be found gravid and capable of spawning throughout the year in the southern portion of its range, while spawning only in the warmest summer months in the northern part of its range (Boolootian *et al.* 1962; Young and DeMartini 1970; Leighton 1974; Tutschulte 1975; Giorgi and DeMartini 1977). Similarly, other species may become gravid and spawn only one yearly, in the period of the spring through fall, whereas others may have two distinct spawning periods during this time (Boolootian *et al.* 1962; Webber and Giese 1969; Leighton 1974; Tutschulte 1975; Hahn 1979).

During the spawning season, the gonad within the conical appendage gradually becomes engorged with sperm or eggs. In the wild, spawning may be triggered in gravid animals by a sudden change in water temperature, exposure to air during a low tide (for intertidal species), or other stimuli that might cause sudden contraction of he foot muscle. Specific environmental triggers of spawning also have been suspected (Boolootian *et al.* 1962; Webber and Giese 1969; Young and DeMartini 1970; Tutschulte 1975; Giorgi and DeMartini 1977; Hahn 1979). Release of eggs or sperm by one animal is observed to trigger the spawning of many gravid animals nearby; this effect has been shown to be caused by a hormone-like substance released into the water with the spawn (Murayama 1935; Carlisle 1945; von Medem 1948; Owen *et al.* 1971; Morse *et al.* 1977a). The act of spawning is accomplished by contraction of the shell muscle; this compresses the gonad against the body, thus squeezing the eggs or sperm out of the gonad, into the branchial chamber, and thence through the respiratory pores.

The number of sperm and eggs released depends upon the size of the abalone as well as its state of gravidity; as with most broadcast spawners, the number usually is very large. A large gravid red abalone may spawn as many as 10 million eggs at a time; it has been estimated that 10^{12} sperm are released by a gravid male of this species in a 30-min spawning (Morse *et al.* 1977a). Spawning may cause weight losses of as much as 5–10% of the body weight (Newman 1968; cf., Boolootian *et al.* 1962). Fertilization of the released gametes occurs in the open water. Eggs (ca. 0.2 mm in diamter) that have become fertilized sink to the bottom and begin embryological development.

Larval Development

Approximately 24 hours after fertilization, depending upon species and water temperature, a free-swimming, trocophore larva (0.2 mm) hatches from the egg membrane. This drifting, planktonic larva swims weakly as a result of the action in a ring of beating, microscopic cilia. After its first day in the plankton, the trocophore larva develops into the veliger larval stage; swimming cilia now are borne by a distinctive collar of cells, known as the velum, surrounding the developing head. During the next week in the plankton, development of the veliger is completed; this larvae, still approximately 0.2 mm in diameter, carries a complete larval shell with a closable operculum.

It is a distinct advantage for cultivation that larval development is completely supported by nutrients from the egg yolk; abalone larvae are fully lecithotrophic, and require no feeding during this time (Morse *et al.* 1979a,b, 1980b). The processes of abalone development are described in more detail by Crofts (1929), Ino (1952), Oba (1964), Shibui (1972), Seki and Kan-no (1977) and Morse *et al.* (1979a,b, 1980a,b).

The planktonic larval stage serves as a means for disperal of abalone. The veliger abalone larvae are positively phototropic; they swim upward toward the surface in a spiraling motion, then cease swimming and sink to the bottom, shortly after which they resume their upward swimming once more. This behavior, coupled with dispersion in coastal currents, distributes the larvae widely. As described in the next section, abalone larvae can remain in this developmentally arrested, dispersive stage for as long as 3 weeks without mortality, developmental impairment, settlement, or metamorphosis. During their time in the plankton, with periodic opportunities to inspect the bottom, the larvae begin a characteristic "searching" behavior that aids in their location of suitable substrate for final settlement and metamorphosis (Morse *et al.* 1980a).

Settlement and Metamorphosis

When larval development is complete (approximately 4–10 days after fertilization, depending upon species and temperature), larval abalone become competent to settle and undergo metamorphosis to the juvenile form (Morse *et al.* 1979a,b, 1980a,b). Further development is arrested at this point, until metamorphosis begins.

Many if not all of the U.S. abalone species, and many species in other countries as well, require larval contact with a specific natural inducer to trigger the settlement response and metamorphosis (Morse *et al.* 1979a,b, 1980-c; 1980, 1981, 1984a–c). The natural inducer of settlement and metamorphosis, for many abalone species, is a biochemical substance present at the surface of the crustose red algae commonly found coating rocks and other substrates in the coastal environment (Morse *et al.* 1979a,b, 1980a; Morse and Morse, 1984). These "recruiting" algae, on which the abalone larvae preferentially settle, include the coralline red algae, such as *Lithothamnium* spp. and *Lithophyllum* spp., and the crustose noncoralline red algae such as *Hildenbrandia* spp. (Morse *et al.* 1979a, 1980a). Recruitment of abalone larvae to encrusting red algae was suspected previously by Crofts (1929) and Shepherd (1973).

In the absence of their natural algal (and biochemical) inducers of settlement and metamorphosis, abalone larvae can postpone settlement and survive as healthy planktonic larvae for more than 3 weeks. This prolonged survival as developmentally arrested, planktonic larvae is dependent upon high-quality, clean seawater; absence of microorganisms and pollutants; species; and water temperature.

When competent abalone larvae come into direct physical contact with their natural recruiting algal substrates (or with the inducing biochemicals extracted from these algae), ciliary swimming quickly ceases; the larvae attach to the algal surfaces with their foot, and begin a characteristic snaillike gliding motion over the substrate. Soon after settlement, the recruited abalones begin their typical grazing behavior, with side-to-side scraping of the algal surface. This behavioral metamorphosis is soon followed by complete developmental or physical metamorphosis to the juvenile abalone form.

Developmental metamorphosis begins with the abscission (shedding) of the specialized larval swimming organ, which consists of the ciliated, columnar epithelial cells of the velum. This profound change irreversibly commits the settled larvae to the benthic or bottom-dwelling mode of life. Within the next few days, the various organs rapidly develop: the heart is completed and begins to beat; the operculum is shed; and a new shell characteristic of the adult abalone begins to be elaborated (Morse *et al.* 1979a, 1980b). Unlike the small, smooth, and symmetrical larval shell, this new shell grows into a flattened, ribbed, and ever-widening right-handed whorl, immediately recognizable as the typical abalone shell. The energy supply necessary for these dramatic metamorphic changes is still provided by the yolk carried by the larval embryo.

The natural requirement of abalone larvae for specific algal inducers of settlement and metamorphosis represents a natural "fail-safe" mechanism protecting the larvae from the very high mortalities that would result from settlement and metamorphosis in suboptimal habitats (Morse *et al.* 1979a, 1980a). The naturally required algae provide the abalone with specific biochemical inducers of settlement and metamorphosis; suitable food supplies capable of supporting early development and rapid growth; protective microhabitats and niches; and a source of camouflaging pigment, which is incorporated into the growing shell, thus reducing the visibility of the young abalone and protecting them from predators. The biochemical, developmental, and ecological mechanisms and significance of this process of induction of metamorphosis, triggered by specifically required biochemical inducers uniquely present at the surfaces of specific algal species, are described in more detail elsewhere (Morse *et al.* 1979a, 1980a,b; Morse 1982c; Morse and Morse, 1984).

Nutrition and Growth

With the completion of metamorphosis, the young abalone begins active feeding by grazing crustose red algal substrates for the sloughing algal cells and carbohydrate-rich mucus secreted from the algal epithelium. Epiphytic diatoms and other microscopic algae become increasingly important sources of nutrients as the abalone grows larger. Growth is relatively slow, but more or less continuous, as long as food is available.

Aquaculturists frequently identify the time of development of the first respiratory pore in the growing shell (at 1–3 months following metamorphosis) as the start of the juvenile stage, and the end of postlarval development. The juvenile stage is generally considered to last until the advent of sexual maturity, at about 1½ years of age, although these distinctions are somewhat arbitrary. Early in juvenile growth, thin-bladed species of macroalgae, such as *Ulva,* may become a significant part of the diet. At about 6 months of age, juvenile abalone have attained sufficient size to begin feeding on those macroalgae that will sustain them as adults; these frequently include the thick, kelplike brown algae, which require the rasping action of a robust radula for efficient ingestion.

Each species of abalone exhibits a unique range of algal preferences, which in some cases may not coincide with the relative efficiencies of the algae in supporting most rapid growth (Leighton 1966, 1968). Common algal foods for the U.S. species of abalone, depending upon location, season, and species, include the following: brown algae—*Macrocystis* (giant kelp), *Nereocystis* (bull kelp), *Egregia* (feather boa kelp), and *Eisenia* (sea palm kelp); red algae—*Gigartina, Gelidium,* and *Plocamium;* and green algae—*Ulva* (sea lettuce).

Whereas small juvenile abalone generally are highly mobile and cryptic, emerging from rock crevices to forage at night, most of the larger juveniles and mature adults of the species that inhabit the North American coast are more sedentary, establishing permanent positions on rock ledges or in crevices. By cleaning the area beneath them, these animals establish scars on the rocks which allow them a firm hold and greater protection from predators. Feeding primarily at night, these abalone catch drifting fragments of macroalgae being moved by the currents. When feeding, the abalone adopts a distinct feeding posture, extending its chemosensory tentacles, elevating the shell from the substrate, and lifting the anterior portion of the foot to face into the current. This activity is stimulated both by water currents (Shepherd 1973, 1975) and by chemical sensing of substances released from nearby kelp or other foods. When a drifting fragment of alga is contacted, either by the foot or the epipodial tentacles, the sides of the foot quickly fold toward the midline, grasping the plant frond securely and pulling it down beneath the foot (thus trapping it). The food then may be eaten immediately, or held for later consumption. When food is scarce, abalone will forage for food at night and then find and clean a new spot rather than return to their original location.

Typical growth rates for juvenile abalone in natural environments or in hatcheries, where food is a principal limitation, are on the order of 20–30 mm (ca. 1 in.) per year for the red abalone; growth rate more than double this rate have been observed when food supplies are significantly increased. Because feeding behavior is suppressed by high illumination, both feeding and growth rates are enhanced by reduced lighting (Shibui 1972; Uki 1981; E. Ebert, personal communication). Somatic growth and gametogensis are under reciprocal control; hence, somatic (meat) growth is slowed considerably after gametogenesis begins (after the first or second year), particularly when gametogenesis is accelerated by seasonal factors such as warming.

Diseases and Parasites

Significant problems of disease or parasitism are virtually unknown in abalone. It can be anticipated, however, that these may constitute problems in unusually stressed populations. Simple microbial overgrowth is a major cause of early mortality but can be readily controlled (Morse *et al.* 1979b; see Conventional Cultivation Technology Larval Development).

Predation

The principal source of mortality of abalone, from the postlarval to adult stage, is predation. Voracious small worms and crustaceans have been observed to decimate localized populations of postlarval abalone; these predators are common and especially cryptic (well camouflaged and hidden) on the natural recruiting algal substrates (Morse *et al.*

1979b). The widespread but patchy distribution of these minute but especially effective predators probably accounts for a major share of the reduction in newly recruited populations. Juvenile abalone of all species are vulnerable to a multitude of fish, crustacean, starfish, and molluscan predators; hence, these small abalone remain limited almost entirely to protected niches on the undersides of rocks or in deep crevices, foraging for food only at night.

Larger abalones become "emergent," occupying open rock faces in areas north or south of the principal range of the sea otter, *Enhydra lutris*. Sea otters are particularly voracious predators of the abalone (Woodhouse *et al.* 1977; Woodhouse 1982; Hardy *et al.* 1982; Cicin-Sain *et al.* 1982). This diving mammal effectively uses its lower jaw to pry abalone from the substrate when the molluscs are surprised in the elevated feeding posture. Even more surprising, the otter is frequently observed to harvest abalone with a tool. Using a rock grasped between both paws, the otter beats repeatedly on the top of the abalone's shell, until a large hole is opened through which the soft tissues are exposed and eaten (Houk and Geibel 1974). The rapid, unchecked growth of sea otter populations off the central California coast, following the establishment of the otter's protected status as an endangered species, has been widely considered to be the principal cause of the virtual extinction of the commercial and recreational fisheries that previously had flourished at Monterey and Morro Bay in California (Burge *et al.* 1975; Woodhouse *et al.* 1977; Hardy *et al.* 1982; Silva 1982; Cicin-Sain *et al.* 1977, 1982).

Predation by otters on the central California and Alaska coasts and the unchecked spread of otter populations both north and south remain the principal barriers to efficient abalone ranching, such as that practiced effectively in Japan (cf., Mottet 1978, 1980), and the principal threat to the development of technology for ocean ranching and fisheries restoration and enhancement in northern and southern California (Cicin-Sain *et al.* 1981). Where otter populations are large or likely to be introduced, ocean ranching of abalones (and other invertebrates) by direct seeding is precluded, and only the less efficient strategy of intensive cultivation in containments (on land or in the water) remains open to prospective abalone producers. This major constraint to efficient abalone cultivation in the United States is discussed in more detail later in this chapter.

Among invertebrate predators, the octopus is a major cause of abalone mortality (Tegner *et al.* 1981). Octopus of several species can pull small abalone from the substrate; large abalone are taken by the octopus drilling a hole through the shell with its radula and then injecting a paralyzing venom (Pilson and Taylor 1961). Starfish, rock crabs, lobsters, and predatory molluscs such as whelks also prey on smaller abalone (Cox 1962; Shepherd 1975). Fish predators of abalone include robust-jawed rockfish such as the cabezon *(Scorpaenichthys),* the sheephead *(Semicossyphus,* formerly *Pimelometopon)* (Cox 1962), and a number of rays (principally the bat ray, *Holorhinus*) (cf., Shepherd 1973). In Japan, the effectiveness of ocean seeding of hatchery-reared abalone is enhanced by communal control and intensive harvest of invertebrate and fish predators prior to planting of abalones in circumscribed areas (see Ocean Ranching section).

CONVENTIONAL CULTIVATION TECHNOLOGY

Most initial efforts to cultivate abalone commercially in the United States have been based on the large red abalone, *H. rufescens.* This species possesses several important advantages: relatively fast growth, excellent meat quality and value, hardiness in hatchery conditions, and ease of reproductive control. Recently, efforts have been made to expand production to the warmer waters of southern California, using the thermophilic but smaller green abalone, *H. fulgens,* (Leighton 1972; Morse *et al.* 1978; Morse 1980, 1981, 1984a,b); Leighton *et al.* 1981) and hatchery-produced hybrids of the red and green abalones (J. McMullen, personal communication).

The basic technology conventionally used for abalone cultivation in the United States is similar to that originally developed in Japan (cf., Ino 1952; Shibui 1972; Kan-no 1975), with some modifications and recent improvements to overcome remaining barriers to efficient production. Only an outline of this modified traditional technology will be present in this section. A more detailed discussion of the factors precluding economically efficient production using such conventional technology and of recent technological developments designed to overcome these problems is presented in the following sections.

Control of Reproduction

Brood stock may be taken from the wild or from hatchery-reared populations that have attained sexual maturity. Brood stock collected from the wild generally are acclimated for a period of a few weeks, during which the animals are held in tanks with good circulation of seawater, at a temperature to initiate and maintain gametogenesis, and with ample supplies of macroalgal food. Gravidity (readiness for spawning) may be judged by inspection of the degree of distention of the gonad (see Biology section), or by determining the inducibility of spawning under precisely defined conditions (Morse *et al.* 1977a, 1978) as described later.

There are two methods generally used for the induction of spawning in gravid abalones. One is the technique developed in Japan by Kikuchi and Uki (1974) in which gravid brood stock are exposed to flowing and heated seawater previously irradiated with ultraviolet light. Spawning may be induced within 3–16 hr. The other method presently in use was developed in our laboratory at the University of California at Santa Barbara (Morse *et al.* 1977a,b, 1978). In this method, gravid abalone are placed in containers of seawater (15°C, 59°F, for red abalone) that is made slightly alkaline (pH of about 9.1) by the addition of base and then adjusted to a concentration of 5-millimolar hydrogen peroxide by the addition of freshly diluted peroxide from a concentrated, stable stock solution. After 2.5 hr, the alkaline seawater containing hydrogen peroxide is replaced with fresh seawater at the original temperature. Approximately 3 hr after the original introduction of hydrogen peroxide, gametes are copiously released into the water, which is clean and free of chemicals.

These two methods apparently are dependent upon a common underlying mechanism (Morse *et al.* 1977b, 1978), which is discussed in more detail in the section on control of reproduction and seed production in thermophilic and other abalone species. A previously developed technique, based upon the prolonged dessication and heat-shock (Carlisle 1945), of brood stock is unsatisfactory because it causes the release of a large proportion of immature and defective gametes (Morse *et al.* 1977a, 1978).

Some workers spawn both male and female abalone in the same container, thus accomplishing fertilization as spawning occurs. Others (ourselves included) prefer to spawn the sexes separately, in order to maintain greater control over fertilization and thus prevent developmental abnormalities resulting from polyspermy (Morse *et al.* 1977b; Vacquier *et al.* 1981). Fertilization occurs rapidly; after a few minutes, the eggs are washed free of excess sperm, either by repeated decantation or by filtration through plastic screens. The fertilized eggs then are placed in containers of clean seawater so that they cover the bottom in a single layer. In our laboratory, embryological development of red abalone is allowed to proceed at 14°–16°C (57°–61°F). Regulation of temperature and water quality (to ensure freedom from microorganisms) is of the greatest importance for ensuring reproducibly high yields, normal organisms, and rapid development and growth of the resulting larvae and juveniles. These methods, utilized successfully by commercial hatcheries in the United States and abroad, have been described in detail elsewhere (Morse *et al.* 1977a,b, 1978, 1979a,b).

Use of hydrogen peroxide for the induction of spawning, and the methods described here for controlled fertilization and development, result in reliable, copious yields of gametes fully competent for normal fertilization and larval development.

Larval Development

In the development of red abalone at 15°C (59°F), hatching of the swimming larval trochophore occurs 18–24 hr after fertilization. At this time, the trochophores are either decanted or filtered free of the remaining egg debris. In our laboratory, the larvae then are placed in clean mesh-bottom containers and cultivated in continuously flowing, fresh, filtered, and UV-sterilized seawater maintained at 14°–16°C (57°–61°F) with a thermostat (Morse *et al* 1979b). Development into competent veliger larvae occurs 7 days after fertilization at 15°C (59°F). As mentioned previously, the larvae are nourished from the original yolk, and thus require no feeding during this time.

Most of the mortality and developmental abnormalities frequently encountered during the development of abalone larvae in hatcheries results from microbial growth in the larval cultures (Morse *et al.* 1979b). When this growth is minimized by precise thermostatic control to retard microbial division and by use of continuously flowing, filtered, and UV-sterilized seawater to reduce microbial input and accumulation, mortality during larval cultivation can be kept virtually nil. Under these conditions, with very good fresh seawater, larvae are routinely cultivated in densities of 20–30/ml, with virtually no losses or retardation of development. A new commercial hatchery near Santa Barbara is employing these procedures with similar results.

Larval Settlement, Metamorphosis, and Early Postlarval Development and Growth

When the larvae have become competent to settle and metamorphose, they generally are transferred in hatcheries to "seasoned" tanks or containers in which a thin layer of benthic diatoms, bacteria and microalgae have been allowed to develop; these often contain patches of crustose coralline and noncoralline red algae, as well. Tanks containing these red algae thus may contain variable supplies of the required natural inducers of settlement and metamorphosis, as well as food supplies ample to support early postlarval development and growth.

The size, shape, and arrangement of the settling tanks varies in different hatcheries. In general, however, these tanks usually are made of fiberglass, are rectangular and placed in a stacked array, are provided with filtered running seawater, and are lighted by fluorescent fixtures during part of each 24-hr period to encourage diatom growth. The density of animals in settling and early grow-out containers also is unique to each of the different hatcheries, depending upon local water quality, the expected mortality, and the particular feeding strategy that is employed. Some hatcheries rely upon natural diatom growth to provide food during this early period; others supplement this regime with additions of intensively cultivated monotypic or mixed microalgae. Postlarval abalone have been successfully reared on a number of small species of diatom, including *Navicula* spp.; *Cocconeis* spp.; *Amphora* spp.; *Nitzschia* spp.; and the small unicellular gree algae, *Tetraselmis* spp. (Sagara 1975; Kan-no 1975; Leighton 1977; Sedgwick 1978; Ino 1980).

Growth of Juveniles

When the juvenile abalone grown on diatoms and microalgae reach approximately 5 mm in size, they usually are transferred to larger "grow-out" tanks where they are introduced to macroalgal species such as *Egregia* and *Macrocystis*. Since diatoms and microalgae also are present on the walls of these tanks, as well as on the fronds of the macroalgae, the transition from microalgal to macroalgal diet is a gradual one. By the time the young abalone reach 10–15 mm (ca. 0.5 in.) they are capable of sustained growth on a diet consisting exclusively of macroalgae.

In the existing commercial hatcheries, mortality of young abalone less than 5 mm in size is known to be very high, although specific data often are not available. Japanese abalone growers have found that typical survival to the 5-mm length is 0.1–1%; subsequent survival from 5 mm to 3 cm is 90% (Sagara 1975). When provided with a good water flow, adequate oxygen, and ample food supply, juvenile abalone are able to thrive in high density. In Japanese culture systems, with gravel-bottom tanks for rearing animals from 5 mm to 3 cm, stocking densities are 2000–2500 abalone m^2 (186–232/ft^2) of bottom area (Inoue 1976).

Commercial abalone growers in California are not attempting to grow animals to the large sizes typical of the commercial wild harvest; efforts are concentrated instead on production of animals that are 5–10 cm (2–4 in.) long. Abalone of this size can be sold both in the Asian market and in the domestic restaurant market. Typical growth rates for young red abalone are approximately 20–30 mm (ca. 1 in.) per year; therefore, growing time to marketable size for this species is 2 to 5 years, using presently available technology. Growth rates vary substantially, depending upon species, age, algal foods provided, and temperature (Leighton 1974; Leighton *et al.* 1981; Uki *et al.* 1981).

Techniques for increasing the survivability and growth rate of postlarval and juvenile abalone are presently being tested. Some commercial hatcheries have successfully adopted new techniques developed specifically for this purpose (Morse *et al* 1979b; Morse 1980, 1981, 1984a,b). Enhancement of growth rate by use of artificial diets is being examined by at least one California firm (J. Weddington and J. Charest, personal communication). Artificial diets for young abalone were developed originally by Japanese investigators. A principal ingredient of these artificial foods is sodium alginate, extracted from the giant kelp, *Macrocystis,* and used as both a binder for the bulk constituents of the diets and as a feeding stimulant for the abalone. Many different ingredients have been tested for the bulk portion of the diet, including fish meal, marine yeast, brewers yeast, terrestrial plant fiber, and the green alga, *Chlorella.* Normal growth was observed when the total crude protein in the diet was greater than 20% (Sagara and Sakai 1974). This work has been reviewed by Mottet (1978).

The aquaculture of abalone is still largely in a research and development phase, although some firms recently have reported net profits from sales of hatchery-produced stock. Three different approaches to grow out presently are being tested by various U.S. growers. Some have chosen on-land strategies for grow out, rearing abalone with intensive feeding in tanks, raceways, or ponds. This approach requires a high initial investment for land and facilities and involves high operating costs for pumping (energy) and maintenance; in return, a high degree of control and relative ease of maintenance are obtained. Other growers now are experimenting with containment systems for grow out in the open ocean or protected harbors. In these systems, abalone are placed in habitats or cages that are suspended from buoys or piers, or anchored to the ocean floor. Containment systems offer the prospect of high-density cultivation, excellent water quality, and reduced costs in comparison with facilities that require constant pumping of seawater. The disadvantages of this approach are the difficulties of feeding contained animals that require macroalgae; maintenance problems; and potential damage to the structures and loss of stock due to storms and the action of boats, fishing nets, etc. An alternative strategy under investigation by both the abalone fishing and aquaculture industries, with assistance from state and federal programs, entails the open-ocean planting of hatchery-reared seed abalone on lease holdings obtained from the state. This form of ocean ranching requires minimal investment in grow-out facilities. Although the final percentage recovery from uncontained plantings is expected to be low (as a result of predation and other losses), the potential economies of scale require further investigation (see Ocean Ranching).

BARRIERS TO EFFICIENT PRODUCTION USING CONVENTIONAL TECHNOLOGY

Attention recently has been focused on legislative constraints on the development of profitable abalone and other marine invertebrate aquaculture industries in the United States (cf., National Academy of Sciences 1978). While frustrations over such bureaucratic impediments may be, to a large extent, well justified, these also have served to distract the attention of new aquaculture entrants, investors, the general public, and governmental officials from the far more serious *technological* barriers and unsolved problems that remain as major impediments to the profitability of abalone aquaculture.

These problems are particularly characteristic of the less-efficient, traditional methods of cultivation developed one or two decades ago, and still in wide use in firms and government programs in the United States and abroad. Significant remaining biological and physical barriers, which have until recently prevented the emergence of profitable abalone production in this country, are discussed in this section. Recent investigations, developments of new solutions to these problems, and the resulting improved prospects for future industrial development of abalone production in the United States then are described in the remaining three sections of this chapter.

Defective Juveniles and Seed with Low Early Survival

As noted previously, abalone larvae require contact with a specific biochemical substance, normally found on the surfaces of the recruiting red algae, for efficient induction of normal settlement, metamorphosis, rapid development, and the start of juvenile growth. Traditional hatchery technology, developed before this natural requirement had been identified or its biological significance recognized by aquaculturists, generally has omitted the regular use of this specific natural inducer of abalone metamorphosis and development.

The low percentage of abalone larvae that settle and metamorphose in the absence of adequate supplies of the required morphogenetic inducer generally reflect a settlement response inefficiently triggered by the presence of bacteria and protozoans in relatively contaminated cultures (Morse *et al.* 1979b). These microorganisms subsequently produce high mortality and developmental abnormalities in the few remaining survivors (Morse *et al.* 1979b). When abalone larvae are cultivated in clean seawater and clean vessels, in the absence of large populations of contaminating microorganisms (or added red algae or other inducers), settlement and metamorphosis do not occur. In some cases, incidental but varying (and generally insufficient) amounts of the required inducer may be present in hatchery systems as a result of the growth of red algae encrusting settlement tanks. However, the use of this alga itself is generally unsuitable for efficient hatchery production, as the alga is invariably infested with large numbers of voracious micropredators (worms and crustaceans), which quickly decimate populations of new abalone recruits (Morse *et al.* 1979b; Morse 1981).

In the absence of sufficient supplies of the required natural inducers, then, survival through metamorphosis and early development is very low. Survival from the planktonic larval phase to 1 month following metamorphosis generally is 0.01–1% when abalone are cultivated without the addition of a sufficient supply of the required biochemical inducer. This low survival rate has been verified in commercial-scale cultivation using traditional technology representative of that in several of the hatcheries now operating in the commercial and governmental sectors in California (Leighton *et al.* 1981).

The importance of a natural requirement for morphogenetic induction in abalone larvae, often insufficiently satisified by traditional hatchery technology, was obscured by the optimism first generated by the high fecundity of the abalone. Aquaculturists practicing traditional methods of cultivation have been encouraged to believe that this fecundity can offset the low survival and low efficiency of metamorphosis and early postlarval developmental obtained with conventional technology. Because the spawning of a single pair of adult abalones will produce as many as a few million competent larvae, a low percentage of survival through metamorphosis and early postlarval develpment might appear, superficially, to provide an adequate supply of juveniles and seed. This is not the case, however.

The true significance, to the commercial cultivation of abalone, of the natural requirement for a specific biochemical inducer of settlement, metamorphosis, and development lies not only in its control over the number of survivors of the metamorphic transition, but in its control of subsequent successful postmetamorphic development, growth, and survivability. Those few juveniles and seed that survive cultivation when inadequate amounts of required morphogenetic inducer are present are developmentally retarded compared to siblings that have received adequate inducer (Morse 1981, 1984a,b). Seed produced in hatcheries with uncontrolled or suboptimal supplies of the required morphogenetic inducer (ie., in hatcheries relying on what is traditionally called "natural settlement") thus show a high incidence of shell and other growth deformities. The resistance of such seed to the trauma of transport and handling is low, and survival of this seed in trial ocean plantings has been disappointingly poor (Morse 1984a, 1984b).

Coastal Pollution

The various problems and expense involved in the location and establishment of hatcheries on coastal sites with access to high-quality seawater are not unique to abalone cultivation, and are discussed in detail elsewhere in this volume. A major barrier to the successful development of commercial abalone cultivation in the United States, however, is the susceptibility of the abalone in its early developmental stages to the high concentrations of pollutants now present in the protected semi-urban bays in which hatcheries first were located by pioneering industrial entrants several years ago.

This problem, now recognized to be severe, was originally underestimated in part because the data first available indicated that abalone larvae and adults are relatively resistant to the presence of toxic pollutants. However, it recently has been found that the critical recognition of and successful response to the required morphogenetic inducing signal by abalone larvae are far more sensitive to interference from heavy metals, chlorinated hydrocarbon pesticides, and other common pollutants than is the simple survival of the larvae or the adults (Morse *et al.* 1979b; Morse 1980, 1981). Thus, previous estimates of sensitivities derived from bioassays measuring only simple survival failed to identify the high sensitivity of the critical developmental processes of abalone to pollutants in hatcheries. Before the extent of this problem and its impact could be assessed, the early hatcheries were built. Levels of interfering pollutants in urban bays and agricultural drainage areas, in which some of these pioneer hatcheries had first been located in California, have increased markedly in recent years. As a result, a number of the present commercial or developing commercial hatcheries now face serious operational limitations imposed by coastal pollutants of industrial, agricultural, and urban origin.

Unreliable Supplies of Seed and Juveniles of Warm-Water Species

In addition to the mounting restrictions on cost-effective hatchery siting resulting from increased coastal pollution, the U.S. abalone fisheries and the related cultivation industry have faced increasing economic pressure resulting in a southward migration in recent years (see Introduction, The Declining Fishery). Unchecked predation by rapidly growing populations of sea otters, protected now by federal and state laws (see Biology, Predation) has virtually eliminated the commercial and recreational fisheries once based in central California. Predation by otters and legal closure of waters in northern California also preclude low-cost ocean ranching of hatchery-produced seed abalone in these areas (Cicin-Sain *et al.* 1982). In such areas, only the more cost-intensive and labor-intensive methods of cultivation in protected containments (on land or in water) remain possible for commercial production of abalone.

These pressures have combined in recent years to force the movement of the U.S. abalone industry southward (Cicin-Sain *et al.* 1977). A number of new abalone aquaculture firms have been established in the Santa Barbara area and other southern California locations in the past few years; many of these companies now are exploring the feasibility of ocean-ranching strategies, using ocean bottom areas leased from the state. This recent trend has produced a shift in the focus of new industrial efforts from the central California area, where these efforts first began, to southern California.

Successful cultivation and ranching of abalone in the warmer waters of southern California, particularly in the most southern portion of this range, will depend upon the ability to produce warm-water-tolerant (thermophilic) species capable of good survival and growth in these areas. The largest thermophilic species of principal commercial value include the pink abalone *(H. corrugata)*, the green abalone *(H. fulgens)*, and the black abalone *(H. cracherodii)* (see Principal U.S. Species). There also has been considerable interest in the possibility of commercial hybridization of these species with the larger red abalone *(H. rufescens)* to produce thermophilic stocks with more desirable meat value, growth, etc. Until recently, however, efforts in the industrial and governmental sectors had failed to produce reliable supplies of seed and juveniles of these warm-water abalones (Tegner *et al.* 1981). The recent successful development of techniques to overcome this problem is described later in this chapter.

Slow Growth and Inefficient Food Conversion

Abalone aquaculture industries in the U.S. and abroad face serious economic difficulties resulting from the slow rate of growth and the inefficient feeding of abalones (Staton 1981; Morse 1981, 1984a,b). Growth rates of abalones are intrinsically and significantly lower than those observed for many faster-growing molluscs such as oysters. Typical growth rates for the relatively large and fast-growing red abalone are 20–30 mm (ca. 1 in.) per year; thus it takes 2–5 years of growth for animals to reach market size, depending upon the market chosen (Staton 1981; J. McMullen, personal communication). Furthermore, the growth rates observed within production crops in hatcheries are extremely heterogeneous, thus seriously complicating handling and marketing procedures (Staton 1981; Morse 1981, 1984a,b). Growth rates two to three times greater than the mean rate occasionally are observed in some individuals grown in hatcheries or in the wild; rates of growth only 20–30% of the mean rate also are common in hatchery situations (Oba *et al.* 1968; Morse 1984b). For reasons discussed later, it can be concluded that these differences are not primarily of genetic origin, but reflect instead the differences between individuals in physiological and nutritional status (Morse 1981, 1984a).

In addition to low and variable growth rates, abalones also show low and variable food-conversion efficiencies. Although efficiencies near 15% have been reported for very small, rapidly growing juveniles (Kikuchi *et al.* 1967), conversion efficiencies of 5–10% are more typical for the greater period of juvenile growth in commercial hatcheries in the United States and Japan (Staton 1981; Kikuchi *et al.* 1967). These low efficiencies may be compared with the maximal value of nearly 30% seen in the nongametogenic Japanese oyster, *Crassostrea gigas,* and other rapidly growing filter-feeding molluscs.

Also basisically related to these problems are the low and variable rates of synthesis of protein and glycogen—the constituents of the meat principally responsible for its nutritional value, flavor, and overall commercial value (Morse 1984a,b). As a consequence, the observed meat:shell ratio of hatchery-produced animals is relatively low and quite variable, thus presenting further significant economic problems to commercial growers (Staton 1981; Morse 1984a,b). It is likely, too, that susceptibility to trauma and stress is increased in slowly growing animals with low rates of accumulation of protein and glycogen stores.

It should be noted that significant progress in the development of feeds and feeding strategies designed to overcome these problems has been claimed by some companies in the private sector. However, as information concerning this development generally is protected as proprietary, few details are available other than the general information already discussed (see Conventional Cultivation Technology, Growth of Juveniles). To balance these claims, it also must be noted that no commercial or government hatcheries have yet brought abalones to market with considerably reduced production time, less heterogeneity of size, or with consistently improved meat weight or value.

Restricted Success of Ocean Grow Out

Growth in the ocean to market size of small juveniles produced in hatcheries is the typical procedure in most commerically successful molluscan shellfish industries throughout the world. This basic strategy uses the capital-intensive and labor-intensive hatchery operation to minimize early mortality and economically produce large numbers of healthy, small seed; final grow out then is accomplished in the ocean, to minimize the long-term investment of intensive effort and capital costs required for production. Although this strategy has had some success (albeit based upon continuing subsidization) in Japan (Inoue 1969; Kan-no and Hayashi 1974; Kan-no 1975; Saito 1979, 1981; Miyamoto *et al.* 1982), major problems have until recently restricted its profitable implementation in the United States.

As mentioned already, heavy predation by sea otters, and legislative closure of northern California waters, continues to preclude open-ocean ranching north of Point Conception, California. In these areas, only the intrinsically more labor-intensive strategy of cultivation in containments is possible for abalone production. While the use of such containments in the ocean may help reduce rquirements for costly coastal land facilities and the continuous pumping of seawater through a land-based production facility, the nutritional requirements of abalones pose a major limitation to the practicability of containment cultivation. Unlike filter-feeding bivalve molluscs, for which such methods first were developed, abalones normally feed on macroalgae; mesh capable of containing the abalones and restricting predator access also prevents efficient access of macroalgal foods. Thus, intermittent labor-intensive feeding of abalone in such containments is required on a continuing basis.

Several commercial firms have experimented with various designs of abalone containments moored, tethered, anchored, suspended from oil-drilling platforms, or otherwise held in the ocean; however, only one, The Ab Lab in Port Hueneme, California, has reported success. This firm now is producing a small but continuous supply of abalone from a series of 208-liter (55-gal) plastic containments moored from a pier in protected water (J. McMullen, personal communication). This location facilitates frequent feeding with kelp, and cleaning, and serves to protect the array from ocean storms and shipping. As a further innovation, The Ab Lab is supplying a new market in California for the small ("gourmet-size") abalone produced after only 2 years of cultivation. Development of this market was accomplished by John McMullen, the proprieter of Ab Lab, with the assistance of John Richards of the University of California Sea Grant Cooperative Extension Service. Other firms attempting containment cultivation of abalone in less-protected ocean environments report several major problems: difficulties in gear-handling and logistics for feeding with kelp and in maintaining feeding schedules (because of weather and unreliability of access by boats and diving); fouling of containment meshes, which restricts water flow; occasional decimation of stock by octopus that invade damaged containments; and damage or loss of containments from storms, shipping, or fishing nets.

In southern California, cost-effective seeding of the ocean bottom (as practiced in Japan, China, France, and other countries) is possible on leaseholdings obtained from the state for this purpose. Seed animals placed on the ocean bottom can obtain food by unrestrained foraging, but must be harvested by diving. Migration rates are relatively low in areas with good food abundance (cf., Saito 1979, 1981; Miyamoto *et al.* 1982); predation is the principal source of losses in this method. Although the existing sea otter range is hundreds of miles north of the commercial abalone beds and planting sites now in use, the potential threat of this predator is a principal concern to the industry. The abalone aquaculture and commercial fishing industries, other aquaculture and commercial fishing industries, recreational fishermen, the California Abalone Association, other industrial and public associations and the California Department of Fish and Game currently are working to obtain a legislative plan for zonal management of the sea otter in order to protect the remaining abalone and other shellfish industries that might utilize ocean ranching in southern California waters (Smith 1982; Cicin-Sain 1982; Cicin-Sain *et al.* 1982).

At the present time, predation by fish and invertebrates (see Biology, Predation) poses the greatest hindrance to development of cost-effective abalone seeding and harvesting operations in southern California. In Japan, communal control of geograhically confined planting sites by the abalone cooperatives makes possible the selective reduction of predators prior to the planting of abalone seed. However, predator control in the more open U.S. waters generally is considered impractical. Additional major limitations to economical ocean ranching include the patchy distributions of suitable macroalgae and the drifting fronds from these plants (Saito 1979); competition for macroalgae from urchins (Kan-no 1975); and the need for labor-intensive harvesting by diving. Strategies for the development of cost-effective abalone seeding now are under investigation by the private sector (see Ocean Ranching, United States).

BIOCHEMICAL AND GENETIC ENGINEERING TECHNIQUES TO IMPROVE PRODUCTION

Recent research at the University of California, conducted in cooperation with interested participants from the growing U.S. abalone aquaculture industry and the abalone fishery, has been aimed at developing practical methods to improve the reliability and cost-effectiveness of control over those basic biological processes that presently limit the efficiency, yield, and profitability of abalone production. Areas in which useful applications have been developed include the control of reproduction, larval development, enhanced survival, success at metamorphosis, rapid juvenile development, and problems of interference from pollutants. Some of these recently developed methods already are in wide use in abalone cultivation in the United States and abroad. These new techniques also are being used to improve production of a number of other species of gastropods, oysters, scallops, mussels, and clams. New techniques developed to overcome the remaining barriers to profitability (see previous section) are discussed in this and the following section. More detailed information is available in other, specialized publications (Morse et al. 1977a,b, 1978, 1979a,b, 1980 a–c; Morse 1980, 1981, 1984a–c).

Biochemical Induction of Settlement, Metamorphosis, and Rapid Juvenile Development

The natural requirement of abalone larvae for a specific, biochemical inducer of normal settlement, metamorphosis, and rapid subsequent juvenile development can be fully and efficiently satisfied by provision of simple and inexpensive substances first identified in the naturally required red algae (Morse *et al.* 1979a,b, 1980 a–c). Of these related small inducers, the simplest, most effective, most reliable, and most inexpensive for use in aquaculture is the readily available amino acid γ-aminobutyric acid, or GABA. The simple addition of a low but precisely measured concentration of GABA to a culture of competent abalone larvae (4–7 days posthatching, dependent upon species and temperature) will induce rapid, synchronous, and completely normal settlement; attachment to the substrate, metamorphosis, and rapid subsequent juvenile development and growth in virtually 100% of the larvae. In the absence of GABA or other related biochemical inducer, no more than a few percent of the larvae show such induction. The potential for improvement of cultivation efficiency and yield that is afforded by this new technique is illustrated directly in Fig. 9.3.

Similar, efficient induction of settlement and metamorphosis can be achieved by providing competent larvae with the crustose coralline red algae that they normally require in their ocean habitat, or with specific proteins purified from these algae; however, production of these more complex inducers is both time-consuming and costly, and high mortality subsequently results from predation by microscopic fauna associated with the intact alga as noted earlier (Morse *et al.* 1979b). It is thus far more convenient, efficient, and inexpensive to induce settlement and metamorphosis—with minimal mortality—through the

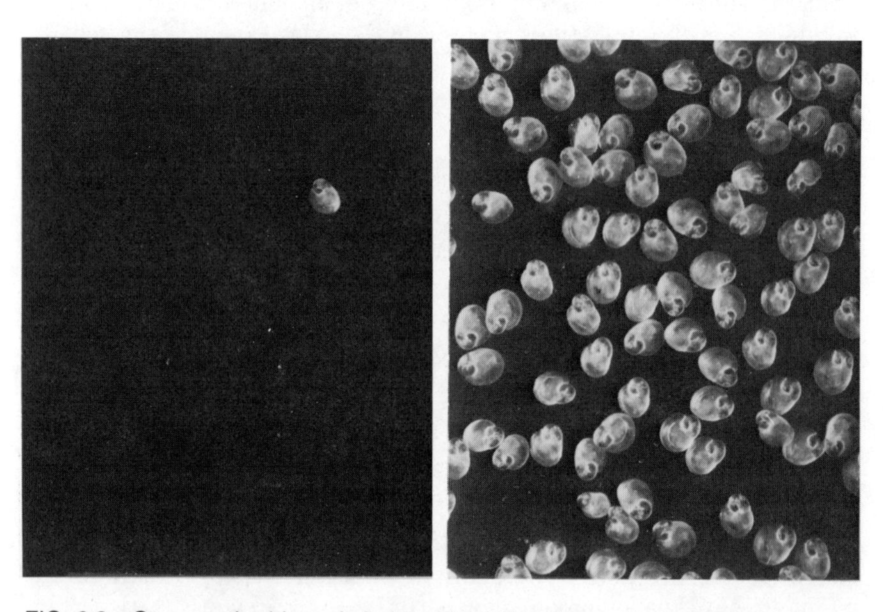

FIG. 9.3 One-month-old seed abalone *(Haliotis rufescens)* produced using conventional hatchery technology *(left)* and new technology with induction by GABA *(right)*. These siblings were grown in parallel, with conventional feeding with diatoms and microalgae. Survival to 1 month with induction by GABA was 93%; without GABA, it was 1%. Average length of seed is approximately 2 mm.
From Morse (1984a) with permission from Elsevier Scientific Publishing Company, Inc.

use of GABA. This and similar amino acid-derived compounds are proving effective and widely useful for the reliable induction of settlement, metamorphosis, and rapid juvenile development in a number of commercially produced abalone species and other valuable molluscs (Morse *et al.* 1979a,b; Morse 1980, 1981, 1984a–c).

The significance of these findings, and the potential advantages of this new technique, extend beyond the increased yield, at lower cost, of very large numbers of seed and juvenile abalone. The maximal activation of biochemically controlled growth achieved through the use of GABA produces healthy and rapidly growing juveniles, in marked contrast to the low survival rates and production of developmentally retarded seed and juveniles that are typical of cultivation using older, traditional hatchery technology without the use of GABA or some other required morphogenetic inducer. Direct applications of this new technique now are being made, particularly in the newer abalone aquaculture firms in the United States (and in governmental programs in Australia, Europe, and other countries), to achieve improvements in the economic efficiency and yield of containment cultivation; to assay for interfering pollutants; to extend the number of species and geographic locations suitable for practicable cultivation; to identify factors that can accelerate abalone growth; and to develop and test new methods for ocean ranching. Each of these applications is discussed in the next few sections.

New Bioassay for Evaluation of Potential Hatchery Locations and Seawater Sources

A continuous supply of fresh, flowing, partially sterilized pollutant-free seawater is needed for optimal production of healthy, developmentally competent larvae (Morse *et al.* 1979b). As described previously (see Conventional Cultivation Technology, Larval Development), the very high densities of healthy abalone larvae produced in our laboratory and in some of the newer U.S. aquaculture firms on a regular and frequent basis only can be supported by continuous flow-through cultivation systems, which are free of pollutants. Interruption of this seawater flow through the developing larval cultures for more than one day at any time during the typical week-long cultivation period (required for larvae to develop from hatching to full competence for metamorphosis) usually produces developmental abnormalities. These abnormalities appear as failures in efficient settlement or normal metamorphosis, with subsequent abnormally high mortality and retarded development.

In addition to this requirement for pollutant-free water during the cultivation of larvae, the induction of normal settlement, metamorphosis, and rapid postlarval development and growth is especially sensitive to interference from pollutants. Heavy metals, chlorinated hydrocarbons, pesticides, and other common pollutants of impacted coastal waters seriously reduce the efficiency and yield of this essential metamorphic transition, and thus all but abolish the potential for cost-effective hatchery operation (Morse *et al.* 1979b; Morse 1981). Even after development of the more resistant juveniles, continuously flowing seawater free of pollutants still is required to stimulate optimal feeding and growth.

Various "scrubbing" systems for the repurification of recirculated seawater have proven only marginally effective in our experience. Thus, there can be no substitute for the continuous availability of pollutant-free seawater for efficient production of healthy juvenile and seed abalones.

As an aid to aquaculturists in evaluating the quality of seawater sources (and, thus, the suitability of potential hatchery or seeding locations), a sensitive new bioassay recently has been developed. This bioassay is based specifically on the high sensitivity of GABA-induced settlement and metamorphosis of abalone larvae to interference from a wide spectrum of common coastal pollutants (Morse *et al.* 1979b; Morse 1980, 1981). This new procedure is far more sensitive, reliable, rapidly quantifiable, and directly relevant to the siting of prospective abalone hatchery or seeding operations than previously available assays based on the survival of the more resistant larvae, juveniles, or adults of abalone or others species. This procedure also is more directly useful than specific chemical assays for individual pollutants, as the bioassay measures the interference from *all* components in the seawater with one of the most sensitive and critical developmental processes in abalone production. The assay thus can be used to test the suitability of any potential hatchery site. Details of this simple new procedure have been published elsewhere (Morse *et al.* 1979b).

Control of Reproduction and Seed Production in Thermophilic and Other Abalone Species

Analysis of the underlying biochemical mechanisms controlling reproduction revealed that hormone-like prostaglandins normally regulate and initiate spawning in abalones and a number of other mulluscan species (Morse *et al.* 1977a,b, 1978). Thus, spawning of gravid adults can be induced simply by the addition of trace amounts of prostaglandins to the surrounding seawater. Even more reliably and inexpensively, spawning can be induced at will by activation with hydrogen peroxide of the natural enzymatic synthesis of prostaglandins in gravid animals. Peroxide activation of prostaglandin-dependent spawning has been found widely useful for obtaining synchronous and copious release of fully competent gametes (both eggs and sperm) in a large number of species of abalones, oysters, scallops, clams, and other valuable molluscs, which are objects of cultivation efforts throughout the world (Morse *et al.* 1977a,b, 1978; Morse 1980, 1981, 1984 a–c).

Because this method has proved to be so reliable, the inducibility of spawning with hydrogen peroxide can be used in studies to identify other requirements for gametogenesis and spawning in those species that previously had proved refractory or difficult to spawn using other methods. Using this new procedure, research conducted in cooperation with several molluscan hatcheries has defined the conditions required for reliable gametogenic conditioning, spawning, larval development, metamorphosis, and efficient seed production in valuable thermophilic (warm-water) species—the green abalone, pink abalone, and black abalone (Morse 1981, 1984a–c; Hooker and D. E. Morse, manuscript in preparation). These species are of major importance to the U.S. industry in its recent move southward; they also are the principal species of remaining industrial harvest in the warm waters of Mexico. Previous efforts using other techniques had failed to achieve reliable control of reproduction in these species (Tegner *et al.* 1981).

It has been found that efficient and accelerated gametogenesis in these thermophilic species requires saturation feeding for 1 to 2 months at 18°–25°C (64°–77°F), depending upon the species, feeding rate, water quality, and water exchange rate. Spawning of these gravid animals then is reliably inducible with the hydrogen peroxide procedure described earlier if the animals are held at 20°–25°C (68°–77°F), depending upon species, rather than at the 15°C (59°F) typically used in spawning the cool-water red abalone. Larval development, induction of settling and metamorphosis, and production of healthy seed, using GABA as the inducer of metamorphosis, also were found to be rapid and efficient, with very high survival, when conducted at elevated temperatures of 17°–22°C (63°–72°F) (Morse, 1982a–c; N. Hooker and D. B. Morse, manuscript in preparation). Conditions for efficient gametogenic conditioning, spawning, larval development, metamorphosis and seed production of a number of other commercially important species, including several cold-water species as well, similarly have been defined using the new hydrogen peroxide and GABA techniques. Recent results with *H. discus hannai* will be described in detail elsewhere (D. E. Morse, A. D. Morse, and F. Zhang, manuscript in preparation). Species in which production has been improved by use of hydrogen peroxide to induce spawning and GABA to induce metamorphosis are listed in Table 9.1.*

Successful application of these techniques to the efficient production of interspecific hybrids, bred for their extended range of cultivation, meat quality, growth rate, or the genetic marking of stock, also has been demonstrated in the industrial sector (J. McMullen and D. Leighton, personal communication) and in related research (Vacquier *et al.* 1981, Leighton and Lewis, 1982). These recent results thus have increased the economic efficiency of production of a large number of abalone species; have increased the number of species and hybrids successfully cultivated; and have extended the geographic range of potentially efficient abalone cultivation in the United Studies and abroad.

TABLE 9.1. ABALONE SPECIES IN WHICH PRODUCTION IS IMPROVED BY HYDROGEN PEROXIDE AND GABA

Haliotis	Common Name	Industry Base
rufescens	Red abalone	USA
fulgens	Green abalone	USA, Mexico
corrugata	Pink abalone	USA, Mexico
cracherodii	Black abalone	USA, Mexico
kamtschatkana	Pinto abalone	USA, Canada
tuberculata	Ormer	Ireland, France, UK
midae	African abalone	South Africa
iris	Blue abalone	New Zealand
australis	Yellowfoot abalone	New Zealand
virginea	Whitefoot abalone	New Zealand
ruber	Blacklip abalone	Australia
discus hannai	Ezo or Hokkaido abalone	Japan, China

Source: Morse 1984a; reproduced with permission from Elsevier Scientific Publishing Company, Inc.

Increased Nutritional Efficiency and Acceleration of Growth

Recent analyses have indicated that the basis of the low and variable rate of abalone growth and the related low efficiency of food conversion, discussed previously, is hormonal, and thus may be alle-

*We are especially grateful to our colleagues in industry and research who have assisted us in developing this information, particularly including Mr. John McMullen (The Ab Lab, Port Hueneme, California), Dr. Colin Sumner (Tasmanian Fisheries Development Authority, Australia), Dr. L. J. Tong (New Zealand Ministry of Agriculture and Fisheries), Dr. Robert LaTouche (Shellfish Research Laboratory, Conway, Ireland), Dr. R. Caldwell (Washington Dept. of Fisheries), Dr. K. Chew (Univ. of Washington), and Mr. J. Steinbeck (Calif. Polytechnic State Univ.).

viated by an approach utilizing modern methods of biochemical and genetic engineering (Morse 1981, 1984a–c). As efforts in this area of research are just beginning, only a preliminary account is given here.

Analysis of the natural requirements for rapid early juvenile growth in abalones recently revealed that two peptide hormones (insulin and growth hormone), which normally are produced by the molluscs in very low and variable concentration, can significantly accelerate the growth of abalones, for a short period of time, when provided exogenously in very low concentration (Morse 1981, 1984a,c). Even more significantly, the extreme heterogeneity of sizes and growth rates generally observed in the cultivation of abalones was shown in these experiments to result not from genetic variation, deficits in sperm or egg, or inadequate nutrition (as frequently cited in the semipopular aquaculture literature, but from physiological deficiencies in cultivation conditions, which may be corrected or enhanced by external biochemical (hormonal) control. These results thus indicate that genetic breeding programs based upon selection of abalones for apparently desirable growth rate properties cannot efficiently be started until the physiological and biochemical requirements for optimal growth are fully defined.

Research now in progress is attempting to further characterize the mechanisms of hormonal control that normally regulate nutrient uptake and utilization, and the synthesis of meat protein and glycogen (Morse 1984a). That prolonged acceleration of abalone growth may be possible is indicated both by the observations of hormone-stimulated accelerated growth (with significantly reduced heterogeneity in size) in young abalones cited above and by the widespread observations that some abalones (in both hatcheries and natural environments) have growth rates at least three times greater than the average rate of their siblings.

Efforts now are being made to identify further the specific peptide hormones normally made by the abalone for endogenous regulation of nutrient utilization, protein and glycogen synthesis, and growth. This information then may be used, in conjunction with recombinant DNA technology, to purify and clone the genes of the abalone that code for these essential, growth-accelerating hormones. Toward this end, the first potentially useful "abalone gene banks," or "libraries," of cloned abalone genes have been established, from which the requisite hormone-producing genes can be identified and purified (Morse 1984a,b).

This genetic engineering approach may make it possible to produce at a reasonable cost sufficient amounts of the precise peptide hormones normally made by the abalone to safely and reliably accelerate abalone growth on a practical, industrial scale. Successful enhancement of nutrient utilization, protein and glycogen synthesis, and rate of growth in the abalone also should significantly reduce the sensitivity to trauma and stress typical of poorly growing animals. Furthermore, use of natural (homologous) hormones should preclude side effects that might result from the use of foreign (heterologous) hormones.

A similar approach has formed the basis of the recent technological revolution in terrestrial agricultural production and research, and in the commercial production of growth hormones and resistance factors in medicine and the pharmaceutical industry. It also may be possible, ultimately, to use cloned, amplified, and modified genes to directly engineer desirable modifications for reintroduction into the genome (chromosomes) of the animal under cultivation. The outcome of these experiments, now in their early stages, may hold particular significance for commercial cultivation of abalone in containments, which is now severely limited by inefficient feeding and slow growth.

OCEAN RANCHING

Ocean ranching, or the planting of hatchery-produced seed for later harvest from the ocean bottom, is the principal method of abalone production under active investigation and semicommercial development in governmentally subsidized programs in Japan, China, France, Australia, Mexico, South Africa, and a number of other countries. The greatest advances in ocean ranching of abalone have been made in Japan, where this practice has been heavily subsidized by long-term governmental support for the mass production of hatchery-reared seed. In the United States, exploratory efforts begun in cooperative research undertaken by the University of California and the California Department of Fish and Game (Tegner *et al.* 1981) now are being followed by a new initiative from the private sector, with governmental assistance being provided at several levels.

Japan

In Japan, governmental subsidies support the mass production of abalone seed for ocean ranching; recoveries are facilitated by communal control over the shallow bays used for planting by local abalone-producing cooperatives (Inoue 1969). Harvesting of selected invertebrates (starfish, octopus, and urchins) prior to the planting of abalone seed reduces losses from predation and competition (Kan-no 1975). The effectiveness of abalone seeding in Japan has been shown to depend upon the availability of suitable local supplies of macroalgae (Saito, 1979); hence, abalone seeding is supported there by a strong and coordinated program of planting of desired macroalgal species, coupled with urchin removal (Kanno 1975).

Rigorous quantitative evaluations of abalone ocean ranching in Japan have been reported recently. These reports document the relatively low mobility and high recoveries of planted abalone *(Haliotis discus hannai)* in Funka Bay and Oshoro Bay, Hokkaido. Using temporary transplantation cages or habitats to reduce early mortality from predation, and to allow acclimation of the transplanted seed to the ocean-bottom environment, Saito (1979, 1981) and Miyamoto *et al.* (1982) found that most of the seed abalone left the temporary habitats to take up residence in rock crevices on the bottom within a few days after ocean planting. In these experimental plantings, seed generally averaged 25 mm (1 in.) in length; after nearly 2 years in the ocean, recoveries averaged approximately 15–20%. Recovery was positively correlated with the size of seed animals planted, averaging 23–31% for seed greater than 22 mm, and somewhat less than 10% for seed less than 22 mm in length. Similar estimates of recoveries from seedling had been made earlier, drawn from data reflecting the enhancement of abalone harvests resulting from ocean ranching in Japan (Inoue 1969; Kan-no and Hayashi 1974; Kan-no 1975; Saito 1979, 1981).

United States

Preliminary abalone seeding efforts in the United States were not encouraging. However, these first experimental plantings, conducted largely with government support and supervision, omitted the use of the temporary transplantation habitats upon which the successful Japanese studies had relied. The first U.S. study also employed relatively small numbers of largely defective and expensive seed animals (most in the range of 20–40 mm, or 1.0–1.5 in.) produced by the older and relatively inefficient hatchery technology. Many of these seed were developmentally retarded, showing conspicuous shell deformities, evidence of arrested growth, and abnormally high sensitivities to stress and transplantation trauma (R. Schmitt, D. Breitburg, R. Burge, and M. Tegner, personal communication). As a probable further consequence of their prolonged conditioning in the artificially protected hatchery environment, these seed animals also proved to be especially susceptible to predation, presumably as a result of their abnormal behavior after transplantation or in the presence of added predators (Tegner *et al.* 1981).

A potentially more promising approach now is under investigation.

This new effort relies upon the use of temporary transplantation habitats such as those that proved essential for the success of ocean ranching in Japan. To offset the lack of governmental subsidies for the mass production of seed and the lack of adequate means for predator reduction (both available to the Japanese), the new thrust in ocean ranching in the United States is based upon those recent technological developments that improve the cost effectiveness of mass production of healthy, rapidly growing, small seed. Although recoveries from plantings of very small seed are expected to be low, the use of very large numbers of such inexpensive "miniseed" and the economies of scale afforded by the new technology for efficient production of such seed may make ocean ranching economically feasible in the United States.

Techniques now are available for the convenient genetic marking of abalone stocks by hybridization and by the breeding and planting of species and variants not preexisting in a designated planting site. This simple expedient, coupled with continued strict enforcement of state regulations governing abalone harvests and sales, will obviate potential problems of competition between ocean ranchers and commercial and recreational fisheries (these fisheries being required by law to harvest larger animals of specific minimum sizes). Genetic marking of stock, and continued state inspections of sales, also will reduce problems of commercial-scale poaching, thus reducing requirements for the patrolling of planted stock.

In southern California, sea otter predation does not presently impinge upon the productive fisheries of the Santa Barbara Channel and the Channel Islands. In these productive waters, ocean ranching of inexpensive, mass-produced, healthy small seed may be a potentially profitable type of abalone aqua culture. The state of California has facilitated the investigation of this means of production by making available leaseholdings of productive bottom areas for regulated abalone planting, growth, and harvest. The California Department of Fish and Game also is working in cooperation with industrial and public interest groups to develop an acceptable plan for zonal management of the sea otter, to safeguard the remaining public and commercial fisheries and the developing ocean-ranching industries in southern California for the future (Cicin-Sain 1982).

Progress recently has been made in the cultivation and successful ocean planting of kelp and other macroalgal species desired for abalone production, with new technology developed in the research and commercial sectors in California (Neushul and Coon 1978; M. Neushul, personal communication). A profitable commercial fishery for urchins also recently has developed in southern California, which may help limit problems of competition for kelp.

SUMMARY AND PROSPECTS FOR FUTURE INDUSTRIAL DEVELOPMENT

Although small-scale industrial efforts at abalone cultivation began in the United States in the early 1960s, only a few of the first pioneer entrants now remain. As in the case of the U.S. oyster seed producing industry, many of these early pioneers in abalone aquaculture began their efforts before the development of the necessary base of information and enabling technology. With assets frozen in large physical plants committed to relatively inefficient, traditional methods of cultivation, and in several cases further hindered by poor locations on partially polluted or otherwise suboptimal waters, a number of these early entrepreneurs failed to develop profitable operations. Although legislative and bureaucratic constraints have been widely blamed for these failures, lack of adequate control of the basic biology of the abalone was at least equally responsible for the early difficulties encountered by the young industry.

The development of new technologies for improved control over those basic biological processes that limit the efficiency and yield of abalone production has infused new life into this nascent industry in the United States. A number of new firms have been started in the past few years, with their operations based, in part, on the use of these new and enabling technologies. The development of these new firms has been centered in southern California, as this is the present hub of the existing commercial abalone harvesting industry and the center of technological development work at the University of California, in cooperation with state and federal agencies and local industrial and public groups.

Successful use of the hydrogen peroxide technique for reliable induction of abalone spawning recently has helped The Ab Lab (Port Hueneme, California) to expand regular production to the point of achieving profitability. This milestone also is due in part to the development of a new domestic market for small, "gourmet-size" abalone, now gaining popularity in southern California restaurants. A number of other California firms are planning to follow this successful lead.

As seen in Table 9.1, the new biochemical technologies for improving abalone production are applicable to a number of abalone species of commercial importance in the U.S. and abroad. The successful application of these new techniques to the efficient production of the thermophillic (warm-water) species of principal commercial importance in southern California and Mexico, and to the production of cold-water species found in Oregon, Washington, Alaska, Australia, New Zealand, Japan, China, and Europe, has extended the range of practical cultivation manyfold over that previously available to the traditional early aquaculturists.

While most U.S. development of abalone cultivation has occurred in California, recent industrial and governmental research and development efforts also indicate feasibility and interest in Oregon, Washington, and Alaska. Abalone brought from the U.S. mainland have been found to grow well in Hawaii (T. Pryor, personal communication), but the absence of suitable supplies of kelp-type species of large brown algae is a further barrier to abalone cultivation in this state and in the eastern and Gulf states as well.

Because of sea otter predation, abalone cultivation in northern and central California apparently will remain restricted to intensive production and grow-out methods in protected containments on land or in the ocean. In view of the high capital and operating costs required for large-scale, intensive cultivation in such containments, further progress in developing techniques to significantly and inexpensively accelerate the relatively slow growth of abalones may be necessary for this method of cultivation to become profitable in the United States.

Whereas the presence of large and protected populations of sea otters will continue to preclude ocean ranching of abalone in northern and central California, this potentially cost-effective method of cultivation is particularly attractive to new industrial entrants and the commercial abalone harvesting industry based in Southern California. The combination of efficient new hatchery technology— to produce large number of inexpensive, healthy, rapidly growing, and genetically marked seed—with ocean planting which requires relatively small labor and capital investments during the relatively long time required for growth of abalone to market size, now is under investigation by the industry in the productive waters of the Santa Barbara Channel and the protected, relatively pollution-free Channel Islands of southern California.

As the traditional, commercial abalone harvesting industry continues its transition toward mariculture-based seed production, planting, and sustained harvesting operations, and as new technologically based entrants into the field of abalone cultivation bring modern methods and needed capital into this developing industry, prospects for the profitable production of abalone in the United States now appear encouraging.

ACKNOWLEDGMENTS

This article is based, in part, on research supported by grants from the U.S. Department of Commerce (NOAA)—University of California Sea Grant Program (Grants R/A-25, -32, -43, -51) to D. E. M.

LITERATURE CITED

BOOLOOTIAN, R. A., FARMANFARMAIAN, A. and GIESE, A. C. 1962. On the reproductive cycle and breeding habits of two western species of *Haliotis*. Biol. Bull. Woods Hole, Mass. **122,** 183–193.

BOWDEN, G. 1981. Coastal aquaculture law and policy: A case study of California. Westview Press, Boulder, Colorado.

BURGE, R., SCHULTZ, S., AND ODEMAR, M. 1975. Report on recent abalone research in California, with recommendations for management. California Department of Fish and Game, Sacramento, California.

CARLISLE, J. G. 1945. The technique of inducing spawning in *Haliotis rufescens* Swainson. Science **102,** 566–567.

CICIN-SAIN, B. 1982. Sea otters and shellfish fisheries in California: the management framework. *In* Social Science Perspectives on Managing Conflicts between Marine Mammals and Fisheries, B. Cicin-Sain, P. M. Grifman, and J. B. Richards (editors), pp. 195–232. Univ. of California Coop. Ext. Service, San Luis Obispo, California.

CICIN-SAIN, B., MOORE, J. E., and WYNER, A. J. 1977. Management approaches for marine fisheries: the case of the California abalone. Sea Grant Publ. No. 54, Institute of Marine Resources, University of California, La Jolla, California.

CICIN-SAIN, B., P. M. GRIFMAN, AND J. B. RICHARDS (editors) 1982. Social Science Perspectives on Managing Conflicts between Marine Mammals and Fisheries. Univ. of California Coop. Ext. Serv., San Luis Obispo, California.

CONTE, F. S. 1981. Abalone aquaculture. California Aquacult. Newsl. Sea Grant College Program Coop. Ext., University of California, La Jolla, California.

COX, K. 1960. Review of the abalone in California. Calif. Fish Game **46,** 381–406.

COX, K. W. 1962. California abalones, family Haliotidae. Calif. Dept. Fish Game Fish Bull. **118,** 1–133.

CROFTS, D. R. 1929. *Haliotis*. Liverpool Mar. Biol. Comm. Mem. Typical Br. Mar. Plants Anim. **29,** 1–174.

FAO. 1975. Yearbook of Fisheries Statistics, Vol. 40. United Nations Food and Agricultural Organization, Rome, Italy.

GIORGI, A., and DEMARTINI, J. D. 1977. A study of the reproductive biology of the red abalone, *Haliotis rufescens* Swainson, near Mendocino, California. Calif. Fish Game **63,** 80–94.

HAHN, K. 1979. The reproductive cycle and gonadal histology of the pinto abalone,

Haliotis kamtschatkana Jonas, and the flat abalone, *Haliotis walallensis* Stearns. Unpublished.

HARDY, R., WENDELL, F. and DEMARTINI, J. 1982. A status report on California shellfish fisheries. *In* Social Science Perspectives on Managing Conflicts between Marine Mammals and Fisheries, B. Cicin-Sain, P. M. Grifman, and J. B. Richards (editors), pp. 328–340. Univ. of California Coop. Ext. Service, San Luis Obispo, California.

HOUK, J. L., and GEIBEL, J. J. 1974. Observations of underwater tool use by the sea otter, *Enhydra lutris* Linnaeus. Calif. Fish Game **60**(4), 207–208.

INO, T. 1952. Biological studies on the propagation of the Japanese abalone (genus *Haliotis*). Bull. Tokai Reg. Fish. Res. Labo. **5,** 1–102.

INO, T. 1980. Fisheries in Japan: Abalone and oysters. Japan Marine Products Photographic Material Association.

INOUE, M. 1969. Mass production and transplantation of abalone. Bull. Kanagawa Exp. Fish. Stn. **131,** 295–307.

INOUE, M. 1976. Abalone. *In* Fisheries Propagation Data Book. Suisan Suppan, Tokyo. (translated by M. Mottet)

KAN-NO, H. 1975. Recent advances in abalone culture in Japan. Proc. 1st Int. Conf. aquacult. Nutr. University of Delaware, pp. 195–211.

KAN-NO, H., and HAYASHI, T. 1974. The present status of shellfish culture in Japan. Proc. 1st U.S.–Japan Meet. Aquacult. Tokyo, Japan. (Nat. Mar. Fish. Serv. Circ. 388, pp. 23–25.)

KIKUCHI, S., SAKURAI, Y., SASAKI, M., and ITO, T. 1967. Food values of certain marine algae for the growth of the young abalone, *Haliotis discus hannai*. Bull. Tohoku Reg. Fish. Res. Lab. **27,** 93–100.

KIKUCHI, S., and UKI, N. 1974. Technical study on artificial spawning of abalone, genus *Haliotis,* II. Effects of seawater irradiated with ultraviolet rays on induction of spawning. Bull. Tohoku Reg. Fish. Res. Lab. **33,** 79–86.

LEIGHTON, D. L. 1971. Observations of the effect of diet on shell coloration in the red abalone, *Haliotis rufescens* Swainson. Veliger **4,** 29–32.

LEIGHTON, D. L. 1966. Studies of food preference in algivorous invertebrates of Southern California kelp beds. Pacific Sci. **20,** 104–113.

LEIGHTON, D. L. 1968. A comparative study of food selection and nutrition in the abalone, *Haliotis rufescens* Swainson, and the sea urchin *Strongylocentrotus purpuratus* Stimpson. Scripps Inst. Oceanogr. Contrib. **38,** 1853–1854.

LEIGHTON, D. L. 1974. The influence of water temperature on larval and juvenile growth in three species of southern California abalones. Calif. Dept. Fish Game Fish Bull. **72,** 1137–1145.

LEIGHTON, D. L. 1977. Some problems and advances in culture of North American abalones *(Haliotis)*. Proc. Symp. Latin Am. Aquacult. Assoc. Maracay, Venezuela, pp. 1–14.

LEIGHTON, D. L., BYHOWER, M. J., KELLY, J. C., HOOKER, G. N., and MORSE, D. E. 1981. Acceleration of development and growth in green abalone, *Haliotis fulgens,* using warmed effluent seawater. World Maric. Soc. **12,** 170–180.

LEIGHTON, D. L. and LEWIS, C. A. 1982. Experimental hybridization in abalones. *Int. J. Invert. Reprod.* **5,** 273–282.

MCALLISTER, R. 1976. California Marine Fish Landings for 1974. California Dept. Fish Game Fish Bull. No. 166.

MIYAMOTO, T., SAITO, K., MOTOYA, S., NISHIKAWA, N., MONMA, H., and KAWAMURA, K. 1982. Experimental studies on the release of the cultured seeds of abalone. *Haliotis discus hannai* Ino in Oshoro Bay, Hokkaido. Sci. Rep. Hokkaido Fish. Exp. Stn. **24,** 59–89.

MORRIS, R. H., ABBOTT, D. P., and HADERLIE, E. C. 1980. Intertidal Invertebrates of California. Stanford University Press, Stanford, California.

MORSE, A. N. C., and MORSE, D. E. 1984. Recruitment and metamorphosis of *Haliotis* larvae are induced by molecules uniquely available at the surfaces of crustose red algae. J. Exp. Mar. Biol. Ecol. **75,** 191–215.

MORSE, D. E. 1980. Recent advances in biochemical control of reproduction, settling, metamorphosis and development of abalones and other molluscs: applicability for more efficient cultivation and reseeding. Proc. Nat. Shellfisher. Assoc. **70,** 132–133.

MORSE, D. E. 1981. Biochemical and genetic control of critical physiological processes in molluscan life-cycles: Basic mechanisms, water-quality requirements and sensitivities to pollutants. Sea Grant College Program 1978–1980 Biennial Rep. pp. 83–87. University of California, La Jolla, California.

MORSE, D. E. 1984a. Recent progress in biochemical and genetic engineering for improved production of abalone and other commercially valuable molluscs. *In* Recent Innovations in the Cultivation of Pacific Molluscs, D. E. Morse, K. Chew, and R. Mann (editors), pp. 263–282. Elsevier, New York.

MORSE, D. E. 1984b. Prospects for the California abalone resource: recent development of new technologies for aquaculture and cost-effective seeding for restoration and enhancement of commercial and recreational fisheries. *In,* Ocean Studies, S. Hansch (editor), pp. 165–189. Calif. Coastal Commis., San Francisco, California.

MORSE, D. E. 1984c. Biochemical control of larval recruitment and marine fouling. *In* Marine Biodeterioration. J. D. Costlow and R. C. Tipper, editors, pp. 134–140. U. S. Naval Institute, Annapolis, Maryland.

MORSE, D. E., DUNCAN, H., HOOKER, N., and MORSE, A. 1977a. Hydrogen peroxide induces spawning in molluscs, with activation of prostaglandin endoperoxide synthetase. Science **196,** 298–300.

MORSE, D. E., DUNCAN, H., HOOKER, N., and MORSE, A. 1977b. An inexpensive chemical method for the control and synchronous induction of spawning and reproduction in molluscan species important as protein-rich food resources. Proc. U. N. Symp. Coop. Invest. Caribb. Adjacent Reg., Caracas, 1976 (FAO UN Fish. Bull. **200,** 291–300).

MORSE, D. E., HOOKER, N., and MORSE, A. 1978. Chemical control of reproduction in bivalve and gastropod molluscs, III: An inexpensive technique for mariculture of many species. Proc. World Maric. Soc. **9,** 543–547.

MORSE, D. E., HOOKER, N., DUNCAN, H., and JENSEN, L. 1979a. γ-Aminobutyric acid, a neurotransmitter, induces planktonic abalone larvae to settle and begin metamorphosis. Science **204,** 407–410.

MORSE, D. E., HOOKER, N., JENSEN, L., and DUNCAN, H. 1979b. Induction of larval abalone settling and metamorphosis by γ-aminobutyric acid and its congeners from crustose red algae, II: applications to cultivation, seed-production and bioassays; principal causes of mortality and interference. Proc. World Maric. Soc. **10,** 81–91.

MORSE, D. E., TEGNER, M., DUNCAN, H., HOOKER, N., TREVELYAN, G., and CAMERON, A. 1980a. Induction of settling and metamorphosis of planktonic molluscan *(Haliotis)* larvae, III: Signalling by metabolites of intact algae is dependent on contact. *In* Chemical Signalling in Vertebrate and Aquatic Animals, D. Muller-Schwarze and R. M. Silverstein (editors), pp. 67–86. Plenum, New York.

MORSE, D. E., DUNCAN, H., HOOKER, N., BALOUN, A., and YOUNG, G. 1980b. GABA induces behavioral and developmental metamorphosis in planktonic molluscan larvae. Fed. Proc. Fed Am. Soc. Exp. Biol. **39,** 3237–3241.

MORSE, D. E., HOOKER, N., and DUNCAN, H. 1980c. GABA induces metamorphosis in *Haliotis,* V: stereochemical specificity. Brain Res. Bull. **5,** Suppl. 2, 381–387.

MOTTET, M. G. 1978. A review of the fishery biology of abalones. Dept. of Fish. Tech. Rep. **37,** 1–81. Seattle, Washington.

MOTTET, M. G. 1980. Factors leading to the success of Japanese aquaculture, with an emphasis on Northern Japan. Fish. Tech. Rep. **52,** 1–83, Seattle, Washington.

MURAYAMA, S. 1935. On the development of the Japanese abalone, *Haliotis gigantea.* J. Coll. Agric., Tokyo Imp. Univ. **13,** 227–232.

NAS. 1978. Aquaculture in the United States: Constraints and Opportunities. National Academy of Sciences, Washington, DC.

NATIONAL MARINE FISHERIES SERVICE. 1982. Southwestern Regional Headquarters, Long Beach, California. Unpublished data.

NEUSHUL, M., and COON, D. 1978. Kelp bed mariculture and resource management. Sea Grant College Program Annu. Rep. 1976–1977, pp. 77–79. University of California, La Jolla, California.

NEWMAN, G. G. 1968. Growth of the South African abalone, *Haliotis midae.* Invest. Rep. **67,** 1–24. Div. Sea Fish. Union of South Africa.

OBA, T. 1964. Studies on the propagation of an abalone, *Haliotis diversicolor supertexta* Lischke: II, On the development. Bull. Jpn. Soc. Sci. Fish. **30,** 809–819.

OBA, T., SATO, H., TANAKA, K., and TOYAMA, T. 1968. Studies on the propagation of an abalone, *Haliotis diversicolor supertexta:* III, On the size of the one-year specimen. Bull. Jpn. Soc. Exp. Fish. **34,** 457–459.

OLIPHANT, M. S. 1979. California Marine Fish Landings for 1976. Calif. Dep. Fish Game Fish Bull. **170,** 1–56.

OLSEN, D. A. 1968a. Banding patterns of *Haliotis rufescens* as indicators of botanical and animal succession. Biol. Bull. Woods Hole, Mass.: **134,** 139–147.

OLSEN, D. A. 1968b. Banding patterns in *Haliotis.* II. Some behavioral considerations and the effect of diet on shell coloration for *Haliotis rufescens, Haliotis corrugata, Haliotis sorenseni,* and *Haliotis assimilis.* Veliger **11,** 135–139.

OWEN, B., McLEAN, J. H., and MEYER, R. J. 1971. Hybridization in the Eastern Pacific abalones *(Haliotis).* Bull. Los Angeles Coty Mus. Nat. History Sci. **9,** 1–37.

PILSON, M. E. Q., and TAYLOR, P. B. 1961. Hole drilling by octopus. Science **134,** 1366–1368.

SANGARA, J. 1975. Abaloine culture. *In* Culture of Marine Life. Japan Int. Coop. Agency, Tokyo.

SAGARA, J. and SAKAI, K. 1974. Feeding experiment of juvenile abalones with four artificial diets. Bull. Tohoku Reg. Fish. Res. Lab. **77,** 1–5.

SAITO, K. 1979. Studies on propagation of Ezo abalone, *Haliotis discus hannai* Ino-I. Analysis of the relationship between transplantation and catch in Funka Bay coast. Bull. Jpn. Soc. Sci. Fish. **45,** 695–704.

SAITO, K. 1981. Coastal ranching of abalone in Northern Japan. Panel Proceedings, Int. Symp. World Maric. Soc., Venice, Italy, pp. 16–17.

SEDGWICK, P. 1978. Replanting the ocean garden: Abalone farming off Santa Barbara. Oceans, pp. 61–62.

SEKI, T. and KAN-NO, H. 1977. Synchronized control of early life in the abalone, *Haliotis discus hannai* Ino, Haliotidae, Gastropoda. Bull. Tohoku Reg. Fish. Res. Lab. **38,** 143–153.

SHEPHERD, S. A. 1973. Studies on the Southern Australian abalone (genus *Haliotis*) I. Ecology of live sympatric species. Aust. J. Mar. Freshwater Res. **24,** 217–257.

SHEPHERD, S. A. 1975. Distribution, habitat and feeding habits of abalone. Austr. Fish. **34,** 12–15.

SHIBUI, T. 1972. On the normal development of the eggs of the Japanese abalone, *Haliotis discus hannai* Ino, and ecological and physiological studies of its larvae and young. Bull. Iwate Prefect. Fish. Exp. St. **2,** 1–69.

SILVA, M. 1982. Management of sea otters and shellfish fisheries in California: who is affected. *In* Social Science Perspectives on Managing Conflicts between Marine Mammals and Fisheries, B. Cicin-Sain, P. M. Grifman, and J. B. Richards (editors). Univ. of California Coop. Ext. Service, San Luis Obispo, California.

SMITH, E. 1982. Legal perspectives on the sea otter conflict. *In* Social Science Perspectives on Managing Conflicts between Marine Mammals and Fisheries, B. Cicin-Sain, P. M. Grifman, and J. B. Richards (editors). Univ. of California Coop. Ext. Service, San Luis Obispo, California.

STATON, H. 1981. Current prospects in abalone aquaculture in California. *In* Transcript of the California Aquaculture Association Meeting, San Diego, January, 1981.

STEIN, J. L. 1981. The role of chemosynthetic bacteria in the diet and distribution of the black abalone *(Haliotis cracherodii)* at subtidal hydrothermal vents. Abstr. West. Soc. Naturalists Annu. Meet., Santa Barbara, California, p. 42.

TEGNER, M. J., CONNELL, J. M., DAY, R. W., SCHMITT, R. J., SCHROETER, S., and RICHARDS, J. B. 1981. Experimental abalone enhancement program. Sea Grant College Program 1978–1980 Biennial Rep., pp. 114–118. University of California, La Jolla, California.

TUTSCHULTE, T. 1975. The comparative ecology of three sympatric abalones. Ph.D. Dissertation, University of California, San Diego, California.

UKI, N. 1981. Feeding behavior of experimental populations of the abalone, *Haliotis discus hannal.* Bull. Tohoku Reg. Fish. Res. Lab. **43,** 53–58.

UKI, N., GRANT, J. F., and KIKUCHI, S. 1981. Juvenile growth of the abalone, *Haliotis discus hannai,* fed certain benthic microalgae, related to temperature. Bull. Tohoku Reg. Fish. Res. Lab. **43,** 59–64.

VAQUIER, V., LEIGHTON, D. L., and LEWIS, C. A. 1981. Assessment of sperm-egg interactions during fertilization and hybrid formation of California abalones. Sea Grant College Program 1978–1980 Biennial Rep., University of California, La Jolla, California.

VON MEDEM, F. G. 1948. Untersuchungen über die Ei-und Spermawikstoffe bei marinen Mollusken. Zool. Jahrb. Abt. Allg. Zool. Physiol. Tiere **61,** 1–44.

WEBBER, H. H., and GIESE, A. C. 1969. Reproductive cycle and gametogenesis in the black abalone *Haliotis cracherodii* (Gastropoda: Prosobranchiata). Mar. Biol. **4,** 152–159.

WOODHOUSE, C. 1982. A summary of the literature on *Enhydra lutris. In* Social Science Perspectives on Managing Conflicts between marine Mammals and Fisheries, B. Cicin-Sain, P. M. Grifman, and J. B. Richards (editors). Univ. of California Coop. Ext. Service, San Luis Obispo, California.

WOODHOUSE, C. D., JR., COWEN, R. K., and WILCOXON, L. R. 1977. A summary of knowledge of the sea otter, *Enhydra lutris* L., in California, and an appraisal of the completeness of biological understanding of the species. Final Rep., U.S. Marine Mammal Commiss., Contr. No. MM6AC008, Washington, D.C.

YOUNG, J., and DeMARTINI, J. D. 1970. The reproductive cycle, gonadal histology, and gametogenesis of the red abalone, *Haliotis rufescens* (Swainson). Calif. Fish Game **56,** 298–309.

10

Water Quality

Robert P. Romaire

Introduction
Physical Variables
  Temperature and Thermal Stratification
  Salinity
  Turbidity and Color
Chemical Variables
  Dissolved Oxygen
  pH
  Carbon Dioxide
  Total Alkalinity and Total Hardness
  Nitrogenous Compounds
  Sulfur
  Hydrogen sulfide
  Chlorine
Biological Variables
  Plankton
  Aquatic Plants
Pesticides
Water Analysis
Literature Cited

INTRODUCTION

Aquatic ecosystems consist of a diverse assemblage of organisms whose interactions between each other and their chemical and physical environment form a complex set of interrelationships. In aquaculture there are many environmental variables that affect the survival, reproduction, growth, and yield of cultured species. Fortunately, management of an aquaculture system does not require a detailed knowledge of all the interactions in the system. A thorough knowledge of only a few important environmental variables is generally sufficient to effectively manage aquaculture systems. These are the variables that aquaculturists should concentrate on and attempt to manage.

More commercial and experimental data on environmental factors affecting commercially cultured finfishes are available than for invertebrates only. Thus, this chapter presents information relative to finfishes as well as invertebrates. Information is presented that will help in determining the potential of a body of water for producing fish and, by inference, invertebrates, in improving water quality, in avoiding stress-related disease problems, in maintaining fish for research purposes, and in increasing fish production. Though brief, this chapter covers the most important points. The relationships between water quality and fish production—including where possible, the ranges of desirable levels of water quality—are discussed. For a more detailed account of water quality management of warm-water aquaculture systems consult *Water Quality in Warmwater Fish Ponds* (Boyd 1979a) and *Water Quality Management for Pond Fish Culture* (Boyd 1982).

PHYSICAL VARIABLES

Temperature and Thermal Stratification

Water temperature has a greater effect on aquatic invertebrates than any other environmental variable. Water temperature has significant effects on respiration, food consumption, digestion, assimilation, growth, and behavior (Schmidt-Nielsen 1979). All fishes are poikilothermic, that is, they have nearly the same body temperature as their surroundings and their body temperature depends on external heat sources, primarily solar radiation. Fish may be grouped arbitrarily into water temperature groups: cold water, cool water and warm water (Klontz 1979). Cold-water fishes are those fish that reside in waters of 15°C (59°F) or less; cool-water fishes in waters of 15°–20°C (59°–68°F); and warm-water fishes in waters above 20°C (68°F). Each species of fish has preferred water temperatures at which growth and other biological functions are optimal. Warm-water fishes grow best at temperatures between 25° and 32°C (77° and 90°F). Water temperatures are in this range the year around at low altitudes in the tropics, but in temperate regions water temperatures are too low in winter for rapid growth of warm-water fish and fish food organisms.

Consequently, management procedures such as fertilization and feedings are often halted or reduced in winter (Boyd 1979a, 1982).

Temperature has a pronounced effect on chemical and biological processes. In general, a temperature increase of 10°C (18°F) causes rates of chemical and biological reactions to double or triple (Schmidt-Nielsen 1979). Thus, fish will consume two to three times as much dissolved oxygen at 30°C (86°F) as at 20°C (68°F), and their biochemical reactions will progress at double or triple the rate at 30°C as at 20°C. Consequently, dissolved oxygen requirements of fish are more critical in warm water than in cooler water. The increase in rate caused by a 10°C (18°F) increase in temperature is called the Q_{10}. If the rate doubles, Q_{10} is 2; if the rate triples, Q_{10} is 3; and so on. Chemical treatments of ponds also are affected by temperature. In warm water, fertilizers dissolve faster, herbicides act quicker, fish toxicants are more effective and degrade more rapidly, and the rate of oxygen consumption by decaying manure is greater (Boyd 1979a).

In natural and impounded waters, heat enters at the surface so surface waters heat faster than deeper waters. Because the density of water decreases with increasing water temperatures above 4°C (39°F), surface waters may become so warm and light that they do not mix with cooler, heavier waters in lower layers. The separation of water into distinct warm and cool layers is called thermal stratification. The upper, warmer layer is called the epilimnion and the lower, cooler layer is known as the hypolimnion. The layer of rapidly changing temperature between the epilimnion and hypolimnion is termed the thermocline (Fig. 10.1). In temperate regions, large ponds may stratify in the spring and remain stratified until fall. In small, shallow ponds in temperate regions and in tropical ponds, stratification may exhibit a diel pattern: during the day, the surface layers warm and form a distinct layer; at night the surface waters cool to the same temperature as the lower waters and the two layers mix.

In some areas surface waters may exceed 35°C (95°F), which is above the optimum for most warm-water fishes (Colt *et al.* 1979; Pope *et al.* 1981). Most aquatic invertebrates are bottom dwellers and are provided haven in cooler, subsurface waters. Many fishes have poor tolerance to sudden changes in temperature. Often, a sudden change in temperature of as little as 5°C (9°F) will stress or even kill fish; the effect is usually worse when fish are moved from cooler to warmer water. Fish can readily tolerate gradual changes in temperature. For example, one could raise the water temperature from 25° to 30°C (77° to 96°F) over several days without harming fish; but fish suddenly removed from 25°C water and placed in water of 30°C might die. The

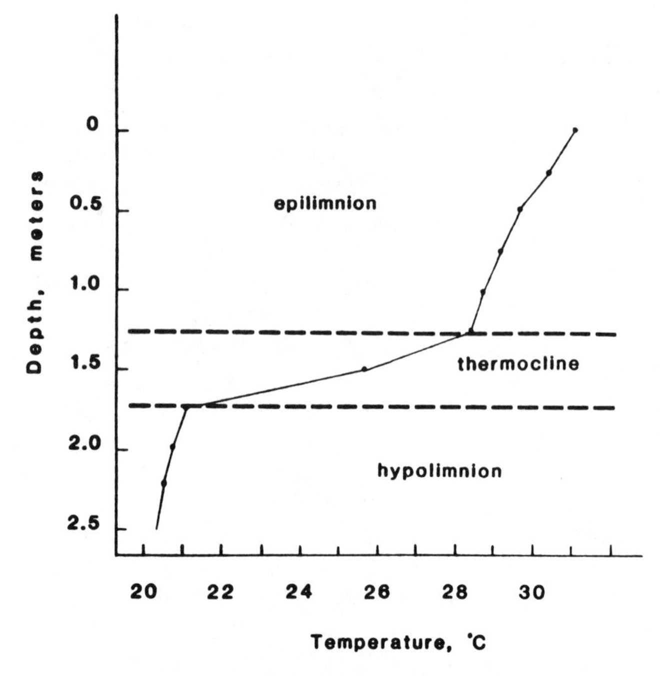

FIG. 10.1 Themal stratification in a fish culture pond.

effects of water temperature on some commercially cultured aquatic invertebrates are summarized in Table 10.1.

Salinity

Salinity is a measure of the total concentration of all dissolved ions in water expressed in grams per liter or parts per thousand (ppt). The major dissolved ions are sodium (Na^+) and chloride (Cl^-), with magnesium (Mg^{2+}), calcium (Ca^{2+}), potassium (K^+), sulfate (SO_4^{2-}) and bicarbonates (HCO_3^-) present in significant amounts. The level of salinity in water reflects geological and hydrological conditions (Hutchinson 1957; Hem 1970). Surface waters in areas of high rainfall where soils are continuously leached usually have low salinity, 0.01 to 0.25 ppt. In arid regions, evaportation exceeds precipitation and salinity increases as a result of evaporation. Salinity in ponds of arid regions often range between 0.5 and 2.5 ppt, and much higher salinities are often found. In areas of high rainfall, groundwater from wells may sometimes have salinity values as high as those encountered in surface waters of arid regions. Seawater has a high salinity of 35 ± 3 ppt;

TABLE 10.1. EFFECTS OF WATER TEMPERATURE ON SELECTED CULTURED INVERTEBRATES

Crawfish *(Procambarus clarkii* and *P. acutus acutus)*—Upper lethal temperature is 36°C or 97°F (Huner and Barr 1980). The optimum growth for most crawfish species occurs at 18°–22°C or 64–72°F (Pope *et al.* 1981).

Freshwater prawn *(Macrobrachium rosenbergii)*—Upper and lower temperature tolerance levels are 36° and 16°C or 97° and 61°F, respectively (Armstrong 1978). Optimum growth occurs at 30°C or 86°F (Farmanfarmaian and Moore 1978).

Brine shrimp *(Artemia salina)*—Upper and lower lethal temperatures are greater than 35°C and lower than 100°C or 95° and 212°F, respectively (McShan *et al.* 1974). The optimum temperature for growth ranges from 20° to 30°C or 68° to 86°F (Hernandorena 1978).

Brown shrimp *(Penaeus aztecus aztecus)*—Growth of postlarvae in laboratory is nil at 11°C or 52°F (Zein-Eldin and Aldrich 1965); rate of growth increases with temperature up to 32.5°C or 91°F; survival greatly reduced at 32.5°C or 91°F; death at 35°C or 95°F; no appreciable growth below 20°C or 68°F; optimum growth between 22.5 and 30°C or 73° and 86°F (Zein-Eldin and Griffith 1966).

Pink shrimp *(Penaeus duorarum)*—Juveniles and adults can tolerate temperatures between 11° and 40°C or 52° and 105°F. Individuals may die when exposed to a low of 10°C or 50°F for 6–10 hours. In ponds, feeding activity is low if temperature drops below 18°C or 64°F (Tabb *et al.* 1972).

American lobster *(Homarus americanus)*—Optimum growth for larval lobsters in stage 1 to 4 is 20° to 23°C or 68° to 73°F (Hughes and Mathiessen 1962); optimum temperature for culture is 24°C or 75°F (Ford *et al.* 1975).

American oyster *(Crassostrea virginica)*—Oysters can live in waters with a minimum of 1°C or 34°F during winter in northern states to a maximum of 36°C or 97°F in Gulf Coast. On flats exposed to sun the temperature of oysters has reached 46°C to 49°C or 115 to 120°F. Inhibition of growth and feeding occurs below 8°C or 46°F (Galtsoff 1964).

the salinity of brackish-water ponds reflects the degree of dilution of seawater with freshwater. High rates of evaporation in brackish-water ponds during periods of low rainfall may cause them to become excessively saline. Salinities in excess of 45 ppt are difficult for marine fishes to tolerate.

The osmotic pressure of water increases with increasing salinity. Fishes differ in their osmotic pressure requirements so that the optimum salinity for acquaculture depends on the species. Some fishes can tolerate wide variations in salinity and are called euryhaline. Other species have a limited tolerance to variations in salinity and are referred to as stenohaline. There is no sharp separation between euryhaline and stenohaline animals, and there is no commonly accepted definition that places a given species in one group or another. Salinity tolerances of some cultured aquatic invertebrates are presented in Table 10.2.

TABLE 10.2. EFFECTS OF SALINITY ON SELECTED CULTURED INVERTEBRATES

Red swamp crawfish *(Procambarus clarkii)*—Newly hatched young die at salinities above 15 ppt, intermediate-sized crawfish die at 30 ppt, and adults tolerate salinities of 30 ppt for one week (Loyacano 1967). In pond studies, crawfish grow and reproduce in salinities of 6 ppt and below (Perry 1971).

Macrobrachium rosenbergii—Larval stages are relatively euryhaline (Fujimura 1966). Best survival for larval stages 1 and 2 is 10 ppt; stages 3 to 5, 14 ppt; stage 6, 10–12 ppt; stages 7 and 8, 10 ppt (Sick and Beaty 1974). Optimum salinity for growth of adults is 0 to 5 ppt.

Brine shrimp *(Artemi* spp.)—Brine shrimp can tolerate 3–300 ppt salinity and 0.8–150 ppt NaCl (McShan et al. 1974). Optimum salinity for hatching is 5–70 ppt.

Brown shrimp *(Penaeus aztecus aztecus)*—Young shrimp have a wide salinity tolerance (2–35 ppt), but there is evidence that juvenile shrimp grow best in ponds with salinities of 15–25 ppt (Broom 1970). Maximum growth of postlarvae is between salinities of 8.5 ppt and 17 ppt under laboratory conditions (Bidwell 1975).

Pink shrimp *(P. duorarum duorarum)*—Regardless of size, pink shrimp can tolerate a wide range of salinity, less than 1 ppt (Gunter and Hall 1965) to a maximum of 60 ppt (Simmons 1957). Juvenile shrimp prefer salinities of 20 ppt or more (Hildebrand 1955).

White shimp *(P. setiferus)*—Salinity tolerances of white shrimp are similar to those of the pink and brown shrimp (Broom 1970).

American oyster *(Crassostrea virginica)*—Growth and development is inhibited outside salinity range of 5 to 30 ppt (Galtsoff 1964).

Small fish of most species are more susceptible than adults to sudden changes in salinity. Sodium chloride (NaCl) or synthetic sea salts may be used to increase the salinity in fish-holding facilities and in small experimental ponds. Conversely, salinity may be lowered in small-scale systems by the addition of water with low salinity. It is usually not practical to adjust the salinity of larger aquaculture systems, except in brackish-water ponds where seawater may be introduced by gravity or tidal movements.

It is seldom practical to measure concentrations of all dissolved ions in water. However, as salinity increases the ability of water to conduct electrical current (conductivity) also rises. A conductivity meter may be used to measure conductivity, from which the salinity can be estimated. Many conductivity meters have scales for reading salinity directly. Another method for obtaining the approximate salinity of water is to measure the concentration of the total dissolved solids (TDS). A sample is filtered through a fine paper, a known volume evaporated, and the residue remaining weighed. The weight of the residue in milligrams per liter is the total dissolved solids concentration, and this closely approximates the salinity. Salinity may be also estimated from the following equation (Wooster *et al.* 1969):

$$\text{salinity (ppt)} = 1.80655 \times [\text{Cl}] \text{ (ppt)}$$

The chloride concentration can be estimated by refractometers, temperature-corrected hydrometers, or by titration of a water sample with standard silver nitrate, using a potassium chromate indicator (American Public Health Association *et al.* 1980).

Turbidity and Color

Turbidity is the degree of opaqueness produced in water by suspensed particulate matter. The concentration of particulate matter determines the transparency of the water by limiting the light transmission within it. The kinds of materials creating turbid conditions in water are as varied as the abiotic and biotic composition of the surrounding watershed. Substances such as humus, silt, organic detritus, colloidal matter, and plants and animals can create turbid conditions (U.S. Environmental Protection Agency 1976).

In aquaculture ponds, turbidity from planktonic organisms is often desirable, whereas that caused by suspended clay particles is generally undesirable. Clay particles, however, are seldom abundant enough in fish pond waters to directly harm fish. If the pond receives runoff that has heavy loads of silt and clay, the silt covers the bottom and can smother fish eggs and fish food organisms (Boyd 1979a, 1982). Moreover, the suspended solids can create thickened gill lamellae on the cultured specie; this impedes oxygen uptake and, in turn, decreases growth rate. Certain species of invertebrates that naturally live in turbid waters have sufficient interlamellar spaces to minimize the contact of suspended solids with gill tissue (Schmidt-Nielsen 1979). The clay particles that remain in suspension restrict light penetration and limit the growth of plants. A persistent clay turbidity which restricts visibility into the water to 30 cm (12 in.) or less may prevent the development of phytoplankton blooms. Water turbidity is often measured with a Secchi disk or by nepholometry (American Public Health Association *et al.* 1980).

Some ponds receive large inputs of vegetative matter either directly (e.g., organic manuring) or indirectly (runoff) from their watersheds. Extracts from this plant material often impart color to the water. Color from vegetative extracts (humates) often appears as a dark stain giving the water the appearance of tea or weak coffee. Pond waters with high concentrations of humates are typically acid and have low total alkalinity. Although color does not affect fish directly, it restricts light penetration and reduces plant growth. Agricultural limestone applications have been used successfully to remove humates from waters (Waters 1956).

In some culture systems it is necessary to remove the turbidity caused by suspended clay particles so that light can penetrate deep enough into the pond for phytoplankton growth. One technique for removing clay turbidity involves application of organic matter. Recommendations vary, but the most popular include the following (Irwin and Stevenson 1951; Swingle and Smith 1947): two or three applications of 2000 kg/ha (1785 lb/Ac) of barnyard manure; one or more applications of 2000–4000 kg/ha (1785–3570 lb/Ac) of hay; and 75 kg/ha (67 lb/Ac) of cottonseed meal plus 25 kg/ha (22 lb/Ac) of superphosphate at 2- to 3-week intervals. The effectiveness of organic matter applications in removing clay turbidity varies, and several weeks must past before success of a particular treatment can be determined.

A better method for removal of clay turbidity is treatment with filter alum (aluminum sulfate, $Al_2(SO_4)_3 \cdot 14H_2O$). Alum will cause suspended clay particles to coagulate and precipitate from the water within a few hours (Ree 1965; Boyd 1979b). The exact application rate for alum may be determined by treating samples of pond water in beakers with concentrations of alum ranging from 10 to 40 mg/liter in increments of 5 mg/liter. The lowest concentration of alum that causes a floc of clay particiles to form in 1 hour is the desirable treatment rate. When applying alum it should be dissolved in water and quickly dispersed over the entire pond surface. Application should be made in calm, dry weather because mixing by wind and rain will break up the floc and prevent it from settling out. Alum has an acid reaction in water, so it destroys total alkalinity and reduces pH. Each 1.0 mg/liter of alum will decrease the total alkalinity by 0.5 mg per liter (Boyd 1979a, 1982). If the total alkalinity is below 20 mg/liter, alum treatment may depress the pH to the point that fish are adversely affected. Hydrated lime [calcium hydroxide, $Ca(OH)_2$] applied simultaneously at the rate of 0.4 mg/liter for each 1.0 mg/liter of alum will prevent unfavorable changes in alkalinity and pH (Boyd 1979a, 1982). Although alum treatment will clear pond water of clay turbidity, it does nothing to correct the cause of turbidity. Unless the source of turbidity (e.g., erosion from an unvegetated watershed) is eliminated, ponds will again become turbid with clay particles.

CHEMICAL VARIABLES

Dissolved Oxygen

Dissolved oxygen (DO) is probably the most critical water-quality variable in aquaculture. Although fish kills in culture ponds may result from excessive concentrations of ammonia, hydrogen sulfide, free carbon dioxide, etc., most fish kills are probably caused by oxygen depletion. The atmosphere is a vast reservoir of oxygen, but atmospheric oxygen is only slightly soluble in water. The solubility of oxygen in water at different temperatures at standard sea level atmospheric pressure is given in Table 10.3. The solubility of oxygen in water decreases as water temperature increases (Hutchinson 1957). When water contains dissolved oxygen at the existing temperature, the water is said to be saturated with oxygen. If water contains more oxygen than it should for a particular temperature, the water is supersaturated. Water can also contain less oxygen than the saturation level. The solubility of oxygen decreases with decreasing atmospheric

TABLE 10.3. SOLUBILITY OF OXYGEN IN PURE WATER AT DIFFERENT TEMPERATURES[a]

°C	mg/liter	°C	mg/liter	°C	mg/liter
0	14.16	12	10.43	24	8.25
1	13.77	13	10.20	25	8.11
2	13.40	14	9.98	26	7.99
3	13.05	15	9.76	27	7.86
4	12.70	16	9.56	28	7.75
5	12.37	17	9.37	29	7.64
6	12.06	18	9.18	30	7.53
7	11.76	19	9.01	31	7.42
8	11.47	20	8.84	32	7.32
9	11.19	21	8.68	33	7.22
10	10.92	22	8.53	34	7.13
11	10.67	23	8.38	35	7.04

Source: After Boyd 1979a.
[a] For an atmosphere saturated with water vapor and at a pressure of 760 mm Hg.

(barometric) pressure. The approximate change in pressure with altitude is as follows (Boyd 1979a): 0 to 600 m—4% decrease in pressure for each 300-m increase in altitude; 600 to 1500 m—3% decrease per 300 m; and 1500 to 3000 m—2.5% decrease per 300 m. The solubility of oxygen in water also decreases as salinity increases. At temperatures of 20°–35°C (68°–95°F), the solubility of dissolved oxygen decreases by about 0.007 mg per liter for each 210-ppm increase in salinity.

Even though dissolved oxygen will diffuse into water, the rate of diffusion is very slow. Thus, photosynthesis by phytoplankton is the primary source of dissolved oxygen in most aquaculture systems (Hepher 1963; Boyd 1973). The primary losses of dissolved oxygen from ponds or other natural waters are caused by respiration by plankton, respiration by benthic organisms, and diffusion of oxygen into the air (Boyd *et al.* 1978; Schroeder 1975). The ranges of expected gains and losses of dissolved oxygen caused by different processes in culture ponds is given in Table 10.4. It is apparent that the major losses of dissolved oxygen are caused by plankton and fish and that photosynthesis is the major source. Diffusion of oxygen into ponds occurs only when waters are undersaturated and diffusion of oxygen out of ponds only occurs when water is supersaturated. The larger the difference between the dissolved oxygen concentration in the pond water and the concentration at saturation, the greater is the rate of diffusion (Schroeder 1975). Wind and wave action enhance diffusion.

TABLE 10.4. RANGES OF EXPECTED GAINS AND LOSSES OF DISSOLVED OXYGEN CAUSED BY DIFFERENT PROCESSES IN FISH PONDS, FOR PONDS OF 1.0–1.5 METERS (39–59 IN.) AVERAGE DEPTH

Process	Range (mg/liter)
Gains	
Photosynthesis by phytoplankton	5–20
Diffusion	1–5
Losses	
Plankton respiration	5–15
Fish respiration	2–6
Respiration by organisms in mud	1–3
Diffusion	1–5

In an aquaculture system, more oxygen must enter or be produced in the water by phytoplankton than is used by the organisms or dissolved oxygen depletion will occur. Since nutrients are normally abundant in culture ponds, light is the primary factor regulating photosynthesis by phytoplankton (Edmondson 1956; Ryther 1956). Light rapidly decreases in intensity as it passes through water. This is true even in pure water, but the decrease is even faster in ponds because planktonic organisms and other suspended and dissolved substances reflect and absorb light. Therefore, the rate of oxygen production by phytoplankton decreases with depth, and below a certain depth no oxygen is produced. Since oxygen is used continuously by the pond biota and only produced by phytoplankton during daylight hours, there is a depth at which the dissolved oxygen produced by the phytoplankton and that entering by diffusion just equal that consumed by the pond biota. This depth is called the compensation point. The compensation point in fish ponds is usually less than 1 m (3.3 ft) and sometimes less than 0.5 m (1.6 ft) (Hepher 1962; Boyd 1973). The stratification of dissolved oxygen in ponds usually corresponds to thermal stratification (Beasley 1963). The epilimnion contains oxygen and the hypolimnion is often depleted of oxygen.

The depth to which light intensity is great enough for photosynthesis to provide surplus dissolved oxygen is related to plankton density. Photosynthesis decreases with decreasing light intensity, and as plankton becomes more abundant, the rate of oxygen consumption by the plankton increases. When plankton abundance is great, dissolved oxygen production is very high near the surface. Because of shading, the rate of oxygen production will decrease rapidly with depth and only a thin layer of surface water, often less than 1 m (3.3 ft), will contain appreciable dissolved oxygen. In waters where plankton is less abundant, rates of dissolved oxygen production are not as high within the illuminated layer of water, but there is appreciable oxygen production and surplus dissolved oxygen at greater depths than in ponds with greater plankton. The influence of plankton turbidity on the depth distribution of dissolved oxygen is illustrated in Fig. 10.2. As a general rule, most waters contain enough dissolved oxygen to support fish to a depth two to three times the Secchi disk visibility (Boyd 1979a).

There is a marked diel fluctuation in DO concentration in ponds and natural waters. Concentrations of dissolved oxygen are lowest at

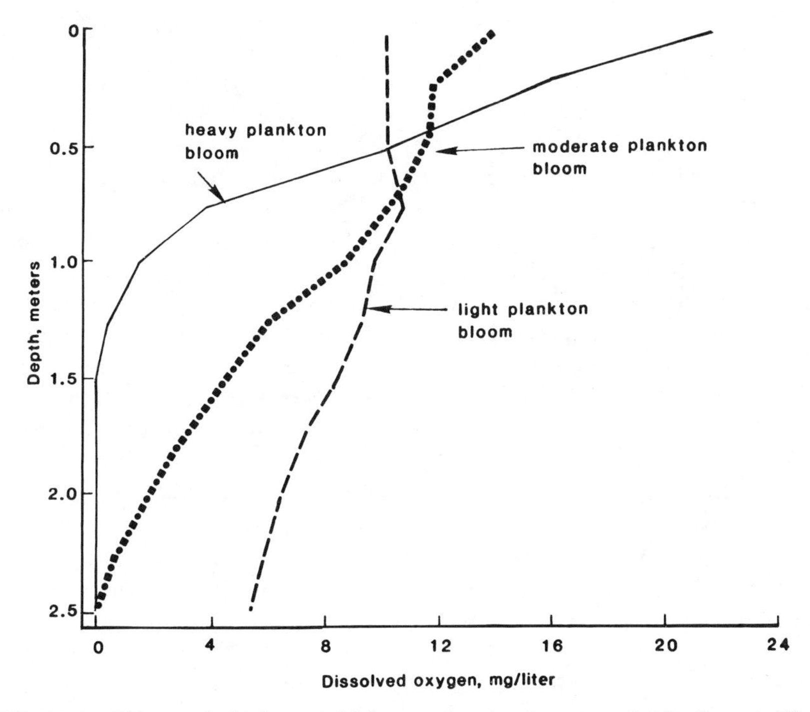

FIG. 10.2 Effects of plankton turbidity on dissolved oxygen distribution at different depths in ponds.

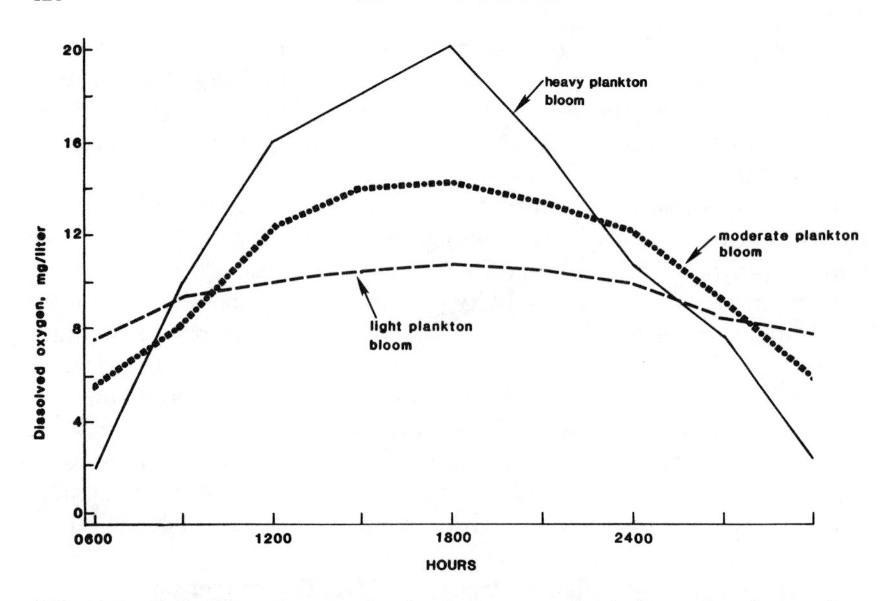

FIG. 10.3 Daily fluctuations in dissolved oxygen concentrations of surface waters in ponds with different densities of plankton.

sunrise, increase during daylight hours to a maximum in late afternoon, and then decline during the night. The magnitude of fluctuation is greatest in waters with heavy plankton blooms and least in ponds with light plankton densities (Fig. 10.3).

Problems with low DO concentrations in waters usually result from prolonged cloudy weather, algal die-offs, overturns (sudden destratification), or treatment of ponds with phytotoxic chemicals (Boyd 1979a, 1982). The production of oxygen is less on a cloudy day than on a clear or partly cloudy day, when DO concentrations do not increase to the usual afternoon levels (Romaire and Boyd 1978). This results in lower than usual dissolved oxygen concentrations the following morning. Extended periods of cloudy weather may result in low dissolved oxygen concentrations even in waters with moderate plankton blooms.

In waters with heavy plankton blooms, scums of algae often form at the surface. These scums of algae may suddenly die, often as a result of light injury, and their decomposition will result in oxygen depletion. Phytoplankton die-offs usually occur during calm, clear, warm weather (Boyd *et al.* 1975, 1978). Dissolved oxygen concentrations do not return to normal following a die-off until a new phytoplankton community is established.

Winds and/or heavy, cold rains may break up thermal stratification in waters, causing a complete mixing (overturn) of the oxygenless water of the hypolimnion and the oxygenated water of the epilimnion. If the pond contains a large volume of oxygenless water, oxygen depletion may occur. Decomposition of plants killed by herbicides may also result in oxygen depletion. Treatment of ponds with phytotoxic substances (e.g., potassium permanganate and formalin) to control diseases may lead to oxygen depletion (Allison 1962).

Fish require adequate concentrations of dissolved oxygen for survival and growth. At low DO concentrations, the respiratory and metabolic activity of fish may be greatly limited. The dissolved oxygen tolerances of species vary widely because of distinct physiological and behavioral differences. The minimum concentration of oxygen for survival varies with the time of exposure. A fish may tolerate a low concentration of dissolved oxygen for several hours without ill effect, but will die if exposed to this same concentration for several days. The oxygen requirements of some commercially important aquatic invertebrates are summarized in Table 10.5. Low DO concentrations affect fish at levels that do not cause mortality by making them more susceptible to diseases. In addition, fish do not feed or grow as well when dissolved oxygen concentrations remain continuously below 25% oxygen saturation. Daily fluctuations of dissolved oxygen in fish ponds apparently have little effect on feeding and growth for some species as long as the minimum dissolved oxygen does not drop below 2 mg/liter in the early morning and then rise to near saturation a few hours after sunrise.

Almost all problems with dissolved oxygen in fish culture are the consequences of heavy plankton blooms (Boyd *et al.* 1978). Suitable

TABLE 10.5. EFFECTS OF DISSOLVED OXYGEN ON SELECTED CULTURED INVERTEBRATES

Red swamp crawfish *(Procambarus clarkii)*—Juveniles 9 to 12 mm long had an LC_{50} of 0.75 to 1.10 mg/liter dissolved oxygen; the LC_{50} for 31 to 35 mm juveniles was 0.49 mg/liter dissolved oxygen (Melancon 1975). Crawfish in burrows tolerated oxygen tensions as low as 1.4–0.2 mg/liter with no mortality but did not remain submerged (Jaspers and Avault 1969).

Freshwater prawn *(Macrobrachium rosenbergii)*—At 33°C *M. rosenbergii* becomes oxygen dependent at 4.65 mg/liter; at 23° and 28°C oxygen dependency is 2.08 and 2.90 mg/liter, respectively (Sharp 1976).

Brine shrimp *(Artemia* spp.)—For salinities of 75–320 ppt the critical oxygen level varies between 1.3 and 0.76 mg/liter (Mitchell and Geddes 1977). Optimum conditions for hatching are oxygen levels greater than 3 mg/liter (Sogeloos and Persoone 1975).

Hardshell clam *(Mercenaria mercenaria)*—Eggs developed at oxygen levels down to 0.5 mg/liter but had 100% mortality at 0.2 mg/liter; larval growth was normal at oxygen levels at or above 4.2 mg/liter but was curtailed at 2.4 mg/liter or below (Morrison 1971).

plankton densities result in Secchi disk visibilities of 30–60 cm (12–24 in.). The probability of problems with low DO concentration increases as Secchi disk visibility decreases below 30 cm (12 in.). In ponds with Secchi disk visibilities of 10–20 cm (4–8 in.), DO concentrations may fall so low at night that fish are stressed or even killed (Romaire and Boyd 1978).

A number of procedures are used to prevent fish kills when DO concentrations are dangerously low. Application of up to 6 or 8 mg/liter of potassium permanganate ($KMnO_4$) has been frequently recommended (Lay 1971). The $KMnO_4$ is supposed to oxidize organic matter and lower the demand for dissolved oxygen in the pond. However, $KMnO_4$ is not effective for this purpose and its application may actually increase the time required for DO concentrations to return to normal levels (Tucker and Boyd 1977). Applications of calcium hydroxide [$Ca(OH)_2$] have been recommended to destroy organic matter in ponds with low DO concentrations, thereby reducing rates of oxygen consumption by bacteria, but research has not shown that applications of $Ca(OH)_2$ will lower concentrations of organic matter. However, when the DO concentration is low, CO_2 is usually quite high. The application of $Ca(OH)_2$ will remove CO_2 and allow fish to better utilize the low concentration of dissolved oxygen (Boyd 1979a, 1982). Each 1.0 mg per liter of CO_2 will require 0.84 mg/liter of $Ca(OH)_2$ for its removal (Boyd 1979a, 1982). For example, if a pond contains 30 mg/liter of CO_2, $Ca(OH)_2$ treatment with 25 mg/liter of $Ca(OH)_2$ would remove the CO_2. Following phytoplankton die-offs, fertilizers have been applied to encourage phytoplankton growth and encourage oxygen production. Research to evaluate the effectiveness of this procedure has not been conducted, but nutrient concentrations are high in ponds following phytoplankton die-offs and it is doubtful that fertilization is necessary.

The only really effective procedure for preventing fish mortality during periods of extremely low dissolved oxygen involves the use of mechanical devices (Boyd and Tucker 1979). Large, tractor-powered pumps may be used to pump fresh, oxygenated water from a nearby water source into a pond with low DO. Alternatively, well water may be released into an oxygen-depleted pond. Well water should be discharged across a baffle or aeration screen because well water is deficient in DO (Fig. 10.4). When oxygenated water is released into a pond with low oxygen, bottom water (which usually contains less oxygen and more carbon dioxide than surface waters) should simultaneously be released from the pond if possible. Pumps may also be used to pump water from a pond with low DO oxygen content. However,

FIG. 10.4 Water aeration screen.

this method is not as effective as pumping fresh, oxygenated water from another pond or well into the pond with low oxygen.

Various types of aeration devices may be used to introduce oxygen into waters with low DO. Small spray-type surface aerators are in common use. These aerators are most effective in small ponds or when several are operated in a large pond. More powerful aerators, such as the portable low-lift pump and sprayer and the paddlewheel aerator, supply considerably more oxygen to ponds than do spray-type surface aerators. However, low-lift pumps and paddlewheel aerators are expensive and must be operated from the power take-off of a farm tractor. The oxygen-delivery capacity and relative efficiency of several types of emergency aeration devices are given in Table 10.6. Large fish farms and research stations can afford to maintain and operate emergency aeration equipment, but small-scale fish culturists have little recourse when faced with dissolved oxygen problems. Problems with dissolved oxygen seldom occur except in ponds where fish are fed at high rates.

TABLE 10.6. AMOUNTS OF OXYGEN ADDED TO POND WATERS BY DIFFERENT TECHNIQUES OF EMERGENCY AERATION

Aeration Technique	Oxygen Added (kg/ha)	Relative Efficiency (%)
Paddlewheel aerator	54.8	100
Crisafulli pump with sprayer	34.9	64
Crisafulli pump to discharge oxygenated water from adjacent pond	21.3	39
Otterbine aerator (3.7 kw)	17.0	31
Crisafulli pump to circulate pond water	13.2	24
Otterbine aerator (2.2 kw)	12.7	23
Rainmaster pump to circulate pond water	12.0	22
Rainmaster pump to discharge oxygenated water from adjacent pond	6.7	12
Air-o-lator aerator (0.25 kw)	4.4	8

Source: After Boyd and Tucker 1979.

Fish culturists often monitor DO concentrations during the night in ponds to determine if emergency aeration is needed. Procedures have been developed for predicting how much DO will fall during the night (Boyd *et al.* 1978; Romaire and Boyd 1978). Such predictions permit the culturist to prepare for emergency aeration in advance. The simplest of these procedures involves the measurement of DO concentrations at dusk and 2 or 3 hours later. These two values are plotted versus time on a graph; a straight line projected through the two points can be used to estimate the DO concentration later in the night (Fig. 10.5).

Dissolved oxygen may be determined by the traditional Winkler procedure (American Public Health Association *et al.* 1980). However, polarographic dissolved oxygen meters provide an easier and more rapid means of determination. When using dissolved oxygen meters, it is important to occasionally check the results against data obtained by the Winkler procedure to verify the accuracy of the meter.

pH

The pH of natural waters is greatly influenced by the concentration of carbon dioxide, an acidic substance (Strumm and Morgan 1970). Phytoplankton and other aquatic vegetation remove carbon dioxide from the water during photosynthesis, so the pH of a body of water rises during the day and decreases at night (Fig. 10.6). Waters with a low alkalinity often have pH values of 5 to 7.5 before daybreak, but when phytoplankton respiration is great, afternoon pH values may

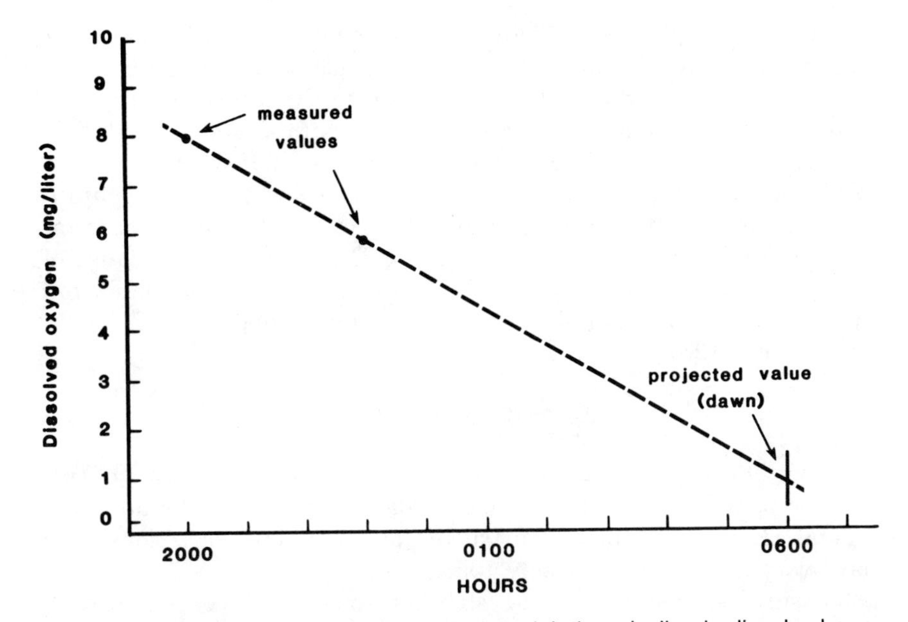

FIG. 10.5 A graphical method for predicting nighttime decline in dissolved oxygen in fish ponds.

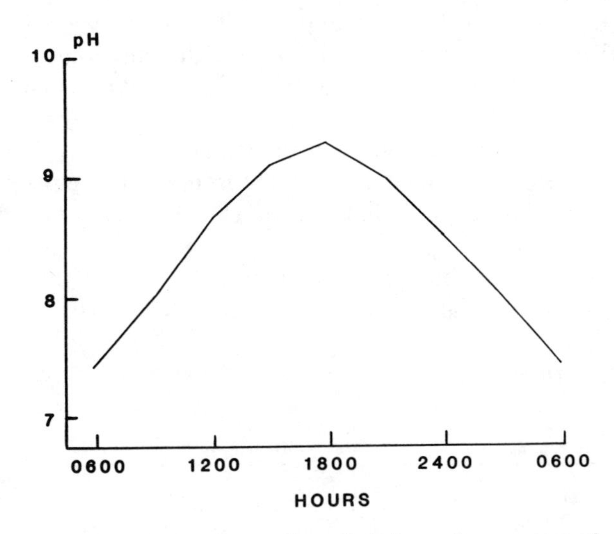

FIG. 10.6 Daily fluctuation of pH in surface waters of fish ponds.

exceed 10 (Swingle 1961). Fluctuations in pH are not as great in waters with higher total alkalinity; in these waters, pH values normally range from 7.5 or 8 at daybreak to 9 or 10 during the afternoon. In some waters with high total alkalinity and very low total hardness, the pH may rise to above 11 during periods of rapid photosynthesis (Boyd 1982). Waters with pH values permanently below 5 usually contain sulfuric acid resulting from the oxidation of sulfide-containing minerals in bottom muds or on watersheds. The most common cause of mineral acidity in natural waters originates from the oxidation of iron pyrite by sulfur-oxidizing bacteria under aerobic conditions (Coleman and Thomas 1967):

$$4FeS_2 + 15\ O_2 + 2H_2O \rightarrow 2Fe_2(SO_4)_3 + 2H_2SO_4$$

Soil that developed from marine sediments containing deposits of sulfide compounds is called "cat's clay." Cat's clay is often found along coastal plains, and ponds constructed on cat's clay formations may have very acidic waters (Neely 1958).

Measurements of pH in ponds should be made in the early morning and again in the afternoon to determine the typical diurnal pH pattern. Waters with a pH values of 6.5 to 9 at daybreak are considered best for fish production (Fig. 10.7). The suitability of waters for fish growth decreases above and below this pH range (Swingle 1961). Reproduction diminishes at pH values below 6. Some ponds that receive drainage from acid soils or swamp may be too acid for fish culture. Waters with extremely high total alkalinities also may not support fish production. The acid and alkaline death points for fishes approximate pH 4 and pH 11 (Fig. 10.7). Even though fish may survive, production will generally be poor in ponds with early morning pH values of 4–6 or 9–10. The afternoon pH in many fish culture systems rises to 9 or 10 for short periods without adverse effect on fish (Boyd 1982).

Applications of lime have proved successful in improving acid waters for fish production (Thomaston and Zeller 1961). However, liming to

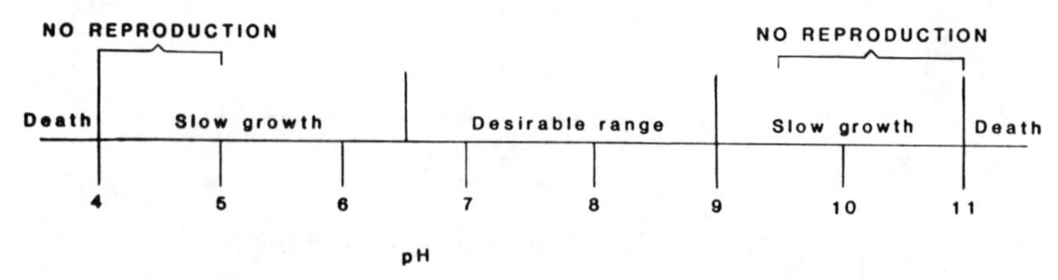

FIG. 10.7 Effects of pH on cultured aquatic species.

increase total hardness is of no value in preventing high pH because lime applications increases total alkalinity and total hardness by roughly the same amount (Boyd 1979a). Agricultural gypsum (calcium sulfate, $CaSO_4$) may be applied to water to increase the total hardness without affecting the total alkalinity. The treatment rate for agricultural gypsum should be the amount that will increase the total hardness to a level that equals total alkalinity. The treatment rate may be determined from the following equation (Boyd 1982):

$$\text{agricultural gypsum (mg per liter)} = [(\text{total alkalinity} - \text{total hardness})] \times [2.2]$$

The agricultural gypsum should be applied in the same manner as liming materials.

Applications of ammonium fertilizers have been recommended to lower the pH of pond waters. The ammonia in fertilizer is nitrified to nitrate with a release of hydrogen ion, which lowers pH (Alexander 1961). However, at a high pH a large percentage of the ammonium ion will immediately be transformed to unionized ammonia, which is highly toxic to fish (Boyd 1979a, 1982). Filter alum (aluminum sulfate, $Al_2(SO_4)_3 \cdot 14\ H_2O$) may be added to ponds to decrease pH. An alum treatment equal in mg per liter to the phenolphthalein alkalinity will reduce pH to approximately 8.3. Although alum treatment may be used to prevent a fish kill when the pH is too high, it does nothing to change the conditions responsible for high pH.

The only reliable method for determining pH is with a pH meter employing the glass electrode (American Public Health Association *et al.* 1980). The meter should be standardized with a buffer solution of pH 7. It should then be verified that the meter will correctly indicate the pH of a buffer solution of some other pH. A pH meter gives more reliable results if it is standardized with a buffer solution of a pH near that of the sample.

Carbon Dioxide

Carbon dioxide is a natural component of all natural waters. It may enter surface waters by absorption from the atmosphere, but only when the partial pressure of carbon dioxide in the water is less than the partial pressure of carbon dioxide in the atmosphere. Carbon dioxide may also be produced in water through biological oxidation of organic materials. If photosynthetic activity is limited, then the partial pressure of carbon dioxide in water may exceed that in the atmosphere and carbon dioxide will escape from the water. Thus, surface waters are constantly absorbing or losing carbon dioxide to maintain equilibrium with the atmosphere. Although carbon dioxide is highly soluble in water the amount that can exist at equilibrium with the atmosphere is very small because of the low partial pressure of carbon dioxide in the atmosphere.

Carbon dioxide (CO_2) acts as an acid in water as shown (Garrels and Christ 1965):

$$H_2O + CO_2 \rightleftharpoons H_2CO_3$$

$$H_2CO_3 \rightleftharpoons H^+ + HCO_3^-$$

Less than 1% of the free carbon dioxide dissolved in water forms carbonic acid (H_2CO_3), which strongly dissociates. Therefore, we may consider CO_2 plus H_2CO_3 as total CO_2 and

$$\text{Total } CO_2 + H_2O \rightleftharpoons H^+ + HCO_3^-$$

Pure water saturated with carbon dioxide at 25°C (77°F) has a total CO_2 concentration of 0.48 mg/liter and theoretically has a pH of 5.68. At greater carbon dioxide concentrations the pH will be less. For example, if the total carbon dioxide concentration is 30 mg/liter at 25°C, the pH will be approximately 4.8. It is usually assumed that carbon dioxie cannot make water more acid than pH 4.5 (Strumm and Morgan 1970). Conversely, the pH at which the free carbon dioxide concentration decreases to an analytically undetectable level is important for both analytical and practical purposes. A pH of 8.34 is often given as the level above which free carbon dioxide is absent.

Ground waters and waters from the hypolimnion of stratified ponds often contain considerable quantities of carbon dioxide (Hutchinson 1957). The carbon dioxide results principally from bacterial oxidation of organic matter in which the water has been in contact. Under these conditions carbon dioxide is not free to escape to the atmosphere. Carbon dioxide is an end product of both anerobic and aerobic bacterial respiration; thus its concentration is not limited by the amount of dissolved oxygen originally present. It is not uncommon to find ground waters with 30–50 mg/liter of carbon dioxide.

Algae use carbon dioxide in photosynthesis and this removal is responsible for increases in pH. As the pH increases, the carbon dioxide forms change, with the result that carbon dioxide can be extracted for algal growth from both bicarbonates (HCO_3^-) and from carbonates (CO_3^{2-}) in accordance with the following equilibrium equations (Boyd 1979a):

$$2HCO_3^- \rightleftharpoons CO_3^{2-} + H_2O + CO_2$$
$$CO_3^{2-} + H_2O \rightleftharpoons 2OH^- + CO_2$$

Thus the removal of carbon dioxide by algae tends to cause a shift in the forms of alkalinity present from bicarbonate to carbonate, and from carbonate to hydroxide. Algae can continue to extract carbon dioxide from water until an inhibitory pH is reached, usually in the range from pH 10 to 11 (Sawyer and McCarty 1978). During darkness algae produce rather than consume carbon dioxide. The carbon dioxide production tends to reduce pH.

Concentrations of free carbon dioxide as high as 60 mg/liter can be tolerated by fish for a very short period. However, waters supporting good fish populations normally contain less than 5 mg/liter free carbon dioxide (Boyd 1979a). In waters used for intensive fish culture, phytoplankton activity usually reduces free carbon dioxide concentrations to very low levels. At night carbon dioxide levels increase (usually by 2 to 10 mg/liter) until daybreak. Continuously high concentrations are seldom found in pond waters except after phytoplankton die-offs or overturns, after destruction of thermal stratification, and during cloudy weather (Boyd 1982). Levels of free carbon dioxide in well water are often high enough to be harmful to fish. Unfortunately, waters with high free carbon dioxide concentrations usually have low concentrations of dissolved oxygen.

Removal of free carbon dioxide may be accomplished by the application of calcium hydroxide ($Ca(OH)_2$), also referred to as slaked or hydrated lime (Boyd 1979a). From the equation below it is seen that one mole of carbon dioxide (44 g)

$$CO_2 + Ca(OH)_2 \rightarrow CaCO_3 + H_2O$$

reacts with one mole of calcium hydroxide (74 g). Therefore, 1 mg per liter of carbon dioxide is removed by 1.68 mg per liter of slaked lime (Boyd 1979a, 1982). Care must be taken to avoid use of more calcium hydroxide than is needed to remove the desired amount of free carbon dioxide because calcium hydroxide causes pH to increase sharply.

Carbon dioxide can be measured by means of standard solutions of an alkaline reagent. The usual method is titration of the water sample to the phenolphthalein end point (pH = 8.3) with a standard base, usually, sodium hydroxide or sodium carbonate (American Public Health Association *et al.* 1980). Special precautions must be taken during the collection, handling, and analysis of water samples for carbon dioxide measurements, regardless of the method used. The sample container should be allowed to overflow to exclude any water that has come in contact with air. If the sample must be transported to the laboratory for analysis, the bottle should be filled completely and capped so as to leave no air pocket.

Total Alkalinity and Total Hardness

Total alkalinity refers to the total concentration of titratable bases in water expressed as mg per liter of equivalent calcium carbonate ($CaCO_3$). In most waters these bases are principally bicarbonate (HCO_3^-) and carbonate (CO_3^{2-}) ions. Bicarbonates represent the major form of alkalinity, since they are formed in considerable amounts from the action of carbon dioxide in water reacting with bases in rocks and soils to form bicarbonates, as illustrated for two alkaline earth carbonates, calcite ($CaCO_3$) and dolomite ($CaMg(CO_3)_2$):

$$CO_2 + CaCO_3 + H_2O \rightleftharpoons Ca^{2+} + 2\ HCO_3^-$$

$$CaMg(CO_3)_2 + 2CO_2 + 2H_2O \rightleftharpoons Ca^{2+} + Mg^{2+} + 4HCO_3^-$$

Other salts of weak acids, such as borates, silicates, and phosphates, may be present in small amounts and contribute to total alkalinity. Ammonia or hydroxides may also contribute to the alkalinity of waters.

Another way to think of alkalinity is in terms of resistance to pH change. The volume of acid required to cause a specified change in pH increases as a function of the total alkalinity of the water. In general, early morning pH is greater in waters with moderate or high total alkalinity than in waters with low total alkalinity. The availability of carbon dioxide for photosynthesis is related to alkalinity. Waters with total alkalinities of less than 15 to 20 mg/liter usually contain little available carbon dioxide (Boyd 1974); waters with total alkalinities of 20 to 150 mg/liter contain suitable quantities of carbon dioxide to permit plankton production for fish culture (Boyd 1982). Carbon dioxide is often low in waters with total alkalinity more than 200 to 250 mg/liter. The afternoon pH in waters with low total alkalinity may often be as great as in waters with moderate or high total alkalinity. Waters of low alkalinity are poorly buffered against pH change, and the removal of carbon dioxide results in rapidly rising pH (Sawyer and McCarty 1978).

The total concentration of divalent metal ions (primarily calcium and magnesium) expressed in mg per liter of equivalent calcium carbonate is termed the total hardness of water. Total alkalinity and total hardness values are normally similar in magnitude because calcium, magnesium, bicarbonate, and carbonate ions in water are derived in equivalent quantities from the solution of limestone in geological depositsd (Hem 1970). However, in some waters total alkalinity may exceed total hardness and vice versa. When the total alkalinity of a water exceeds its total hardness, some of the bicarbonate and carbonate is associated with potassium and sodium rather than calcium and magnesium. Likewise, if the total hardness exceeds the total alkalinity, some of the calcium and magnesium is associated with sulfates, chlorides, silicates, or nitrates rather than bicarbonates and carbonates. If total alkalinity is high and total hardness low, pH may rise to very high levels (greater than 10.5) during periods of rapid photosynthesis.

Waters are categorized according to degrees of hardness (Sawyer and McCarty 1978) as follows:

0–75 mg per liter	Soft
75–150 mg per liter	Moderately hard
150–300 mg per liter	Hard
over 300 mg per liter	Very hard

This classification has no biological meaning, but is important in water treatment. However, the classification is often used by fish culturists.

Desirable levels of total hardness and total alkalinity for fish culture generally fall within the range of 20–300 mg/liter (Boyd and Walley 1975). If total alkalinity and total hardness are too low, they may be raised by liming. However, there is no practical way of decreasing total alkalinity and total hardness when they are above a desirable level. As a general rule, the most productive waters for fish culture have a total hardness and total alkalinity of approximately the same magnitude. For example, a water with a total alkalinity of 100 mg/liter and total hardness of 10 mg/liter is not as good for fish culture as a water in which the total alkalinity is 100 mg/liter and the total hardness is 100 mg/liter. Greater production does not result directly from higher levels of total hardness and total alkalinity per se, but from the higher concentrations of phosphorus and other essential elements that increase along with hardness and alkalinity (Boyd 1979a).

Total alkalinity is normally determined by titration to the methyl orange end point (pH = 4.5) with standard acid, generally sulfuric or hydrochloric (American Public Health Association *et al.* 1980). Some workers prefer to use a pH electrode rather than a chemical indicator to detect the end point of a total alkalinity titration. Total hardness is usually determined by titration with standard ethylenediamine tetraacetic acid (EDTA). Eriochrome—T black is often used to detect the end point of this titration.

Nitrogenous Compounds

The primary source of inorganic nitrogen for aquatic systems comes from the fixation of atmospheric nitrogen (N_2) by biological, meterological, and industrial processes. The chemistry of nitrogen is complex because of the several oxidation states that nitrogen can assume (Sawyer and McCarty 1978). The relationships that exist between the various forms of nitrogen compounds and the changes that occur in nature are best illustrated by a diagram of the nitrogen cycle (Fig. 10.8).

A primary source of nitrogenous compounds in aquaculture systems is from organic matter, for example, plankton, detritus, feeds, etc. Most of the nitrogen in organic matter exists as amino groups in protein. Proteins are deaminated through biological activity and ammonia nitrogen is produced. This process is called ammonification and the am-

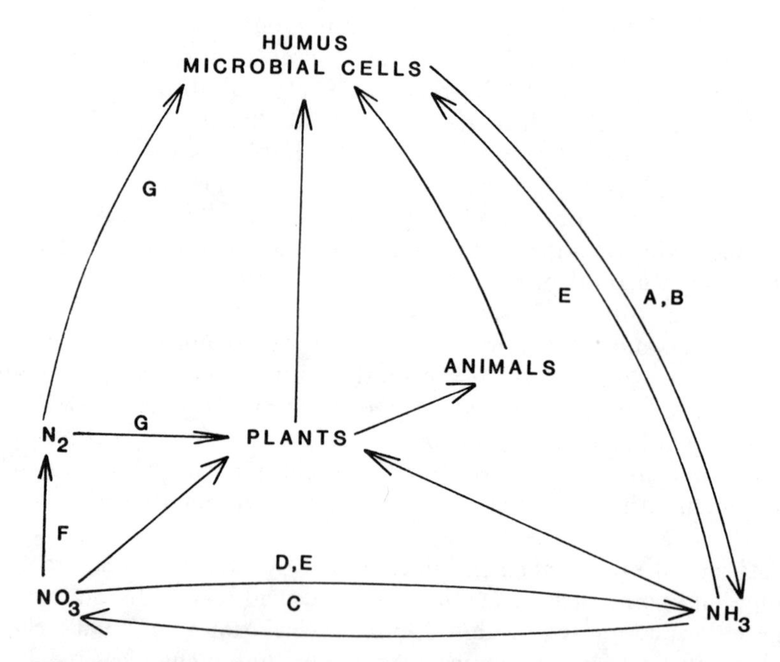

FIG. 10.8 Nitrogen cycle in fish ponds. A. Ammonification; B, Mineralization; C. Nitrification; D. Nitrate reduction; E. Immobilization; F. Denitrification; G. N_2 Fixation

monia is released into the environment or assimilated into microbial tissue (Alexander 1961). The ammonia released into the environment establishes the following equilibrium:

$$NH_3 + H_2O \rightleftharpoons NH_4OH \rightleftharpoons NH_4^+ + OH^-$$

In most environments, NH_4^+ will be the predominate form. Ammonification is a heterotrophic process, which can occur under either aerobic or anaerobic conditions.

Ammonia nitrogen may be used by acquatic plants or nitrified to nitrate, which can also be used by aquatic plants (Boyd 1979a, 1982). Nitrification occurs in two steps:

$$2NH_4^+ + 3O_2 \rightarrow 2NO_2^- + 2H_2O$$

$$2NO_2^- + O_2 \rightarrow 2NO_3^-$$

The oxidation of ammonia to nitrate is by chemoautophic bacteria, primarily *Nitrosomonas* in the first step and principally *Nitrobacter* in the second step (Alexander 1961). Nitrification is most rapid at pH 7 to 8 and at temperatures of 25°–35°C (77°–95°F). The oxidation of ammonia to nitrate is a potential source of acidity in aquatic systems.

Under anaerobic conditions nitrates and nitrities are both reduced by a process called denitrification. Gaseous forms of nitrogen (NH_3, N_2O, N_2) released as metabolites by heterotrophic microorganisms are lost to the atmosphere. Denitrification occurs in the hypolimnion of eutrophic bodies of water or when oxidized nitrogen compounds diffuse into anaerobic layers of mud (Patrick and Tusneem 1972; Bouldin *et al.* 1974). Certain free-living bacteria and blue-green algae are able to fix considerable quantities of nitrogen in aquatic systems (Fogg *et al.* 1973). Nitrogen fixation is assumed to be of appreciable magnitude in fish ponds (Swingle *et al.* 1963). When nitrogen-fixing organisms die and decompose, the nitrogen that was contained in their tissues is mineralized.

Atmospheric nitrogen is soluble in water to the extent of about 12 mg/liter at 25°C (77°F). The other forms of inorganic nitrogen in usual order of increasing abundance in fish ponds are nitrite (NO_2^-), un-ionized ammonia (NH_3), nitrate (NO_3^-) and ionized ammonia (NH_4^+) (Boyd 1979a). Concentrations of nitrite are seldom appreciable in fish ponds except when dissolved oxygen levels are low. The proportion of un-ionized and ionized ammonia varies with pH (Trussell 1972). Concentrations of inorganic nitrogen in natural, unpolluted waters are seldom great, but in ponds or other systems used for intensive fish culture inorganic nitrogen is generally much higher. Nitrogen is also present in soluble organic compounds and as a constituent of living and dead particulate organic matter.

Ammonia reaches pond water principally as a product of fish metabolism and decomposition of organic matter by bacteria. In water, ammonia nitrogen occurs in two forms, un-ionized ammonia (NH_3) and ionized ammonia (NH_4^+). Un-ionized ammonia is toxic to fish, but the ammonium ion is harmless except at extremely high concentrations (Dowing and Merkens 1955). High ambient levels of un-ionized ammonia can affect osmoregulation and oxygen transport in aquatic species, and sublethal ammonia levels can cause pathological changes in fish organs and tissues (Smith and Piper 1975). The toxic levels for un-ionized ammonia for short-term exposure usually lie between 0.6 and 2.0 mg/liter for pond fish, and sublethal levels may occur at 0.1–0.3 mg/liter (Boyd 1979a). The pH, temperature, and salinity of the water regulate the proportion of total ammonia that occurs in the un-ionized form. The pH is most important. A pH increase of 1 unit causes roughly a tenfold increase in the proportion of un-ionized ammonia. Table 10.7 contains factors for calculating un-ionized ammonia at selected temperatures and pH values. For example, at 25°C, the percentages of total ammonia in un-ionized form are as follows: pH 7, 0.59; pH 8, 5.32 and pH 9, 35.98. Ammonia concentrations are seldom high enough in earthen fish ponds to affect fish growth. However, ammonia can build up to appreciable levels in high-density culture systems such as raceways, silos, aquaria, and other closed-loop systems (Spotte 1979). The greatest concentrations of total ammonia nitrogen usually occur after phytoplankton die-offs or sudden destratification of ponds (Boyd *et al.* 1975). However, the high concentrations of free carbon dioxide that occur during such events depress pH and the proportion of un-ionized ammonia.

In ponds or flow-through systems, ammonia is generally the principal toxic metabolic by-product, but in recycled systems both ammo-

TABLE 10.7. UN-IONIZED AMMONIA FACTORS[a]

	pH					
°C	6.5	7.0	7.5	8.0	8.5	9.0
10	0.0006	0.0019	0.0058	0.1082	0.0556	0.1567
15	0.0009	0.0027	0.0086	0.0268	0.0801	0.2159
20	0.0001	0.0040	0.0124	0.0383	0.1118	0.2847
25	0.0002	0.0059	0.0175	0.0532	0.1510	0.3598
30	0.0026	0.0081	0.0251	0.0751	0.2045	0.4482

Source: American Public Health Association *et al.* 1980.

[a] Multiply appropriate factor by the total ammonia nitrogen concentration to determine the amount of NH_3.

nia and nitrite may occur at toxic levels (Spotte 1979). Nitrite is the ionized form of nitrous acid (HNO_2). The level of nitrous acid depends on total nitrite, pH, temperature, and ionic strength. The toxicity of nitrite is due principally to its effects on oxygen transport and tissue damage. Nitrite in the blood oxidizes hemoglobin to methemoglobin, which is not capable of transporting oxygen (Tomasso *et al.* 1979). Methemglobin in fish can be detected by the color of the blood and gills, which turn brown. The same reaction is thought to occur with the hemocyanin of crustaceans. Apparently, the toxicity of nitrite decreases markedly with increasing salinity (Spotte 1979). It is postulated that the increased concentration of cations in saline water combines with nitrite ion and preventions its uptake from solution. The toxic levels for nitrite (NO_2–N) for short-term exposure lie between 8.6 and 15.4 mg/liter for shrimp to 500 to 750 mg/liter for mollusk (Pope *et al.* 1981). Nitrate is the terminal product of nitrification and accumulates to very high levels in closed systems. Fortunately, nitrate is not acutely toxic to acquatic animals even in large concentrations, although its effects over extended periods of time have not been determined (Boyd 1979a, 1982).

The nesslerization technique has been widely used for the determination of total ammonia nitrogen in water (American Public Health Association *et al.* 1980). This procedure is time-consuming because distillation of the water sample is required. The phenate method for measuring total ammonia nitrogen does not require distillation. Ammonia electrodes may also be used to determine ammonia nitrogen on undistilled samples. Nitrite is often determined by the phenoldisulfonic method (American Public Health Association *et al.* 1980). However, the cadium reduction procedure has been found to be more sensitive for nitrate determination of pond water (Boyd and Hollerman 1981). In the cadium reduction procedure, nitrate is reduced to nitrite. The nitrite is then determined by the highly sensitive diazotization technique that employs sulfanilamide as a diazotizing reagent and *N*-(1-naphthyl)-ethylenediamine as a coupling reagent. The diazotization procedure is also used to determine nitrite in pond waters. Nitrate and nitrite may be determined with ion specific electrodes; however, nitrate and nitrite electrodes are usually not sensitive enough for highly accurate analysis of pond waters because of the low concentrations of nitrate and nitrite commonly present (Boyd 1982).

Sulfur

Sulfate. The sulfate (SO_4^{2-}) ion is one of the major ions occurring in natural waters (Hem 1970). Concentrations in ponds vary with the nature of the geological materials in the watershed. In regions with waters of low salinity, concentrations of sulfate often range from 1 to 5 mg/liter as sulfur. However, in regions with waters of higher salinity, and particularly in arid regions, sulfate concentrations are much greater.

Hydrogen Sulfide. Hydrogen sulfide in unionized form (H_2S) is toxic to fish at concentrations less than 1 mg/liter (Smith *et al.* 1976). The level of undissociated hydrogen sulfide assumed to be safe for most aquatic species is 0.002 mg/liter (Boyd 1979a). Under anaerobic conditions in the hypolimnion certain heterotrophic bacteria use sulfate and other oxidized sulfur compounds in metabolism and excrete sulfide as illustrated (Boyd 1979a):

$$SO_4^{2-} + \text{organic matter} \xrightarrow[\text{bacteria}]{\text{anaerobic}} S^{2-} + H_2O + CO_2$$

The sulfide excreted is an ionization product of hydrogen sulfide (H_2S). Hydrogen sulfide dissociates in water as follows:

$$H_2S \rightleftharpoons HS^- + H^+$$

$$HS^- \rightleftharpoons S^{2-} + H^+$$

As pH decreases, concentrations of H_2S increase while those of HS^- and S^{2-} decrease (Hutchinson 1957). The amount of H_2S in water may be calculated if the total sulfide content and pH is known (American Public Health Association *et al.* 1980). For example, at 25°C and pH 9, 99% of the sulfide is in the form of HS^-; at pH 7 it is equally divided between HS^- and H_2S; and at pH 5 about 99% is present as H_2S. The percentages of un-ionized hydrogen sulfide in water at different pH values are given in Table 10.8. Low pH obviously favors the presence of un-ionized hydrogen sulfide, and acid bodies of water that contain high concentrations of hydrogen sulfide may be improved for fish culture by liming (Boyd 1982). Hydrogen sulfide accumulates in the hypolimnion of ponds and mixes with surface waters during overturns. Un-ionized hydrogen sulfide can be oxidized quickly with potassium permanganate ($KMnO_4$) according to the following reaction (Willey *et al.* 1964):

$$4KMnO_4 + 3H_2S \rightarrow 2\ K_2SO_4 + S + 3MnO + MnO_2 + 3H_2O$$

The amount of $KMnO_4$ needed to remove a given amount of un-ionized hydrogen sulfide from pond water cannot be calculated directly because $KMnO_4$ reacts with other reduced inorganic compounds and

TABLE 10.8. HYDROGEN SULFIDE FACTORS[a]

pH	Factor	pH	Factor
5.0	0.99	7.4	0.24
5.4	0.97	7.6	0.10
6.0	0.89	7.8	0.11
6.4	0.76	8.0	0.072
6.6	0.66	8.2	0.046
6.8	0.55	8.4	0.03
7.0	0.44	8.8	0.012
7.2	0.33	9.2	0.0049

Source: American Public Health Association *et al.* 1980.

[a] Multiply appropriate factor by the total dissolved sulfide concentration to determine the amount of H_2S. Values are for 25°C (77°F).

organic matter. Fortunately, hydrogen sulfide is seldom a problem in aquaculture.

Soils in some areas contain sulfide deposits. Such soils are usually found where coal is mined or along coastal plains. When exposed to air, the sulfide is oxidized to sulfuric acid and the runoff from these soils may have an extremely low pH. Construction of fish ponds should not be encouraged on watersheds where sulfide-bearing minerals occur at or near the surface unless adequate lime is applied to neutralize the acidity.

Total dissolved sulfide may be determined by the methylene blue method (American Public Health Association *et al.* 1980). The procedure is based on the reaction of sulfide, ferric chloride, and dimethyl-*p*-phenylenediamine to produce methylene blue. Ammonium phosphate is added after color development to remove ferric chloride color. The un-ionized hydrogen sulfide may be calculated from the concentration of total dissolved sulfide, the sample pH, and the ionization constant of H_2S. Total dissolved sulfide may also be determined with a silver–sulfide electrode.

Chlorine

Chlorination is often used to disinfect facilities used for holding fish. Moreover, chlorinated water is sometimes used for small-scale culture systems such as aquariums and hatcheries. Chlorine is usually added to water at hatcheries as molecular chlorine (Cl_2) or as calcium hypochlorite, $Ca(OCl)_2$. Chlorine combines with water to form hypochlorous acid (HOCl) and hydochloric acid which dissociates (Sawyer and McCarty 1979):

$$Cl_2 + H_2O \rightleftharpoons HOCl + H^+ + Cl^-$$

$$HOCl \rightleftharpoons H^+ + OCl^-$$

Both free chlorine (Cl_2) and hypochlorite ion from calcium hypochlorite react in water to form free chlorine and hypochloric acid in proportions determined by the pH. These chlorine species are commonly referred to as free chlorine residuals. Chlorine and hypochlorous acid also react with ammonia to form chloramines. Chloramines are called combined chlorine residuals.

Both free and combined chlorine residuals are extremely toxic to acquatic species at concentrations less than 0.1 mg/liter (Boyd 1982). If measurable concentrations of chlorine residuals are present in water, the water may not be safe for holding fish. Chlorine can be removed from tap water by letting it age in an open container for several days, or by passing the water through an activated charcoal filter (Spotte 1979). The most effective method of chlorine removal is treatment with sodium thiosulfare ($Na_2S_2O_3{\cdot}5H_2O$), which reacts with chlorine residuals as illustrated for free chlorine below (Boyd 1979a):

$$Cl_2 + 2Na_2S_2O_3 \cdot 5H_2O \rightarrow Na_2S_4O_6 + 2NaCl + 10\ H_2O$$

According to this reaction, 6.99 mg/liter of sodium thiosulfate is required to remove 1 mg/liter of chlorine.

Several procedures are available for determining chlorine and its reaction products in water. The DPD method provides a rapid and reliable determination of chlorine (American Public Health Association *et al.* 1980). In the presence of excess iodine, free chlorine residuals and chloramines react with *N, N*-dimethyl-*p*-phenylenediamine (DPD) to produce a red color. This color may be destroyed by titration with standard ferrous ammonium sulfate, which reduces chlorine and its reaction products to chloride.

BIOLOGICAL VARIABLES

Plankton

Plankton comprises all the microscopic organisms that are suspended in water and includes small plants (phytoplankton), small animals (zooplankton), and bacteria. When there is enough plankton in the water to discolor it and make it appear turbid, the water is said to contain a plankton "bloom." The phytoplankton uses inorganic salts, carbon dioxide, water, and sunlight to produce its own food. The zooplankton feeds on living or dead phytoplankton and other tiny particles of organic matter in the water. Bacteria utilize organic matter in the water for food. In most fish culture systems where fish are not provided supplemental feed, plankton forms the base of the food web.

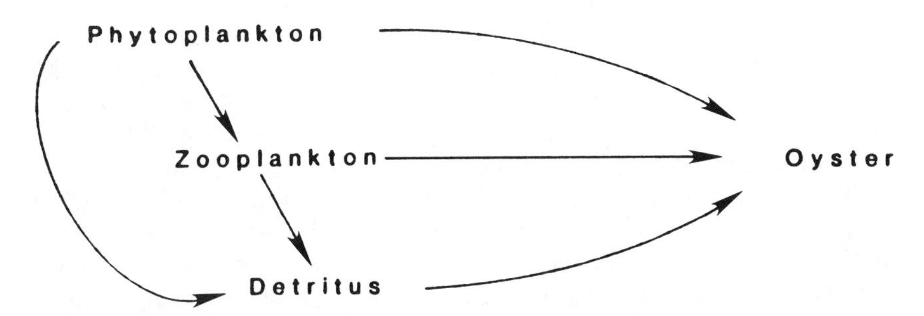

FIG. 10.9 A representative food web in an oyster culture system.

Examples of food webs in aquatic invertebrate culture systems are given in Figs. 10.9 and 10.10. Since each step in the food web is rather inefficient, an aquaculture system with a more direct food web will produce a greater biomass of fish per unit area.

Because plankton forms the base of the food web, there is a strong relationship between plankton abundance and fish production. In addition to encouraging fish growth, plankton makes water turbid and prevents the growth of undesirable aquatic weeds through shading (Boyd 1982). Heavy phytoplankton blooms usually contain large numbers of blue-green algae, which can form scums at the surface. These scums absorb heat during the day and cause shallow thermal stratification. During the night, heavy plankton blooms consume large amounts of dissolved oxygen and may cause oxygen depletion before the next morning (Romaire and Boyd 1978). Scums of plankton may suddenly die, decompose, and cause oxygen depletion (Boyd *et al.* 1978). In addition to causing dissolved oxygen problems, organisms in heavy plankton blooms often produce substances that impart a strong off-flavor to fish flesh (Lovell and Sackey 1973).

There are many techniques for measuring plankton abundance, but

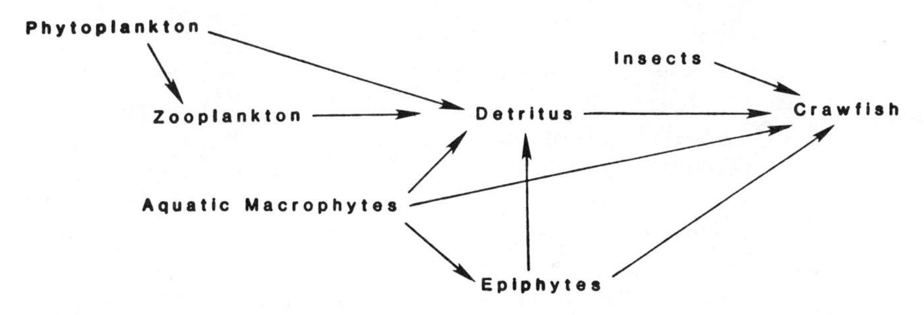

FIG. 10.10 A representative food web in a crawfish culture system.

most are too tedious for use by a commercial fish culturist. The most practical technique for use in ponds that do not contain appreciable clay turbidity is to measure the Secchi disk visibility. The Secchi disk visibility is the depth at which a disk 20 cm in diameter with alternate black and white quadrants disappears from view. There is generally a high correlation between Secchi disk visibility and plankton abundance. The ideal plankton turbidity for fish culture has not been established. However, a Secchi disk visibility of 30–60 cm (12–24 in.) is generally adequate for good fish production and for shading underwater weeds. As Secchi disk visibilities decrease below 30 cm (12 in.), there is an increase in the frequency of dissolved oxygen problems. At values above 60 cm (24 in.), light penetrates to greater depths encouraging macrophyte growth.

Plankton communities are constantly changing in species composition and in total abundance. This results in corresponding fluctuations in Secchi disk visibility and in the appearance of pond water. However, unless plankton becomes so dense that dissolved oxygen problems occur or so thin as to encourage underwater weeds, the changes do not affect fish production appreciably (Boyd 1979a). By monitoring Secchi disk visibility on a regular schedule (once or twice weekly) and observing the appearance of the water, the fish culturist can obtain information on the continuing condition of the plankton community in a pond and on the supply of fish food organisms.

The ability of water to produce plankton depends on many factors, but the most important is usually the availability of inorganic nutrients for phytoplankton growth. Essential elements for phytoplankton growth include carbon, oxygen, hydrogen, phosphorus, nitrogen, sulfur, potassium, sodium, calcium, magnesium, iron, manganese, copper, zinc, boron, cobalt, chloride, and possibly others. Phosphorus is most often the element limiting phytoplankton growth in ponds. The addition of phosphate fertilizer will cause an increase in plankton production and a subsequent increase in fish production in most ponds (Boyd 1979a, 1982). Inadequate supplies of nitrogen, potassium, and carbon also limit phytoplankton in some ponds (Boyd 1972).

In general, the level of plankton production in unmanaged ponds is related to the basic soil fertility of the surrounding watersheds. Therefore, the basic levels of plankton and fish production are greater in ponds located on watersheds with fertile soils than in ponds located on watersheds with poor soils (Boyd 1976). Even though the basic fertility of ponds differs greatly depending on the management and soils of their watersheds, the level of plankton production in most ponds can be raised to that needed for good fish production. Inorganic fertilizers may be added to ponds with low basic fertility to increase plankton production. In some ponds, both lime and fertilizer applications may be required to increase plankton production. Manures also increase plankton production.

Aquatic Plants

Large aquatic plants (aquatic macrophytes) which may grow in aquaculture systems are usually undesirable. They interfere with fish management operations such as seining, feeding, and fish harvest, compete with phytoplankton for nutrients, provide havens for prey fish to escape predatory fish and thus encourage unbalanced fish populations, favor mosquito production, and contribute to water loss through evapotranspiration. One notable exception to this, however, is commercial crawfish culture operations in the southeastern U.S. where the growth of aquatic and semi-aquatic plants is often encouraged (Huner and Barr 1980).

Aquatic macrophytes include filamentous algae (e.g., *Spirogyra, Rhizoclonium, Pithophora, Cladophora, Oedogonium, Hydrodictyon*) and submersed (e.g., *Najas, Ceratophyllum, Myriophyllum, Caboma, Elodea, Potamogeton, Vallisneria*), floating and floating-leafed (e.g., *Lemna, Wolffia, Eichornia, Nuphar, Nymphaea*), and emergent (e.g., *Typha, Scirpus, Juncus, Polygonum, Sagittaria, Alternanthera, Ludwigia*) macrophytes. Aquatic macrophytes, which begin their growth at the pond bottom, are limited to relatively transparent waters. Therefore, management procedures that favor plankton turbidity will often eliminate macrophytes. Floating or floating-leafed macrophytes must be controlled by other methods.

One effective method of controlling macrophytes is fertilization to produce plankton turbidity and shade the pond bottom (Boyd 1979a, 1982). This technique is effective especially if no area of the pond is shallower than 60 cm (24 in.). Grass carp *(Ctenopharygodon idella)* eat great quantities of aquatic vegetation and provide a biological method of controlling macrophytes. When stocked at 60–80 per ha (24–32/Ac), grass carp will control most species of macrophytes that can not be controlled by plankton turbidity (Avault 1965).

Herbicides are also used in fish culture to control macrophytes. Usually, the concentrations of aquatic herbicides used to kill macrophytes are safe to fish. Decay of macrophytes killed by herbicides can cause dissolved oxygen depletion. If ponds have extensive areas of macrophytes, one-fourth to one-fifth of the pond should be treated at 1- to 2-week intervals to reduce the chance of oxygen depletion. The major limitation to the use of herbicides for controlling macrophytes is that once the concentration of a herbicide falls below a non-toxic level macrophytes will regrow. Common herbicides used for control of aquatic macrophytes include copper sulfate, 2,4-D, endothall, diquat, and simazine.

Algicides are sometimes used to control phytoplankton in ponds. Copper sulfate ($CuSO_4 \cdot 5\ H_2O$) is widely used to control phytoplankton. The effectiveness of copper sulfate in controlling algae is dependent upon the alkalinity of the water (Boyd 1982). Cupric ions react with bicarbonate ions to form insoluble copper carbonate, which precipitates. Copper sulfate will kill most species of phytoplankton at concentrations of 0.1–0.5 mg/liter in waters with total alkalinities of 40–50 mg/liter (Toth and Riemer 1968). In waters with higher alkalinities, copper sulfate concentrations of 1.0 mg/liter or higher may be required to kill phytoplankton. Phytoplankton killed by copper sulfate decomposes rapidly and may result in low dissolved oxygen concentrations. Copper sulfate has no appreciable residual toxicity and phytoplankton growth will resume soon after treatment. (Tucker and Boyd 1978). Fish are susceptible to copper sulfate, and in waters with alkalinities less than 20 mg/liter, treatment with 0.5–1.0 mg/liter of copper sulfate may kill fish (Boyd 1979a, 1982).

Synthetic algicides such as diuron [3-(3,4-dichlorophenyl)-1,1-dimethyl urea] and simazine [2-chloro-4,6-*bis*(ethylamino)-triazine] are sometimes used to kill phytoplankton. These algicides are extremely toxic to algae, have a long residual action, and are not toxic to fish at concentrations used to kill phytoplankton (Tucker and Boyd 1978). As with copper sulfate, extensive mortality of phytoplankton following applications of synthetic algicides may result in depletion of dissolved oxygen. Some fish culturists have attempted to "thin" phytoplankton blooms by small, periodic applications of synthetic algicides to ponds receiving heavy applications of feed. However, research has demonstrated that this practice results in prolonged periods of low dissolved oxygen concentrations and reduced fish yields (Tucker and Boyd 1978).

PESTICIDES

Fish ponds are usually constructed in areas where industrial pollution is not prevalent. However, agricultural pollutants, especially pesticides, may reach ponds in runoff or drift. Many pesticides, insecticides in particular, are extremely toxic to invertebrates. Crustaceans are particularly sensitive to insecticides becasue of their close phylogenetic relationship to insects. Acute toxicity values for many commonly used insecticides range from 0.005 to 10 mg/liter and much lower concentrations may be toxic upon longer exposure. Even if adult fish are not killed outright, long-term damage to fish populations may occur in environments contaminated with pesticides. The abundance of food organisms may decrease, fry and eggs may suffer mortality, and growth rates of fish may decline.

Pesticides sprayed onto fields may drift over considerable areas and reach ponds. Therefore, ponds in agricultural areas are often contaminated to some degree with pesticides. The key variables that influence the degree of pesticide contamination of fish ponds are distance from pesticide-treated fields, tree and other vegetative cover between ponds and fields, topographic barriers to drift or runoff from treated fields, and methods of application of pesticides to fields (Boyd 1979a). If watersheds receive heavy applications of persistent pesticides, ponds in the area may not be suitable for fish production. Cotton and other nonfood crops are often treated with especially toxic and persistent pesticides. In some regions, pesticides that contain heavy metals, such as arsenic and lead, are still used. Heavy metals may reach ponds and kill fish or adversely affect production.

WATER ANALYSIS

Water analysis is a highly specialized field, and methods for measuring the concentration of almost any possible constituent of water are available. These methods may be found in several standard water analysis manuals. The most widely used of these manuals is the *Standard Method for the Examination of Water and Wastewater* (American Public Health Association *et al.* 1980). Other widely used manuals include *Methods for Chemical Analysis of Water and Wastes* (U.S. Environmental Protection Agency 1974), *A Practical Book of Seawater Analysis* (Stickland and Parsons, 1972) and *Water Quality in Warmwater Fish Ponds* (Boyd 1979a).

To conduct water analyses according to standard procedures, a water analysis laboratory and a well-trained analyst are necessary. In practical fish culture, however, only limited water-quality data are needed in making water quality management decisions. These normally include pH, total alkalinity, total hardness, dissolved oxygen and carbon dioxide concentrations, and plankton abundance. Water analysis kits are available at a modest cost. The kits provide sufficiently accurate data on which to base management decisions. A Secchi disk, which may be constructed from common items or purchased for a small cost, may be used to estimate plankton abundance.

Water samples for dissolved oxygen or carbon dixoide analyses must

FIG. 10.11 Water sampler used for collecting water from depths up to several meters.

be collected so that they do not come in contact with the atmosphere. If a sample is supersaturated with dissolved gases, the gases are lost to the atmosphere; samples below saturation will gain gases from the atmosphere. A number of samplers are available for collecting water for dissolved-gas analyses, but the least expensive types may be obtained from the manufacturers of water analysis kits. Samples for total alkalinity, total hardness, or pH may come in contact with the air without introduction of appreciable errors in measurement. Samples of surface water may be secured by simply immersing an open-mouthed bottle and allowing it to fill. Samplers may also be purchased for obtaining water from greater depths (Figure 10.11). Once the water sample has been collected it should be analyzed as soon as possible to prevent changes in concentrations of the constituents.

LITERATURE CITED

ALEXANDER, M. 1961. Introduction to Soil Microbiology. John Wiley & Sons, New York.

ALLISON, R. 1962. The effect of formalin and other parasiticides upon oxygen consumption in ponds. Proc. Southeastern Assoc. Game Fish Commiss. **16,** 446–449.

APHA, American Water Works Association, and Water Pollution Control Federation 1980. Standard methods for the examination of water and wastewater, 15th ed. American Public Health Association, New York.

ARMSTRONG, D. A. 1978. Responses of *Macrobrachium rosenbergii* to extremes of temperature and salinity. Pap. No. 4507, pp. 18–51. Department of Water Sciences and Engineering, University of California, Davis, California.

AVAULT, J. W., JR. 1965. Preliminary studies with grass carp for aquatic weed control. Prog. Fish Cult. **27,** 207–209.

BEASLEY, P. G. 1963. The penetration of light and the concentration of dissolved oxygen in fertilized pond waters infested with *Microcystis.* Proc. Southeastern Assoc. Game Fish Commiss. **17,** 222–226.

BIDWELL, J. P. 1975. Brown shrimp culture at Gulf Coast Research Laboratory. Seascope **5**(3/4), 1, 6.

BOULDIN, D. R., JOHNSON, R. L., BURDA, C., and KAO, C. 1974. Losses of inorganic nitrogen from aquatic systems. J. Environ. Qual. **3,** 107–114.

BOYD, C. E. 1972. Sources of CO_2 for nuisance blooms of algae. Weed Sci. **20,** 492–497.

BOYD, C. E. 1973. Summer algal communities and primary productivity in fish ponds. Hydrobiologia **41,** 357–390.

BOYD, C. E. 1974. Lime requirements of Alabama fish ponds. Alabama Agric. Exp. Stn., Bull. 459, Auburn University, Auburn, Alabama.

BOYD, C. E. 1976. Water chemistry and plankton in unfertilized ponds in pastures and woods. Trans. Am. Fish. Soc. **105,** 634–636.

BOYD, C. E. 1979a. Water quality in warmwater fish ponds. Alabama Agric. Exp. St., Auburn University, Auburn, Alabama.

BOYD, C. E. 1979b. Aluminum sulfate (alum) for precipitating clay turbidity from fish ponds. Trans. Am. Fish. Soc. **108,** 307–313.

BOYD, C. E. 1982. Water Quality Management for Pond Fish Culture. Elsevier Scientific Publ. Co., New York.

BOYD, C. E., and HOLLERMAN, W. D. 1981. Determination of nitrate in waters from fish ponds. Alabama Agric. Exp. Stn. leaflet 99. Auburn University, Auburn, Alabama.

BOYD, C. E., and TUCKER, C. 1979. Emergency aeration of fish ponds. Trans. Am. Fish. Soc. **108,** 299–306.

BOYD, C. E., and WALLEY, W. W. 1975. Total alkalinity and hardness of surface waters in Alabama and Mississippi. Alabama Agric. Exp. Stn. Bull. 465. Auburn University, Auburn, Alabama.

BOYD, C. E., DAVIS, J. A., and JOHNSTON, E. 1978. Die-offs of the blue green alga, *Anabaena variabilis,* in fish ponds. Hydrobiologia **61,** 129–133.

BOYD, C. E., PRATHER, E. E., and PARKS, R. W. 1975. Sudden mortality of a massive phytoplankton bloom. Weed Sci. **23,** 61–67.

BOYD, C. E., ROMAIRE, R. P., and JOHNSTON, E. 1978. Predicting early morning dissolved oxygen concentrations in channel catfish ponds. Trans. Am. Fish. Soc. **107,** 484–492.

BROOM, J. G. 1970. Shrimp culture. Proc. World Maric. Soc. **1,** 63–68.

COLEMAN, N. T., and THOMAS, G. W. 1967. The basic chemistry of soil acidity. *In* Soil Acidity and Liming, R. W. Pearson and F. Adams (editors). Monogr. 12, American Society of Agronomy, Madison, Wisconsin.

COLT, J., MITCHELL, S., TCHOBANOGLOUS, G., and KNIGHT, A. 1979. The environmental requirements of fish. *In* The Use and Potential of Aquatic Species for Wastewater Treatment. California State Water Resources Control Board, Sacramento, California.

DOWNING, K. M., and MERKENS, J. C. 1955. The influence of dissolved oxygen concentrations on the toxicity of un-ionized ammonia to rainbow trout (*Salmo gairdneri:* Richardson) Ann. Appl. Biol. **43,** 243–246.

EDMONDSON, W. T. 1956. The relationship of photosynthesis by phytoplankton to light in lakes. Ecology. **37,** 161–174.

FARMANFARMAIAN, A., and MOORE, R. 1978. Diseasonal thermal aquaculture-1. Effect of temperature and dissolved oxygen on survival and growth of *Macrobrachium rosenbergii.* Proc. World Maric. Soc. **9,** 55–66.

FOGG, G. E., STEWART, W. D., FAY, P., and WALSBY, A. E. 1973. The Blue-Green Algae. Academic Press, New York.

FORD, R. F., VAN OLST, J. C., CARLBERG, J. M., DORBAND, W. R., and JOHNSON, R. L. 1975. Beneficial use of thermal effluent in lobster culture. Proc. World Maric. Soc. **6,** 156–173.

FUJIMURA, T. 1966. Notes on the development of a practical mass culturing techniques of the giant prawn *Macrobrachium rosenbergii.* Indo-Pacific Fisheries Council, 12th Session. FAO Regional Office for Asia and the Far East, Bangkok, Thailand.

GALTSOFF, P. S. 1964. The American oyster, *Crassostrea virginica.* Fish Wildl. Serv. Fish. Bull. **64.**

GARRELS, R. M., and CHRIST, C. L. 1964. Solutions, Minerals, and Equilibria. Freeman, Cooper, and Company. San Francisco, California.

GUNTER, G., and HALL, G. E. 1965. A biological investigation of the Caloosahatchee estuary of Florida. Gulf Res. Rep. **2.**

HART, J. S. 1944. The circulation and respiratory tolerance of some Florida freshwater fishes. Proc. Florida Acad. Sci. **7,** 221–246.

HEM, J. D. 1970. Study and interpretation of the chemical characteristics of natural waters. U.S. Geol. Surv., Water-Supply Pap. 1473. U. S. Government Printing Office, Washington, D.C.

HEPHER, B. 1962. Primary production in fish ponds and its applications to fertilization experiments. Limnol. Oceanog. **7,** 131–135.

HERNANDORENA, A. 1978. The effects of temperature on the nturitional requirements of *Artemia salina.* Biol. Bull. Woods Hole, Mass. **151,** 314–321.

HILDEBRAND, H. H. 1955. A study of the fauna of the pink shrimp (*Penaeus duorarum* Burkenroad) grounds in the Gulf of Campeche. Inst. Mar. Sci., University of Texas Ser. **4,** 169–232.

HUGHES, J. T. and MATTHIESSEN, G. C. 1962. Observations on the biology of the American lobster. Limnol. Oceanogr. **7,** 414–421.

HUNER, J. V., and BARR, J. 1980. Red swamp crawfish: biology and exploitation. Louisiana Sea Grant Program, Publ. No. LSU-T-80-001, Louisiana State University, Baton Rouge, Louisiana.

HUTCHINSON, G. E. 1957. A Treatise on Limnology, Vol. I. Geography, Physics, and Chemistry. John Wiley & Sons, New York.

IRWIN, W. H., and STEVENSON, J. H. 1951. Physiochemical nature of clay turbidity with special reference to clarification and productivity of impounded wates. Oklahoma Agric. Mech. College Bull. Arts and Science Studies, Biol. Serv. **48,** 1–54.

JASPERS, E., and AVAULT, J. W., JR. 1969. Environmental conditions in burrows and ponds of the red swamp crawfish *Procambarus clarkii*. Proc. Southeastern Assoc. Game Fish Commiss. **23,** 592–605.

KLONTZ, G. 1979. Fish health management: Concepts and methods of intensive aquaculture. Fish. Resources Office, Cont. Educ., University of Idaho, Moscow, Idaho.

LAY, B. A. 1971. Applications for potassium permanganate in fish culture. Trans. Am. Fish. Soc. **100,** 813–815.

LOVELL, R. T., and SACKE, L. A. 1973. Absorption by channel catfish of earth-musty flavor compounds synthesized by cultures of blue-green algae. Trans. Am. Fish. Soc. **102,** 774–777.

LOYACANO, H. 1967. Some effects of salinity on two populations of red swamp crawfish, *Procambarus clarkii*. Proc. Southeastern Assoc. Game Fish Commiss. **21,** 423–434.

McSHAN, M., TRIEFF, N. M., and GRAJCER, D. 1974. Biological treatment of wastewater using algae and *Artemia*. J. Water Pollut. Control Fed. **46,** 1742–1743.

MELANCON, E. J., JR. 1975. Design and use of a continuous-flow system for determining the oxygen tolerance of juvenile red swamp crawfish, *Procambarus clarkii*. M.S. Thesis, Louisiana State University, Baton Rouge, Louisiana.

MITCHELL, B. D., and GEDDES, M. C. 1977. Distribution of brine shrimps *Parartemia zietziana* Sayce and *Artemia salina* (L.) along a salinity and oxygen gradient in a south Australian saltfield. Freshwater Biol. **7,** 461–467.

MORRISON, G. 1971. Dissolved oxygen requirements for embryonic and larval development of the hardshell clam, *Mercenaria mercenaria*. J. Fish. Res. Board Can. **28,** 379–381.

NEELY, W. W. 1958. Irreversible drainage—a new factor in waterfowl management. North Am. Wildl. Conf. **23,** 342–348.

PATRICK, W. H., and TUSNEEM, M. E. 1972. Nitrogen loss from flooded soil. Ecology **53,** 735–737.

PERRY, W. G., JR. 1971. Salt tolerance and factors affecting crawfish production in coastal marshes. Presented 2nd Annu. Meet. Louisiana Crawfish Farmers Association, September 14, 1971, University of Southwestern Louisiana, Lafayette, Louisiana.

POPE, P., COLT, J., and LUDWIG, R. 1981. The Environmental Requirements of Crustaceans. *In* The use and potential of Aquatic Species for Wastewater Treatments. California State Water Resources Control Board, Sacramento, California.

REE, W. R. 1965. Emergency alum treatment of open reservoirs. J. Am. Water Works Assoc. **55,** 275–289.

ROMAIRE, R. P., and BOYD, C. E. 1978. Predicting nighttime oxygen depletion in catfish ponds. Alabama Agric. Exp. Stn. Bull. 505. Auburn University, Auburn, Alabama.

RYTHER, J. H. 1956. Photosynthesis in the ocean as a function of light intensity. Limnol. Oceanogr. **1,** 61–70.

SAWYER, C. N., and McCARTY, P. L. 1978. Chemistry for Environmental Engineering, 3rd Ed. McGraw-Hill Book Co., New York.

SCHMIDT-NIELSEN, K. 1979. Animal Physiology: Adaptation and Environment. Cambridge University Press, New York.

SCHROEDER, G. L. 1975. Nighttime material balance for oxygen in fish ponds receiving organic wastes. Bamidgeh **27,** 65–74.

SHARP, J. 1976. The effects of dissolved oxygen, temperature, and weight on respiration of *Macrobrachium rosenbergii*. Department of Water Science and Engineering, University of California, Davis, Pap. 4501. California.

SICK, L., and BEATY, H. 1974. Culture techniques and nutrition studies for larval stages of the giant prawn, *Macrobrachium rosenbergii*. Tech. Rep. Ser. No. 74-5. Georgia Marine Science Center, University System of Georgia, Skidway Island, Georgia.

SIMMONS, E. G. 1957. An ecological survey of the upper Laguna Madre of Texas. Institute of Marine Science University of Texas Ser. **4,** 156–200.

SMITH, C. E., and PIPER, R. G. 1975. Lesions associated with chronic exposure to ammonia. *In* The Pathology of Fishes. W. E. Ribelin and G. Migaki (editors), University of Wisconsin Press, Madison, Wisconsin.

SMITH, L. L., JR., OSEID, D. M., KIMBALL, G. L., and EL-KANDELGY, S. M. 1976. Toxicity of hydrogen sulfide to various life history stages of bluegill *(Lepomis macrochirus)*. Trans. Am. Fish. Soc. **105,** 442–449.

SOGELOOS, P., and PERSOONE, G. 1975. Technological improvements for the cultivation of invertebrates as food for fishes and crustaceans. II. Hatching and culturing of the brine shrimp, *Artemia salina* L. Aquaculture **6,** 303–317.

SPOTTE, S. 1979. Fish and Invertebrate Culture. John Wiley & Sons, New York, New York.

STRICKLAND, J. D. H., and PARSONS, T. R. 1972. A practical handbook of seawater analysis, 2 ed. Fisheries Res. Board Can. Bull. 167.

STRUMM, W., and MORGAN, J. J. 1970. Aquatic Chemistry. John Wiley & Sons, New York, New York.

SWINGLE, H. S. 1961. Relationships of pH of pond waters to their suitability for fish culture. Proc. Pacific Sci. Congr. 9, 1957, **10,** 72–75.

SWINGLE, H. S., and SMITH, E. V. 1947. Management of farm fish ponds. Alabama Agric. Exp. Stn. Bull. 254. Auburn University, Auburn, Alabama.

SWINGLE, H. S., and GOOCH, B. C., and RABANAL, H. R. 1963. Phosphate fertilization of ponds. Proc. Southeastern Assoc. Game Fish Commiss. **17,** 213–218.

TABB, D. C., and YANG, W. T. 1972. A manual for culture of pink shrimp *Penaeus duorarum,* from eggs to postlarvae suitable for stocking. Sea Grant Spec. Bull. No. 7. University of Miami, Miami, Florida.

TOMASSO, J. R., SIMCO, B. A., and DAVIS, K. 1979. Chloride inhibition of nitrite induced methemoglobinemia in channel catfish *(Ictalurus punctatus)* J. Fish. Res. Board Can. **36,** 1141–1144.

TOTH, S. J., and RIEMER, D. N. 1968. Precise chemical control of algae in ponds. J. Am. Water Works Assoc. **60,** 367–371.

THOMASTON, W. W., and ZELLER, H. O. 1961. Results of a six-year investigation of chemical soil and water analysis and lime treatment in Georgia fish ponds. Proc. Southeastern Assoc. Game Fish Commiss. **15,** 236–245.

TRUSSEL, R. P. 1972. The percent of un-ionized ammonia in aquaeous ammonia solutions at different pH levels and temperatures. J. Fish. Res. Board Can. **29,** 1505–1507.

TUCKER, C. S., and BOYD, C. E. 1977. Relationships between potassium permanganate treatment and water quality. Trans. American Fish. Soc. **106,** 481–488.

TUCKER, C. S., and BOYD, C. E. 1978. Consequences of periodic applications of copper sulfate and simazine for phytoplankton control in catfish ponds. Trans. Am. Fish. Soc. **107,** 316–320.

U.S. ENVIRONMENTAL PROTECTION AGENCY. 1974. Methods for chemical analysis of water and wastes. EPA-625-/6-74-003a. Environmental Monitoring and Support Laboratory, Environmental Research Center, Cincinnati, Ohio.

U.S. ENVIRONMENTAL PROTECTION AGENCY 1976. Quality criteria for water. Superintendent of Documents, Washington, D.C.

WATERS, T. F. 1956. The effects of lime application to acid bog lakes in northern Michigan. Trans. Am. Fish. Soc. **86,** 329–344.

WILLEY, B. F., JENNINGS, H., and MUROSKI, F. 1964. Removal of hydrogen sulfide with potassium permanganate. J. Am. Water Works Assoc. **56,** 475–479.

WOOSTER, W. S., LEE, A. J., and DIETRICH, G. 1969. Redefinition of salinity. J. Mar. Res. **27,** 358–360.

ZEIN-ELDIN, Z. P., and ALDRICH, D. V. 1965. Growth and survival of postlarval *Penaeus aztecus* under controlled conditions of temperature and salinity. Biol. Bull., Woods Hole, Mass. **129,** 199–216.

ZEIN-ELDIN, Z. P. and GRIFFITH, G. W. 1966. The effect of temperature upon the growth of laboratory-held postlarval *Penaeus aztecus*. Biol. Bull. Woods Hole, Mass. **131,** 186–196.

Appendix

The Brine Shrimp, Genus *Artemia**

Jay V. Huner

INTRODUCTION

Brine shrimp, genus *Artemia*, are small crustaceans that are indispensable to the culture of most finfish and crustacean species with complicated larval development. The newly hatched larvae, called nauplii, are the most common larval food, but adult brine shrimp also constitute valuable foods for several species. There is some interest in culturing adults as human food supplements.

Brine shrimp belong to a complex of phyllopodus crustaceans, the Anostraca, referred to commonly as fairy shrimps. Although this cosmopolitan group of organisms has long been considered monospecific, many taxonomists believe that the genus comprises several species in addition to *Artemia salina*. The common name is based on the typical environment, hypersaline lakes and salterns, inhabited by this group of fairy shrimps.

Fairy shrimps are small, 12–30 mm (0.5–1.25 in.), crustaceans. All species have a delicate, feathery appearance with 11 pairs of fine, oarlike legs called phyllopods. The phyllopods beat almost continu-

*The editors had originally intended to devote an entire chapter to brine shrimp biology and culture but were unable to obtain a manuscript prior to completion of the rest of the text. In this brief appendix, sources of information are not referenced in the conventional manner, but readers who wish to pursue the subject are directed to the three-volume series entitled *The Brine Shrimp Artemia* edited by G. Persoone, P. Sorgeloos, O. Roels, and E. Jaspers and published by Universa Press, Watternen, Belgium, in 1980. Virtually every aspect of brine shrimp biology and culture is discussed in these volumes by the premier authorities in their respective fields.

ously, moving the tiny shrimps gracefully through the water. They normally swim upside down.

All fairy shrimps are found in very harsh environments such as brine lakes and temporary, seasonal freshwater ponds and puddles. They are totally defenseless and must live in places where there are no predators or where they can complete their life cycles before predators appear. Because their habitats are so unstable, often being filled with water less than 3 weeks, various species complete their life cycles, egg to egg, in 2.5–6.5 weeks. Fairy shrimps produce two types of eggs; thin-shelled ones that often hatch while still inside the mother's brood pouch and thick-shelled ones, called cysts, that can withstand drying and freezing, remaining viable for up to 4 years.

BRINE SHRIMP PRODUCTS

The brine shrimp cyst is the product most commonly encountered. Cysts are incubated to produce nauplii that are fed to larval fishes and crutaceans. Brine shrimp nauplii resemble tiny tear drops with three pairs of erector-set limbs. They contain a droplet of yolk material packed with nutrients that make them excellent feeds.

In the United States, most brine shrimp come from the Great Salt Lake in Utah and salterns near San Francisco, California. Two forms of brine shrimp are harvested, adults and cysts. Adults are frozen or dried; the cysts are collected after they wash ashore and dry. Although cysts are not overly abundant, there is hardly a shortage; however, cyst sales are tightly controlled and prices are quite high at the retail level. For example, 5 gr (0.18 oz) cost around $2 but will yield as many as 1 million nauplii occupying 30 ml of space. This is equivalent to about $330 per kg ($150 per lb). Bulk prices are far more reasonable.

High prices attest to the great demand for brine shrimp cysts. Much research has gone into the development of artificial larval foods that can replace brine shrimp nauplii, but success has been limited. Another solution to the problem has been the planting of brine shrimp in suitable lakes and salterns where they did not/do not exist. This has been done in several areas of the world and has been responsible for the stabilization of bulk prices for cysts.

Cysts are covered by a very tough "shell." Hatching success can be limited by the failure of nauplii to escape from the shell. The outer cyst wall can be removed chemically without harm to the inactive embryo within. This product can be purchased.

Other brine shrimp products besides cysts are available. Both live and frozen adult brine shrimp are sold as feeds for juvenile and adult fishes and crustaceans. Although it is very easy to grow brine shrimp, few people do it commercially, so prices are high. Live brine shrimp retail at prices of about $5 for several milliliters. Because frozen brine shrimp are harvested in large volumes, prices are much more modest, around $11 per kg ($5 per lb).

LIFE CYCLE

The life cycle usually begins with the hydration (absorption of water) of the inactive, dry cyst. Within hours, the outer walls burst and the embryo appears, surrounded by a hatching membrane. As the embryo leaves the cyst shell, the hatching membrane bursts, releasing the nauplius. The brine shrimp develops gradually, molting 15 times before assuming the mature adult form. The nauplius has only three pairs of appendages. New appendages are added with each molt through the 10th molt, at which time all segments and appendages are present. The nauplius beings to eat after the 3rd molt (3 or 4 days after hatching) when its digestive tract becomes functional. Food includes yeast cells, bacteria, and simple algae cells. Structural changes leading to sexual differentiation begin after the 10th molt. The first pair of limbs (antennae) on the male assume a grotesque fang-shaped appearance at maturity. These are used to cling to the female while breeding takes place. A male and a female will swim about, clasped together for many hours, oblivious to everything taking place around them.

A female may produce 10 or 11 broods over a 50- to 60-day life span, although many brine shrimp live as long as 6.5 months. Average brood size is 134. Females produce two kinds of eggs: thin-shelled eggs that yield free-swimming nauplii or heavy-shelled cysts that contain partly developed embryos when laid. The partly developed embryos can remain dormant for many months. The cysts are produced when conditions become harsh such as the onset of cold weather or the evaporation of a pond or pool. When waters are evaporating rapidly, oxygen concentrations vary widely from day to night and salinities reach their highest levels. In some areas, females may produce viable offspring in the absence of males by parthenogenesis.

Brine shrimp are very tolerant of adverse conditions. For example, they can thrive in salinities of 80–100 ppt. They survive oxygen levels of 1–2 ppm. In fact, if they are subjected to low oxygen levels for any length of time, they turn red. This is because they produce hemoglobin in its molecular form and it circulates freely in the hemolymph.

Brine shrimp are filter feeders. Initially, setae on the antennae filter food particles from the environment. However, a very efficient filtering system replaces the antennal system as development progresses. In more advanced juveniles and adults, the space between the legs widens as the legs move forward on the forestroke. Water is drawn into this space from the area below the midline of the body. Small filtering setae collect particles, including food items, from the incoming stream. On the backstroke, water is forced out through the space between the legs, but the food is concentrated in a food groove below the bases of the legs. The food groove leads to the mouth. Trapped food particles are transferred to the mouth by several complex mechanisms. Glands along the groove secrete adhesive material that clumps the particles together into balls. The rates of food capture and consumption are relatively constant. When food is present at high concentrations, digestion is extremely inefficient and brine shrimp can actually starve because the food passes through the gut too quickly to properly remove nutrients. The positive phototaxis of all brine shrimp life stages may cause problems in culturing them; bright lights can attract so many brine shrimp in a culture that feeding efficiency is reduced significantly.

As previously mentioned, brine shrimp may take on a red color from the synthesis of hemoglobin. They will also take on the color of the dominant food item and may become green, brown, or red depending on the principal pigment of algal blooms on which they are feeding.

BRINE SHRIMP CULTURE

Some consider the management and harvest of brine shrimp and their cysts from salinas in southern California to be a form of extensive aquaculture. This is certainly a lucrative business, but there are very few places in the United States where such systems can be duplicated on a commercial scale. Most attempts to culture brine shrimp to the preadult/adult stages have involved intensive static or flow-through systems based on the work of P. Sorgeloos, G. Persoone, and co-workers at the State University of Ghent, Ghent, Belgium.

An array of feeds can be used in the culture of brine shrimp. These include live and frozen algae, yeast suspensions, flour, and rice bran (defatted is preferred). Ideal particle size is 50μm. Feed suspensions should be about 10–15 g/m^3. Algal cells densities should exceed 6×10^4

cells/ml. Cysts may or may not be incubated directly in the culture system but require constant illumination and periodic agitation to insure maximal hatching success. Excellent results have been obtained at salinities of 28–32 ppt and temperatures of 28°–30°C (82°–86°F). Food conversion values range from 1:1 to 2:1.

A 2-week batch culture in an airlift pump-operated raceway can produce 2–3 preadults/ml after innoculation with 10 g of cysts. This is equivalent to about 2 kg of live weight biomass/m^3. A flow-through raceway system might produce as much as 20 kg of live weight biomass/m^3 from an innoculation of 15 g of cysts in 2 weeks. (One kg of dry brine shrimp biomass is equivalent to about 10 kg of fresh brine shrimp biomass.) For comparison purposes, continuous outdoor static cultures with 400 adults/liter have generated 5 g (live weight) of brine shrimp/m^3/day, while indoor, intensive systems with daily medium renewal have produced 100 g (live weight)/m^3/day.

Index

A

Abalone
- biochemical production techniques, 397-404
  - growth acceleration, 403-404
  - juvenile development, 397-399
  - nutritional efficiency, 402-404
  - seawater bioassay, 399-400
  - thermophilic species, 400-402
- biology, 376-385
  - anatomy and function, 376-378
  - diseases, 383
  - growth, 382, 383
  - larval development, 379-380
  - metamorphosis, 380-382
  - nutrition, 382-383
  - parasites, 383
  - reproduction, 378-379
  - settlement, 380-382
- black, 372, 374, 393, 401
- declining fishery, 367-369
- definition, 366
- developing aquaculture industry, 369-371
- distribution, 371, 373
- flat, 372, 376
- genetic engineering techniques, 403-404
- green, 372, 374-375, 393, 401
- grow-out systems, 388-390
- *Haliotis discus hannai,* 405
- industry outlook, 407-409
- interspecific hybrids, 376
- larvae
  - development, 379-380, 387-388
  - metamorphosis, 380-382, 387
  - settlement, 380-382, 387-388
- ocean ranching, 389-390, 404-407
  - Japan, 404-405
  - United States, 405-407
- pink, 372, 373-374, 393, 401
- pinto, 373, 375
- predators, 368, 383-385, 393, 395, 396-397, 406, 408
- principal species, 371-376
- problems, 390-397
  - coastal pollution, 392-393
  - defective juveniles, 390-392
  - inefficient food conversion, 394-395
  - juvenile supply, 393-394
  - ocean grow out, 395-397
  - seed supply, 393-394
  - seed survival, 390-392
  - slow growth, 394-395
- red, 372, 373, 393, 398, 401
- reproduction control, 385-387
- retail prices, 366
- threaded, 372, 375
- white, 372, 375
- world market, 366
- worldwide production, 366-367

Aeration devices, 24, 428-430

Aerators, 23, 429, 430

Aerococcus viridans, 190

Aeromonas, 16

Alabama, penaeid shrimp culture, 145

Alaska
- abalone fisheries, 367
- oyster culture, 252

Algae, *see also* Phytoplankton
- biotoxins, 256, 337, 344
- blooms
  - dissolved oxygen effects, 425, 426
  - in prawn culture, 86, 98-99, 222
  - red tide, 256, 337, 344
  - in shrimp culture, 142, 146
- blue-green, 445
- brown, 382, 388, 389, 408

Algae *(continued)*
  carbon dioxide removal, 434-435
  crustose, 380-381
  culture methods
    Milford, 284-286
    tanks, 262, 263, 264
    Wells-Glancy, 283, 284, 285
  epibionts, 105-106
Algal feeds
  abalone, 382, 388, 389
  oysters, 262-264
  penaeid shrimp, 142
  prawns, 87
Algicides, 448
Alkalinity, 430, 432-433
  measurement, 437-438
  total, 436-438
Alligator weed, 25, 44
Alum, 422, 433
Amino acids, 184
γ-Aminobutyric acid, 398-399, 400, 401, 402
Ammonia, 439-441
  ionized, 439, 440
  toxicity, 84-85
  un-ionized, 439, 440
Ammonification, 438-439
Ammonium fertilizer, 433
Amphora, 388
Anax junius, 34
Ancistrocama, 251, 296
Anostraca, 457
Antimycin, 34
Anuenue system, 89-91, 93
Aphanomyces astaci, 6, 15
Aquacell, 144
Aquacop system, 89-93
Aquaculture, *see also* individual species
  world yield, 300
Arasan, 35
Archaeogastropods, 376
Arctica islandica, see Ocean quahog
Arkansas
  crawfish culture, 2, 46, 47
  prawn culture, 110
Artemia, see Brine shrimp
Arthropoda, 163
Artificial insemination, 128, 170-171
Astacidae, 3, 5
Asterias, see Starfish

B

Backswimmer, 34
Bacteria
  chemoautopic, 39
  chitinivorous, 15-16
  chitinoclastic, 106
  planktonic, 444
Bacterial disease
  clams, 296
  crawfish, 15-16
  lobsters, 189, 190
  oysters, 251
  penaeid shrimp, 146-147
  prawns, 105
Baits, 2, 27-30
Balanus glandula, 217
Barnacle, 25
Bat ray fish, 251
Baytex, 34
Beetles, as predators, 34
Belon oyster, *see* Oyster, European
Belostoma lutaricum, 34
Benlate, 35
Bicarbonates, 434-435, 436, 437
Bigmouth buffalo, 37-38
Birds, as predators, 34, 298, 342-343
Black drum, 245
"Black spot" disease, 105, 106, 107
Blue crab, 204-215
  cannibalism, 207
  crab pounds, 207-208
  as crawfish predator, 33
  eggs, 206
  feeds, 205
  hard-shelled, 204, 214
  larval development, 207
  life span, 205
  market value, 214-215
  megalopal stage, 206-207
  molting, 208, 209
  as oyster predator, 246
  peeler stage, 207, 208, 209, 214, 215
  range, 204-205
  reproduction, 205-206
  shedding systems
    floats, 207, 208-210
    on-shore tables, 208, 211-213
    open flow-through, 208, 212
    operation, 214
    water management, 212-213
  soft-shelled, 204

spawning, 206
zoeae, 206
Bluegill, 2
Bodo, 16
Bottom culture, 314, 315, 321, 326
Bouchot culture, 314-316
Box dredge, 243
Brachiobdellid worms, 16-17
Brackish water, 418-419
Brine shrimp, 457-461
as crab feed, 217
culture, 178-179, 460-461
dissolved oxygen effects, 427
life cycle, 459-460
as lobster feed, 176-177, 178-179, 183
as penaeid shrimp feed, 141-142
as prawn feed, 87, 88, 93-94
products, 458-459
salinity range, 461
temperature range, 419
Brooding tray, 175
Burrow's golden eye, 342
Busycon, 298
Butter clam, 277, 278, 279

C

Cabezon, 385
Cadmium reduction procedure, 441
Calcite, 436
Calcium carbonate, 10, 86
Calcium hydroxide, 428, 435
Calcium hypochlorite, 443
Calcium sulfate, 433
California
abalone
culture, 385, 388-389, 391, 392-394, 395-397, 406-407, 408
fisheries, 367-370, 373, 374-376
clam culture, 278, 302
crawfish culture, 5
oyster culture, 250, 251, 252, 257
prawn culture, 110
spiny lobster culture, 227
vibriosis occurrence, 297
Callinectes sapidus, see Blue crab
Cambaridae, 3
Cancer boralis, see Crab, jonah
C. magister, see Crab, Dungeness
C. irroratus, see Crab, rock
Cannibalism, 167, 176, 178, 207
Captan, 35
Carapace length, 163, 165
Carbon dioxide, 433-436
acidic action, 434
algal removal, 434-435
analysis, 449-450
measurement, 435-436
pH, effects, 430
sources, 433-434
Carbonates, 434-435, 436, 437
Carotenoids, dietary, 187
Carp, 447
Catfish, 2, 37-38
Cat's clay, 432
Cestodes, 296
Channel catfish, *see* Catfish
Cherax destructor, 6
C. tenuimanus, 6, 42
Chionoecetes, see Crab, snow
Chlamydiae, 296
Chlorine, 443-444
Ciliate(s), 251, 296
Ciliate gill disease, 146
Citrobacter freudii, 16
Clam, 275-310, *see also* individual species
biology, 280-282
candidate species, 276, 278-280
commercial fisheries, 276-278
cultural constraints, 297-300
diseases, 295-297
economics, 303
grow-out systems, 293-295
hatchery systems, 282-288
algal culture, 283-286
brood stock maintenance, 283-284
larval culture, 286-288
larval cycle, 282
nurseries, 288-293
field, 288-290
onshore, 290-292
parasites, 295-297
predators, 298
production, 278
seed production, 301-302
status, 300-302
yield, 275-276
Climatic restrictions
clams, 297-298
oysters, 251, 256, 259
penaeid shrimp, 147-148
prawns, 107-109
Coccidia, 296

Cocconeis, 388
Conch, 245, 298
Concholepas concholepas, 367
Connecticut, vibriosis occurrence, 297
Copepod, parasitic, 251, 259, 296, 343
Copper sulfate, 448
Corixidae, *see* Water boatmen
Corthunia, 16, 105
Crab
  blue, *see* Blue crab
  canceroid, 203-204, 215-219
    culture techniques, 217-219
    fecundity, 216-217
    feeds, 217-218
    larval development, 216, 217
    life histories, 215-216
    soft-shelled, 219
  Dungeness, 215, 216, 217-219
  jonah, 215, 217, 218, 219
  king, 204
  parasitic, 296
  pinnotherid, 296
  predaceous, 251, 255, 298, 343, 384-385
  rock, 215, 216, 217, 218, 219
  snow, 204
  stone, 204
Crab pounds, 207-208
Crassostrea gigas, see Oyster, Pacific
Crassotrea virginica, see Oyster, American
Crawfish, 1-61, *see also* individual species
  annual yield, 2
  Australian, 6, 42
  as bait, 2
  biology, 7-15
    environmental requirements, 11-14
    food habits, 10-11
    genetics, 11
  growth, 9-10
  life cycle, 7
  molting, 9-10
  reproduction, 7-9, 25
  species relationship, 14-15
brood, 24, 25
diseases, 15-18
economics, 49-52
eggs, 8-9, 44
feeds and feeding, 2-3, 25-27, 43-44
food web, 445
harvesting, 3, 27-32
  baits, 27-30
  size effects, 11
    techniques, 30-32
    traps and trapping, 27, 29
  important species, 3-6
  intensive culture, 40-44
  marketing, 51-52
  marron, 6, 42
  pesticide sensitivity, 14
  pests, 15-18
  pond culture, 18-38
    forages, 25-27
    polyculture, 37-38
    pond construction, 18-22
    population monitoring, 32-33
    predator control, 33-35
    rice/crawfish double cropping, 35-37
    stocking, 24-25
    water management, 22-24
  processing, 47-48
  salinity ranges, 420
  signal, 3, 4, 5-6, 7, 43, 44, 45, 53
  soft-shelled, 38-40
  status, 44-47
    climatic restrictions, 44-45
    current acreage, 45-47
    expansion possibilities, 47
  stress, 13
  stunting, 11, 12
  temperature ranges, 11, 12, 419
  yabbie, 6
Crayfish, *see* Crawfish
Crustaceans, 203-204, *see also* individual species parasitic, 296
Ctenopharygodon idella, see Grass carp
Cultch, 240-243, 247-248, 258, 264, 265
Cyprinodon variegatus, see Killifish

D

Daphnia, 204
Dasyatus, 298
Delaware Bay disease, 244
Denitrification, 439
"Dermo", 244, 296
Detrital food chain, 3, 26
Diatoms, 142, 388
2,4-Dichlorophenoxyacetic acid, 35-36
Difolatan, 35
N,N-Dimethyl-*p*-phenylenediamine, 444
Dinoflagellates, 339, 344
Diseases, *see also* under individual species

bacterial, 15-16, 105, 146-147, 189, 190, 251, 297
fungal, 6, 15, 18, 189, 297
nutritional, 146
protozoal, 16, 146, 244
viral, 146, 147, 245
Dissolved oxygen, 422-430
aeration, 24, 428-430
analysis, 449-450
compensation point, 424
diel fluctuation, 425-426
losses, 423-424
low concentration, 426-430
monitoring, 430, 431
phytoplankton effects, 423, 424, 425, 426
sources, 423, 424
species ranges
crawfish, 11, 12, 13, 23, 24
lobsters, 177
penaeid shrimp, 136
prawns, 83-84
Diuron, 448
Diving beetle, 34
Dolomite, 436
Dorosoma cepedianum, see Gizzard shad
Double cropping, 35-37, 45
Dragonfly, 34
Ducks
as predators, 34, 298, 342-343
as trematode hosts, 339
Dunaliella, 284

E

Ecdysis, *see* Molting
Ecdysone, 10, 40
Economics, *see under* individual species
Ectocommensals, 105-106
Egg(s), *see under* individual species
Egg collector, 140
Egregia, see Kelp
Ehydra lutris, see Sea otter
Eider duck, 339, 342
Eisenia, see Kelp
Electro-trawl, 30-31
Endotoxins, 16
Ensis, see Razor clam
Epibionts, 105-106
Epicommensals, 146
Epilimnion, 424, 427
Epistylis, 16, 105
Eupleura, see Oyster drill
Euryhaline, 419
Evasterias troschelii, 343
Extensive culture, 128-129, 136, 145
Eyestalk ablation
blue crab, 214
lobster, 179
penaeid shrimp, 139-140

F

Fairy shrimp, *see* Brine shrimp
Feeds, *see also* under individual species
cover crops, 3
detritus food chain, 3, 26
microencapsulated, 182-183
water stability, 88
Fertilizers, 428, 433, 446-447
Filter feeders, 238-239
Fish, *see also* individual species
carbon dioxide effects, 435
dissolved oxygen effects, 427-430
osmotic pressure effects, 419
oxygen depletion effects, 423
pH effects, 432
as predators, 107, 245, 298, 343, 385, 396
temperature effects, 416-418
Fish crow, 298
Fish kills, 422-423
Flatworms, as predators, 246, 251, 259
Flavobacterium, 16
Floating boxes, 207, 208-210
Florida
penaeid shrimp culture, 147
prawn culture, 110
soft-shell crab prices, 214
Flume, 261
Food web, 444-445
Forages, 25-27
Formalin, 18, 427
Fouling organisms, 224, 251
Fungal diseases
clams, 297
crawfish, 6, 15, 18
lobsters, 189
Fungicides, 35
Furadan, 36

G

GABA, *see* γ-Aminobutyric acid
Gaffkemia, 189-190

Galveston hatchery method, 128, 141, 142
Gastrolith, 10
Gastropods, 245
Gelidium, 382
Genetic engineering, 403-404
Genetic marking, 406
Geoducks, 277, 278
Georgia
  clam culture research, 302
  grass shrimp culture, 229
  prawn culture, 110
Gigartina, 382
Gizzard shad, 27-28, 29
Glair, 8, 25
Gonyaulax catenella, 337, 344
G. excavata, 344
G. polyedra, 344
G. tamarensis, 256
Grass carp, 447
Green crawfish, *see* Crawfish, soft-shelled
Gregarines, 146, 296
Grow-out systems
  abalone, 388-390
  clams, 293-295
  lobsters, 177-181
  penaeid shrimp, 129, 143-145
  prawns, 98-104
Growth, hormone-stimulated, 402-404
Gymnophallus bursicola, 339-340
Gypsum, 433

H

Habitat traps, 227
Haliotis, see Abalone
H. corrugata, see Abalone, pink
H. cracherodii, see Abalone, black
H. fulgens, see Abalone, green
H. kamtschatkana, see Abalone, pinto
H. kamtschatkana assimillis, see Abalone, threaded
H. rufescens, see Abalone, red
H. sorenseni, see Abalone, white
H. walallensis, see Abalone, flat
Haplosporidians, 296, 343-344
Hard clam, 217, 276
  commercial hatcheries, 301-302
  diseases, 296
  distribution, 280-281
  field nursery, 288
  grow-out system, 293-294
  larval development, 287
  onshore nursery, 290
Harvesters, 242
Harvesting, *see under* individual species
Hatchery methods
  clams, 282-288
  crabs, 217
  mussels, 337-338
  oysters, 236, 238, 239, 259-266
  penaeid shrimp, 128-129, 141-142, 148
  prawns, 64, 89-94
Hawaii
  crawfish culture, 47
  oyster culture, 252
  penaeid shrimp culture, 129, 147, 149, 150, 151
  prawn culture, 64
    economics, 111-112, 115
    methods, 86, 89, 90, 99
    size distribution, 79
    status, 109-110
Hay, as detrital substrate, 26
Heated effluent culture, 97, 328, 344-345
Heavy metals, 449
Helminths, 16-17, 296
Hemipterans, 34
Herbicides, 35-36, 427, 447-448
Heterodonta, 280
Hildenbrandia, 381
Hippolysmata wurdemanni, see Shrimp, candy-striped dancing
Holorhinus californicus, see Bat ray fish
Homarus, see Lobster
H. americanus, see Lobster, American
H. gammarus, see Lobster, European
Hormones
  genetically-engineered, 403-404
  growth-accelerating, 403-404
  molt-inhibiting, 165
  molt-stimulating, 10
  spawning-inducing, 400-401, 402
Hyaloklossia, 296
Hybridization
  abalone, 376
  lobsters, 174-175
  penaeid shrimp, 128
Hydrogen peroxide, as spawning inducer, 400-402, 407
Hydrogen sulfide, 442-443
Hymenocera elegans, see Prawn, painted
Hypolimnion, 424, 427, 434
Hyporochlorous acid, 443

I

Ictalurus, see Catfish
Ictiobus cyprinellus, see Bigmouth buffalo
Insect control, 34
Insecticide toxicity, 14
Intensive culture
  penaeid shrimp, 128, 137, 143-144, 151
  prawns, 102-104
Intermolt, 76
Isochrysis, 284

J

Jellyfish, as predators, 107

K

Kelp, 382, 388, 406-407
Killifish, 146
Kocide, 35

L

Lagenidium, 189
Lagenophrys, 105
Largemouth bass, 2
Larviculture, *see* Hatchery methods
Lepomis macrochirus, see Bluegill
Leucothrix, 105, 146
Levees, 18, 20, 21-22
Life cycle, *see under* individual species
Lime, 24, 432-433, 435
Lipids, dietary, 185-186
Lithophyllum, 381
Lithothamnium, 381
Littleneck clam, 277, 278, 279
Live car, 207
Lobster, 159-201
  as abalone predator, 385
  American, 159, 164
    consumer demand, 193
    egg development, 175
    market size, 161
    metabolic activity, 185
    reproductive cycle, 168, 172, 174
  biology, 162-167
    anatomy, 163-164
    aggressive behavior, 166-167
    feeding behavior, 166-167
    growth, 164-166, 174
    molting, 164-166
    morphology, 163-164
  brood stock, 167-175
    artificial insemination, 170-171
    controlled mating, 169-170
    egg development rate, 172-173
    egg extrusion, 169, 170, 171-173, 174-175
    nutrition, 174
    photoperiod effects, 171-174
    reproductive cycle, 168-169, 171
    sources, 167-168
    temperature regulation, 172-174
  cannibalism, 167, 176, 178
  coral, 160
  culture facilities
    design, 191, 192-193, 195
    site selection, 191-192, 195-196
  diseases, 189-191
  European, 159, 161
    digestion, 184
    egg hatching, 175
    feeding pattern, 183
    reproduction, 168, 169, 174
  eyestalk ablation, 179
  feeds and feeding, 166, 178-179, 181-189
    consumption, 183-184
    conversion efficiency, 188-189
    digestion, 183-184
    physical characteristics, 181-183
  grow-out systems, 177-181
    associated growth factors, 179, 181
    early juvenile, 178-179
    larger juvenile, 179
    water quality, 177
  habitat, 165-166
  hybridization, 174-175
  industry outlook, 194-196
  larvae
    development, 160
    rearing, 175-177
  marketing, 193-194
  mating behavior, 167
  molting, 164-166, 172
  Norway, 160-161, 184
  nutritional requirements, 184-188
    carotenoids, 187
    energy, 185
    lipids, 185-187
    minerals, 188

Nutritional requirements *(cont.)*
  protein, 184-185
  vitamins, 187-188
slipper, 160
sperm banks, 171
spiny, 225-228
  culture, 226-228
  life history, 226
temperature ranges, 160, 191-192, 419
Long-line culture, 317-319, 321, 323, 324, 325
Louisiana
  crawfish culture, 2, 7
    double cropping, 35, 36-37
    economics, 49, 50
    harvesting, 32
    marketing, 51-52
    polyculture, 37-38
    pond culture, 18, 19, 21, 22, 26, 27
    processing, 47-48
    shedding system, 39
    total acreage, 45, 46
  oyster culture, 246
  penaeid shrimp culture, 137
  prawn culture, 110
  soft-shell crab prices, 214
Ludvigia, see Water primrose
Lysmata grabhami, see Shrimp, lady

M

Macoma balthica, 296
Macrobrachium, see Prawn
M. rosenbergii, see Prawn
Macrocystis, see Algae, brown
Macrophytes, 446, 447-448
Mahogany clam, *see* Ocean quahog
Maine
  mussel culture, 344
    experimental culture, 331, 332, 333, 334, 335
    feasibility, 319-320
    heated effluent culture, 344-345
    methods, 321, 323, 324, 326
    pearl incidence, 338, 339, 341
  oyster culture, 236, 242, 252-256, 264
  vibriosis occurrence, 297
Malathion, 36
Manila clam, 277, 278, 279
  field nursery, 288
  grow-out system, 294
Marketing, *see under* individual species
Marsh culture, 21-22
Maryland
  clam culture research, 302
  oyster culture, 242, 244, 246
Mating behavior
  lobsters, 167
  prawns, 72-75
Maturation tanks, 138, 140
Megalopa, 206-207, 216
Melanitta, see Scoter
Menippe mercenaria, see Crab, stone
Mercenaria arenaria, see Soft-shell clam
M. campechiensis, see Hard clam
M. mercenaria, see Hard clam
Metapenaeus joyneri, 128
M. monoceros, 128
Methemoglobin, 441
Methyl parathion, 34
Methylene blue method, 443
Microbial disease, 296, *see also* specific diseases
Microprosthema semilaevis, see Shrimp, scarlet flame
Micropterus salmoides, see Largemouth bass
Microsporodians, 146
Minchinia costalis, see Seaside disease
M. nelsoni, see Delaware Bay disease
Minerals, dietary, 188
Mink, as predators, 34
Mississippi
  crawfish culture, 2, 32, 46, 47, 52
  prawn culture, 110
"Molt death" syndrome, 186
Molt-inhibiting hormone, 165
Molt-stimulating hormone, 10
Molting, *see also under* individual species
  induction, 10, 40
  vitellogenesis and, 172
Monochrysis, 284
Moon snail, 259, 298
Moss-back condition, 105-106
MSX, *see* Delaware Bay disease
Mussel, 311-363
  algal biotoxin, 337, 344
  annual landings, 312, 313-314
  bottom culture, 314, 315, 321, 326
  carrying-capacity model, 345-350
  commercial operations, 320-327
  economics, 350-354
  eggs, 328, 329, 330

European culture technology, 314-320
bottom culture, 314, 315
bouchot culture, 314-316
feasibility in U.S., 319-320
long-line, 317-319
raft, 316-317
experimental culture
East Coast, 331-337
West Coast, 337-338
growth rate, 313, 327, 333, 334, 335, 336, 337
hatcheries, 337-338
heated effluent culture, 344-345
industry status, 314
life cycle, 328-331
long-line culture, 317-319, 321, 323, 324, 325
marketable size, 312-313, 327, 333
maximum sustainable yield, 312
pearl incidence, 326, 338-342
predators, 342-343
production, 321
raft culture, 316-317, 321, 322, 323, 324, 332-333
research, 327-328
salinity ranges, 351
seed mussels, 333, 335-337, 338
temperature ranges, 345, 351
Mya arenaria, see Soft-shell clam
Mycoplasma, 296
Mycosis, larval, 297
Myliobatis, 298
Mysidopsis, 204
Mytilicola, 296
M. intestinalis, 343
M. orientalis, 251, 259
Mytilus edulis, see Mussel
M. galloprovincialis, 340

N

National Aquaculture Development Plan, 147, 300
National Sea Grant Program, 162
Nauplii, 457, 458
Navicula, 388
Nematopsis, 296
Neomysis, 204
Neoplasia, 296
Nephropidae, 160-161
Nephrops norvegicus, see Lobster, Norway
Nereocystis, see Kelp
Nesslerization technique, 441
New Jersey, clam culture research, 302
New York
crawfish culture, 2, 4
oyster culture, 241
vibriosis occurrence, 297
Nitrate, 39, 441
Nitrification, 84, 438, 439, 441
Nitrite, 84, 85, 439, 441
Nitrobacter, 439
Nitrogen cycle, 438
Nitrogen fixation, 439
Nitrogenous compounds, 438-441
measurement, 441
as prawn culture waste, 84-85
sources, 438-439
Nitrosomonas, 439
Nitrous acid, 441
Nitzschia, 388
North Carolina, oyster culture, 242
Nursery methods
clams, 288-293
prawns, 94-98
Nursery tanks, 96, 97
Nutritional diseases, 146

O

Ocean quahog, 277, 281
Ocean ranching, 389-390
Japan, 404-405
U.S., 405-407
Ocenebra japonica, see Oyster drill
Octopus, as predator, 384
Ohio, crawfish culture, 7
Oidemia nigra, see Scoter
Oncorhynchus, see Salmon
Orconectes, 39, 52-53
O. immunis, 3, 4, 7
O. nais, 44
O. rusticus, 7, 44
O. virilis, 5, 7, 43, 44
Ordram, 35, 36
Oregon
abalone fisheries, 367
clam culture, 278
crawfish culture, 5
oyster culture, 250, 251-252, 256-257
Ortmannicus, 6
Osmotic pressure, 419

Ostrea, see Oyster, cupped
O. edulis, see Oyster, European
O. lurida, see Oyster, Olympia
Oxygen solubility, 423, *see also* Dissolved oxygen
Oyster, 235-273
  American, 236, 237, 239-247
    culture, 240-244
    diseases, 244-245
    economics, 246-247
    harvesting, 242, 243, 244
    parasites, 245-246
    seeding, 240-243
    spawning, 240
  cupped, 236, 238, 239, 261-262
  European, 236, 237, 252-256, 257
    culture, 252-255
    parasites, 255
    predators, 255
    feeds and feeding, 238-239
  flat, 236, 238, 239, 261, 262
  food web, 445
  habitat, 239
  hatcheries, 259-266
    larval culture, 262-266
    methods, 260-262
  industry outlook, 267-268
  industry problems, 266-267
  landings, 236, 238, 246
  larval development, 236, 238, 239
  natural history, 236, 238-239
  Olympia, 236, 237, 256-259
    climatic restrictions, 259
    culture, 257-258
    diseases, 258-259
    economics, 259
    parasites, 259
    predators, 259
  Pacific, 236, 237, 247-252
    climatic restrictions, 251, 256
    culture, 247-250
    diseases, 250-251
    economics, 251-252, 256
    parasites, 251
    predators, 251
  per capita consumption, 267
  predators, 245-246
  reproduction, 236, 238
  salinity ranges, 420
  spawning, 261-262
  temperature ranges, 419
Oyster drill, 245, 251, 255, 259, 298

P

Pacifastacus, 10, *see also* Crawfish
P. leniusculus, see Crawfish, signal
Paddlefish, 37-38
Palaemonetes, see Shrimp, grass
Palaemonidae, 65
Pandalus platyceros, see Spot prawn
Panulirus, see Lobster, spiny
P. argus, see Lobster, spiny
P. interruptus, see Lobster, spiny
Paragonemus kellcoti, 16, 17
P. westermani, 16, 17
Paralithodes camtschatica, see Crab, king
Paralytic shellfish poisoning, 256, 278
Parasites, *see also* names of specific parasites
  abalone, 383
  clams, 295-297
  crawfish, 16-18
  oysters, 245-246, 251, 255, 259
  penaeid shrimp, 145-147
Parastacidae, 6
Pavlova lutheri, 284
Pearls, 326, 338-342
Pediveliger larvae, 286, 330, 338
Pen, bottom-contact, 218
Penaeid shrimp, 65, 127-157, *see also* individual species
  artificial insemination, 128
  biology, 135-136
  culture requirements, 136-137
  diseases, 145-147
  economics, 150-151
  extensive culture, 136, 145
  feeds and feeding, 136, 138-139
  Galveston culture methods, 141, 142
  grow-out systems, 129, 143-145
  hatchery methods, 141-142
  hybridization, 128
  important species, 130-135
  industry outlook, 151-152
  industry status, 147-149
    companies, 148-149
    constraints, 147-148
  intensive culture, 137, 143-144, 149, 151
  Japanese culture method, 141-142
  low-technology culture, 151
  marketing, 151
  maturation, 138-140
  parasites, 145-147
  pond culture, 129, 136-137, 143-145

production, 151
reproduction, 135
salinity ranges, 136, 138, 142, 147
semi-intensive culture, 143, 149, 150
spawning, 138, 140
temperature ranges, 136, 138, 142, 147, 419
Penaeus aztecus, 135, 140
P. duorarum, 128, 129, 134, 419
P. indicus, 128
P. japonicus, 127, 128, 133, 138
P. kerathurus, 131
P. merguiensis, 128
P. monodon, 128, 130, 131, 132, 138, 143, 147
P. orientalis, 131
P. plebejus, 128
P. semisulcatus, 128, 131
P. setiferus, 128, 129, 134, 135
P. stylirostris, 128, 130-131, 132, 138, 145, 147
P. teraoi, 128
P. vannamei, 128, 130, 132, 145, 147
Perca flavescens, see Yellow perch
Perkinsus marinus, see "Dermo"
Pesticide sensitivity, 14, 86-87
Pesticides, 448-449
pH, 430-433
alkalinity and, 430, 432, 436
biological effects, 432
diel fluctuation, 430-431, 432
measurement, 432, 433
Phenolodisulfonic method, 441
Phospholipids, dietary, 186-187
Photoperiod, 171-174
Photosynthesis
carbon dioxide removal, 430, 434, 435
dissolved oxygen source, 423, 424, 425
Phyllopod, 457-458
Phyllosoma, 160, 161, 226
Phyllospadix torreyi, see Surfgrass
Phytoplankton
blooms, 444, 445
die-offs, 426
dissolved oxygen effects, 423, 424, 425, 426
nutrients, 447-448
Pimelometopon, see Sheephead
Pink shrimp, *see Penaeus duorarum*
Pisaster ochraceus, 343
Plankton, 444-447, *see also* Phytoplankton; Zooplankton
Plankton kriesel, 175, 176-177
Plants, aquatic, 446, 447-448
Platymonas, 284
Plocamium, 382
Pogonias cromis, see Black drum
Polinices, see Moon snail
Pollutant sensitivity, 86-87
Pollution, 392-393
Polychaetes, 139
Polyculture
crawfish, 37-38
grass shrimp, 229
prawn, 220, 224
Polygonum, see Smartweed
Polyodon spathula, see Paddlefish
Pond(s)
fertility, 446-447
marsh, 21-22
open, 18-20, 45
ricefield, 18, 45
wooded, 20-21, 45
Pond culture, *see under* individual species
Population monitoring (crawfish), 32-33
Postmolt, 76
Potassium permanganate, 18, 427, 428, 442-443
Prawn, 63-125
biology, 65-89
behavior, 81
distribution, 65-67
growth, 75-81, 102
life cycle, 66-67
molting, 75-77
diseases, 104-106
economics, 111-117
eggs, 66, 75, 93
environmental requirements, 81-87
alkalinity, 85-86
dissolved oxygen, 82-83, 427
hardness, 85, 86
nitrogenous wastes, 84-85
pH, 85
salinity ranges, 82-83, 420
temperature ranges, 81-82, 83, 108-109
toxic materials, 86-87
feeds and feeding, 87-89, 93-94, 95, 98, 100, 101, 103
grow-out systems, 98-104
intensive, 102-104
temperate zone, 100-102
tropics, 98-100

grow-out systems *(continued)*
harvesting, 80-81, 95-96, 99-100, 101, 108
hatchery methods, 89-94
important species, 65
industry constraints, 107-109
industry outlook, 111-117
industry status, 109-111
life cycle, 66-67
marketing, 108
nursery methods, 94-98
temperate zone, 96-98
tropics, 94-96
painted, 230
pH range, 85
predators, 106-107
processing, 104
production rates, 91, 102, 103-104, 109-110
reproduction, 67-75
mating behavior, 72-75
sexual characteristics, 67-72
seed stock, 93, 99, 109
size distribution, 78-80, 102
spawning, 75
spot, *see* Spot prawn
stocking rates, 91, 93, 94, 97, 99, 102
taxonomy, 65
Predators, *see also* names of predators
abalone, 368, 383-385, 393, 396-397, 406, 408
clams, 298
crawfish, 33-35
mussels, 342-343
oysters, 245-246, 251, 255, 259
prawns, 106-107
Premolt, 10, 39, 40, 76
Procambarus, 3, 10, 53-54, *see also* Crawfish
P. acutus acutus, 4, 5, 7, 11, 14-15, 45, 47, 52
P. clarkii, 3, 4-5
biology, 7, 8, 9, 11, 12, 13, 14-15
culture, 41-42, 43, 44, 45, 47, 52
dissolved oxygen effects, 427
wasting disease, 18
P. hayi, 6
P. troglodytes, 6, 7
Propanil, 35
Prostaglandins, 400-401
Protein, dietary, 184-185
Protothaca, see Littleneck clam
Protozoal disease, 16, 105, 106, 146, 244
Pseudoklossia, 296
Pseudomonas, 16
Pueruli, 226, 227, 228
Pumps, 213, 428-429, 430
Purina Marine Rations, 88-89
PVC trays/troughs, 169, 170, 180

R

Raccoon, as predator, 34-35, 298
Raceways
clams, 290
penaeid shrimp, 137, 144, 145, 149, 151
Rack culture
clams, 288
oysters, 248, 249, 250
Raft culture
clams, 289
mussels, 316-317, 321, 322, 323, 324, 332-333
oysters, 248, 249-250
Rays, 298, 385
cownose, 245
Razor clam, 277, 278
Recombinat DNA technology, 403-404
Red swamp crawfish, *see Procambarus clarkii*
"Red tail" disease, *see* Gaffkemia
Red tide, 256, 337, 344
Regulatory constraints, 298-300, 390
Reproduction, *see under* individual species
Rhinoptera, see Rays
Rhode Island, mussel culture, 320, 325, 332
Rice pond culture, 18, 35-37
Rice straw, as detrital substrate, 25, 26
Rickettsiae, 296
Rockfish, 385

S

Salinity, 418-421, *see also under* individual species
control, 420
nitrite toxicity effects, 441
oxygen solubility effects, 423
seawater, 418
Salmon, 220, 224
Saprolegnia, 18
Saxidomus, see Butter clam
Scapulicambarus, 6

Scorpaenichthys, see Cabezon
Scoter, 339, 342
Scyllarides nodifer, see Lobster, slipper
Sea gull, 298
Sea lettuce, 382
Sea otter, 368, 384, 393, 395, 406, 408
Seaside disease, 244
Seawater
  bioassay, 399-400
  salinity, 418
Secchi disk, 421, 428, 446, 449
Semicossyphus, see Sheephead
Semi-intensive culture, 143, 149, 150
Shedding systems, 207-215
  deionized water, 39
  floats, 207, 208-210
  mortality rates, 213-214
  on-shore tables, 208, 211-213
  open flow-through, 208, 212
  operation, 214
  troughs, 39
  water management, 212-213
Sheephead, 385
Shell abnormalities, 296
Shell disease, 190-191
Shrimp
  brown, *see Penaeus aztecus*
  candy-striped dancing, 230
  caridean, *see* Prawn
  coral, 229-230
  fairy, *see* Brine shrimp
  ghost, 259
  grass, 204, 228-229
  lady, 230
  mud, 259
  ornamental, 229-230
  penaeid, *see* Penaeid shrimp
  pink, *see Penaeus duorarum*
  scarlet flame, 230
  stenopidean, 229-230
  white, *see Penaeus setiferus*
Siliqua, see Razor clam
Simazine, 448
Siphon abnormalities, 296
Sirolpidium zoophthorum, 297
Skeletonema costatum, 217
Smartweed, 25
Sodium chloride, therapeutic use, 18
Soft-shell clam, 277, 281
South Carolina
  clam culture, 294, 295, 302
  crawfish culture, 2, 46
  oyster culture, 242
  penaeid shrimp culture, 137, 149, 150
  prawn culture
    economics, 112, 116-117
    methods, 86, 89, 97, 101, 103-104
    status, 110
Spawning, *see also under* individual species
  induction, 400-402, 407
Sperm banks, 171
Spermatophores
  extrusion, 170, 174
  lobster, 170-171, 174
  penaeid shrimp, 140
  prawns, 71, 72, 74
Spisula solidissima, see Surf clam
Sporozoal disease, 15
Spot prawn, 203, 219-225
  culture constraints, 224-225
  feeds and feeding, 220, 221, 222, 224
  larval history, 220-221
  postlarval development, 221-224
  reproduction, 220
  salmon polyculture, 220, 224
Stake culture, 248, 249, 250
Starfish, as predator, 245, 251, 255, 259, 298, 343, 384
Stenohaline, 419
Stenopus hispidus, see Shrimp, coral
Stylochus ellipticus, see Oyster leech
Sulfate, 441-442
Sulfur, 441-443
Surf clam, 277
  distribution, 281
  grow-out system, 294
  onshore nursery, 290
Surfgrass, 227
Swimmerets, 7, 9
Synaxidae, 160

T

Tanner crab, *see* Crab, snow
Tapes, see Manila clam
Temperature, 416-418, 419, *see also under* individual species
Tetraselmis, 388
Texas
  crawfish culture, 2, 19, 46, 47, 50, 52
  penaeid shrimp culture, 145, 147, 149, 150
  prawn culture, 110, 111
Thais, see Oyster drill

T. haemostoma, see Conch
T. lamellosa, see Whelk
T. lapillus, see Whelk
Thelonia, 15
Thermal stratification, 417-420, 445
Total alkalinity, 432, 433, 436-438
Total dissolved solids, 420
Total hardness, 432, 433, 436-438
Traps and trapping, 27, 29
  Witham habitat, 227
Tray culture, 180, 288, 290
Trematodes, 339-340
Trichodina, 251, 296
Trocophore
  abalone, 379-380
  clam, 280
  mussel, 328
Trough culture, 169, 170
Tumors, 296
Turbidity, 421-422, 444, 445

U

Umbo, 328
Up-flow culture, 290-293
Upogebia pugettensis, see Shrimp, mud
Urosalpinx, see Oyster drill

V

Veliger larvae
  abalone, 380
  clam, 280, 281-282, 286
  mussel, 328, 330, 331
Vibrio/vibriosis, 190, 251, 297
Virginia
  clam culture, 294, 302
  oyster culture, 240, 241, 243, 244, 246
Virus diseases
  oysters, 245
  penaeid shrimp, 146, 147
Vitamins, leaching, 187-188
Vitellogenesis, molting and, 172
Vorticella, 105

W

Washington state
  abalone fisheries, 367
  clam culture, 294, 302
  crawfish culture, 5
  mussel culture, 320-321, 322-323, 327, 338
  oyster culture, 236
    methods, 247, 248, 249, 258, 262, 265
    production, 251, 252
"Wasting disease", 18
Water analysis, 449-450
Water boatmen, 17-18
Water bugs, 34
Water management systems
  clam culture, 286-287
  crawfish culture, 22-24
  lobster culture, 177
  prawn culture, 91, 93, 96, 103
  shedding systems, 212-213
Water primrose, 25
Water quality, 415-455
  biological variables, 444-448
  chemical variables, 422-444
    carbon dioxide, 433-436
    dissolved oxygen, 422-430
    nitrogenous compounds, 438-441
    pH, 430-433
    sulfur, 441-443
    total alkalinity, 432, 433, 436-438
    total hardness, 432, 433, 436-438
  pesticides, 448-449
  physical variables, 416-422
    color, 421
    salinity, 418-421
    temperature, 416-418, 419
    thermal stratification, 417-420
    turbidity, 421-422
Water scorpion, 34
Well water, 428-429
Whelk, 259, 298, 343, 385
White river crawfish, *see Procambarus acutus acutus*
White shrimp, *see Penaeus setiferus*
Winkler procedure, 430
Wisconsin, crawfish culture, 5

Y

Yellow perch, 2

Z

Zoeae, 206, 216
Zooplankton, 444, 445
Zoothamnium, 16, 105